HANDBOOK OF
SOLUBILITY DATA for PHARMACEUTICALS

HANDBOOK OF
SOLUBILITY DATA for PHARMACEUTICALS

Abolghasem Jouyban

CRC Press
Taylor & Francis Group
Boca Raton   London   New York

CRC Press is an imprint of the
Taylor & Francis Group, an **informa** business

CRC Press
Taylor & Francis Group
6000 Broken Sound Parkway NW, Suite 300
Boca Raton, FL 33487-2742

© 2010 by Taylor and Francis Group, LLC
CRC Press is an imprint of Taylor & Francis Group, an Informa business

No claim to original U.S. Government works

Printed in the United States of America on acid-free paper
10 9 8 7 6 5 4 3 2 1

International Standard Book Number: 978-1-4398-0485-8 (Hardback)

Library of Congress Cataloging-in-Publication Data

Jouyban, Abolghasem.
   Handbook of solubility data for pharmaceuticals / Abolghasem Jouyban.
     p. ; cm.
   Includes bibliographical references and indexes.
   ISBN 978-1-4398-0485-8 (hardcover : alk. paper)
   1. Drugs--Solubility--Handbooks, manuals, etc. I. Title.
   [DNLM: 1. Drug Discovery. 2. Pharmaceutical Preparations--chemistry. 3. Organic Chemicals--chemistry. 4. Solutions. 5. Solvents. QV 744 J86h 2010]

   RS201.S6J68 2010
   615'.19--dc22
                                        2009027798

Visit the Taylor & Francis Web site at
http://www.taylorandfrancis.com

and the CRC Press Web site at
http://www.crcpress.com

*To my parents
and to my family*

Contents

Preface

Aqueous solubility is one of the major challenges in the early stages of drug discovery, and any attempt to increase solubility is of great importance in the pharmaceutical industry. One of the most common and effective methods for enhancing solubility is the addition of an organic solvent to the aqueous solution. In addition to experimental efforts to determine the solubility in mono and mixed solvents, a number of mathematical models are also proposed. The available experimental solubility data in mono and mixed solvents are collected and a brief discussion on the mathematical models is also provided. The solvents considered are the pharmaceutical cosolvents and other organic solvents that could be used in syntheses, separations, and other pharmaceutical processes.

The solutes included are the available data for official drugs, drug candidates, precursors of drugs, metabolites, and degradation products of pharmaceuticals. The aqueous solubility reported by Yalkowsky and He is excluded from the data in this book and readers are instead referred to the *Handbook of Aqueous Solubility*. The solubilities of amino acids are also considered as they play an important role in peptide drug properties and could be used in the pharmaceutical industry.

Abolghasem Jouyban
Faculty of Pharmacy and Drug Applied Research Center
Tabriz University of Medical Sciences

Acknowledgments

I would like to thank all the people who helped in collecting data, especially Dr. Shahla Soltanpour, Dr. Elnaz Tamizi, Ali Shayanfar, and Mohammad A.A. Fakhree. My deepest appreciation goes to Prof. William E. Acree Jr. (University of North Texas) for his contributions to the solubility investigations, careful review of Chapter 1, and very helpful comments and suggestions. Many thanks to Prof. Mohammad Barzegar-Jalali (Tabriz University of Medical Sciences), Prof. Pilar Bustamante (University of Alcala), Prof. Hak-Kim Chan (University of Sydney), and Prof. Brian J. Clark (University of Bradford) for their contributions to the solubility investigations over the last 12 years. Special thanks are also due to Mohammad A.A. Fakhree for his efforts in developing the "Solvomix" software that is available online at http://www.crcpress.com/product/isbn/9781439804858. In addition the software can be found at http://www.kimiaresinst.org/ (Service section).

Abolghasem Jouyban
Tabriz University of Medical Sciences

Author

Abolghasem Jouyban (-Gharamaleki) obtained his PharmD from Tabriz University of Medical Sciences in 1988, his first PhD in pharmaceutical analysis from Bradford University (United Kingdom) in 2001, and his second PhD in pharmaceutics from Sydney University (Australia) in 2004. He is a professor of pharmaceutical analysis at Tabriz University of Medical Sciences and has published more than 140 research articles in international and national peer-reviewed journals. Dr. Jouyban received two gold medals from the Iranian Razi Research Festival in 2000 and 2006 for his academic achievements.

1 Introduction

Before a drug becomes available to its receptors, it should be dissolved in the biological fluids surrounding the receptors; therefore, solubility is an important subject in pharmaceutical sciences. There is a solubility problem with nearly 40% of the drug candidates, and any attempt to predict the solubility is quite important in drug discovery investigations. The oldest rule for solubility prediction is "like dissolves like." This rule is applied to a well-known solubility equation, i.e., the Hildebrand equation, in which the solubility of a solute reaches the maximum value when the Hildebrand solubility parameters of the solute and solvent are equal. However, it has been shown that the Hildebrand equation is valid only for solutions with nonspecific interactions. Water is the unique biological solvent and aqueous solubility is one of the most important physicochemical properties (PCPs) of drugs; it is therefore obvious that simple equations like the Hildebrand equation cannot represent the aqueous solubility of solutes consisting of various functional groups, such as pharmaceuticals. More accurate equations are needed to predict the aqueous solubility of drugs, which will be discussed in detail later on (see Section 1.2). Aqueous solubility is also a key factor in the design of oral, parenteral, and ophthalmic formulations of poorly water-soluble drugs. A comprehensive database of aqueous solubilities of chemicals and pharmaceuticals was collected by Yalkowsky and He [1].

Solubility is defined as the maximum quantity of a drug dissolved in a given volume of a solvent/solution. For ionizable drugs, the solubility could be affected by the pH of the solution, and the intrinsic solubility (S_0) is defined as the concentration of a saturated solution of the neutral form of the drug in equilibrium with its solid. *The United States Pharmacopeia* [2] classified the solubilities of drugs into seven classes, as listed in Table 1.1.

Aqueous solubility has an essential role in the bioavailability of oral drug formulations. There is an established classification, namely, the biopharmaceutical classification system (BCS), which divides drugs into four classes in terms of their solubility and permeability [3]. The BCS classification correlates the in vitro solubility and permeability to the in vivo bioavailability. Soluble and permeable drugs are class I drugs with oral bioavailability being limited by their ability to reach the absorption sites. Class II drugs are poorly soluble but permeable drugs through the gastrointestinal tract (GI), meaning that their oral absorption is limited by the drug's solubility and, as a consequence of the Noyes–Whitney equation, by their dissolution rate. Class III drugs are soluble but poorly permeable and their oral bioavailability is limited by the barrier properties of the GI tract. Drugs of class IV are low soluble and poorly permeable compounds with the limitations of classes II and III. The drug candidates of class I possess suitable bioavailabilities and show appropriate drugability potentials; the bioavailability of drug candidates of class II can potentially be improved by developing suitable formulation designs, while those that belong to classes III and IV are most likely to return to the lead optimization phase for the improvement of their PCP [4]. Table 1.2 lists the details of the BCS along with a number of examples for each class [3,5].

Alsenz and Kansy [6] reviewed various solubility determination approaches in drug discovery and early development investigations. The poor solubility of a drug/drug candidate could be overcome using various solubility enhancement methods including salt formation; crystal engineering techniques; and the addition of cosolvents, surface active agents, complexing compounds, or hydrotrops to the solution. Preparation of soluble prodrugs is an alternative method to increase low aqueous solubility. In addition to the experimental efforts, as well as using solubilization techniques, to determine the solubility in water, a number of articles reported the practical possibility of solubility prediction methods. Lipinski et al. [7] reviewed the experimental determination of drugs' solubilities and

TABLE 1.1
Descriptive Classification of Drug Solubility

	Parts of Solvent Required for Dissolving One Part of the Solute
Very soluble	<1
Freely soluble	1–10
Soluble	10–30
Sparingly soluble	30–100
Slightly soluble	100–1,000
Very slightly soluble	1,000–10,000
Practically insoluble	>10,000

TABLE 1.2
Biopharmaceutical Classification System (BCS) of Drugs and Examples for Each Class

	High Solubility	Low Solubility
High permeability	Class I	Class II
	Acetaminophen	Flurbiprofen
	Caffeine	Warfarin
	Prednisolone	Carbamazepin
	Phenobarbital	Grisofulvin
	Diazepam	Phenytoin
Low permeability	Class III	Class IV
	Acyclovir	Acetazolamide
	Allopurinol	Azathioprine
	Atenolol	Chlorthiazide
	Captopril	Furosemide
	Famotidine	Mebendazole

computational aspects of aqueous solubility prediction methods. Jorgensen and Duffy [8] reviewed various aqueous solubility estimation methods from the chemical structures of the drugs. Ren and Anderson [9] reviewed the solubility prediction methods in lipid-based drug delivery systems. These systems were introduced in 1974 and have attracted more attention in recent years because of the high frequency of low aqueous solubility and dissolution rate problems in drug discovery investigations.

Date and Nagarsenker [10] reviewed the formulations and applications of parenteral microemulsions for delivery of hydrophobic drugs. Brewster and Loftsson [11] reviewed the applications of cyclodextrins as solubilizing excipients in pharmaceutical formulations along with their possible toxicities as well as the theoretical aspects of the complex formations between drugs and cyclodextrins. These compounds are able to improve the aqueous solubilities and dissolution rates of drugs, which result in the improved bioavailability of BCS classes II and IV drugs.

1.1 SOLUBILITY DETERMINATION METHODS

The solubility of a drug/drug candidate could be determined by experimental procedures mainly classified into two groups, namely, the thermodynamic and kinetic solubility determination methods. Combinatorial techniques in the pharmaceutical industry have significantly increased the demand for rapid and efficient solubility determination methods to be used in high-throughput screening. Most drugs are intended for oral administration, and aqueous solubility is essential as it influences the bioavailability of the drug. The ability to acquire solubility determinations at a comparable rate would help identify the soluble and more efficient drug candidates. In addition to the cost effect of such information in the drug discovery investigations, there are severe restrictions on sample amount, time, and also resources. Thermodynamic solubility determination methods are not feasible at the early discovery stage because of the large sample requirement, low throughput, and laborious sample preparation. Kinetic solubility determinations could be used as an alternative method at this stage. However, this method could not be substituted for thermodynamic solubility values, since the crystal lattice energy and the effects of the polymorphic forms of the drug are lost when the drug is dissolved in dimethyl sulfoxide (DMSO); therefore, the kinetic method is suitable for drug discovery purposes, whereas the thermodynamic solubility values are required in the drug development stage. More details of the most common methods used for determining thermodynamic and kinetic solubilities are presented in this book.

1.1.1 SHAKE-FLASK METHOD

The shake-flask method proposed by Higuchi and Connors [12] is the most reliable and widely used solubility measurement method. This method determines thermodynamic solubility and could be carried out in five steps.

Sample preparation

An excess amount of drug is added to the solubility medium. The added amount should be enough to make a saturated solution in equilibrium with the solid phase. In the case of acidic or basic drugs dissolved in an unbuffered solubility medium, further addition of the solid could change the pH of the solution and consequently the solubility of the drug.

Equilibration

Depending on the dissolution rate and the type of agitation used, the equilibration time between the dissolved drug and the excess solid could be varied. Equilibration is often achieved within 24 h. To ensure the equilibration condition, the dissolution profile of the drug should be investigated. The shortest time needed for reaching the plateau of drug concentration against time could be considered a suitable equilibration time. Any significant variation of the dissolution profile after reaching the equilibration should be inspected, since there are a number of possibilities including degradation of the drug as well as its polymorphic transformation. Both of these affect the solubility values of a drug dissolved in the dissolution media. Vortexing or sonicating the sample prior to equilibration could reduce the equilibration time. To overcome the poor wettability of low soluble drugs, one may use small glass microspheres or sonication.

Separation of phases

Two common methods used for phase separations of saturated solutions are filtration and centrifugation. Filtration is the easiest method; however, the possible sorption of the solute on the filter should be considered as a source of error in solubility determinations, especially for very low soluble drugs. Prerinsing the filter with the saturated solution could reduce the sorption of the solute on the filter by saturating the adsorption sites. Centrifugation or ultracentrifugation is preferred in some cases, and the higher viscosity of the saturated solutions, i.e., in mixed solvents, should be considered as a limitation. A combination of filtration and centrifugation could also be used [13].

Analysis of saturated solution and the excess solid

UV spectrophotometric analysis is the most common and the easiest analytical method in solubility determination experiments. The next is the high-performance liquid chromatography (HPLC) methods in both the isocratic and gradient elution modes. The HPLC analysis could also detect the possible impurities or degradation products if a highly selective method was used. X-ray diffraction (XRD) and differential scanning calorimetry (DSC) of the residual solid separated from the saturated solution confirm the possible solid-phase transformations during equilibration.

Data analysis

The collected data could be compared with the previously reported data to ensure the accuracy of the experimental procedure employed. Any mistake in the dilution step, any miscalculations, or using uncalibrated instruments, such as uncalibrated balances, temperature variations, and some other factors, could result in different solubility values for a given drug dissolved in a solvent at a fixed temperature.

1.1.2   KINETIC SOLUBILITY DETERMINATION

To determine the solubility of a drug/drug candidate using kinetic methods, a given concentration of the solute, which was previously dissolved in DMSO, is added to the dissolution media with a fixed rate and the precipitation of the solute is detected by the turbidimetric [7] or nephelometric [14] method. In the first method, aliquots of 1 μL of a 10 μg/μL DMSO solution are added at 1 min intervals to a phosphate buffer (pH 7.0) at room temperature. Precipitation is detected by an increase in UV absorbance from light scattering at 660–820 nm. A total of 14 dilutions are made, resulting in a solubility range of 4–56 μg/mL [7]. In the second method, a 10 mM DMSO solution is diluted in phosphate-buffered saline (PBS) at pH 7.4 and is then diluted 10 times across a 96-well plate with PBS containing 5% DMSO. The concentration at which the compound precipitates is detected by light scattering using a nephelometer with a laser light source at 633 nm.

As described above, the advantages of using the kinetic solubility determination methods are their speed, the small amount of drug required, and the fact that they estimate the solubility under the same conditions as those used in biological assays. It should be noted that a number of drug candidates are low soluble in DMSO and this may cause in solubility determination using kinetic methods.

A flow injection analysis (FIA) was also reported in the literature [15], which could be considered as a semithermodynamic method in which a predissolved drug is equilibrated with problems phosphate buffer (pH 6.5) overnight prior to analysis.

1.2   AQUEOUS SOLUBILITY PREDICTION METHODS

The general solubility equation (GSE) developed by Yalkowsky and coworkers [16] is the most commonly used method and is expressed as

$$\log S_w = 0.5 - 0.01(mp - 25) - \log P \tag{1.1}$$

in which
S_w is the aqueous solubility of drug expressed in mol/L [17]
mp is the melting point (°C)
$\log P$ is the logarithm of the water-to-octanol partition coefficient [16,17]

This equation predicts the aqueous solubilities of solutes with the root mean square error (RMSE) of 0.7–0.8 log unit. The GSE is a combination of the relationships proposed by Irmann [18] and Hansch et al. [19]. Irmann proposed a group contribution method for predicting the aqueous solubility of liquid hydrocarbons and halocompounds with an additional correction term as $\Delta S_m(T_m - T)/1364$, in which ΔS_m is the entropy of fusion and T_m is the melting point temperature of the solute in Kelvin. The correction term was simplified to $-0.0095(mp - 25)$ by replacing ΔS_m with 13 cal/deg/mol [18]. Hansch et al. proposed a linear relationship between $\log S_w$ and $\log P$ of liquids [19]:

$$\log S_w = 0.978 - 1.339 \log P \qquad (1.2)$$

and this equation was extended to calculate the aqueous solubilities of crystalline solutes as [20]:

$$\log S_w = 1.170 - 1.380 \log P \qquad (1.3)$$

Yalkowsky and coworkers combined and applied these relationships to predict the aqueous solubilities of liquid and solid solutes [16,21]. The $(mp - 25)$ term of Equation 1.1 is set to zero for liquid $(mp < 25°C)$ solutes [17].

Jain et al. [22] tested the applicability of Equation 1.1 for predicting the solubility of weak electrolytes (acids and bases, $N = 367$) and concluded that the GSE could be extended to predict the aqueous solubility of weak electrolytes. The reported average absolute error (AAE) for the predicted solubility of weak electrolytes was 0.64, whereas the corresponding value for nonelectrolytes was 0.43 ($N = 582$). The overall AAE for both classes of the solutes was 0.58 ($N = 949$). The GSE cannot be used to predict the solubilities of ampholytes and polyprotic acids and bases [22].

The linear solvation energy relationship (LSER) model developed by Abraham et al. is

$$\log S_w = 0.518 - 1.004E + 0.771S + 2.168A + 4.238B - 3.362A \cdot B - 3.987V \qquad (1.4)$$

where

E is the excess molar refraction
S is the dipolarity/polarizability of the solute
A denotes the solute's hydrogen bond acidity
B stands for the solute's hydrogen bond basicity
$A \cdot B$ is a term representing the acid–base interactions between solute molecules
V is the McGowan volume of the solute in unit of 0.01 (cm^3/mol) [23]

Equation 1.4 has been updated as

$$\log S_w = 0.395 - 0.955E + 0.320S + 1.155A + 3.255B - 0.785A \cdot B - 3.330V \qquad (1.5)$$

in another report from Abraham's research group [24]. It has also been shown that Equation 1.5 provided more accurate predictions compared with Equation 1.4 [25].

The basis of LSER equations dates back to the 1930s, when Hammett established a procedure for understanding the effects of electronic structure on the reaction rates of series of similar molecules [26]. Hammett's linear free energy relationships (LFERs) were extended by Taft to include a steric parameter and to separate the resonance and inductive effects of aromatic ring substitutes [27]. In 1976, Kamlet and Taft developed the solvatochromic equations [28], which were the extended versions of LFERs and were later expanded to include solute properties [29] and known as LSERs. In the original LSER of Kamlet and Taft, a solvent was characterized by its solvatochromic parameters, i.e., dipolarity/polarizability (π^*), and its ability to accept (β) and donate (α) hydrogen bonds [28], and it was designed to calculate the properties of a single solute in multiple solvents. Abraham and coworkers [29] extended the approach to calculate the properties of various solutes in a single solvent. Later comprehensive investigations by Abraham resulted in a model consisting of solvent and solute parameters, which is known as the Abraham solvation model [30]. The Abraham model has found a large number of applications for modeling of various physicochemical and biological properties including the aqueous solubilities of drugs [23].

Although the GSE and LSERs are the golden models for predicting aqueous solubility, they suffer from a lack of accuracy and provide a rough estimate of the aqueous solubility of drugs/drug-like molecules. Efforts to provide better models are still ongoing. The AAE of Equation 1.4 was 0.408 (log unit) for calculating the solubility of 659 solutes, most of which were nonpharmaceutically related compounds [23]. To show the prediction capability of the amended solvation energy relationship (ASER), i.e., Equation 1.4, Abraham and Le trained the model using experimental data for 594 solutes and then predicted the aqueous solubility of a test set consisting of 65 solubility data. The reported AAE for the predicted data was 0.5 [23]. The mean percentage deviation (MPD) of the test set was 134 ± 373% [31]. The corresponding MPD of the correlative equation for the entire set of 659 data points was 244 ± 850%.

Yang et al. [21] compared the accuracies of Equations 1.1 and 1.4 using solubility data for 662 compounds and found that both models provide better results for liquid solutes when compared with the accuracies of the predictions for solid solutes. The reported AAEs for Equations 1.1 and 1.4 for all solutes were 0.446 and 0.431, respectively. The same pattern was observed using residual mean square error (RMSE) as an accuracy criterion in which the RMSEs were 0.622 and 0.615 for Equations 1.1 and 1.4, respectively. In a more recent study, Jain et al. [17] compared the prediction capabilities of GSE and aqueous functional group activity coefficient (AQUAFAC) methods using 1642 pharmaceutically or environmentally related solutes. The authors reported AAEs of 0.576 and 0.543 for GSE and AQUAFAC methods, respectively [17]. By converting the results of the predicted solubilities to the MPD scale, one could deal with the practical applicability of the predictions. The MPDs (±SDs) of the GSE and AQUAFAC methods for reported S_w values were 1199 ± 11272% and 432 ± 2755%, respectively. Very large SD values reveal that there are some outlier data points with very large deviations. By ignoring these outliers, the MPDs were reduced to 817% and 338% [32].

Hou and coworkers [33] proposed a multiple linear regression (MLR) model for aqueous solubility predictions based on atomic contributions in which each atom of the drug molecule has its own contribution. Atom types, 76 of them, hydrophobic carbon, and square of the molecular weight (MW^2) of the drugs were used to fit to their model, which is expressed as

$$\log S_w = C_0 + \sum_i a_i n_i + \sum_j b_j B_j \tag{1.6}$$

where
  C_0 is a constant
  a_i and b_j are the regression coefficients
  n_i is the number of the atom type i in the drug structure
  B_j denotes the correction factors, i.e., MW^2 and hydrophobic carbon

Equation 1.6 was trained using 1290 solubility data points and then the solubility of the commonly used test set of 21 pharmaceutically and environmentally compounds of interests was predicted. The accuracy of this prediction is presented in Table 1.3. For further evaluation of the prediction capability, it was trained using 1207 data points and the solubility of additional 120 solutes was predicted. The obtained results were comparable with those of the test set of 21 compounds [33].

TABLE 1.3
The Accuracies of Various Models to Predict the Solubility of 21 Data Set

Model	Descriptors	MPD (SD)	RMSE	AAE	Reference
Klopman	2D substructures	194 (324)	0.54	0.97	[101]
Kuhn		943 (2532)	0.96	0.86	[102]
Yan-MLR	3D descriptors	4689 (18440)	1.06	0.54	[103]
Yan-ANN	3D descriptors	218 (293)	0.52	0.75	[103]
Hou	Atomic	242 (591)	0.53	0.65	[33]
Huuskonen-MLR (Equation 1.10)	Topologicals	95 (104)	0.53	0.79	[38]
Huuskonen-ANN	Topologicals	101 (86)	0.53	0.62	[38]
Duchowicz[a]	Dragon	617 (940)	100[a]	0.95	[35]
Tetko		105 (153)	0.52	0.51	[104]
Liu		132 (127)	0.51	0.61	[105]
Wegner		240 (394)	0.53	0.71	[106]

[a] The solubility of antipyrine was predicted as 0 whereas its experimental log(g/L) was 2.665, and this causes a very large RMSE value.

Faller and Ertl [34] proposed a three-parameter model for calculating the aqueous solubility of pharmaceuticals and tested its performance using solubility data of 60 generic drugs. Their model was

$$\log S_w = 9.5942 + 0.7555 c \log P + 0.0088 \, PSA - 0.6438 \, revVol \tag{1.7}$$

where
$c \log P$ is the calculated log P
PSA is the polar surface area
revVol is the ratio of molecular volume to the number of nonhydrogen atoms in the molecule

The effect of tautomerism on the numerical values of $c \log P$ was shown for one of the Novartis compounds in which nine possible tautomers existed and the calculated $c \log P$ of the different tautomeric forms varied between −0.03 and 3.43 [34]. This point should be considered in all solubility prediction methods employing $c \log P$ and some other descriptors as independent variables.

Duckowicz et al. [35] proposed an MLR model for calculating the aqueous solubility of organic compounds expressed as g/L ($S_{g/L}$). The model is

$$\log S_{g/L} = 2.970 - 0.435\Omega(\chi_1 sol) - 0.503\Omega(M \log P) + 0.0767\Omega(RDF\,060u) \tag{1.8}$$

where
Ω is the orthogonalized values of ($\chi_1$sol), the solvation connectivity index
($M \log P$) is the Moriguchi octanol–water partition coefficient
($RDF\,060u$) is the radial distribution function 6.0 unweighted

These descriptors were calculated using DRAGON software [36]. Equation 1.8 was trained using 97 experimental solubilities validated using 47 data points. Then the trained and validated model was used to predict the solubility of the test set of 21 compounds. The accuracy of the predictions using RMSE criterion revealed that their equation produces reasonable accuracies with the advantage of employing three independent variables.

Huuskonen et al. [37] employed the neural network models for modeling the aqueous solubility amounts of data sets of structurally related drugs using topological descriptors as input parameters. The architecture of the artificial neural network (ANN) was 5-3-1 and the results were cross-validated using the leave-one-out method. The applicability of the proposed ANN was examined using the experimental solubility of three sets of drugs, i.e., 28 steroids, 31 barbiturates, and 24 reverse transcriptase inhibitors. The RMSEs of steroids, barbiturates, and reverse transcriptase inhibitors were 0.288, 0.383, and 0.401, respectively. The main disadvantage of their treatment is that it is only applicable to drug sets included in the training step and no generalization is possible. In another comparative study, Huuskonen et al. [38] trained three MLR methods and an ANN model using 675 aqueous solubility data points taken from databases. The MLR models were

$$\log S_w = -1.01 \log P - 0.01 \, mp + 0.50 \tag{1.9}$$

$$\log S_w = \sum a_i S_i + 1.52 \tag{1.10}$$

$$\log S_w = 0.005 \, mp + \sum a_i S_i + 1.39 \tag{1.11}$$

in which
$\log P$ is the calculated values of the partition coefficients
a_is are the model constants
S_is are 31 atom-type electrotopological state indices and 3 indicator variables

The same independent variables (S_i) used in the MLR models were employed as input data for the ANN models. The structure of the ANN was 34-5-1, and apparently the bias term, which is usually used in ANN computations, was ignored in the work. The validity of the models was tested using aqueous solubility data of 38 pharmaceuticals. The MPD values for Equations 1.9 through 1.11 were $2466 \pm 12877\%$, $117 \pm 125\%$, and $181 \pm 327\%$, and the corresponding value for the AAN model was 154 ± 193 [39].

The mobile order and disorder (MOD) model that was presented by Huyskens and Haulait-Pirson [40] leads to a general predictive equation for the solubility of solutes in various solvents. The MOD is based on the equilibrium between fractions of times during which protic molecules are involved in or escape from the hydrogen bonding. The model calculates the volume fraction solubility (Φ_B) of a solute in a solvent considering the physicochemical phenomena accompanying the fusion and solution processes—solvent–solvent, solute–solute, and solute–solvent interactions—and is expressed as

$$\ln \Phi_B = A + B + D + F + O + OH \tag{1.12}$$

All contributions to the free energy change are taken into account when a solute is dissolved in a solvent by the fluidization of the solute (term A), with the exchange entropy correction resulting from the difference in molar volumes of the solute and solvent (term B), the change in nonspecific cohesion forces in the solution (term D), the hydrophobic effect of the self-associated solvents (term F), the hydrogen bond formation between proton acceptor solutes and proton donor solvents (term O), and the hydrogen bond formation between amphiphilic solutes and proton acceptor and/or proton donor solvents as well as the autoassociation of the solute in the solution (term OH). Equation 1.12 could be simplified for proton donor solvents (e.g. alcohols) as

$$\ln \Phi_B = A + B + D + F + O \tag{1.13}$$

and for nonproton donor solvent, e.g., alkanes [41], as

$$\ln \Phi_B = A + B + D \tag{1.14}$$

The terms of the MOD model are defined as
 The fluidization term:

$$A = -\frac{\Delta H_m}{R}\left(\frac{1}{T} - \frac{1}{T_m}\right) - \sum\left[\frac{\Delta H_{trans}^i}{R}\left(\frac{1}{T} - \frac{1}{T_{trans}^i}\right)\right] \tag{1.15}$$

where
 ΔH_m is the molar enthalpy of fusion
 R is the gas constant
 T is the absolute temperature
 T_m is the solute's melting point (in K)
 ΔH_{trans}^i and T_{trans}^i are the enthalpy and temperature of the solid-phase transition for the solid solutes
  undergoing solid–solid or solid–liquid crystal transitions

When experimental values of the enthalpy of fusion are not available, the fluidization term could be estimated from [42]

$$A = -0.02278(T_m - 298.15) \tag{1.16}$$

The correction factor of the entropy of mixing or the B term:

$$B = 0.5\Phi_S\left(\frac{V_B}{V_S} - 1\right) + 0.5\ln\left(\Phi_B + \Phi_S\frac{V_B}{V_S}\right) \tag{1.17}$$

where
 Φ_S is the ideal volume fraction of the solvent
 V_B and V_S are the molar volumes of the solute and solvent, respectively

The D term:

$$D = -\left[\frac{1}{1.0 + \max(K_{O_i}, K_{OH_i})\dfrac{\Phi_S}{V_S}}\right]\frac{\Phi_S^2 V_B}{RT}(\delta_B' - \delta_S') \tag{1.18}$$

where
 $\max(K_{O_i}, K_{OH_i})$ is the stability constant governing the strongest specific association between solute–solvent functional groups
 K_{O_i} and K_{OH_i} are the stability constants of the hydrogen bonds formed in the solution
 δ_B' and δ_S' are the modified nonspecific cohesion parameters of the solute and solvent, respectively [42]

Table 1.4 lists the molar volumes and modified nonspecific cohesion parameters of the solvents.

TABLE 1.4
The Molar Volumes (V_S) and Modified Nonspecific Cohesion Parameters (δ_S')
of the Solvents

Solvent	V_S (cm³/mol)	δ_S' (MPa$^{1/2}$)
γ-Butyrolactone	76.8	22.24
1,1-Dibromoethane	92.9	18.77
1,1-Dichloroethane	84.8	18.51
1,2-Dibromoethane	87.0	20.75
1,2-Dichloroethane	78.8	20.99
1,3-Butanediol	89.4	19.40
1,3-Propanediol	72.5	19.80
1,4-Butanediol	88.9	20.24
1,4-Dichlorobutane	112.1	19.78
1-Butanol	92.0	17.16
1-Decanol	191.6	16.35
1-Heptanol	141.9	16.39
1-Hexanol	125.2	16.40
1-Nonanol	174.3	16.37
1-Octanol	158.3	16.38
1-Pentanol	108.6	16.85
1-Propanol	75.1	17.29
2,3-Butanediol	91.3	19.40
2-Butanol	92.4	16.60
2-Butyl acetate	132.0	19.61
2-Propanol	76.9	16.80
3,3,5-Trimethyl hexanol	175.2	16.36
3-Methyl 3-pentanol	124.0	16.40
Acetone	74.0	21.91
Acetonitrile	52.9	23.62
Anisole	108.6	20.20
Benzene	89.4	18.95
Benzyl alcohol	103.8	17.00
Bromobenzene	105.3	21.22
Carbito acetatel acetate	175.5	16.98
Carbitol	136.1	18.60
Carbon disulfide	60.0	18.48
CCl$_4$	97.1	17.04
Cellosolve	97.4	18.63
Cellosolve acetate	136.6	16.98
CH$_2$Cl$_2$	64.5	20.53
CHCl$_3$	80.7	18.77
Chlorobenzene	102.1	19.48
Chlorocyclohexane	120.3	18.45
cis-1,2-Dichloroethylene	76.0	18.61
Cyclohexane	108.8	14.82
Cyclohexanol	106.0	17.88
Cyclooctane	134.9	15.40
Decaline	156.9	18.90
Dibutyl ether	170.3	17.45
Diethyl adipate	202.2	18.17
Diethyl carbonate	121.2	16.89
Diethyl ether	104.8	18.78

TABLE 1.4 (continued)
The Molar Volumes (V_S) and Modified Nonspecific Cohesion Parameters (δ'_S) of the Solvents

Solvent	V_S (cm³/mol)	δ'_S (MPa$^{1/2}$)
Diethyl ketone	106.4	20.13
Diethylene glycol	94.9	21.90
Dimethyl isosorbide	151.5	20.66
Dimethylcyclohexane	145.4	15.50
Dioxane	85.8	20.89
Dipentyl ether	204.0	16.16
Dipropyl ether	141.8	17.96
Dipropyl ketone	140.4	19.49
Ethanediol diacetate	132.4	18.50
Ethanol	58.7	17.81
Ethyl acetate	98.5	20.79
Ethyl butyrate	132.2	19.61
Ethyl decanoate	231.6	18.30
Ethyl hexanoate	165.6	18.99
Ethyl isopropyl ether	124.3	16.12
Ethyl nonanoate	215.2	18.43
Ethyl octanoate	198.2	18.59
Ethyl oleate	356.9	17.69
Ethyl pentanoate	148.5	19.27
Ethyl propionate	115.5	20.05
Ethyl propyl ether	120.4	18.40
Ethylbenzene	123.1	18.02
Ethylene glycol	55.8	19.90
Furfuryl alcohol	86.5	18.99
Glycerol	73.3	21.19
Glyceryl diacetate	148.8	22.50
Glyceryl triacetate	188.7	18.88
Iso-Butanol	92.8	16.14
Isooctane	166.1	14.30
Isopentyl acetate	148.8	19.27
Isopropyl acetate	117.0	20.01
Isopropyl myristate	317.0	17.83
Mesitylene	139.6	17.00
Methanol	40.7	19.25
Methyl acetate	79.8	21.67
Methyl benzoate	125.0	19.62
Methyl cellosolve	79.1	18.63
Methyl cellosolve acetate	118.2	17.67
Methyl ethyl ketone	90.2	20.90
Methyl formate	62.1	22.96
Methyl isobutyl ketone	125.8	19.95
Methylcyclohexane	128.3	15.00
Monoglyme	104.4	20.36
m-Xylene	123.2	17.20
N,N-Dimethylformamide	77.0	22.15
n-Butyl acetate	132.5	19.66
n-Butyl acrylate	143.4	19.28
n-Butyl butyrate	165.9	18.99
n-Butyl carbitol	170.1	17.40

(continued)

TABLE 1.4 (continued)
The Molar Volumes (V_S) and Modified Nonspecific Cohesion Parameters (δ'_S)
of the Solvents

Solvent	V_S (cm³/mol)	δ'_S (MPa$^{1/2}$)
n-Butyl lactate	150.4	19.84
n-Butyl propionate	149.5	18.95
n-Chlorobutane	105.0	17.12
n-Chlorohexane	138.1	18.00
n-Chlorooctane	171.1	18.00
n-Chlorotetradecane	270.2	18.00
n-Decane	195.9	15.14
n-Dodecane	228.6	15.34
n-Heptane	147.5	14.66
n-Hexadecane	294.1	15.61
n-Hexane	131.6	14.56
n-Hexyl acetate	165.8	19.15
Nitrobenzene	102.7	21.77
Nitroethane	71.8	22.44
n-Nonane	179.7	15.07
n-Octane	163.5	15.85
n-Pentadecane	277.7	15.56
n-Pentane	116.1	14.18
n-Pentyl acetate	148.7	19.27
n-Propyl acetate	115.7	20.05
n-Tetradecane	260.3	15.49
n-Tridecane	245.0	15.42
n-Undecane	213.3	15.25
o-Dichlorobenzene	113.1	18.77
o-Xylene	121.4	17.50
p-Cymene	156.0	20.20
Propylene glycol	73.7	19.23
Propylene glycol monomethyl ether acetate	137.1	19.50
p-Xylene	123.9	17.30
Pyridine	80.9	20.94
Squalane	525.0	16.25
tert-Butanol	94.3	15.78
tert-Butylcyclohexane	173.9	15.50
Tetrahydrofurane	81.4	19.30
Tetraline	137.1	19.43
Thiophene	79.6	18.70
Toluene	106.9	18.10
trans-1,2-Dichloroethylene	78.0	18.41
Water	18.1	20.50

Source: Ruelle, P. et al., *Perspect. Drug Discov. Des.*, 18, 61, 2000.

The hydrophobic effect term or *F* term:

$$F = -r_S \Phi_S \frac{V_B}{V_S} + \sum v_{OH_i} \Phi_S (r_S + b_i) \tag{1.19}$$

where r_S is the mobile order factor of the solvent. The numerical value for r_S is constant for a given class of the solvents (0 for nonassociated solvents, i.e., hydrocarbons, esters, ethers, ketones, and nitriles; 1 for strongly associated solvents with a single H-bonded chain (alcohols); and 2 for water and diols) [42].

The O term:

$$O = \sum_i v_{O_i} \ln\left[1 + K_{O_i}\left(\frac{\Phi_S}{V_S} - v_{O_i}\frac{\Phi_B}{V_B}\right)\right] \qquad (1.20)$$

where

> i are the different proton acceptor sites (ketone, ester, nitrile) on the solute molecule capable of a specific interaction with the solvent
>
> v_{O_i} is the number of associations of a given type [42]

The OH term:

$$OH = \sum_i v_{OH_i}\left[\ln\left(1 + K_{OH_i}\frac{\Phi_S}{V_S} + K_{BB_i}\frac{\Phi_B}{V_B}\right) - \ln\left(1 + \frac{K_{BB_i}}{V_B}\right)\right] \qquad (1.21)$$

where

> i are the different proton donor sites (hydroxyl, primary or secondary amine, and amide) on the solute molecule able to interact specifically with the solvent
>
> v_{OH_i} denotes the number of associations between the functional groups of the solute and the solvent
>
> K_{OH_i} is the stability constants for the association of solute–solvent functional groups
>
> K_{BB_i} is the stability constants of the self-associating groups of the solute in the solution [38]

Ruelle et al. estimated the solubility of 38 nonhydroxysteroids [43] and 62 hydroxysteroids [42] in common solvents using the MOD model. These estimations require a number of PCPs as noted above. The authors reported the calculated and experimental volume fraction solubilities of the solutes and discussed the differences between the calculated and experimental solubilities using the ($\Phi_B^{Calculated}/\Phi_B^{Experimental}$) values. The ratios vary from 0.001 (for the solubility of 17-α-hydroxyprogestrone in water) to 152.2 (for solubility of fluocinolone acetonide in water), as reported in the original article [42]. The authors did not report the very large ratio of 9019.6 (for the solubility of triamcinolone in water), revealing that the aqueous solubility of 0.0000051 was predicted as 0.046. When these differences were translated into MPD values, the overall MPD for the solubility of hydroxysteroids in water was 21573.4 ± 135820.1% ($N = 44$). As noted above, the MPDs of a number of solutes were very high (e.g., triamcinolone [901861%], fluocinolone acetonide [14900%], fluocinolone [9043%]). After excluding the MPDs > 1000, the overall MPD was 198.2 ± 222.8% [32].

Cheng and Merz [44] presented a quantitative structure–property relationship (QSPR) to predict the aqueous solubilities of solutes. The authors claimed the employment of two-dimensional (2D) structural information and reduced the number of descriptors in their method. A descriptor association was defined as being proportional to the product of the number of hydrogen bond donors (HBD) and hydrogen bond acceptors (HBA) and was used to describe the intermolecular hydrogen bonding between solute molecules in solid state. This interaction term was used by Abraham et al. [23], who employed a similar interaction term in their developed LSER model (see Equation 1.4). Cheng and Merz claimed that their QSPR method may be applied to different physical conditions (e.g., different temperatures, pressures, pH ranges) and to different solution compositions (e.g., blood, mixed compound solutions, organic and inorganic solutions). An example of the developed QSPR equations developed by Cheng and Merz [44] is

$$\log S_w = -0.7325 A \log P\,98 - 0.4985\,\text{HBD}\cdot\text{HBA} - 0.5172\,\text{Zagreb}$$
$$+ 0.0780\,S\_\text{aaaC} + 0.1596\,\text{Rotlbonds} + 0.2057\,\text{HBD}$$
$$+ 0.1834\,S\_\text{sOH} + 0.2539\,\text{Wiener} \tag{1.22}$$

where

$A \log P\,98$ is the water–octanol partition coefficient calculated using the Cerius $A \log P$ module

Zagreb is the sum of the square of the vertex for the molecule

S_aaaC is the summation of the aromatic carbon atoms in the molecule

Rotlbonds is the number of rotatable bonds

S_sOH is the summation of the electrotopological state values for the functional group type –OH

Wiener is the sum of the chemical bonds between all pairs of heavy atoms (nonhydrogen atoms)

The reported RMSE of various sets were 0.62 to 1.15 log units.

Fillikov [45] presented predictive MLR models trained by solubilities of 53 drugs taken from a reference [46]. Three of the best-performing equations are

$$\log S_w = 1.243 + 0.0421\,\text{VDW} + 0.202\,\text{NoRotWs} - 0.0126\,\text{PSA}$$
$$- 0.0055\,\text{SA} - 0.638 c \log P \tag{1.23}$$

$$\log S_w = 0.94 + 0.0443\,\text{VDW} + 0.182\,\text{NoRot} - 0.0116\,\text{PSA}$$
$$- 0.00545\,\text{SA} - 0.626 c \log P \tag{1.24}$$

$$\log S_w = 1.068 + 0.0482\left(\text{VDW} + \text{HB} + \text{TO}\right) + 0.187\,\text{NoRot}$$
$$- 0.0115\,\text{PSA} - 0.00568\,\text{SA} - 1.6 c \log P \tag{1.25}$$

where

VDW is the van der Waals energy

NoRotWs is the number of rotatable bonds without symmetrical groups

PSA and SA refer to the polar surface area and surface area

$c \log P$ denotes the calculated octanol–water partition coefficient

NoRot is the number of rotatable bonds

HB refers to the hydrogen bond interaction energy

TO is the torsion angle strain energy

The accuracy of Equations 1.23 through 1.25 was compared with that of the predicted solubilities by ACD Labs, i.e., Equation 1.26:

$$\log S = 0.92 - 0.834 c \log P - 0.0084\,\text{MW} \tag{1.26}$$

using RMSE that varied from 0.557 to 0.569 for the three equations and was 1.09 for Equation 1.26. The results show the better performance of the Fillikov equations when compared with Equation 1.26; however, one should keep in mind that Equation 1.26 employs just two descriptors while the Fillikov equations used five descriptors, and hence, from a statistical point of view, it should provide better predictions.

In addition to the journal articles and patents discussed above, a lot of computer software has been developed for predicting drug solubility in water. Select programs and software packages were briefly reviewed elsewhere [47].

1.3   SOLUBILITY PREDICTION IN ORGANIC SOLVENTS

Compared to water, very few papers have been published on the calculation of the solubility of drugs in nonaqueous monosolvents. Yalkowsky et al. [48] calculated the mole fraction solubility of different solutes in octanol (X_{Oct}) at 30°C using

$$\log X_{Oct} = -0.012\,mp + 0.26 \tag{1.27}$$

Equation 1.27 was modified for predicting the mole fraction solubility of weak electrolytes and nonelectrolytes as

$$\log X_{Oct} = -0.011\,mp + 0.15 \tag{1.28}$$

$$\log X_{Oct} = -0.013\,mp + 0.44 \tag{1.29}$$

Equation 1.27 was transformed as

$$\log S_{Oct} = -0.01\,(mp - 25) + 0.8 \tag{1.30}$$

for predicting the molar solubility of drugs in octanol at 30°C [48].

Dearden and O'Sullivan [49] proposed the following equation for calculating the mol/L solubility of drugs in cyclohexane (S_{Cyc}):

$$\log S_{Cyc} = -0.0423\,mp + 1.45 \tag{1.31}$$

which was tested on the solubility of 12 pharmaceuticals and the MPD was 85.1(±21.6)% [32].

Sepassi and Yalkowsky [50] proposed another version of Equation 1.27 to compute the mol/L solubility of drugs in octanol as

$$\log S_{Oct} = -0.01\,(mp - 25) + 0.5 \tag{1.32}$$

Equation 1.32 uses Walden's rule for the entropy of melting (i.e., ΔS_m = 56.5 J/mol K or 13.5 [cal/mol °C] for rigid molecules) and predicts the solubility of crystalline solutes with the solubility parameter of 15–28 $(J/cm^3)^{0.5}$ and mp > 25°C. For the solutes with mp < 25°C, the first term on the right-hand side of Equation 1.32 is set equal to zero. The accuracy of Equation 1.32 was discussed using AAE, and the maximum value of AAE was 3.59 (for the solubility of mirex) and the average value was 0.39 (±0.40) [50]. When these values were translated into MPD, the overall value was 147 (±247)% [32]. Using experimental values of the entropy of melting for 68 solutes, Sepassi and Yalkowsky [50] reported an AAE of 0.29 (±0.04). The corresponding value for Equation 1.32 was 0.35 (±0.04) and the mean difference was not statistically significant (paired t-test, $p > 0.05$).

TABLE 1.5
The Overall MPDs, Standard Deviations (SDs), and Number of Solubility Data Points in Each Set (N) for the Predicted Solubilities Using Equation 1.12

Organic Solvent	MPD	SD	N
Acetone	556.7	814.1	19
Benzene	206.3	294.9	10
CHCl$_3$	122.7	198.9	35
Cyclohexane	41.6	52.9	9
Diethyl ether	1944.8	2202.5	19
Dioxane	160.3	183.6	9
Ethanol	916.6	1454.1	37
Hexane	44.0	50.1	12
Isooctane	30.9	12.6	9
Methanol	890.0	1582.4	11
Octanol	369.4	681.4	9
Propylene glycol	132.0	222.5	12

The MOD model (i.e., Equation 1.12) is also able to calculate the solubility of drugs in organic solvents. Table 1.5 lists the overall MPDs and SDs of Equation 1.12 for a number of organic solvents reported by Reulle et al. [42].

Martin and coworkers [51] calculated the solubility of naphthalene (X_2) in 24 different organic solvents at 40°C using the extended Hansen and extended Hildebrand approaches, the universal functional group activity coefficient (UNIFAC), and the regular solution theory. The obtained models for the extended Hansen and the extended Hildebrand models are

$$\log\left(\frac{X_2^i}{X_2}\right) = 0.0451 + \left\{\left(\frac{V_2\phi_1^2}{2.303RT}\right)\left[1.0488\left(\delta_{1d} - \delta_{2d}\right)^2 - 0.3148\left(\delta_{1p} - \delta_{2p}\right)^2\right]\right\} + 0.2252\left(\delta_{1h} - \delta_{2h}\right)^2$$
(1.33)

$$\log\left[\frac{\left(X_2^i/X_2\right)}{\left(V_2\phi_1^2/2.303RT\right)}\right] = 197.8011 + 6.1130I - 52.2568\delta_1 + 4.6506\delta_1^2 - 0.12908\delta_1^3$$ (1.34)

where
  X_2^i is the ideal solubility of the solute
  V_2 is the molar volume of the solute
  ϕ_1 is the volume fraction of the solvent in the solution
  δ_{1d} and δ_{2d} are the dispersions; δ_{1h} and δ_{2h} are the hydrogen bonds; and δ_{1p} and δ_{2p} are the polar solubility parameters of the solvents and the solute, respectively
  R is the molar gas constant
  T is the absolute temperature

Table 1.6 lists the solubility parameters and the molar volume of water and the common organic solvents [52–54]. The I term is an indicator variable (1 for alcohols and 0 for other solvents investigated in their works) and δ_1 is the Hildebrand solubility parameter of the solvents. Martin et al. [51] reported the accuracy of four investigated models by residual values. The computed

TABLE 1.6
The Solubility Parameters (MPa$^{1/2}$) and the Molar Volumes of the Common Solvents

Solvent	Molar Volume (cm³/mol)	Dispersion	Dipolar	Hydrogen Bonding	Acidic	Basic	Total	Reference
Acetic acid	57.6	14.52	7.98	13.50	14.32	6.34	21.36	[52]
Acetic anhydride	—	15.34	11.05	9.61	—	—	21.21	[54]
Acetone	74.0	15.55	10.43	6.95	4.91	4.91	19.96	[52]
Acetophenone	117.4	19.64	8.59	3.68	2.25	3.07	21.74	[52]
Aniline	91.5	19.43	5.11	10.23	3.89	13.30	22.52	[52]
Benzaldehyde	—	18.72	8.59	5.32	—	—	21.27	[54]
Benzene	89.4	18.41	1.02	2.05	1.43	1.43	18.55	[52]
Benzyl alcohol	103.9	18.41	6.34	13.70	12.07	7.77	23.81	[52]
Butandiol (1,4-)	88.6	16.77	16.57	23.73	37.23	7.57	33.46	[52]
Butane (n-)	—	14.10	0.00	0.00	—	—	14.10	[54]
Butanol (1-)	92.0	15.95	5.73	15.75	13.09	9.41	23.09	[52]
Butanol (iso)	92.4	15.14	5.73	15.95	12.27	10.43	22.77	[52]
Butanol (sec)	92.5	15.75	5.73	14.52	13.50	7.77	22.15	[52]
Butanol (tert)	94.3	14.93	5.11	13.91	19.64	4.91	21.03	[52]
Butyl acetate	132.6	15.75	3.68	6.34	5.73	3.48	17.37	[52]
Butyric acid	91.9	14.93	4.09	10.64	13.09	4.30	18.76	[52]
Butyrolactone (gama)	—	19	16.6	7.4	—	—	—	[53]
Carbon disulfide	60.0	20.46	0.00	0.61	0.41	0.41	20.46	[52]
Carbon tetrachloride	97.1	17.80	0.00	0.61	0.10	1.84	17.82	[52]
Chlorobenzene	102.2	19.02	4.30	2.05	2.05	1.02	19.62	[52]
Chloroform	80.8	17.80	3.07	5.73	6.14	2.66	18.94	[52]
Cyclohexane	108.8	16.77	0.00	0.00	0.00	0.00	16.77	[52]
Cyclohexanol	106.0	17.39	4.09	13.50	14.93	6.14	22.42	[52]
Decane	195.9	15.75	0.00	0.00	0.00	0.00	15.75	[52]
Decane (n-)	—	15.80	0.00	0.00	—	—	15.80	[54]
Dibutyl ether	170.4	15.55	1.64	3.07	0.61	7.98	15.93	[52]
Dichlorobenzene (o-)	—	19.2	6.3	3.3	—	—	—	[53]
Diethyl ether	104.8	14.52	2.86	5.11	1.02	12.89	15.67	[52]

(continued)

TABLE 1.6 (continued)
The Solubility Parameters (MPa$^{1/2}$) and the Molar Volumes of the Common Solvents

Solvent	Molar Volume (cm³/mol)	Dispersion	Dipolar	Hydrogen Bonding	Acidic	Basic	Total	Reference
Diethylamine	—	14.93	2.25	6.14	—	—	16.30	[54]
Diethylene glycol	—	16.6	12	20.7	—	—	—	[53]
Diethylene glycol	—	16.08	14.73	20.46	—	—	29.86	[54]
Diisobutyl ketone	—	15.89	3.68	4.09	—	—	16.71	[54]
Dimethylformamide	—	17.4	13.7	11.3	—	—	—	[53]
DMSO	71.3	18.41	16.36	10.23	4.50	11.66	26.67	[52]
Dioxane (1,4-)	85.7	19.02	1.84	7.36	2.05	13.30	20.48	[52]
Dipropyl ether	139.4	14.93	2.25	4.09	0.82	10.23	15.65	[52]
Dipropylene glycol	—	15.89	20.25	18.41	—	—	31.75	[54]
Ethanol	58.7	15.75	8.80	19.43	16.98	11.25	26.51	[52]
Ethanol amine	—	17	15.5	21.2	—	—	—	[53]
Ethyl acetate	98.5	15.14	5.32	9.20	10.84	3.89	18.49	[52]
Ethyl benzene	—	17.80	0.61	1.43	—	—	17.86	[54]
Ethyl lactate	—	15.95	7.57	12.48	—	—	21.62	[54]
Ethylene dibromide	87.0	19.02	3.48	8.59	22.91	1.64	21.19	[52]
Ethylene dichloride	79.4	19.02	7.36	4.09	4.09	2.05	20.80	[52]
Ethylene glycol	55.9	16.98	11.05	25.77	36.61	9.00	32.71	[52]
Ethylene glycol monobutyl ether	—	16	5.1	12.3	—	—	—	[53]
Ethylene glycol monoethyl ether	—	16.2	9.2	14.3	—	—	—	[53]
Ethylene glycol monomethylether	—	16.2	9.2	16.4	—	—	—	[53]
Formamide	39.9	17.18	26.18	19.02	11.66	15.55	36.66	[52]
Formic acid	—	14.32	11.86	16.57	—	—	24.91	[54]
Furan	—	17.80	1.84	5.32	—	—	18.59	[54]
Glycerol	73.2	17.39	12.07	29.25	40.91	10.43	36.08	[52]
Heptane	147.5	15.34	0.00	0.00	0.00	0.00	15.34	[52]

Heptanol (1-)	141.9	16.57	4.09	12.27	10.84	6.95	21.03	[52]
Hexane	131.6	14.93	0.00	0.00	0.00	0.00	14.93	[52]
Hexanol (1-)	125.2	16.36	4.30	12.89	11.66	7.16	21.29	[52]
Hexyl acetate	164.5	15.95	3.07	5.93	3.89	4.50	17.28	[52]
Isophorone	—	16.57	8.18	7.36	—	—	19.88	[54]
Mesityl oxide	—	16.30	7.16	6.14	—	—	18.84	[54]
Methanol	40.7	15.14	12.27	22.30	17.18	14.52	29.64	[52]
Methyl ethyl ketone	—	16	9	5.1	—	—	—	[53]
Methyl isobutyl ketone	—	15.3	6.1	4.1	—	—	—	[53]
Methyl-2-pyrrolidone	—	18	12.3	7.2	—	—	—	[53]
Methylene chloride	—	18.23	6.34	6.14	—	—	20.31	[54]
Methylene dichloride	—	18.2	6.3	6.1	—	—	—	[53]
Methylethyl ketone	—	15.89	9.00	5.11	—	—	18.96	[54]
Morpholine	—	18.82	4.91	9.20	—	—	21.52	[54]
N,N-Diethylacetamide	126.6	16.77	8.39	7.57	4.30	6.55	20.19	[52]
N,N-Diethylformamide	112.0	16.77	11.45	8.80	5.52	6.95	22.11	[52]
N,N-Dimethylacetamide	93.0	16.77	11.45	10.23	5.93	8.80	22.73	[52]
N,N-Dimethylformamide	77.4	17.39	13.70	11.25	6.95	9.00	24.81	[52]
Nitrobenzene	102.7	20.05	8.59	4.09	4.09	2.05	22.19	[52]
Nitroethane	—	15.95	15.55	4.50	—	—	22.68	[54]
Nitromethane	—	15.75	18.81	5.11	—	—	25.06	[54]
Nitropropane (2-)	—	16.2	12.1	4.1	—	—	—	[53]
N-Methylformamide	59.1	17.18	20.66	12.48	9.82	7.98	29.64	[52]
Nonane	179.7	15.75	0.00	0.00	0.00	0.00	15.75	[52]
Octane	—	15.60	0.00	0.00	—	—	15.60	[54]
Octanol (1-)	158.4	16.98	3.27	11.86	10.64	6.55	20.93	[52]
Pentane	116.1	14.52	0.00	0.00	0.00	0.00	14.52	[52]
Pentanol (1-)	108.6	15.95	4.50	13.91	11.05	8.80	21.66	[52]
Propandiol (1,2-)	73.7	16.77	9.41	23.32	28.84	9.41	30.21	[52]
Propandiol (1,3-)	72.5	16.57	10.84	25.98	22.30	15.14	32.67	[52]
Propanol (1-)	75.1	15.95	6.75	17.39	15.34	9.82	24.53	[52]

(continued)

TABLE 1.6 (continued)
The Solubility Parameters (MPa$^{1/2}$) and the Molar Volumes of the Common Solvents

Solvent	Molar Volume (cm^3/mol)	Dispersion	Dipolar	Hydrogen Bonding	Acidic	Basic	Total	Reference
Propanol (2-)	76.9	15.75	6.14	16.36	14.52	9.20	23.52	[52]
Propionic acid	75.0	14.73	7.77	12.27	12.27	6.14	20.68	[52]
Propyl acetate	115.7	15.75	4.30	6.75	7.36	3.07	17.65	[52]
Propylene carbonate	—	20.11	18.00	4.09	—	—	27.29	[54]
Propylene glycol	—	16.86	9.41	23.32	—	—	30.27	[54]
Pyridine	80.9	19.02	8.80	6.14	2.86	6.55	21.83	[52]
Styrene	—	18.55	1.02	4.09	—	—	19.02	[54]
Tetrahydrofuran	—	16.81	5.73	7.98	—	—	19.47	[54]
Toluene	106.9	18.00	1.43	2.05	1.64	1.23	18.16	[52]
Toluene	—	18	1.4	2	—	—	—	[53]
Toluene	—	18.04	1.43	2.05	—	—	18.26	[54]
Trichloroethylene	—	18	3.1	5.3	—	—	—	[53]
Trichloroethylene	—	17.96	3.07	5.32	—	—	18.98	[54]
Water	18.1	15.55	15.95	42.34	13.70	65.46	47.86	[52]
Xylene (p-)	—	17.69	1.02	3.07	—	—	18.00	[54]

Sources: Beerbower, A. et al., *J. Pharm. Sci.*, 73, 179, 1984; Gharagheizi, F., *J. Appl. Polym. Sci.*, 103, 31, 2007; Ho, D.L. and Glinka, C.J., *J. Polym. Sci. B Polym. Phys.*, 42, 4337, 2004.

MPD values for the models were 13.6%, 23.8%, 12.6%, and 95.0% for the extended Hansen, the extended Hildebrand, the UNIFAC, and regular solution models [32], respectively. In a follow-up study [55], the extended Hansen solubility approach was used to estimate the partial solubility parameters of naphthalene. The same procedures were employed to calculate the partial solubility of a number of solid crystalline solutes [52,55–64]. The computed parameters are listed in Table 1.7.

The extended Hansen equations were presented to reproduce the solubility of a solid solute in individual solvents [51,55] as

$$\log X_S = C_0 + C_1 \delta_{1d}^2 + C_2 \delta_{1d} + C_3 \delta_{1p}^2 + C_4 \delta_{1p} + C_5 \delta_{1h}^2 + C_6 \delta_{1h} \tag{1.35}$$

$$\log X_S = C_0 + C_1 \delta_{1d}^2 + C_2 \delta_{1d} + C_3 \delta_{1p}^2 + C_4 \delta_{1p} + C_5 \delta_{1h}^2 + C_6 \delta_{1h} + C_7 \delta_{1a} \delta_{1b} \tag{1.36}$$

Equations 1.35 and 1.36 could also be used to calculate the partial solubility parameters of the solutes using the following expressions:

$$\delta_{2d} = \frac{-C_2}{2C_1} \tag{1.37}$$

$$\delta_{2p} = \frac{-C_4}{2C_3} \tag{1.38}$$

$$\delta_{2h} = \frac{-C_6}{2C_5} \tag{1.39}$$

$$\delta_{2a} = \frac{-C_6}{C_7} \tag{1.40}$$

$$\delta_{2b} = \frac{-C_5}{C_7} \tag{1.41}$$

Slightly different partial solubility parameters were reported for drugs using Equations 1.35 and 1.36, as noted by the numerical entries in Table 1.7.

Bustamante et al. [65] proposed an MLR model employing ideal solubility (X_2^i), molar volume (V_2), and partial solubility parameters of the solutes to represent the solubility of multiple drugs in different solvents. The model was

$$\log X_S = b_1 \log X_2^i + b_2 V_2 + b_3 \delta_{1d} + b_4 \delta_{1d}^2 + b_5 \delta_{1p} + b_6 \delta_{1p}^2 + b_7 \delta_{1a} + b_8 \delta_{1b} + b_9 \delta_{1a} \delta_{1b} \tag{1.42}$$

where b_1–b_9 are the model constants. Equation 1.42 was trained using the experimental solubilities of sulfadiazine, sulfamethoxypyridazine, benzoic acid, p-hydroxybenzoic acid, and methyl p-hydroxybenzoate at 25°C, as well as the solubility of naphthalene at 40°C. There is no independent variable representing the temperature effects in the model; therefore, there is no logical basis to employ the solubility at 25°C and 40°C altogether. The trained model was reported as [65]

TABLE 1.7
The Solubility Parameters (MPa$^{1/2}$) and the Molar Volumes of the Common Solutes

Solute	Reference	Dispersion (δ_{2d})	Dipolar (δ_{2p})	Hydrogen Bonding (δ_{2h})	Acidic (δ_{2a})	Basic (δ_{2b})	Total (δ_2)
Benzoic acid	[52]	17.26	16.13	4.23	—	—	24.01
Benzoic acid	[52]	17.22	15.05	—	9.04	3.25	24.11
Benzoic acid	[52]	17.75	12.17	—	9.84	6.54	24.34
Benzoic acid	[64]	17.63	10.10	—	9.27	8.23	22.98
Benzoic acid	[64]	17.26	12.17	11.34	9.83	6.54	22.33
Citric acid	[190]	16.49	14.19	16.72	—	—	27.44
Citric acid	[190]	16.24	13.54	—	17.17	8.73	27.35
Diazepam	[61]	15.83	13.35	8.23	—	—	—
Diazepam	[61]	15.90	13.29	8.57	—	—	—
Haloperidol	[60]	17.12	10.65	8.43	—	—	21.85
Haloperidol	[60]	17.16	9.30	9.24	5.29	8.08	21.57
Ibuprofen	[64]	16.44	6.39	8.89	—	—	19.75
Ibuprofen	[64]	16.37	7.67	—	5.36	4.85	19.46
Lactose	[59]	17.57	28.67	18.99	14.50	12.43	38.61
Lorazepam	[61]	15.75	15.81	13.10	—	—	—
Lorazepam	[61]	15.92	15.64	13.83	—	—	—
Mannitol	[59]	16.15	24.53	14.56	8.71	12.18	32.78
Methyl p-hydroxybenzoate	[58]	18.59	14.89	—	9.57	1.49	24.42
Naphthalene	[52]	21.49	3.64	—	5.89	1.12	22.11
Naphthalene	[55]	20.61	4.01	1.88	—	—	21.08
Naproxen	[57]	17.35	12.14	9.86	12.31	3.95	23.35
Naproxen	[57]	18.70	13.01	11.51	12.17	5.46	25.51
Naproxen	[57]	18.70	11.39	11.11	14.11	4.38	24.55
Niflumic acid	[62]	16.75	12.11	—	13.90	4.96	23.77
Oxazepam	[61]	15.44	16.13	12.78	—	—	—
Oxazepam	[61]	15.62	16.17	13.64	—	—	—
Paracetamol	[190]	17.63	14.50	8.08	—	—	24.22
Paracetamol	[190]	16.60	13.78	—	19.98	7.69	27.80
Paracetamol	[190]	16.58	14.26	—	22.29	7.57	28.56
p-Hydroxybenzoic acid	[58]	16.99	14.48	—	11.92	8.00	26.26
Pimozide	[63]	16.53	11.05	16.51	—	—	25.77
Pimozide	[63]	23.73	11.46	5.50	—	—	26.85
Piroxicam	[62]	16.76	21.36	—	3.03	7.12	27.93
Piroxicam	[62]	16.78	21.22	—	3.01	7.08	27.83
Piroxicam	[62]	16.95	18.97	—	4.81	6.45	26.62
Piroxicam	[62]	17.00	16.03	8.74	—	—	24.94
Prazepam	[61]	15.79	13.01	6.73	—	—	—
Prazepam	[61]	15.65	12.60	6.79	—	—	—
Saccharose	[59]	17.09	18.52	13.05	11.26	7.57	28.38
Sodium benzoate	[64]	16.28	29.19	13.04	—	—	35.87
Sodium diclofenac	[57]	16.47	16.62	18.54	—	—	29.85
Sodium diclofenac	[57]	16.27	18.05	13.48	9.82	9.26	27.79
Sodium ibuprofen	[64]	17.5	17.89	29.31	—	—	38.54
Sulfamethoxypyridazine	[56]	25.52	17.01	10.49	8.96	6.13	32.31
Temazepam	[61]	15.56	15.92	9.71	—	—	—
Temazepam	[61]	15.66	15.89	11.24	—	—	—
Temazepam	[191]	21.90	9.80	8.22	—	—	—

$$\log X_S = 0.850 \log X_2^i - 0.0112V_2 - 0.221\delta_{1d} + 0.021\delta_{1d}^2 + 0.415\delta_{1p}$$
$$- 0.025\delta_{1p}^2 + 0.167\delta_{1a} + 0.230\delta_{1b} - 0.050\delta_{1a}\delta_{1b} \tag{1.43}$$

A similar task was also achieved using the solvatochromic approach as

$$\log X_S = -0.01(\text{mp} - 25) - 0.002V_2 + 4.211\beta_m \tag{1.44}$$

where β_m is the hydrogen bonding acidity of the solute.

The Abraham solvation model provides a more comprehensive solubility prediction method for organic solvents. The Abraham model written in terms of solubility is

$$\log\left(\frac{S_S}{S_W}\right) = c + e \cdot E + s \cdot S + a \cdot A + b \cdot B + v \cdot V \tag{1.45}$$

where

S_S and S_W are the solute solubilities in the organic solvent and water (in mol/L), respectively

E is the excess molar refraction

S is the dipolarity/polarizability of the solute

A denotes the solute's hydrogen bond acidity

B stands for the solute's hydrogen bond basicity

V is the McGowan volume of the solute

In Equation 1.45, the coefficients c, e, s, a, b, and v are the model constants (i.e., solvent's coefficients), which depend on the solvent system under consideration. Table 1.8 lists the Abraham solvent coefficients for a number of solvents taken from a reference [66]. These coefficients were computed by regression analysis of the measured $\log(S_S/S_W)$ values, the infinite dilution activity coefficients, and the partition coefficients of various solutes against the corresponding solute parameters [67]. The Abraham solvent coefficients (c, e, s, a, b, and v) and the Abraham solute parameters (E, S, A, B, and V) represent the extent of all known interactions between solutes and solvents in the solution [24]. It is obvious that Equation 1.45 could be rewritten as

$$\log S_S = \log S_W + c + e \cdot E + s \cdot S + a \cdot A + b \cdot B + v \cdot V \tag{1.46}$$

to calculate the solubility in organic solvents.

1.4   THE ACCURACY CRITERIA IN SOLUBILITY CALCULATION METHODS

Li [68] compared the accuracy of the extended log-linear model with the log-linear model using the average error of prediction (AEP) calculated by

$$\text{AEP} = \frac{\sum \left| \log X_m^{\text{Calculated}} - \log X_m^{\text{Observed}} \right|}{N} \tag{1.47}$$

The AEP term has also been used in other reports with the AAE term [22,23,69–71].

TABLE 1.8
The Abraham Solvent Coefficients

Solvent	c	e	s	a	b	v
1-Butanol (dry)	0.152	0.437	−1.175	0.098	−3.914	4.119
1-Decanol (dry)	−0.062	0.754	−1.461	0.063	−4.053	4.293
1-Heptanol (dry)	−0.026	0.491	−1.258	0.035	−4.155	4.415
1-Hexanol (dry)	0.044	0.470	−1.153	0.083	−4.057	4.249
1-Octanol (dry)	−0.034	0.490	−1.048	−0.028	−4.229	4.219
1-Pentanol (dry)	0.080	0.521	−1.294	0.208	−3.908	4.208
1-Propanol (dry)	0.147	0.494	−1.195	0.495	−3.907	4.048
2-Butanol (dry)	0.106	0.272	−0.988	0.196	−3.805	4.110
2-Methyl-1-propanol (dry)	0.177	0.355	−1.099	0.069	−3.570	3.990
2-Methyl-2-propanol (dry)	0.197	0.136	−0.916	0.318	−4.031	4.113
2-Pentanol (dry)	0.115	0.455	−1.331	0.206	−3.745	4.201
2-Propanol (dry)	0.063	0.320	−1.024	0.445	−3.824	4.067
3-Methyl-1-butanol (dry)	0.123	0.370	−1.243	0.074	−3.781	4.208
Acetone (dry)	0.335	0.349	−0.231	−0.411	−4.793	3.963
Acetonitrile (dry)	0.413	0.077	0.326	−1.566	−4.391	3.364
Benzene	0.142	0.464	−0.588	−3.099	−4.625	4.491
Chlorobenzene	0.040	0.246	−0.462	−3.038	−4.769	4.640
Decane	0.160	0.585	−1.734	−3.435	−5.078	4.582
Dibutyl ether (dry)	0.203	0.369	−0.954	−1.488	−5.426	4.508
Diethyl ether (dry)	0.330	0.401	−0.814	−0.457	−4.959	4.320
Ethyl acetate (dry)	0.358	0.362	−0.449	−0.668	−5.016	4.155
Gas to water	−0.994	0.577	2.549	3.813	4.841	−0.869
Heptane	0.325	0.670	−2.061	−3.317	−4.733	4.543
Hexadecane	0.087	0.667	−1.617	−3.587	−4.869	4.433
Hexane	0.361	0.579	−1.723	−3.599	−4.764	4.344
Isooctane	0.288	0.382	−1.668	−3.639	−5.000	4.461
Methanol (dry)	0.329	0.299	−0.671	0.080	−3.389	3.512
Methyl *tert*-butyl ether (dry)	0.376	0.264	−0.788	−1.078	−5.030	4.410
Methylcyclohexane	0.246	0.782	−1.982	−3.517	−4.293	4.528
Octane	0.223	0.642	−1.647	−3.480	−5.067	4.526
Tetrahydrofuran (dry)	0.207	0.372	−0.392	−0.236	−4.934	4.447
Toluene	0.143	0.527	−0.720	−3.010	−4.824	4.545

Source: Flanagan, K.B. et al., *Phys. Chem. Liq.,* 44, 377, 2006.

The RMSE (in log scale) is calculated by the following equation:

$$\text{RMSE} = \sqrt{\frac{\sum \left(\text{Observed} - \text{Predicted}\right)^2}{N}} \tag{1.48}$$

which was used in many publications [37,72–75].

The MSD is the mean squared deviation and is calculated by the equation

$$\text{MSD} = \sqrt{\frac{\left(X_m^{\text{Calculated}} - X_m^{\text{Observed}}\right)^2}{N - 1 - p}} \tag{1.49}$$

and was employed by various investigators [76–78].

The solubility ratio of the experimental and calculated values was used by Ruelle et al. [42] and defined as

$$SR = \left(\frac{\Phi_B^{\text{Calculated}}}{\Phi_B^{\text{Experimental}}} \right) \tag{1.50}$$

where Φ_B is the volume fraction solubility of the solute.

To provide a meaningful error term, which is comparable to relative standard deviation (RSD) values, the deviations between the predicted and experimental values were calculated using

$$MPD = \frac{100}{N} \sum \left(\frac{|S_{\text{Calculated}} - S_{\text{Observed}}|}{S_{\text{Observed}}} \right) \tag{1.51}$$

in which N is the number of data points in each set. This criterion was used in nearly all publications from our group with various terminologies. The MPD reflects the overall calculation error and the individual percentage deviation (IPD) could also be computed as

$$IPD = 100 \left(\frac{|S_{\text{Calculated}} - S_{\text{Observed}}|}{S_{\text{Observed}}} \right) \tag{1.52}$$

1.5 ACCEPTABLE MPD RANGE IN SOLUBILITY CALCULATIONS

Solubility calculations can be studied from the point of view of calculation error. In order to provide a comprehensive approach to evaluate the process of predicted solubilities, these could be divided into three groups: predictive, semipredictive, and correlative calculations.

Pure predictive calculations, such as group contribution methods, which were widely employed to predict the aqueous solubility of pharmaceutical and environmentally important compounds, do not need any experimental data. In the pharmaceutical area, Yalkowsky and coworkers have presented a large number of publications. A method using the AQUAFAC has been presented for predicting aqueous solubility of 168 nitrogen-containing compounds including some pharmaceuticals [70]. The results are reported as the AAE on a logarithmic scale as 0.42.

The calculated MPD value based on tabulated aqueous solubility data of 165 compounds [70] is 2108%. The IPDs span between 0% for nonylamine and 308219% for dipropalin (its logarithm of the experimental aqueous solubility is −2.966 and the predicted log S_w is −6.455). The MPD of AQUAFAC is still very high (MPD = 284%) when five outliers that produced very high error were removed [39].

In a chemical engineering application, the solubilities of three polycyclic aromatic hydrocarbons (anthracene, fluoranthene, and pyrene) were predicted with the modified UNIFAC (Dortmund) model. The MPD were reported to be 36%, 36%, and 35%, for anthracene, fluoranthene, and pyrene, respectively [79].

Semipredictive calculations, such as the GSE, were proposed by Yalkowsky and Valvani [16]. The basis of these calculations was to employ a training set to compute the model constants and then to use the computed coefficients to predict other data points. In a comparative study reported by Huuskonen et al. [38], three MLR methods (including GSE) and an ANN model were trained using 675 aqueous solubility data points taken from published databases. In this study, the prediction capabilities of the models were assessed using aqueous solubility data of 38 pharmaceuticals. Here the MPD values for MLR models are 2466 ± 12877%, 117 ± 125%, and 181 ± 327%, and the corresponding value for AAN model is 154 ± 193. As noted previously, Abraham and Le [23] proposed the ASER to calculate the aqueous solubility of solutes. The authors employed a training set, which includes the aqueous solubility of 659 compounds, and then from this they predicted the aqueous solubility of an additional 65 compounds. The reported AAE for the 65 predicted solubility values on a logarithmic scale was 0.50. The corresponding MPD value obtained was 134 ± 373%. These MPD values are so large as to be meaningless and therefore it is suggested that such models would not be of any use in the industry [39]. However, even with such enormous errors, both types of the solubility calculations are in high demand in the pharmaceutical area, where fast, reliable, and generally applicable methods are needed for the prediction of aqueous solubility of new drugs until a promising drug candidate becomes available through development. From this information it is clear that current knowledge and expertise are insufficient to guarantee the reliability of the predictions. Thus, they are not acceptable as a basic procedure prior to developing an optimized process, but they do serve as a rough guide for further testing and for priority setting of parameter range.

Correlative calculations such as simple least square methods provide a means to screen the experimental data to detect the possible outliers where redetermination is required. In addition, the correlative trained models can be employed to predict undetermined data. As an example, the GSE model is proposed for correlating aqueous solubility of different compounds in water. Its correlation ability and that of a revised form of GSE have been studied by Jain and Yalkowsky [80]. The MPD values of the original and revised forms of GSE have been calculated employing aqueous solubility of 582 nonelectrolytes; the MPDs obtained are 2881 ± 30716% and 1417 ± 15393%. The reported error percentage for the correlative form of an ASER employing 659 data is 0.408 and the corresponding MPD value is 244 ± 850% [23,32].

When a drug has a low aqueous solubility as is found from either experimental measurements or theoretical computations, its solubility often needs to be increased by using different techniques including cosolvency. The reported MPDs in the calculation of aqueous solubility of solutes are too large and cannot be extensively used in industry; however, these models could be considered as the first generation of solubility models.

A summary of different types of solubility calculations collected from the published chemical and pharmaceutical literature [25,38,69,73,74,76–78,80–88] is shown in Table 1.9. The differences between predicted and experimental solubilities have been presented by MPD. In cases where it was not possible to calculate the MPD, the MSD, RMSE, or AAE values reported in the original papers were employed. There are linear relationships between various accuracy criteria as shown in Figure 1.1 for MPD, RMSE, and AAE [25].

As noted earlier, most of the models provided predictions with relatively high error values. In trying to explain why prediction errors are so high, a number of sources can be discussed. In addition to the error related to the nature of the models and calculation procedure, the quality of solubility data is one possible contribution to error sources in solubility data modeling. Experimental solubility data for a given solute can vary from laboratory to laboratory. As representative examples, published solubility data for paracetamol, sulfadiazine, sulfadimidine, sulfapyridine, ibuprofen, theophylline, and naphthalene [23,80,81,89–95] are compared in Table 1.10. The tabulated experimental data clearly illustrate that there can be a significant amount of variability in experimental solubilities. In an interesting work by Kishi and Hashimoto [96], solubility data of anthracene and fluoranthene reported by 17 different laboratories using a standard method by the environmental agency of Japan have been summarized. The results showed that even when all variables were kept constant, interlaboratory difference can still be very significant. The mean solubilities of anthracene and fluoranthene from the study spanned between 0.17 and 0.36 log unit and MPD value was 51% [96]. It is obvious that the ranges of individual solubilities are even greater where the range of logarithm of mol/L solubility for anthracene is −7.08 to −5.23. In addition to the variation of the results for different laboratories, the RSD values for the repeated experiments from the same laboratory are significantly high. As examples, the reported RSD values are up to 9.2% [95], 4.4% [97], 10% [98], and 28% [99].

TABLE 1.9
The Reported Deviations for Solubility Calculations Collected from the Literature

System[a]	Model	Reference	Correlative Calculations		Predictive Calculations	
			N[b]	Deviation	N[b]	Deviation
I	UNIFAC	[82]	5–17	MPD[c] = 55	—	—
I	Equation 1.65	[82]	5–17	MPD = 36	—	—
I	Equation 1.28	[82]	5–17	MPD = 49	—	—
I	Equation 1.127	[74]	11	RMSE[d] = 0.049	—	—
I	Equation 1.98	[74]	11	RMSE = 0.063	—	—
I	Ruckenstein and Shulgin	[74]	11	RMSE = 0.064	—	—
I	Equation 1.98 ($i = 2$)	[73]	—	—	460	RMSE = 0.120
I	Equation 1.57	[69]	294	AAE[e] = 0.48	—	—
I, II	Equation 1.72	[85]	12–13	MPD = 6.9	7–8	MPD = 7.3
I, II	Equation 1.76	[85]	12–13	MPD = 11.4	7 to 8	MPD = 14.6
I, II	Equation 1.87	[85]	12–13	MPD = 4.9	7–8	MPD = 7.9
I, II	Equation 1.98 ($i = 4$)	[85]	12–13	MPD = 7.5	7–8	MPD = 8.0
I, II	Mean predicted solubility	[85]	12–13	MPD = 5.9	7–8	MPD = 6.4
I, II	Equation 1.87	[83]	11–21	MPD = 5.56	6–16	MPD = 10.33
I, II	ANN	[83]	11–21	MPD = 0.90	6–16	MPD = 9.04
II	Equation 1.72	[85]	—	—	6–8	MPD = 8.7
II	Equation 1.76	[85]	—	—	6–8	MPD = 13.3
II	Equation 1.87	[85]	—	—	6–8	MPD = 7.2
II	Equation 1.98 ($i = 4$)	[85]	—	—	6–8	MPD = 7.4
II	Mean predicted solubility	[85]	—	—	6–8	MPD = 7.6
II	Equation 1.98 ($i = 3$)	[86]	—	—	6–16	MPD = 13.1
II	Equation 1.76	[86]	—	—	6–16	MPD = 14.2
II	Equation 1.87	[86]	—	—	6–16	MPD = 12.5
II	Equation 1.95	[86]	—	—	6–16	MPD = 15.0
II	Mean predicted solubility	[86]	—	—	6–16	MPD = 10.6
II	Equation 1.127	[74]	—	—	455	RMSE = 0.035
II	Equation 1.98	[74]	—	—	455	RMSE = 0.050
II	Equation 1.57	[73]	—	—	460	RMSE = 0.423
II	Equation 1.127	[73]	—	—	460	RMSE = 0.069
II	Equation 1.88	[69]	294	AAE = 0.19	—	—
II	Equations 1.88 and 1.91	[25]	—	—	467	MPD = 18.5
II	Equation 1.57	[25]	—	—	467	MPD = 76.7
III	Equation 1.87	[83]	496	MPD = 90.42	—	—
III	ANN	[83]	496	MPD = 24.76	—	—
IV	Equation 1.87	[83]	—	—	236	MPD = 81.10
IV	ANN	[83]	—	—	236	MPD = 55.97
V	Equation 1.87	[83]	454	MPD = 20.36	—	—
V	ANN	[83]	454	MPD = 2.02	—	—
VI	Equation 1.65 modified	[81]	142	MPD = 35	—	—
VI	Equation 1.87	[84]	142	MPD = 18	—	—
VI	Equation 1.87	[83]	278	MPD = 18.37	—	—
VI	ANN	[83]	278	MPD = 4.70	—	—
VII	Equation 1.87	[83]	120	MPD = 67.19	—	—
VII	ANN	[83]	120	MPD = 3.36	—	—
VIII	Least square	[76]	331	MSD[f] = 0.30	17	MSD = 0.34
VIII	AQUAFAC	[77]	873	MSD = 0.56	97	MSD = 0.56
VIII	GSE	[77]	873	MSD = 0.80	97	MSD = 0.80

(continued)

TABLE 1.9 (continued)
The Reported Deviations for Solubility Calculations Collected from the Literature

System[a]	Model	Reference	Correlative Calculations		Predictive Calculations	
			N[b]	Deviation	N[b]	Deviation
VIII	ANN	[78]	123	MSD = 0.22	13	MSD = 0.23
VIII	Least square	[78]	123	MSD = 0.28	13	MSD = 0.28
VIII	AQUAFAC	[87]	133	MSD = 0.33	25	MSD = 0.324
VIII	GSE	[87]	—	—	25	MSD = 0.423
VIII	Mobile order	[192]	—	—	531	MSD = 0.37
VIII	Least square	[38]	675	—	38	MPD = 2466
VIII	Least square	[38]	675	—	38	MPD = 117
VIII	Least square	[38]	675	—	38	MPD = 181
VIII	ANN	[38]	675	—	38	MPD = 154
VIII	GSE	[80]	582	MPD = 2881	—	—
VIII	GSE-revised form	[80]	582	MPD = 1417	—	—

[a] System, I: solubility of a given solute in binary aqueous solvent mixture, II: solubility of a given solute in binary aqueous solvent mixture using the trained model employing a minimum number of experimental data points, III: correlation of solubility of various solutes in different water–cosolvent mixtures, IV: prediction of solubility of various solutes in different water–cosolvent mixtures, V: correlation of solubility of various solutes in a given water–cosolvent mixture, VI: correlation of solubility of structurally related solutes in a given water–cosolvent mixture, VII: correlation of solubility of a given solute in different water–cosolvent mixtures, and VIII: correlation of aqueous solubility of different solutes.

[b] N is the number of correlated/predicted data points in each set.

[c] MPD is calculated by the equation:

$$MPD = \frac{100}{N} \sum \left(\frac{|S_{Calculated} - S_{Observed}|}{S_{Observed}} \right)$$

[d] RMSE (in log scale) is calculated by the equation:

$$RMSE = \sqrt{\frac{\sum (Observed - Predicted)^2}{N}}$$

[e] AAE (in log scale) is calculated by the equation: $AAE = \dfrac{\sum |Observed - Predicted|}{N}$

[f] MSD is calculated by the equation: $MSD = \sqrt{\dfrac{(S_{Calculated} - S_{Observed})^2}{N - 1 - p}}$

The possible reasons for such differences in solubilities arise from

1. Solute purity
2. Lack of equilibration
3. Temperature (small differences)
4. Analysis method
5. Laboratory technique [99]
6. Typographical error
7. Polymorphism [100]
8. Enantiomeric forms [94]

From a computational chemist's point of view, the ideal model would have an MPD value equal to zero. However, this is impossible because of variations in experimental results (uncertainty) that appear in the training process of the model. In practice, experimental chemists look for a model,

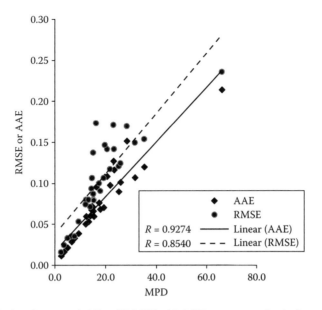

FIGURE 1.1 Correlations between AAE and RMSE with MPD accuracy criteria for predicted solubilities.

TABLE 1.10
The Differences between Aqueous Solubility of Drugs from Different Research Groups

Solute	Reference 1	Solubility 1	Reference 2	Solubility 2	Difference (%)
Ibuprofen	[23]	0.000174	[94]	0.00943 (racemate)	98
Naphthalene	[80]	0.000251	[95]	0.000201	25
Paracetamol	[89]	0.001	[90]	0.001915	191
Sulfadiazine	[91]	0.00000493	[133]	0.00000327	51
Sulfadimidine	[91]	0.0000274	[81]	0.00000302	804
Sulfapyridine	[91]	0.0000194	[93]	0.00000178	988
Theophylline	[23]	0.0407	[95]	0.0340	17

which enables them to calculate the solubility with MPD values lower than the RSD values that are obtained from repeated experiments.

There is a need for an improved solubility prediction method, which is reasonably accurate, capable of rapid analysis even for large data sets or complex molecules, and capable of being performed using simple computational methods. As shown in Table 1.9, the error levels for aqueous solubility calculations are between 117% and 2500%. For solubility calculations in mixed solvent systems, the calculation error is not more than 18%–55% for correlative models [39]. Dickhut et al. [82] proposed that a mixed solvent solubility estimation method is considered acceptable if solubility calculations are, on average, within 30% of the experimentally determined values. This has also been confirmed by other research groups [52,93]. From a practical point of view, any model providing more accurate correlations/predictions in comparison with the previous models could be considered as a step forward and one should also keep in mind that a single and simple model will not be able to produce the most accurate results for all data sets studied.

While comparing the limitations and predictive accuracies of the various models for aqueous solubility data, it is important to use the same experimental data sets in order to make the results compatible with each other and to avoid any bias in the conclusion. A 21-test data set was used in some of the reports and Table 1.3 listed the various models that were investigated [35,38,101–106], the employed descriptors, and the references of the models along with the calculated MPD values [32]. Although this data set provided comparable accuracy criteria across various models, it suffers from a shortage of the experimental solubility data points. In addition, the diversity of data is insufficient and some of the employed data are not pharmaceutically related. For pharmaceutical applications, the solubility of all official drugs should be used as the test set for all coming aqueous solubility prediction methods [47].

1.6   SOLUBILIZATION OF DRUGS BY COSOLVENCY

There are several methods to enhance the aqueous solubility of drugs including cosolvency, hydrotropism, complexation, ionization, and the use of surface active agents. These methods have been discussed in detail in the literature [107]. In cosolvency, mixing a permissible nontoxic organic solvent with water is the most common technique to enhance the aqueous solubility of drugs. The common cosolvents that are used in the pharmaceutical industry are ethanol, propylene glycol, glycerine, glycofural, polyethylene glycols (PEGs) (mainly 200, 300, and 400), N, N-dimethyl acetamide, DMSO, 2-propanol, dimethyl isosorbide, N-methyl 2-pyrrolidone, and room temperature ionic liquids [2,108–110]. Their applications and possible side effects are discussed in the literature [108,111–114]. The cosolvency phenomenon has a wide range of applications in different fields. In addition to their application in drug formulation, solubilizing agents like *tert*-butyl ether are used to dissolve cholesterol gallstones [115]. From the environmental scientist's viewpoint, cosolvency is an important subject, because the organic solvents can change distribution and movement of hydrophobic contaminants in the environment [116]. The solubility behavior of a solute in mixed solvents provides some theoretical basis for the chemist. From this, it is possible to have some idea about solute–solvent and solvent–solvent interactions in the solution. The addition of a cosolvent to the aqueous phase can also affect the chemical stability of the solute [117], which is of importance from a pharmaceutical point of view. This phenomenon is also applicable to analytical techniques in different kinds of chromatographies such as thin layer chromatography (TLC), liquid chromatography (LC), HPLC, or capillary electrophoresis (CE) where the cosolvent is employed to alter the separation parameters.

1.7   REVIEW OF COSOLVENCY MODELS

Apart from experimental determinations of solute solubility in water–cosolvent mixtures, there are many mathematical models that describe the solubility in mixed solvents [95,118–127]. Mathematical models can be based on thermodynamic and theoretical considerations describing how molecules are believed to interact in the solid state or in fluid solution, or they can be semitheoretical or strictly empirical in nature. Theoretical models provide some evidence for better understanding of solubility behavior for drugs in mixed solvents, while semitheoretical or empirical approaches are very useful models for correlating experimental solubilities to independent variables such as volume fraction of the cosolvent.

One can categorize these models into two groups, i.e., predictive models and correlative equations. The advantage of predictive models, such as UNIFAC [128], is that these models do not employ any experimental data points. However, the low prediction capability of the UNIFAC model for solubility of biphenyl in several binary mixed solvents (as a simple model system) has been reported [129]. The correlative equations, such as the Jouyban–Acree model, employ the curve-fitting parameters to correlate experimental solubility data with respect to the concentration of the cosolvent. In order to calculate these curve-fitting parameters, one has to determine a set of

experiments in mixed solvents to train the model. From a practical point of view, a model containing a minimum number of the curve-fitting parameters is the best cosolvency model.

The final goal of developing cosolvency equations is to enable researchers to predict the solute solubility in mixed solvents from a minimum number of experiments or even without experimental data. However, it has been shown that when experimental data points are insufficient, solubility prediction in binary solvents using correlative equations can suffer from low accurate predictions [130]. To cover this limitation, Bustamante and coworkers [81] employed a modified form of the extended Hildebrand solubility (EHS) approach to correlate structurally related drug solubilities in binary solvent mixtures. The authors used the solute solubility in water and cosolvent, the solute solubility parameter, the Hildebrand solubility parameter of the solvent, and the basic solubility parameter of the mixed solvent as independent variables. This approach was a useful solution to the solubility problem of chemically similar solutes. In a paper from the Jouyban and Acree group [84], the applicability of the combined nearly ideal binary solvent/Redlich–Kister model (renamed as the Jouyban–Acree model) for reproducing solubility data of structurally related drugs in binary solvents was presented. It was also shown that the prediction error of the Jouyban–Acree model was less than that of the modified form of the EHS approach [84]. A review of available cosolvency models is provided here and the predictive abilities of the various models are briefly discussed.

1.7.1 HILDEBRAND SOLUBILITY APPROACH

A solution obeying Raoult's law is known as an ideal solution and its solubility can be calculated from the enthalpy of fusion of the solute (ΔH_m^f) and ΔC_p (the difference of the heat capacities of this solid and its supercooled liquid) using the Hildebrand and Scott equation expressed as

$$\log X = -\frac{\Delta H_m^f}{4.575}\left(\frac{T_m - T}{T \cdot T_m}\right) + \frac{\Delta C_p}{4.575}\left(\frac{T_m - T}{T}\right) - \frac{\Delta C_p}{1.987}\log\left(\frac{T_m}{T}\right) \tag{1.53}$$

where

X is the mole fraction solubility at temperature T
T_m is the melting point of the solute
$\Delta C_p = C_p^l - C_p^s$, where C_p^l and C_p^s are the molal heat capacities of the liquid and solid forms, respectively

For regular solutions, the equation was modified as

$$\log X = -\frac{\Delta H_m^f}{4.575}\left(\frac{T_m - T}{T \cdot T_m}\right) + \frac{\Delta C_p}{4.575}\left(\frac{T_m - T}{T}\right) - \frac{\Delta C_p}{1.987}\log\left(\frac{T_m}{T}\right) - \frac{V_s\varphi_m}{4.575T}\left(\delta_m - \delta_s\right)^2 \tag{1.54}$$

in which

V_s is the molar volume of the solute
φ_m is the volume fraction of the solvent
δ_m and δ_s are the solubility parameters of the solvent and the solute, respectively

Although Hildebrand restricts the application of the model to nonpolar solvents, Chertkoff and Martin [131] used the model for calculating the solubility of benzoic acid in binary mixtures of hexane, ethyl acetate, ethanol, and water. These mixtures provided a wide polarity range of solvent from 7.3 H (for hexane) to 23.4 H (for water), and the maximum solubility of benzoic acid was observed in $\delta_m = 11.5$ H (H: Hildebrand unit, 1 H = 1 (cal/cm^3)$^{1/2}$ = 0.489 MPa$^{1/2}$). The numerical values of δ_m of binary solvents were calculated using $\delta_m = f_1\delta_1 + f_2\delta_2$ in which f_1 and f_2 are the volume fractions of solvents 1 (cosolvent) and 2 (water), and δ_1 and δ_2 are the solubility parameters of solvents 1 and 2. Table 1.6 lists the total or Hildebrand solubility parameters and partial solubility parameters of the common solvents.

1.7.2 SOLUBILITY–DIELECTRIC CONSTANT RELATIONSHIP MODEL

Paruta and coworkers [132] tried to correlate the solubility of salicylic acid as a function of dielectric constant (ε) of the solvent mixture where ε values were determined using a resonance method. The model predicts the maximum solubility of a solute in binary solvents; however, there are different solubilities observed for a given ε value. This means that the dielectric constant cannot adequately represent the solvent effects on the solubility of solutes. This approach was employed later on as a polynomial of ε values to correlate the logarithm of solubilities in mixed solvents [133]. The numerical values of the ε_m were calculated using $f_1 \cdot \varepsilon_1 + f_2 \cdot \varepsilon_2$ [134], which are not correct calculations since there is a nonlinear relationship between experimental ε_m and f_1 values as was shown in a paper [135]. However, the main reason for the accurate correlation of the log X_m using the ε_m polynomial is that the polynomial could be arranged as a polynomial of f_1 values by simple algebraic manipulations [136].

1.7.3 LOG-LINEAR MODEL OF YALKOWSKY

Yalkowsky is one of the pioneers of systematic solubility investigations in water–cosolvent mixtures and developed the log-linear model, which was used for many pharmaceutical cosolvents. The log-linear or algebraic mixing rule [126] is expressed as

$$\log X_m = f_1 \log X_1 + f_2 \log X_2 \tag{1.55}$$

where
  X_m is the solubility of the solute
  f_1 and f_2 are the volume fractions of cosolvent (solvent 1) and water (solvent 2) in the absence of the solute
  X_1 and X_2 denote the solubility in neat cosolvent and water, respectively

The main assumptions on which this log-linear model is based are as follows:

1. The Gibbs energy of transferring a solute to an ideal solvent mixture is the sum of the corresponding energies in pure solvents
2. The solvent molecules behave in a mixture the same way as they do in neat solvents
3. The ratio of the solvent to cosolvent surrounding a solute molecule is the same as volume fraction of the solvents in the mixture
4. The molar volume of solute in the solution is not very different from the molar volumes of the solvent and cosolvent
5. No degradation, solvation, or solvent-mediated polymorphic transitions of the solute occur [137]

Most of these assumptions are not present and applicable to the solubility of drugs in aqueous binary mixtures, and the model thus produces large deviations from the true measured experimental data.
  The log-linear model could be rearranged as

$$\log X_m = \log X_2 + \left(\log \frac{X_1}{X_2} \right) f_1 = \text{Intercept} + \text{Slope} \cdot f_1 \tag{1.56}$$

Equation 1.56 is a correlative model; however, it has been demonstrated that a log-linear relationship between the solubility of a nonpolar solute and the fraction of the cosolvent exists as [126]

$$\log X_m = \log X_2 + \sigma \cdot f_1 \tag{1.57}$$

where σ is the solubilization power of the cosolvent and is theoretically equal to $\log(X_1/X_2)$. Valvani et al. [138] reported a linear relationship between σ and the logarithm of drugs' partition coefficient $(\log P)$, which is a key relationship and could improve the prediction capability of the log-linear model. The relationship was expressed as

$$\sigma = M \cdot \log P + N \tag{1.58}$$

where M and N are the cosolvent constants and are not dependent on the nature of the solute. The numerical values of M and N were reported for most of the common cosolvents earlier [139] and have been summarized in Table 1.11. This version of the log-linear model could be considered as a predictive model, and it has provided the simplest solubility estimation method and requires the aqueous solubility of the drug and its experimental/calculated log P value as input data. Updated M and N values for ethanol (0.93, 0.40), propylene glycol (0.77, 0.58), PEG 400 (0.74, 1.26), and glycerol (0.35, 0.26) were reported by Millard et al. [71] employing published data and the data selection criteria of the following: (a) adequate time for equilibration (24 h) or test for equilibration, (b) room-temperature experiments, i.e., 22°C–27°C, and (c) at least duplicated experiments. These slight variations of M and N values could not affect the prediction capability of the log-linear model as was shown in a recent work [69].

By using the known values of M and N along with the log P of a drug, it is possible to predict the cosolvent concentration for solubilization of the desired amount of the drug employing only the

TABLE 1.11
Numerical Values of M and N of Common Cosolvents for Calculating the Slope (σ) and Half-Slope ($\sigma_{0.5}$) of the Log-Linear Model

	σ		$\sigma_{0.5}$	
	M	N	M	N
Acetone	1.14	−0.10	1.25	0.21
Acetonitrile	1.16	−0.49	1.04	0.44
Butylamine	0.64	1.86	0.67	3.83
Dimethylacetamide	0.96	0.75	0.89	1.28
Dimethylformamide	0.83	0.92	0.65	1.70
DMSO	0.79	0.95	0.72	0.78
Dioxane	1.08	0.40	0.99	1.54
Ethanol	0.95	0.30	0.81	1.14
Ethylene glycol	0.68	0.37	0.52	0.28
Glycerol	0.35	0.28	0.38	0.14
Methanol	0.89	0.36	0.73	0.70
PEG 400	0.88	0.68	0.78	1.27
1-Propanol	1.09	0.01	1.03	1.76
2-Propanol	1.11	−0.50	0.96	1.00
Propylene glycol	0.78	0.37	0.55	0.87

Source: Li, A. and Yalkowsky, S.H., *Ind. Eng. Chem. Res.*, 37, 4470, 1998.

experimental aqueous solubility data [71]. This prediction method produces AAE of ~0.5 (in log scale) for the solubility of drugs in water–ethanol mixtures and can be used employing experimental values of log P or computed values from various commercial software packages [69]. The AAE of ~0.5 (for 26 data sets reported in Ref. [69]) is equal to the MPD of ~300% [32]. The produced MPDs are relatively high when compared with the experimentally determined RSD values (usually <10%) and acceptable percentage error range (~30%) in the pharmaceutical area [52,93]. However, the method is straightforward and easy to use and employs only aqueous solubility data of the drug (one data point for each drug).

Li [68] extended the log-linear model using the activity coefficients of water and cosolvent computed by the UNIFAC method. The most accurate extended model of Li was

$$\log X_{\mathrm{m}} = \log X_2 + \sigma \cdot f_1 + \ln \alpha_2 + \varphi_1 \ln\left(\frac{\alpha_1}{\alpha_2}\right) \tag{1.59}$$

where
  α_2 and α_1 are the activity coefficients of solvents 2 (water) and 1 (cosolvent) in the mixture
  φ_1 is the mole fraction of the cosolvent in the binary solvent mixture in the absence of the solute

Table 1.12 lists the AEP of the log-linear and the extended model for the predicted solubilities of various solutes dissolved in 13 different water–cosolvent mixtures. This extension improved the prediction capability of the log-linear model for 7 of the 13 cosolvents [68].

As discussed previously, solute solubilities in monosolvents, i.e., log X_1 and log X_2 (aqueous solubility), could be calculated using

TABLE 1.12
The AEP and Standard Deviation (SD) of Various Models for Calculating the Solutes Solubility in the Common Cosolvents and Their Overall Values

Cosolvent	N^a	Log-Linear		Extended Log-Linear	
		AEP	SD	AEP	SD
Acetone	220	0.306	0.303	0.246	0.310
Acetonitrile	103	0.416	0.284	0.208	0.256
DSMO	131	0.186	0.253	0.316	0.311
Dimethylacetamide	97	0.263	0.312	0.412	0.357
Dimethylformamide	133	0.216	0.260	0.221	0.262
Dioxane	349	0.608	0.334	0.386	0.295
Ethanol	1631	0.328	0.379	0.261	0.357
Ethylene glycol	111	0.186	0.148	0.265	0.164
Glycerol	124	0.096	0.119	0.095	0.117
Methanol	688	0.274	0.375	0.280	0.371
1-Propanol	116	0.570	0.397	0.356	0.336
2-Propanol	188	0.413	0.413	0.291	0.366
Propylene glycol	503	0.193	0.295	0.212	0.292
Overall	—	0.318	0.392	0.269	0.362

Source: Li, A., *Ind. Eng. Chem. Res.*, 40, 5029, 2001.
[a] N is the number of data points.

$$\log X_i = \frac{-\Delta S_{\mathrm{m}}(\mathrm{mp} - 25)}{1364} - \log P_{\substack{\mathrm{Octanol} \\ \mathrm{solvent}\ i}} + C_i t \tag{1.60}$$

where

ΔS_{m} is the entropy of fusion

mp is the melting point of the drug (°C)

$P_{\substack{\mathrm{Octanol} \\ \mathrm{solvent}\ i}}$ is the octanol–solvent i partition coefficient of the solute

C_i is a constant, which depends on the unit of the solubility [140]

The accuracy of the full predictive version of the log-linear model where $\log X_2$ was calculated using Equation 1.60 has not been reported in the literature.

The solubility profiles of drugs are often linear up to $f_1 = 0.5$, and it has been shown that the log-linear model could be modified as

$$\log X_{\mathrm{m}} = \log X_2 + \sigma_{0.5} \cdot f_1 \tag{1.61}$$

which provides more accurate predictions and is more practical in the pharmaceutical area since most of the cosolvents employed in $f_1 < 0.5$ [108] fractions. There is also a linear relationship between $\sigma_{0.5}$ and $\log P$ of the solutes and a summary of the M and N values is listed in Table 1.11.

The model was extended to the ternary solvents [141] as

$$\begin{aligned} \log X_{\mathrm{m}} &= f_1 \log X_1 + f_2 \log X_2 + f_3 \log X_3 \\ &= B_0 + B_1 f_1 + B_2 f_3 \end{aligned} \tag{1.62}$$

and for quaternary solvent mixtures [142,143] as

$$\begin{aligned} \log X_{\mathrm{m}} &= f_1 \log X_1 + f_2 \log X_2 + f_3 \log X_3 + f_4 \log X_4 \\ &= B_0 + B_1 f_1 + B_2 f_3 + B_3 f_4 \end{aligned} \tag{1.63}$$

where

X_3 and X_4 are the solubilities in neat solvents 3 and 4

f terms denote the fractions of the solvents in the mixture

The general form of the log-linear model for multicomponent solvent systems could be written as

$$\log X_{\mathrm{m}} = \log X_2 + \sum \sigma_i f_i \tag{1.64}$$

where σ_i and f_i are the solubilization power and the fractions of cosolvent i [68].

The accuracy of the log-linear model and its extended version for calculating the solubilities in ternary and quaternary solvent mixtures were tested using three solubility data sets for each solvent system. The overall AEP of the log-linear and the extended model for ternary solvents were 0.247 and 0.203, respectively, and the corresponding values for quaternary solvents were 0.330 and 0.268 [68].

1.7.4 EXTENDED HILDEBRAND SOLUBILITY APPROACH

Using the EHS approach, Martin and coworkers [120] expanded the applicability of the regular solution theory to solubility in water–cosolvent mixtures by avoiding Hildebrand's geometric assumption for the interaction term. In the original Hildebrand equation the solute–solvent interaction term is assumed equal to $(\delta_m \times \delta_s)$ in which δ_m and δ_s are the solubility parameters of mixed solvent and solute, respectively, and the model can describe the regular behavior of the solution. Instead the group used an empirical solute–solvent interaction parameter (WW). This modification widened the applications of the model to semipolar crystalline drugs in irregular solutions involving self-association and hydrogen bonding, as occurs in polar binary mixtures. Using the EHS model, the negative logarithm of the mole fraction solubility $(-\log X_m)$ can be expressed as

$$-\log X_m = -\log X_s^i + \frac{V_s \varphi_m^2 \left(\delta_m^2 + \delta_s^2 - 2WW \right)}{2.303RT} \tag{1.65}$$

where

X_s^i denotes the ideal mole fraction solubility of the solute

V_s is the molar volume of the solute

φ_m represents the volume fraction of the solvent in solution and because of very low solubility of the solute it can be assumed to be equal to 1 [129,134,144,145]

R is the molar gas constant

T denotes absolute temperature

WW is the interaction term, which is calculated by a power series of δ_m:

$$WW = \sum_{i=0}^{p} A_i \delta_m^i \tag{1.66}$$

where A_i denotes the curve-fit parameter and δ_m is calculated by using Equation 1.67:

$$\delta_m = f_1 \delta_1 + f_2 \delta_2 \tag{1.67}$$

in which δ_1 and δ_2 are the solubility parameters of pure cosolvent and water, respectively. However, to obtain an estimation of ideal solubility based on experimentally determined entropy or enthalpy of fusion, a sophisticated high-cost instrument, such as a differential scanning calorimeter, is required for measurements. In addition to experimentally measured solubility, solution density and an estimation of physical parameters of V_s and δ_s are required. All of these are essential to calculate solute solubility by the EHS equation. However, dependence of δ_s values on solvent polarity restricts the applications of EHS for predictive purposes [84].

The model was modified to directly relate the solubility of solutes to the solubility parameters of the solvent mixtures [81,93,146] as

$$\log X_m = C_0 + C_1 \delta_m + C_2 \delta_m^2 + C_3 \delta_m^3 + \cdots + C_n \delta_m^n \tag{1.68}$$

where C terms are the curve-fitting parameters. When this modified version is used, there is no need for experimental determination of ideal solubility of the solute and other terms required in the EHS approach [81]. This polynomial, i.e., Equation 1.68, was converted to the general single model (GSM) using simple algebraic manipulations [121]. This version of the EHS is a correlative model and did not provide accurate predictions.

An extended version of Martin's EHS model was proposed for describing the multiple solubility maxima of solutes in solvent mixtures where the WW term was correlated with a power series of the solvent compositions. This extension was applied to describe the multiple solubility maxima of five drugs in water–ethanol and ethanol–ethyl acetate mixtures with an MPD of ~11% [147].

Bustamante et al. [81] made several modifications to the EHS approach to enable it to predict the solubility of structurally related drugs in a given water–cosolvent mixtures. The modified model was

$$\log X_m = b_0 + b_1 \log X_1 + b_2 \log X_2 + b_3 \delta_m \delta_s + b_4 \delta_m^2 + b_5 \delta_m^3 + b_6 \delta_{m,b} \tag{1.69}$$

where

b_0–b_6 are the model constants
$\delta_{m,b}$ is the partial basic solubility parameter of the solvent mixture

Equation 1.69 was tested on two sets of structurally related drugs, i.e., xanthines and sulfonamides in water–dioxane mixtures. In a separate report from Bustamante's group [146], the trained version of Equation 1.69 was used to predict the solubility of sulfapyridine in water–dioxane mixtures. In another article from the same research group [100], the experimental solubility of two polymorphs of mefenamic acid in water–ethanol and ethanol–ethyl acetate mixtures was correlated using a single model as

$$\log X_m = -18.6826 b_0 + 0.5452 \log X_2 + 2.3687 \delta_m - 0.0836 \delta_m^2 + 0.0008337 \delta_m^3 \tag{1.70}$$

in which $\delta_m = f_{water} \delta_{water} + f_{ethanol} \delta_{ethanol} + f_{ethyl\ acetate} \delta_{ethyl\ acetate}$. More accurate predictions could be achieved using a trained version of the Jouyban–Acree model [84,148] in comparison with Equations 1.69 and 1.70.

1.7.5 Williams–Amidon Model

A slightly different set of solubility expressions, based on the excess free energy model, was developed by applying the excess Gibbs energy of Wohl and was expressed as [125]

$$\log X_m = f_1 \log X_1 + f_2 \log X_2 + A_{1\text{-}2} f_1 f_2 \left(\frac{V_s}{V_1} \right) \tag{1.71}$$

$$\log X_m = f_1 \log X_1 + f_2 \log X_2 - A_{1\text{-}2} f_1 f_2 (2f_1 - 1) \left(\frac{V_s}{V_1} \right) + 2A_{2\text{-}1} f_1^2 f_2 \left(\frac{V_s}{V_2} \right) + C_s f_1 f_2 \tag{1.72}$$

$$\log X_m = f_1 \log X_1 + f_2 \log X_2 - A_{1\text{-}2} f_1 f_2 (2f_1 - 1) \left(\frac{V_s}{V_1} \right) + 2A_{2\text{-}1} f_1^2 f_2 \left(\frac{V_s}{V_2} \right)$$
$$+ 3D_{12} f_1^2 f_2^2 \left(\frac{V_s}{V_2} \right) + C_2 f_1 f_2^2 \left(\frac{V_s}{V_2} \right) + C_1 f_1^2 f_2 \tag{1.73}$$

where

$A_{1\text{-}2}$, $A_{2\text{-}1}$, C_s, D_{12}, C_2, and C_1 are solvent–solvent or solute–solvent interaction terms
V_s, V_1, and V_2 represent the molar volumes of the solute, cosolvent, and water, respectively [125]

The molar volume differences between solvents, cosolvents, and solutes are considered as the V terms. In the derivation of the log-linear model, an equal size for the solute, solvent, and cosolvent was assumed to be of comparable molecular size. This is not necessarily the case for solute solubility in binary solvents; therefore, the addition of the V terms could improve the accuracy of the model when compared with the log-linear model. The solvent–cosolvent interaction constants (D_{12} and A terms) are obtained from vapor–liquid equilibrium data. Williams and Amidon [125] reported different values of A for two-suffix (Equation 1.71), three-suffix (Equation 1.72), and four-suffix (Equation 1.73) excess free energy models with respect to a given binary solvent system. This fact is due to different definitions of the A terms in these equations. The solute–solvent interaction terms are estimated from experimental solubility data. This model is able to improve the predictability of the log-linear model by employing additional terms.

1.7.6 MIXTURE RESPONSE SURFACE MODEL

Statistically based mixture response surface methods [124] have been proposed for correlative purposes and these models are as follows:

$$\log X_m = \beta_1 f_1' + \beta_2 f_2' + \beta_3 f_1' f_2' \tag{1.74}$$

$$\log X_m = \beta_1' f_1' + \beta_2' f_2' + \beta_3' \left(\frac{1}{f_1'}\right) + \beta_4' \left(\frac{1}{f_2'}\right) \tag{1.75}$$

$$\log X_m = \beta_1'' f_1' + \beta_2'' f_2' + \beta_3'' \left(\frac{1}{f_1'}\right) + \beta_4'' \left(\frac{1}{f_2''}\right) + \beta_5'' f_1' f_2' \tag{1.76}$$

in which
  β_1–β_3, β_1'–β_4', and β_1''–β_5'' are the model's parameters and
  f_1' and f_2', are given by $f_1' = 0.96 f_1 + 0.02$ and $f_2' = 0.96 f_2 + 0.02$ [124].

By converting f values to f', the model can cover the whole range of volume fraction of the cosolvent (f_1: 0–1). The authors showed the superiority of Equations 1.74 through 1.76 to the EHS model by examining the xanthine derivatives solubilities in binary dioxane–water mixtures.

1.7.7 KHOSSRAVI–CONNORS MODEL

Khossravi–Connors model [95] has been formulated as the summary of the free energy changes of the three steps involved in the dissolution of a crystalline solute in a solvent system:

$$\Delta G_{total}^0 = \Delta G_{crystal}^0 + \Delta G_{cavity}^0 + \Delta G_{solvation}^0 \tag{1.77}$$

where
  ΔG_{total}^0 is the total free energy change
  $\Delta G_{crystal}^0$ is the crystal lattice energy and any solute–solute interactions in the solution representing the conversion of the crystalline solute to the gaseous solute
  ΔG_{cavity}^0 is the free energy change of cavity formation
  $\Delta G_{solvation}^0$ is the free energy changes of insertion of the gaseous solute in the cavity and its solvation processes

Khossravi and Connors called $\Delta G_{crystal}^0$ the intersolute effect, ΔG_{cavity}^0 the medium effect, and $\Delta G_{solvation}^0$ the solvation effect. The total free energy change could be related to the experimental solubility of a solute in binary solvent mixtures (X_m) using

$$\Delta G^0_{\text{total}} = -2.303 kT \log X_{\text{m}} \tag{1.78}$$

where k is Boltzmann's constant [95].

The crystal composition of a solute is independent of the solvent composition; therefore, $\Delta G^0_{\text{crystal}}$ (intersolute effect) is a solvent-independent term. The authors correlated the $\Delta G^0_{\text{cavity}}$ and $\Delta G^0_{\text{solvation}}$ to the solvent composition of the aqueous-organic solvent mixtures. The $\Delta G^0_{\text{cavity}}$ was formulated as

$$\Delta G^0_{\text{cavity}} = g \cdot A \cdot \left[\gamma_1 + \gamma' \left(\frac{\beta_1 f_1 f_2 + 2\beta_2 f_2^2}{f_1^2 + \beta_1 f_1 f_2 + \beta_2 f_2^2} \right) \right] \tag{1.79}$$

where

g is the curvature effect factor

A is the area defined by the van der Waals radii of the solute atoms and is assumed as a solvent-independent quantity

γ_1 is the surface tension of pure solvent 1

$\gamma' = \frac{\gamma_2 - \gamma_1}{2}$ in which γ_2 is the surface tension of the solvent 2

β_1 and β_2 are functions of the equilibrium constants (K_1 and K_2) of the solvation of the solute in the binary solvent mixture, i.e., $\beta_1 = K_1$ and $\beta_2 = K_1 K_2$

The solvation effect was presented as

$$\Delta G^0_{\text{solvation}} = \Delta G^0_{11} F_{11} + \Delta G^0_{12} F_{12} + \Delta G^0_{22} F_{22} \tag{1.80}$$

where

F_{11}, F_{12}, and F_{22} are fractions of the solute in the solvated forms by two molecules of solvent 1, one molecule of solvent 1 and one molecule of solvent 2, and two molecules of solvent 2, respectively

ΔG^0_{11}, ΔG^0_{12}, and ΔG^0_{22} are the free energy changes of the solvation processes by 1-1, 1-2, and 2-2 molecules of the solvents

Equation 1.80 can be rewritten as

$$\Delta G^0_{\text{solvation}} = S_1 F_{12} + S_2 F_{22} + \Delta G^0_{11} \tag{1.81}$$

where

$S_1 = \Delta G^0_{12} - \Delta G^0_{11}$

$S_2 = \Delta G^0_{22} - \Delta G^0_{11}$.

By considering the equilibrium constants of the exchange equilibria of water by the cosolvent and further simplifications, the following equation can be obtained:

$$\Delta G^0_{\text{solvation}} = \frac{S_1 \beta_1 f_1 f_2 + S_2 \beta_2 f_2^2}{f_1^2 + \beta_1 f_1 f_2 + \beta_2 f_2^2} + \Delta G^0_{11} \tag{1.82}$$

Khossravi and Connors [95] used the Leffler–Grunwald delta operator symbolism and defined the total solvent effects ($\Delta G^0_{\text{total}}(f_2)$) as

$$\delta_M \Delta G^0 = \Delta G^0_{\text{total}}(f_2) - \Delta G^0_{\text{total}}(f_2 = 0) \tag{1.83}$$

and all solvent composition independent quantities were vanished:

$$\delta_M \Delta G^0 = \frac{\left(-kT \ln \beta_1 + g \cdot A \cdot \gamma'\right)\beta_1 f_1 f_2 + \left(-kT \ln \beta_2 + 2g \cdot A \cdot \gamma'\right)\beta_2 f_2^2}{f_1^2 + \beta_1 f_1 f_2 + \beta_2 f_2^2} \tag{1.84}$$

In Equation 1.84, the terms, k, T, $gA\gamma'$, $\ln \beta_1$, and $\ln \beta_2$ possess constant values and it is possible to rewrite Equation 1.84 as

$$\delta_M \Delta G^0_{Solution} = \frac{a\beta_1 f_1 f_2 + b\beta_2 f_2^2}{f_1^2 + \beta_1 f_1 f_2 + \beta_2 f_2^2} \tag{1.85}$$

where a and b are unconstrained parameters. Since ΔG^0_{total} is equal to $-2.303kT \log X_m$ in which X_m is the mole fraction solubility of the solute, a combination of Equations 1.78 and 1.85 yields

$$-\log X_m = kT \log X_2 + \frac{a\beta_1 f_1 f_2 + b\beta_2 f_2^2}{f_1^2 + \beta_1 f_1 f_2 + \beta_2 f_2^2} \tag{1.86}$$

1.7.8 Jouyban–Acree Model

The Jouyban–Acree model, formerly known as the combined nearly ideal binary solvent/Redlich–Kister equation, was derived from a thermodynamic mixing model that includes contributions from both two-body and three-body interactions [119]. The basic model was presented by Hwang et al. [149] and an extension of the model was introduced for calculating various PCPs in mixed solvents by our group. The Jouyban–Acree model for calculating the solute solubility in binary solvents at a fixed temperature is expressed as [119]

$$\log X_m = f_1 \log X_1 + f_2 \log X_2 + f_1 f_2 \sum_{i=0}^{n} S_i \left(f_1 - f_2\right)^i \tag{1.87}$$

where S_i stands for the model constants. These constants can be calculated by two procedures:

1. Regressing $\left(\dfrac{\log X_m - f_1 \log X_1 - f_2 \log X_2}{f_1 f_2}\right)$ against $(f_1 - f_2)$ and $(f_1 - f_2)^2$ by a classical least square analysis [118].
2. Regressing $(\log X_m - f_1 \log X_1 - f_2 \log X_2)$ against $f_1 f_2$, $f_1 f_2 (f_1 - f_2)$, and $f_1 f_2 (f_1 - f_2)^2$ by a no intercept least squares analysis. This procedure produced more accurate correlations than the one above for the solute's solubility in aqueous binary solvent [130].

The Jouyban–Acree model is able to adequately represent the spectrum of solution behavior from ideal to highly nonideal systems [150–152]. The model contains as many curve-fitting parameters (usually 3, $n = 2$) as is necessary to accurately describe the actual measured data.

The Jouyban–Acree model was used to calculate multiple solubility maxima as well as solute solubility in mixed solvents at various temperatures [153]. The model was also used to correlate other PCPs in mixed solvent systems, including the electrophoretic mobility of analytes in mixed solvent electrolyte systems [154–156], the instability rate constants in binary solvent systems [157], the acid dissociation constants in water–organic solvent mixtures at fixed and at various temperatures [158,159], the retention factor of analytes in HPLC [160], the dielectric constant [135], surface tension [161], viscosity [162], density [163], solvatochromic parameter [164], refractive index [165], and ultrasound velocity [166] in the solvent mixtures. The theoretical basis of the model for describing the chemical potential of solutes dissolved in mixed solvents [119] and the acid dissociation constants in aqueous-organic mixtures [158] have been provided in earlier papers. The constants of the Jouyban–Acree model represent differences in the various solute–solvent and solvent–solvent interactions in the mixture [119]. Therefore, the model should be able to calculate any PCP in mixed solvents, which is a function of solute–solvent and/or solvent–solvent interactions. The general form of the Jouyban–Acree model is

$$\log PCP_{m,T} = f_1 \log PCP_{1,T} + f_2 \log PCP_{2,T} + f_1 f_2 \sum_{i=0}^{2} \frac{J_i (f_1 - f_2)^i}{T} \tag{1.88}$$

where

$PCP_{m,T}$, $PCP_{1,T}$, and $PCP_{2,T}$ are the numerical values of the PCP of the mixture and solvents 1 and 2 at temperature T, respectively

f_1 and f_2 are the volume (or weight or mole) fractions of solvents 1 and 2 in the mixture

J_i represents the model constants

The model has been successfully extended for the representation of the PCPs in ternary solvents as

$$\log PCP_{m,T} = f_1 \log PCP_{1,T} + f_2 \log PCP_{2,T} + f_3 \log PCP_{3,T} + f_1 f_2 \sum_{i=0}^{2} \frac{J_i (f_1 - f_2)^i}{T}$$

$$+ f_1 f_3 \sum_{i=0}^{2} \frac{J_i' (f_1 - f_3)^i}{T} + f_2 f_3 \sum_{i=0}^{2} \frac{J_i'' (f_2 - f_3)^i}{T} \tag{1.89}$$

where subscript 3 is the solvent 3 characteristics, and J_i' and J_i'' are the subbinary model constants. It is possible to extend the model to represent the PCPs in multicomponent solvents as

$$\log PCP_{m,T} = \sum_{I} f_I \log PCP_{I,T} + \sum_{I} \sum_{J>I} \left\{ f_I f_J \sum_{i=0}^{2} \left[\frac{J_{IJ,i} (f_I - f_J)^i}{T} \right] \right\} \tag{1.90}$$

The main limitations of the Jouyban–Acree model for predicting drug solubilities in water–cosolvent mixtures are (a) requirement of two data points of solubilities in monosolvent systems at each temperature and (b) numerical values of the model constants. To overcome the first limitation, the solubility prediction methods in monosolvent system should be improved. A number of articles reviewed the recent progresses in this field, especially with the aqueous solubility prediction methods [34,50,167–169]. To address the second limitation, a number of solutions were examined:

1. The model constants could be obtained using the solubility of structurally related drugs in a given water–cosolvent system, and then could predict the unmeasured solubility of the related drugs where the expected MPD was ~17% [84].
2. The model constants could be calculated using a minimum number of experimental data points, i.e., three data points, and then could predict the solubilities in the other solvent compositions where the expected prediction MPD was <15% [86]. The reduction in the number of experimental data was also considered in solubility prediction in mixed solvents at various temperatures using Equation 1.88. Using this version of the Jouyban–Acree

TABLE 1.13
Numerical Values of the Jouyban–Acree Model (J_0, J_1, and J_2) for Commonly Studied Cosolvents, the Number of Data Sets (NDS) Employed in the Training Process of the Model, and the References

Cosolvent	J_0	J_1	J_2	NDS	Reference
Dioxane	958.44	509.45	867.44	36	[171]
Ethanol	724.21	485.17	194.41	26	[69]
PEG 400	394.82	−355.28	388.89	79	[172]
Propylene glycol	37.03	319.49	0	27	[173]

model, the solubility data in monosolvents at temperatures of interest was calculated using the van't Hoff plot employing the solubility data at the highest and lowest temperatures of interest [170].

3. The trained versions of the Jouyban–Acree models could be employed for the solubility prediction of drugs in the aqueous mixtures of dioxane [171], ethanol [69], PEG 400 [172], and propylene glycol [173] at various temperatures and the expected MPDs were ~27%, ~48%, ~40%, and ~24%, respectively. Table 1.13 gives the numerical values of the Jouyban–Acree model constants for the four cosolvents studied. Further experimental data sets are required to train similar models for the other cosolvents of pharmaceutical interest. In the latest work, the model constants of PEG 400 were used to predict the solubility of drugs in aqueous mixtures of ethylene glycol and PEG 200 at various temperatures in which the overall MPD was 23.2% [174].

4. In the trained versions of the Jouyban–Acree model, the extent of the solute–solvent interactions is assumed to be the same. This is not necessarily the case, however, because various solutes possess different functional groups leading to different types of solute–solvent interactions. To address this concern, the deviated solubilities from the trained versions of the Jouyban–Acree model were correlated using QSPR models, which resulted in reduced MPD values. The MPD values for dioxane, ethanol, PEG 400, and propylene glycol were 18%, 33%, 38%, and 16%, respectively [175,176].

5. A generalized version of the Jouyban–Acree model was proposed using its combination with the Abraham parameters (experimental values) where the model constants of the Jouyban–Acree model were correlated with the functions of the Abraham solvent coefficients and the solute parameters as

$$J_i = A_{0,i} + A_{1,i}(c_1 - c_2)^2 + A_{2,i}E(e_1 - e_2)^2 + A_{3,i}S(s_1 - s_2)^2$$
$$+ A_{4,i}A(a_1 - a_2)^2 + A_{5,i}B(b_1 - b_2)^2 + A_{6,i}V(v_1 - v_2)^2 \tag{1.91}$$

where
   A terms were the model constants [25]
   c, r, s, a, b, and v are the solvents coefficients
   Subscripts 1 and 2 denote cosolvent and water, respectively
   E is the excess molar refraction
   S is the dipolarity/polarizability of the solute

TABLE 1.14
The Numerical Values of the QSPR Models (Equation 1.91) for Predicting the Model Constants of the Jouyban–Acree Model

	A_0	A_1	A_2	A_3	A_4	A_5	A_6
J_0	2113.119	−1093.783	3380.661	−13.865	−4.921	−5.659	15.250
J_1	−2001.561	1142.780	−2735.160	−38.541	13.176	0.811	38.508
J_2	1474.963	−1507.479	4421.302	17.981	−21.196	6.595	−13.386

Source: Jouyban, A. et al., *J. Pharm Pharmaceut. Sci.*, 10, 263, 2007.

A denotes the solute's hydrogen bond acidity

B stands for the solute's hydrogen bond basicity

V is the McGowan volume of the solute in unit of 0.01 (cm³/mole)

All solute and solvent parameters were adopted from the LSER approach of Abraham et al. [24]. The J_i (i.e., J_0, J_1, and J_2) terms of the studied solubility data sets were regressed against $(c_1 - c_2)^2$, $E(e_1 - e_2)^2$, $S(s_1 - s_2)^2$, $A(a_1 - a_2)^2$, $B(b_1 - b_2)^2$, and $V(v_1 - v_2)^2$ values to compute $A_{j,i}$ terms, and then the $A_{j,i}$ terms were replaced in Equation 1.88 and the solubility of drugs in binary solvents were predicted employing experimental values of X_1 and X_2. The applicability of the generalized model was evaluated employing 30 data sets including various cosolvents and the expected MPD for the predicted solubilities was ~18.5% [25]. Using this version, the required data is the solubility in water and cosolvent systems along with the experimental values of the Abraham solute parameters. Table 1.14 lists the numerical values of the model constants of Equation 1.91.

6. In the latest attempt, another generalized version of the Jouyban–Acree model was proposed for calculating the solubility of drugs using theoretically computed Abraham parameters using the PharmaAlgorithm software [177]. The trained model was

$$\log X_{m,T} = f_1 \log X_{1,T} + f_2 \log X_{2,T}$$

$$+ \left(\frac{f_1 f_2}{T}\right) \left\{ \begin{array}{l} 1639.07 - 561.01\left[(c_1 - c_2)^2\right] - 1344.81\left[E(e_1 - e_2)^2\right] - 18.22\left[S(s_1 - s_2)^2\right] \\ -3.65\left[A(a_1 - a_2)^2\right] + 0.86\left[B(b_1 - b_2)^2\right] + 4.40\left[V(v_1 - v_2)^2\right] \end{array} \right\}$$

$$+ \left(\frac{f_1 f_2 (f_1 - f_2)}{T}\right) \left\{ \begin{array}{l} -1054.03 + 1043.54\left[(c_1 - c_2)^2\right] + 359.47\left[E(e_1 - e_2)^2\right] - 1.20\left[S(s_1 - s_2)^2\right] \\ +30.26\left[A(a_1 - a_2)^2\right] - 2.66\left[B(b_1 - b_2)^2\right] - 0.16\left[V(v_1 - v_2)^2\right] \end{array} \right\}$$

$$+ \left(\frac{f_1 f_2 (f_1 - f_2)^2}{T}\right) \left\{ \begin{array}{l} 2895.07 - 1913.07\left[(c_1 - c_2)^2\right] - 901.29\left[E(e_1 - e_2)^2\right] - 10.87\left[S(s_1 - s_2)^2\right] \\ +24.62\left[A(a_1 - a_2)^2\right] + 9.79\left[B(b_1 - b_2)^2\right] - 24.38\left[V(v_1 - v_2)^2\right] \end{array} \right\}$$

$$(1.92)$$

The training data set consists of 47 drugs in eight different water–cosolvent mixtures at fixed and/or various temperatures with a total number of 1927 points. The overall MPD for the back-calculated solubility data using Equation 1.92 was 42.4%, which was reduced to 29.5% when eight outlier data sets were excluded from the calculations. This model was cross-validated using the leave-one-out method. The overall MPD of leave one drug out was 37.1% [178].

There were good correlations between experimental and computed values of the Abraham solute parameters, as shown in Figure 1.2. The high correlation coefficients revealed that the computed parameters could be replaced with the experimental values.

The overall MPD for the predicted solubilities of 91 data sets with known experimental Abraham parameters was 24.5% and the corresponding MPD for the computed Abraham parameters was 27.4%. Although the mean difference between MPDs was statistically significant, the computed parameters are preferred since their computations are straightforward and no further experiments are needed [178].

The accuracy of Equation 1.92 was compared with those of the trained versions of the log-linear model (Equations 1.57 and 1.58 with M and N values from Table 1.11) and the trained versions of the Jouyban–Acree model for a given cosolvent (Equation 1.88 with J_i terms from Table 1.13). The overall MPDs of Equations 1.92 and the log-linear model were 32.6% and 65.4%, respectively. The overall MPDs of Equation 1.92 were 49.5% and 28.7% for ethanol and dioxane data sets, respectively. The corresponding MPDs for the previous versions of the Jouyban–Acree model (i.e., Equation 1.88 and the model constants of ethanol and dioxane reported in Table 1.13) were 66.1% and 36.5%, respectively. The reductions in MPDs for Equation 1.92 were statistically significant for both models. Figure 1.3 illustrated the IPD distribution of the calculated solubilities sorted in three subgroups.

Equation 1.92 could be simplified to the following equation when a given cosolvent was considered in the computations. The trained model for ethanol was

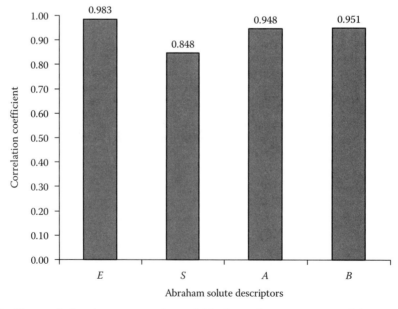

FIGURE 1.2 The correlations between experimental Abraham solute parameters and the computed parameters by the PharmaAlgorithms software.

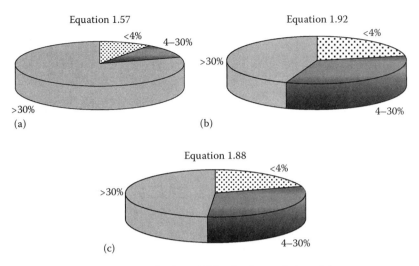

FIGURE 1.3   The relative frequency distribution of IPDs for the studied models.

$$\log X_{m,T} = f_1 \log X_{1,T} + f_2 \log X_{2,T}$$

$$+ \left(\frac{f_1 f_2}{T}\right)\{558.45 + 358.60E + 22.01S - 352.97A + 130.48B - 297.10V\}$$

$$+ \left(\frac{f_1 f_2 (f_1 - f_2)}{T}\right)\{45.67 - 165.77E - 321.55S + 479.48A - 409.51B + 827.63V\}$$

$$+ \left(\frac{f_1 f_2 (f_1 - f_2)^2}{T}\right)\{-493.81 - 341.32E + 866.22S - 36.17A + 173.41B - 555.48V\} \quad (1.93)$$

The overall MPD of Equation 1.93 was 34.3% [178]. These versions of the models provided better predictions compared to all previously presented models. It is obvious that any variations in the test data sets could be reflected in the MPD values and the best comparison could be provided using the same data sets for all models investigated. The main disadvantage of the generalized versions of the Jouyban–Acree model using Abraham parameters is the unavailability of the solvent coefficients for some of the pharmaceutically relevant cosolvents, such as propylene glycol or PEGs.

In practice, when the binary solvents are not able to dissolve the desired amount of a drug in a given volume, ternary solvents are often used. The applicability of the Jouyban–Acree model for calculating the solubility of drugs in ternary solvents was shown using solubility data of 19-nor 1α, 25-dihydroxyvitamin D_2 in water–ethanol–propylene glycol mixtures where the MPD was ~17% [179]. To provide a predictive model for ternary solvents based on solubility data in binary solvents, the subbinary constants of the Jouyban–Acree model were used to predict the solubility of paracetamol in ternary solvents where the MPD was <10% [180]. Using a minimum number of solubility data of salicylic acid in aqueous mixtures of ethanol and propylene glycol as well as ethanol–propylene glycol mixtures (three data points from each binary system), a trained version of the Jouyban–Acree model was presented to predict the solubility of salicylic acid in binary and ternary solvents where the MPD was ~7% [181]. These capabilities should be further investigated employing more experimental data sets.

In the latest work from our group, a new definition of the solubilization power of a cosolvent was presented. Based on the new definition, the solubilization power of a cosolvent is defined as $\omega = \left(\log \dfrac{X_{m,max}}{X_2} \right)$ in which the $X_{m,max}$ could be calculated using the Jouyban–Acree model employing the experimental values of X_1 and X_2 for a drug of interest. The solubilizing power of the cosolvent increases with ω value. A large ω value means that a desired amount of a drug could be dissolved at the minimum concentration of the cosolvent [182]. Table 1.15 lists the observed and predicted values of $X_{m,max}$ and the corresponding f values of the cosolvent [183]. The concentration of the cosolvent in the pharmaceutical formulations is an important factor from both the toxicological and economical points of views.

1.7.9 Modified Wilson Model

An additional procedure is the modified Wilson model (MWM), which is derived from an expression for the excess free energy of mixing of nonelectrolyte solutions based on the Flory–Huggins theory for thermal mixtures. Comor and Kopecni [184] modified the Wilson model and one can estimate solute solubility in binary solvent mixtures from measured values in the pure solvent and cosolvent as well as excess Gibbs free energies for the binary solvent mixture [185]. The MWM is expressed as

$$\log\left(\frac{X_s^i}{X_m}\right) = 1 - \frac{f_1\left[1 - \log\left(\dfrac{X_s^i}{X_1}\right)\right]}{f_1 + f_2 \Lambda_{12}^{adj}} - \frac{f_2\left[1 - \log\left(\dfrac{X_s^i}{X_2}\right)\right]}{f_1 \Lambda_{21}^{adj} + f_2} \tag{1.94}$$

With the adjustment of this model it was shown that a simplified form of the MWM, SMW [122], was able to calculate the solute solubility in water–cosolvent mixtures more accurately than MWM, although this simplification was not successful in the case of solubility prediction in nonaqueous binary solvents [118]. Thus, the SMW is

$$-\log X_m = 1 - \frac{f_1\left(1 + \log X_1\right)}{f_1 + f_2 \lambda_{12}^{adj}} - \frac{f_2\left(1 + \log X_2\right)}{f_1 \lambda_{12}^{adj} + f_2} \tag{1.95}$$

where Λ_{12}^{adj}, Λ_{21}^{adj}, λ_{12}^{adj}, and λ_{21}^{adj} are adjustable parameters of the models, which can be evaluated via a nonlinear least squares analysis or by developing a simple computer program, which calculates the solute solubility at each composition of the cosolvent and employs preselected values for the adjustable parameters. By using Equation 1.95, there is no need to determine the ideal mole fraction solubility of the solute, which is one of the advantages of Equation 1.95. Nevertheless, Equations 1.94 and 1.95 are mathematically equivalent with the log-linear equation for which the adjustable parameters equal unity.

1.7.10 Margules Equation

Margules equation is expressed as

$$-\log X_m = \frac{\Delta S_m}{2.303R}\left(\frac{T_m}{T} - 1\right) + G_1 f_1^2 + G_2 f_2^2 + \left(G_1 + G_2 - G_3 + G_4\right)f_1 f_2 \tag{1.96}$$

where ΔS_m is the entropy of fusion and G_1–G_4 are the constant values. G_1 and G_2 are determined from measured solubilities in water and pure cosolvent; G_3 is a constant that is independent from the solute and is determined using cosolvent vapor pressure above water–cosolvent mixture, and G_4 denotes

TABLE 1.15
The Predicted and Observed log $X_{m(max)}$ at Volume Fraction of Solvent 1 ($f_{1,max}$) and Absolute Error (AE) for Solubility of Drugs in Water–Ethanol Mixtures at Various Temperatures (T, K)

Solute	Solvent 1	T	Log X_1	Log X_2	Log $X_{m(max)}$ Predicted	Observed	AE	$f_{1,max}$ Predicted	Observed	AE
Acetaminophen	Ethanol	298	2.15	1.18	2.55	2.32	0.23	0.76	0.80	0.04
Acetaminophen	Ethanol	303	2.31	1.32	2.69	2.39	0.30	0.76	0.80	0.04
Acetanilide	Ethanol	298	−3.07	−5.14	−2.88	−3.05	0.17	0.85	0.90	0.05
Acetanilide	Ethanol	293	1.35	−0.28	1.62	1.42	0.20	0.82	0.84	0.02
Acetanilide	Ethanol	298	−1.09	−3.10	−0.90	−1.09	0.19	0.85	1.00	0.15
Acetanilide	Ethanol	303	1.46	−0.16	1.71	1.50	0.21	0.82	0.85	0.03
Alanine (Beta)	Water	298	−0.82	−3.96	−0.76	−0.82	0.06	0.92	1.00	0.08
Alanine (DL)	Water	298	−1.49	−4.35	−1.40	−1.49	0.09	0.90	1.00	0.10
Aminocaproic acid (ε)	Water	298	−0.97	−3.95	−0.89	−0.97	0.08	0.91	1.00	0.09
Amobarbital	Ethanol	298	2.34	−0.25	2.46	2.36	0.10	0.88	0.96	0.08
Asparagine (L)	Water	298	−2.47	−5.87	−2.43	−2.47	0.04	0.93	1.00	0.07
Aspartic acid (L)	Water	298	−3.17	−6.17	−3.10	−3.17	0.07	0.91	1.00	0.09
Barbital	Ethanol	298	1.97	0.86	2.33	2.08	0.25	0.77	0.88	0.11
Benoz [a] pyrene	Ethanol	296	−2.24	−7.90	−2.24	−2.24	0.00	1.00	1.00	0.00
Benzocaine	Ethanol	298	−0.82	−3.22	−0.68	−0.82	0.14	0.87	1.00	0.13
Benzoic acid	Ethanol	288	0.35	−1.70	0.56	0.35	0.21	0.84	1.00	0.16
Benzoic acid	Ethanol	293	0.40	−1.62	0.60	0.40	0.20	0.84	1.00	0.16
Benzoic acid	Ethanol	298	0.44	−1.55	0.64	0.44	0.20	0.84	1.00	0.16
Butabarbital	Ethanol	298	1.92	−0.05	2.12	1.96	0.16	0.84	0.90	0.06
Butyrine (DL)	Water	298	−1.44	−3.82	−1.30	−1.44	0.14	0.87	1.00	0.13
Caffeine	Water	298	−2.68	−2.77	−2.03	−1.83	0.20	0.67	0.60	0.07
Cholordiazepoxide	Ethanol	303	−2.47	−5.21	−2.38	−2.30	0.08	0.90	0.90	0.00
Chrysene	Ethanol	296	−2.78	−8.46	−2.78	−2.78	0.00	1.00	1.00	0.00
Clonazepam	Ethanol	303	−2.99	−6.05	−2.93	−2.98	0.05	0.92	0.90	0.02

(continued)

TABLE 1.15 (continued)
The Predicted and Observed log $X_{m(max)}$ at Volume Fraction of Solvent 1 ($f_{1,max}$) and Absolute Error (AE) for Solubility of Drugs in Water–Ethanol Mixtures at Various Temperatures (T, K)

Solute	T	Solvent 1	Log X_1	Log X_2	Log $X_{m(max)}$			$f_{1,max}$		
					Predicted	Observed	AE	Predicted	Observed	AE
Diazepam	303	Ethanol	-2.12	-5.48	-2.08	-2.07	0.01	0.93	0.90	0.03
Furosemide	298	Ethanol	-2.62	-5.64	-2.55	-2.62	0.07	0.91	1.00	0.09
Glycine	298	Water	-1.25	-4.64	-1.21	-1.25	0.04	0.93	1.00	0.07
Glycylglycine	298	Water	-1.52	-5.89	-1.52	-1.52	0.00	0.98	1.00	0.02
Hexachlorobenzene	296	Ethanol	-2.50	-7.74	-2.50	-2.50	0.00	1.00	1.00	0.00
Ketoprofen	298	Ethanol	2.96	-0.97	2.97	2.96	0.01	0.96	1.00	0.04
Ketoprofen	310	Ethanol	2.97	-0.88	2.98	2.97	0.01	0.97	1.00	0.03
Leucine (L)	298	Water	-2.50	-4.13	-2.24	-2.50	0.26	0.82	1.00	0.18
Lorazepam	303	Ethanol	-2.71	-5.46	-2.62	-2.57	0.05	0.90	0.90	0.00
Mefenamic acid (I)	298	Ethanol	-2.74	-5.49	-2.64	-2.74	0.10	0.89	1.00	0.11
Mefenamic acid (II)	298	Ethanol	-2.60	-5.35	-2.50	-2.60	0.10	0.89	1.00	0.11
Methobarbital	298	Ethanol	1.62	0.30	1.94	1.71	0.23	0.79	0.88	0.09
Methyl p-hydroxybenzoate	298	Ethanol	0.37	-1.84	0.53	0.37	0.16	0.86	1.00	0.14
Nalidixic acid	298	Ethanol	-3.69	-5.62	-3.48	-3.56	0.08	0.84	0.85	0.01
Niflumic acid	298	Ethanol	-1.79	-5.26	-1.75	-1.79	0.04	0.94	1.00	0.06
Norleucine (DL)	298	Water	-2.80	-4.21	-2.50	-2.80	0.30	0.80	1.00	0.20
Octadecanoic acid	298	Ethanol	0.92	-1.47	1.06	0.92	0.14	0.87	1.00	0.13
Oxolinic acid	293	Ethanol	-5.16	-6.06	-4.74	-4.76	0.02	0.75	0.80	0.05
Oxolinic acid	298	Ethanol	-5.09	-5.97	-4.67	-4.68	0.01	0.75	0.80	0.05
Oxolinic acid	303	Ethanol	-4.98	-5.87	-4.57	-4.59	0.02	0.76	0.80	0.04
Oxolinic acid	308	Ethanol	-4.89	-5.79	-4.50	-4.48	0.02	0.76	0.80	0.04
Oxolinic acid	313	Ethanol	-4.79	-5.69	-4.41	-4.39	0.02	0.76	0.80	0.04

Paracetamol	298	Ethanol	−1.27	−2.72	−0.98	−1.10	0.12	0.80	0.85	0.05
Paracetamol	293	Ethanol	−1.28	−2.76	−0.98	−1.14	0.16	0.80	0.85	0.05
Paracetamol	298	Ethanol	−1.27	−2.72	−0.98	−1.10	0.12	0.80	0.85	0.05
Paracetamol	303	Ethanol	−1.21	−2.64	−0.92	−1.06	0.14	0.80	0.85	0.05
Paracetamol	308	Ethanol	−1.18	−3.02	−0.98	−1.01	0.03	0.84	0.85	0.01
Paracetamol	313	Ethanol	−1.15	−2.55	−0.88	−0.97	0.09	0.80	0.85	0.05
Pentachlorobenzene	296	Ethanol	−1.08	−6.12	−1.08	−1.08	0.00	1.00	1.00	0.00
Pentobarbital	298	Ethanol	2.40	−0.30	2.50	2.40	0.10	0.89	1.00	0.11
Perylene	296	Ethanol	−3.33	−8.83	−3.33	−3.33	0.00	1.00	1.00	0.00
Phenacetin	298	Ethanol	−1.84	−5.00	−1.78	−1.76	0.02	0.90	0.90	0.02
Phenobarbital	298	Ethanol	2.07	0.08	2.27	2.12	0.15	0.84	0.92	0.08
Phenyl salicylate	298	Ethanol	1.54	−1.82	1.59	1.54	0.05	0.93	1.00	0.07
Propyl p-hydroxybenzoate	298	Ethanol	0.43	−2.68	0.49	0.43	0.06	0.92	1.00	0.08
Salicyclic acid	298	Ethanol	−0.85	−3.70	−0.76	−0.85	0.09	0.90	1.00	0.10
Salicylic acid	298	Ethanol	−0.89	−3.62	−0.79	−0.89	0.10	0.89	0.90	0.01
Sulfamethiazine	298	Ethanol	−3.13	−5.52	−2.99	−2.83	0.16	0.87	0.80	0.07
Sulfanilamide	298	Ethanol	−2.12	−3.19	−1.75	−1.90	0.15	0.77	0.80	0.03
Thimylal	298	Ethanol	2.21	−1.30	2.25	2.21	0.04	0.94	1.00	0.06
Thiopental	298	Ethanol	1.75	−1.10	1.84	1.99	0.15	0.90	0.94	0.04
Triglycine	298	Water	−2.24	−7.21	−2.24	−2.24	0.00	1.00	1.00	0.00
Tyrosine	298	Ethanol	−1.35	−3.28	−1.13	−1.35	0.22	0.84	1.00	0.16
Valine (DL)	298	Water	−1.97	−4.13	−1.80	−1.97	0.17	0.86	1.00	0.14
Vinbarbital	298	Ethanol	1.79	−0.16	2.00	1.80	0.20	0.84	0.94	0.10

the solute–solvent–cosolvent interaction parameter, which is computed for each data set by a least square analysis [127]. Since ΔS_m, T_m, T, and G_1–G_4 are constant for a given binary system, it is possible to rewrite Equation 1.96 as

$$-\log X_m = H_0 + H_1 f_1^2 + H_2 f_2^2 + H_3 f_1 f_2 \tag{1.97}$$

where H_0–H_3 are the model constants. By replacing f_2 with $1 - f_1$, Equation 1.97 could be made equivalent to Equation 1.98.

1.7.11   GENERAL SINGLE MODEL

As a polynomial equation, the GSM [121] is derived from theoretically based cosolvency models, i.e., Williams–Amidon and Jouyban–Acree models, by algebraic manipulations. It has been used as an empirical equation to correlate a solute solubility in the pharmaceutical literature [186,187]. The GSM is expressed as a single power series of the solute-free cosolvent volume fractions by

$$\log X_m = K_0 + K_1 f_1 + K_2 f_1^2 + K_3 f_1^3 + \cdots \tag{1.98}$$

where K_0–K_3 denote the model constants, which are calculated using least squares analysis. The model is derived from the Williams–Amidon and Jouyban–Acree models as explained here. By substituting f_2 with $(1 - f_1)$ and $A_{1\text{-}2}(V_s / V_1)$ with α_1 in Equation 1.71 we get

$$\log X_m = f_1 \log X_1 + \log X_2 - f_1 \log X_2 + f_1 \alpha_1 - f_1^2 \alpha_1 \tag{1.99}$$

Rearrangements yield

$$\log X_m = \log X_2 + f_1 \left[\log X_1 - \log X_2 + \alpha_1\right] - f_1^2 \alpha_1 \tag{1.100}$$

Since the numerical values of $\log X_1$, $\log X_2$, and α_1 are constant, it is possible to rewrite Equation 1.100 as

$$\log X_m = K_0 + K_1 f_1 + K_2 f_1^2 \tag{1.101}$$

A similar manipulation of Equation 1.72 yields

$$\log X_m = f_1 \log X_1 + \log X_2 - f_1 \log X_2 + f_1 \alpha_1' - f_1^2 \alpha_1' + f_1^2 \alpha_2' - f_1^3 \alpha_2' \tag{1.102}$$

It should be noted that in Equation 1.102, the α terms are $\alpha_1' = \left[A_{1\text{-}2}\left(\dfrac{V_s}{V_1}\right) + C_2\right]$ and $\alpha_2' = 2\left[A_{2\text{-}1}\left(\dfrac{V_s}{V_2}\right) - A_{1\text{-}2}\left(\dfrac{V_s}{V_1}\right)\right]$. By further modifications

$$\log X_m = \log X_2 + \left[\log X_1 - \log X_2 + \alpha_1'\right]f_1 + \left[\alpha_2' - \alpha_1'\right]f_1^2 - \alpha_2' f_1^3 \tag{1.103}$$

or

$$\log X_m = K_0 + K_1 f_1 + K_2 f_1^2 + K_3 f_1^3 \tag{1.104}$$

A similar manipulation of Equation 1.73 yields

$$\log X_{\mathrm{m}} = f_1 \log X_1 + \log X_2 - f_1 \log X_2 + f_1 \alpha_1'' - f_1^2 \alpha_1'' + f_1^2 \alpha_2'' - f_1^3 \alpha_2'' \\ + \alpha_3'' f_1 + \alpha_3'' f_1^3 - 2\alpha_3'' f_1^2 + \alpha_4'' f_1^2 + \alpha_4'' f_1^4 - 2\alpha_4'' f_1^3 \tag{1.105}$$

or

$$\log X_{\mathrm{m}} = \log X_2 + \left[\log X_1 - \log X_2 + \alpha_1'' + \alpha_3'' \right] f_1 \\ + \left[\alpha_2'' - \alpha_1'' - 2\alpha_3'' + \alpha_4'' \right] f_1^2 + \left[\alpha_3'' - 2\alpha_4'' \right] \tag{1.106}$$

or

$$\log X_{\mathrm{m}} = K_0 + K_1 f_1 + K_2 f_1^2 + K_3 f_1^3 + K_4 f_1^4 \tag{1.107}$$

Replacing f_2 with $(1 - f_1)$ in Equation 1.87 ($i = 2$) yields

$$\log X_{\mathrm{m}} = f_1 \log X_1 + (1 - f_1)\log X_2 + f_1(1 - f_1)S_0 \\ + f_1(1 - f_1)\left[f_1 - (1 - f_1) \right]S_1 + f_1(1 - f_1)\left[f_1 - (1 - f_1) \right]^2 S_2 \tag{1.108}$$

or:

$$\log X_m = f_1 \log X_1 + \log X_2 - f_1 \log X_2 + \left(f_1 - f_1^2 \right)S_0 \\ + \left(f_1 - f_1^2 \right)\left[2f_1 - 1 \right]S_1 + \left(f_1 - f_1^2 \right)\left[2f_1 - 1 \right]^2 S_2 \tag{1.109}$$

or

$$\log X_{\mathrm{m}} = \log X_2 + f_1 \log X_1 - f_1 \log X_2 + \left(f_1 - f_1^2 \right)S_0 \\ + \left[f_1^2 - f_1 - 2f_1^3 \right]S_1 + \left[8f_1^3 - 5f_1^2 + f_1 - 4f_1^4 \right]S_2 \tag{1.110}$$

Arrangements in Equation 1.110 yield

$$\log X_{\mathrm{m}} = \log X_2 + f_1 \log X_1 - f_1 \log X_2 + \left(f_1 - f_1^2 \right)S_0 \\ + \left[-f_1 + f_1^2 - 2f_1^3 \right]S_1 + \left[f_1 - 5f_1^2 + 8f_1^3 - 4f_1^4 \right]S_2 \tag{1.111}$$

Rearrangements in Equation 1.111 yield

$$\log X_{\mathrm{m}} = \log X_2 + \left[\log X_1 - \log X_2 + S_0 - S_1 + S_2 \right]f_1 \\ + \left[-S_0 + S_1 - 5S_2 \right]f_1^2 + \left[-S_1 + 8S_2 \right]f_1^3 + \left[-4S_2 \right]f_1^4 \tag{1.112}$$

Since the terms in the brackets are constant, Equation 1.112 is the same as Equation 1.107. Similar substitutions in nonlinear Equations 1.75, 1.76, 1.94, and 1.95 yield

$$\log X_m = \frac{J_0 + J_1 f_1 + J_2 f_1^2 + J_3 f_1^3 + \cdots}{K_0 + K_1 f_1 + K_2 f_1^2 + K_3 f_1^3 + \cdots} \tag{1.113}$$

where J_0–J_3 and K_0–K_3 are the model constants computed using a nonlinear least square analysis. Since $\log X_m$ in the left-hand side of Equations 1.98 and 1.113 are the same, it is possible to write

$$M_0 + M_1 f_1 + M_2 f_1^2 + M_3 f_1^3 + \cdots = \frac{J_0 + J_1 f_1 + J_2 f_1^2 + J_3 f_1^3 + \cdots}{K_0 + K_1 f_1 + K_2 f_1^2 + K_3 f_1^3 + \cdots} \tag{1.114}$$

By multiplying $(M_0 + M_1 f_1 + M_2 f_1^2 + M_3 f_1^3 + \cdots)$ in $(K_0 + K_1 f_1 + K_2 f_1^2 + K_3 f_1^3 + \cdots)$ in Equation 1.114 and by rearranging further we get

$$K_0 \left(M_0 + M_1 f_1 + M_2 f_1^2 + M_3 f_1^3 + \cdots \right)$$

$$+ K_1 f_1 \left(M_0 + M_1 f_1 + M_2 f_1^2 + M_3 f_1^3 + \cdots \right)$$

$$+ K_2 f_1^2 \left(M_0 + M_1 f_1 + M_2 f_1^2 + M_3 f_1^3 + \cdots \right)$$

$$+ K_3 f_1^3 \left(M_0 + M_1 f_1 + M_2 f_1^2 + M_3 f_1^3 + \cdots \right) + \cdots$$

$$= J_0 + J_1 f_1 + J_2 f_1^2 + J_3 f_1^3 + \cdots \tag{1.115}$$

Further rearrangements of Equation 1.115 produce

$$K_0 M_0 + \left(K_0 M_1 + K_1 M_0 \right) f_1 + \left(K_0 M_2 + K_1 M_1 + K_2 M_0 \right) f_1^2$$

$$\left(K_0 M_3 + K_1 M_2 + K_2 M_1 + K_3 M_0 \right) f_1^3 + \cdots$$

$$= J_0 + J_1 f_1 + J_2 f_1^2 + J_3 f_1^3 + \cdots \tag{1.116}$$

Since K and M terms are constant values for a given binary system, it is possible to rewrite Equation 1.116 as

$$A_0 + A_1 f_1 + A_2 f_1^2 + A_3 f_1^3 + \cdots = J_0 + J_1 f_1 + J_2 f_1^2 + J_3 f_1^3 + \cdots \tag{1.117}$$

As an example, Equation 1.95 could be rewritten as

$$\log X_m = -1 + \frac{f_1 \left(1 + \log X_1 \right)}{f_1 + f_2 \lambda_{12}^{adj}} + \frac{f_2 \left(1 + \log X_2 \right)}{f_1 \lambda_{21}^{adj} + f_2} \tag{1.118}$$

By replacing f_2 with $(1 - f_1)$, $(1 + \log X_1)$ and $(1 + \log X_2)$ with λ_3 and λ_4 and by rearranging further we get

$$\log X_{\mathrm{m}} = -1 + \frac{f_1\lambda_3}{f_1 + (1-f_1)\lambda_{12}^{\mathrm{adj}}} + \frac{(1-f_1)\lambda_4}{f_1\lambda_{21}^{\mathrm{adj}} + 1 - f_1}$$

$$= -1 + \frac{f_1\lambda_3}{f_1 + \lambda_{12}^{\mathrm{adj}} - f_1\lambda_{12}^{\mathrm{adj}}} + \frac{\lambda_4 - f_1\lambda_4}{f_1\lambda_{21}^{\mathrm{adj}} + 1 - f_1} \tag{1.119}$$

or

$$\log X_{\mathrm{m}} = -1 + \frac{f_1\lambda_3\left(f_1\lambda_{21}^{\mathrm{adj}} + 1 - f_1\right)}{\left(f_1\lambda_{21}^{\mathrm{adj}} + 1 - f_1\right)\left(f_1 + \lambda_{12}^{\mathrm{adj}} - f_1\lambda_{12}^{\mathrm{adj}}\right)} + \frac{\left(\lambda_4 - f_1\lambda_4\right)\left(f_1 + \lambda_{12}^{\mathrm{adj}} - f_1\lambda_{12}^{\mathrm{adj}}\right)}{\left(f_1\lambda_{21}^{\mathrm{adj}} + 1 - f_1\right)\left(f_1 + \lambda_{12}^{\mathrm{adj}} - f_1\lambda_{12}^{\mathrm{adj}}\right)} \tag{1.120}$$

or

$$\log X_{\mathrm{m}} = -1$$

$$+ \left[\frac{\left(f_1^2\lambda_3\lambda_{21}^{\mathrm{adj}} + f_1\lambda_3 - f_1^2\lambda_3\right) + \left(f_1\lambda_4 + \lambda_4\lambda_{12}^{\mathrm{adj}} - f_1\lambda_4\lambda_{12}^{\mathrm{adj}} - f_1^2\lambda_4 - f_1\lambda_4\lambda_{12}^{\mathrm{adj}} + f_1^2\lambda_4\lambda_{12}^{\mathrm{adj}}\right)}{\lambda_{12}^{\mathrm{adj}} + \left(1 + \lambda_{12}^{\mathrm{adj}}\lambda_{21}^{\mathrm{adj}} - 2\lambda_{12}^{\mathrm{adj}}\right)f_1 + \left(-1 - \lambda_{12}^{\mathrm{adj}}\lambda_{21}^{\mathrm{adj}} + \lambda_{21}^{\mathrm{adj}} - \lambda_{12}^{\mathrm{adj}}\right)f_1^2} \right] \tag{1.121}$$

$$\log X_{\mathrm{m}} =$$

$$\left\{ \frac{\left[-\lambda_{12}^{\mathrm{adj}} + \left(-1 - \lambda_{12}^{\mathrm{adj}}\lambda_{21}^{\mathrm{adj}} + 2\lambda_{12}^{\mathrm{adj}}\right)f_1 + \left(1 + \lambda_{12}^{\mathrm{adj}}\lambda_{21}^{\mathrm{adj}} - \lambda_{21}^{\mathrm{adj}} + \lambda_{12}^{\mathrm{adj}}\right)f_1^2\right] + \begin{bmatrix} \lambda_4\lambda_{12}^{\mathrm{adj}} + \left(\lambda_3 + \lambda_4 - 2\lambda_4\lambda_{12}^{\mathrm{adj}}\right)f_1 + \\ \left(\lambda_3\lambda_{21}^{\mathrm{adj}} - \lambda_3 - \lambda_4 + \lambda_4\lambda_{12}^{\mathrm{adj}}\right)f_1^2 \end{bmatrix}}{\left[\lambda_{12}^{\mathrm{adj}} + \left(1 + \lambda_{12}^{\mathrm{adj}}\lambda_{21}^{\mathrm{adj}} - 2\lambda_{12}^{\mathrm{adj}}\right)f_1 + \left(-1 - \lambda_{12}^{\mathrm{adj}}\lambda_{21}^{\mathrm{adj}} + \lambda_{21}^{\mathrm{adj}} - \lambda_{12}^{\mathrm{adj}}\right)f_1^2\right]} \right\} \tag{1.122}$$

Since λ terms in Equation 1.122 are constant for a given solute in a binary solvent system, one could summarize Equation 1.122 as

$$\log X_{\mathrm{m}} = \frac{J_0 + J_1 f_1 + J_2 f_1^2}{K_0 + K_1 f_1 + K_2 f_1^2} \tag{1.123}$$

where

$J_0 = \left(-\lambda_{12}^{\mathrm{adj}} + \lambda_4\,\lambda_{12}^{\mathrm{adj}}\right)$

$J_1 = \left(-1 - \lambda_{12}^{\mathrm{adj}}\,\lambda_{21}^{\mathrm{adj}} + 2\lambda_{12}^{\mathrm{adj}} + \lambda_3 + \lambda_4 - 2\lambda_4\,\lambda_{12}^{\mathrm{adj}}\right)$

$J_2 = \left(1 + \lambda_{12}^{\mathrm{adj}}\,\lambda_{21}^{\mathrm{adj}} - \lambda_{21}^{\mathrm{adj}} + \lambda_{12}^{\mathrm{adj}} + \lambda_3\,\lambda_{21}^{\mathrm{adj}} - \lambda_3 - \lambda_4 + \lambda_4\,\lambda_{12}^{\mathrm{adj}}\right)$

$K_0 = \lambda_{12}^{\mathrm{adj}}$, $K_1 = \left(1 + \lambda_{12}^{\mathrm{adj}}\,\lambda_{21}^{\mathrm{adj}} - 2\lambda_{12}^{\mathrm{adj}}\right)$, and $K_1 = \left(-1 - \lambda_{12}^{\mathrm{adj}}\,\lambda_{21}^{\mathrm{adj}} + \lambda_{21}^{\mathrm{adj}} - \lambda_{12}^{\mathrm{adj}}\right)$.

For this, we can summarize all cosolvency models as a power series of volume fraction of the cosolvent, the GSM. These results are however not unexpected as it is generally the case that a definite experimental phenomenon, like drug solubility in water–cosolvent mixtures, would have a single mathematical representation. Here it has been shown that this is in fact true in the case of the cosolvency models. The main difference in these models is that the accuracies of most models are different from each other. This is the case because the models employed a different arrangement of the independent variables.

1.7.12 Mobile Order and Disorder Theory

Van den Mooter et al. [188] applied the MOD model to calculate the solubility of temazepam in an aqueous mixture of PEG 6000 in volume fractions of 0.00, 0.05, 0.10, and 0.15 at 24°C, 34°C, and 46°C. The model was derived from the basic model of MOD (Equation 1.12) and was presented as

$$
\begin{aligned}
\ln \Phi_B =& \left\{ -\frac{27400}{R}\left(\frac{1}{T} - \frac{1}{432.6}\right) \right\} \\
&+ \left\{ \frac{1}{2}\left[\left(V_B \frac{\varphi_w}{V_w} + V_B \frac{\varphi_{PEG\,6000}}{V_{PEG\,6000}} - 1\right)\Phi_S \right] + \ln\left[(1-\Phi_S) + \Phi_S\left(V_B \frac{\varphi_w}{V_w} + V_B \frac{\varphi_{PEG\,6000}}{V_{PEG\,6000}}\right)\right] \right\} \\
&+ \left\{ -\frac{V_B}{V_w}\Phi_S\varphi_w \right\} \\
&+ \left\{ \ln\left(1 + K_{O_1}\frac{\varphi_w}{V_w}\right) + \ln\left(1 + K_{O_2}\frac{\varphi_w}{V_w}\right) \right\} \\
&+ \left\{ -\ln\left(1 + \frac{K_{BB}}{V_B} + \ln\left(1 + K_{BhO}^{water}\frac{\varphi_w}{V_w} + K_{BhO}^{PEG\,6000}\frac{\varphi_{PEG\,6000}}{V_{PEG\,6000}}\right)\right) \right\} \\
&+ \left\{ \frac{V_B}{RT}(\delta_B' - \delta_S')^2 \Phi_S^2 \right\}
\end{aligned}
\tag{1.124}
$$

where

Φ_B is the volume fraction solubility of temazepam

φ_w and $\varphi_{PEG\,6000}$ are the volume fractions of water and PEG 6000, respectively

V_B, V_w, and $V_{PEG\,6000}$ are the molar volumes of temazepam, water, and PEG 6000, respectively

Φ_S is the volume fraction of the solvent in the solution

K_{O_1}, K_{O_2}, K_{BB}, K_{BhO}^{water} and $K_{BhO}^{PEG\,6000}$ are the constants' values

δ_B' and δ_S' are the modified solubility parameters of the solute and solvent, respectively

The authors have presented the capability of Equation 1.124 using graphical representation, in which the volume fraction solubility of temazepam in 15% v/v of PEG 6000 in aqueous solution was calculated as 0.0007, whereas the experimental value was 0.0002 [188].

1.7.13 QSPR Model of Rytting

Rytting et al. [189] determined the solubility of 122 solutes in water, PEG 400, and three binary aqueous mixtures containing 25%, 50%, and 75% of PEG 400 and treated the solubility in binary solvents as a separate system. These models are slightly different from other cosolvency models that treated the binary solvents as a continuous system. The general form of the Rytting model is

$$
\log X = c_0 + c_1 MW + c_2 V_s + c_3 RB + c_4 HBA + c_5 HBD + c_6 RG + c_7 D_s
\tag{1.125}
$$

where

X is the molar solubility of the solute

MW is the molecular weight (g/mol)

V_s is the molecular volume (Å^3)
RB is the number of rotatable bonds
HBA is the number of hydrogen bond acceptors
HBD is the number of hydrogen bond donors
RG is the radius of gyration (Å)
D_s is the molecular density (MW/V_s)
$c_0 - c_7$ are the model constants

The investigated solutes were divided into 84 data sets (training set) and 38 data sets (test set). The models (for each solvent composition) were trained using 84 data points and the prediction capabilities of the models were not satisfactory [189]. To provide better models, genetic algorithm was used to divide the training set into subsets 1 and 2 based on the similarities of the molecular descriptors. Two sets of the model constants ($c_0 - c_7$) were reported and the solubilities of the test compounds were predicted using the most similar model of 1 or 2 to the descriptors of the test compound. The similarities between compounds were determined using the d_{ij} term defined by

$$d_{ij} = \sum \left(X_{ik} - X_{jk} \right)^2, \quad k = 1 - 7 \tag{1.126}$$

where
$X_{i1}, X_{i2}, \ldots X_{i7}$ are the seven descriptors of compound i
$X_{j1}, X_{j2}, \ldots X_{j7}$ are the seven descriptors of compound j
d_{ij} is the Euclidean distance between compounds i and j

Rytting et al. [189] evaluated the accuracy of their models by computing the residual ranges in log unit for the test set that was sorted into five ranges, i.e., $<\pm0.5$, $\pm0.5-1.0$, $\pm1.0-1.5$, $\pm1.5-2.0$, and $>\pm2.0$. The results were also compared with those of the log-linear model of Yalkowsky. Figure 1.4

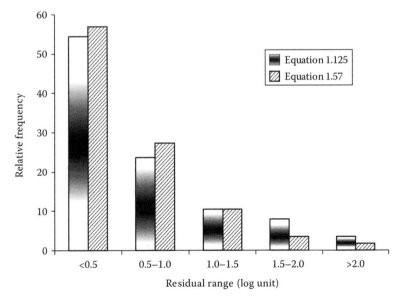

FIGURE 1.4 The relative frequencies of the residuals of the QSPR and log-linear models sorted in five groups.

showed the relative frequencies of the residuals of the QSPR model of Rytting et al. [189] and the log-linear model of Yalkowsky [126] for three water–PEG 400 compositions.

The relative frequencies of both models are similar; however, it should be noted that the log-linear model employs the aqueous solubility data of each solute as an input value and the QSPR model treats each solvent composition as a separate system. In addition, the required computations are more complicated compared to the straightforward calculations of the log-linear model.

1.7.14 ARTIFICIAL NEURAL NETWORK MODEL

An attempt was made using ANN models and different numerical analyses that are mandatory in pharmaceutical applications to predict solubility in mixture of water–cosolvents [83]. The optimized topology of the ANN was 6-5-1 and the network used $f_1, f_2, -\log X_1, -\log X_2, \delta_1, \delta_2$ as input variables, and $-\log X_m$ as its output. The accuracy of the ANN method for computing the solubility data of drugs was checked by calculating the MPD and IPD for 35 data sets of various drugs in eight cosolvent systems. The results were compared with the best MLR models (i.e., the Jouyban–Acree model).

In numerical analysis I, all data points from each set was used to train the models and the back-calculated solubilities were used to compute the MPD and IPD values. The overall MPDs were 0.9% and 5.6% for ANN and MLR models, respectively. In analysis II, five data points from each data set were used as training sets and the solubility at other solvent compositions was predicted. The overall MPDs for ANN and MLR models were 9.0% and 11.3%, respectively. A single ANN model (analysis III) was trained using all data points of 35 data sets and the solubilities were back-calculated where the overall MPD was 24.8%. In the numerical analysis IV, solubility data sets with odd numbers were used as a training set for ANN and the sets with even numbers were used as a prediction set and the overall MPD for predicted solubilities was 56.0% [83].

The correlation ability of ANN and MLR models for the solubility of various drugs in aqueous mixtures of a given cosolvent was investigated (analysis V), where the overall MPDs were 2.0 and 20.4 for ANN and MLR models, respectively. In numerical analysis VI, the correlation ability of the models for structurally related drugs in a given water–cosolvent mixtures was studied. The ANN model produced the overall MPD of 4.7% in comparison with 18.4% for MLR models. In the last numerical analysis (VII), the correlation ability of a given drug in various water–cosolvent systems was investigated. The ANN model produced an overall MPD of 3.4%, whereas the corresponding value for the MLR model was 67.2%. The mean differences between the overall MPDs of the ANN and MLR models were statistically significant in all numerical analyses except analysis IV where there was no significant difference between the two models. The ANN method produced better IPD distribution compared to the MLR model, as shown in Figure 1.5 [83].

1.7.15 COSMO-RS MODEL

The conductor-like screening model for real solvents (COSMO-RS) is a predictive model that integrates the concepts of quantum theory, dielectric continuum models, and surface interactions. Ikeda et al. [72] predicted the solubility of 15 drugs in water, ethanol, acetone, and chloroform and compared the predicted solubilities with the experimental values by computing RMSE whose overall value was 0.64 in log unit. The authors also tested the prediction capability of the COSMO-RS model on oxolinic acid solubility data in water–ethanol and sulfadiazine solubility data in water–dioxane mixtures where the RMSE values were 2.0 and 1.6 log unit for oxolinic acid and sulfadiazine data, respectively [72]. In addition to these deviations, the model applications require relatively sophisticated computations, which are not favored methods for most pharmaceutical scientists.

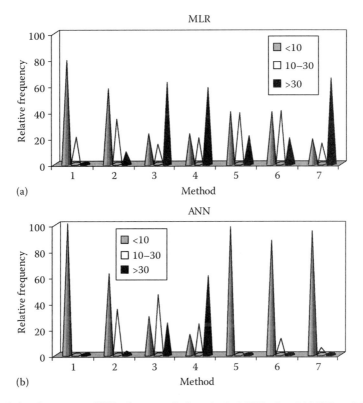

FIGURE 1.5   Relative frequency of IPDs for numerical methods I–VII using (a) MLR and (b) ANN.

1.7.16   NEW MODELS PROPOSED BY YALKOWSKY'S GROUP

A new model was proposed by Machatha et al. [74], which it is claimed is a better predictor of solubility of drugs in water–ethanol mixtures compared to the log-linear model. The model is

$$\log X_{\mathrm{m}} = \frac{\log X_2 + a \cdot f_1}{1 + b \cdot f_1 + c \cdot f_1^2} \tag{1.127}$$

where a, b, and c are the model constants. The authors compared the accuracy of Equation 1.127 with that of GSM model (the third-order polynomial of cosolvent fractions) employing the solubility of 51 compounds in water–ethanol mixtures. The accuracy criteria used in the comparison were RMSE and AAE, where the average RMSE values for Equation 1.127 and the third-order polynomial were 0.035 and 0.049, respectively. The corresponding RMSE for the fluctuation model of Ruckenstein was 0.064 [74].

A bilinear function that accounts for the disparity between log-linear and parabolic models was also presented by Machatha and Yalkowsky [75], and its accuracy was checked using the RMSE criterion by employing the solubility data of 52 sets in water–ethanol mixtures. The proposed model for water–ethanol mixtures is

$$\log\left(\frac{X_{\mathrm{m}}}{X_2}\right) = \sigma_A \cdot f_1 + \frac{(\sigma_A - \sigma_B)f_1}{1 + 10^{-3.6(f_1 - 1)}} \tag{1.128}$$

TABLE 1.16
The RMSE of Various Equations for Calculating Solute Solubility
in Water–Ethanol Mixtures

Solute	N^a	Equation 1.127	Equation 1.98 up to Power 3	Equation 1.98 up to Power 2	Equation 1.128	Equation 1.57
Acetanilide	13	0.021	0.027	0.070	0.041	0.248
Alanine	10	0.010	0.055	0.195	0.050	0.982
Alprazolam	9	0.022	0.035	0.103	0.063	0.365
p-Aminobenzoic acid	6	0.023	0.036	0.044	0.056	0.256
Aminocaproic acid	10	0.029	0.105	0.271	0.047	0.988
Amino-isobutyric acid	5	0.004	0.007	0.025	0.017	0.179
Amino-n-butyric acid	6	0.005	0.044	0.140	0.017	0.574
Anthracene	11	0.061	0.091	0.129	0.123	0.267
Asparagine	5	0.010	0.019	0.128	0.003	0.627
Aspartic acid	9	0.062	0.104	0.220	0.092	0.759
Barbital	11	0.018	0.014	0.061	0.035	0.214
Benzamide	14	0.019	0.012	0.061	0.031	0.324
Benzocaine	11	0.038	0.048	0.110	0.083	0.345
Benzoic acid	11	0.029	0.083	0.103	0.112	0.543
Biphenyl	11	0.069	0.092	0.152	0.204	0.200
Caffeine	6	0.026	0.032	0.135	0.000	0.582
Camphoric acid	12	0.047	0.052	0.069	0.201	0.414
Diazepam	11	0.054	0.057	0.122	0.086	0.513
Didanosine	11	0.050	0.062	0.081	0.075	0.456
Beta-estradiol	6	0.034	0.085	0.155	0.134	0.645
5-Ethylhydantoin	7	0.071	0.019	0.071	0.023	0.296
Formyl-aminobutyric acid	7	0.051	0.011	0.069	0.024	0.211
Formylglycine	9	0.008	0.027	0.074	0.020	0.288
Formylleucine	8	0.043	0.063	0.083	0.076	0.282
Furosemide	13	0.148	0.133	0.361	0.420	0.364
Glutamic acid	6	0.038	0.102	0.217	0.076	0.790
Glutamine	5	0.006	0.007	0.007	0.034	0.623
Glycine	10	0.016	0.072	0.158	0.035	0.148
Glycylglycine	7	0.016	0.074	0.213	0.025	0.854
Histidine	8	0.018	0.033	0.087	0.005	0.275
Hydantoic acid	6	0.016	0.020	0.094	0.014	0.380
Hydantoin	7	0.020	0.027	0.093	0.018	0.356
Ibuprofen	8	0.120	0.123	0.249	0.197	0.462
Indomethacine	10	0.058	0.083	0.172	0.147	0.419
Leucine	5	0.022	0.014	—	—	—
Metharbital	11	0.018	0.028	0.078	0.053	0.279
Methylhydantoic acid	6	0.020	0.020	0.095	0.023	0.294
Naphthalene	6	—	—	0.000	0.000	0.000
Norleucine	10	0.035	0.055	0.137	0.057	0.316
Oxolinic acid	11	0.049	0.063	0.065	0.132	0.147
Paracetamol	13	0.083	0.034	0.057	0.053	0.349
Phenobarbital	12	0.017	0.015	0.092	0.087	0.488
Phenylalanine	8	0.018	0.040	0.097	0.052	0.457
Phenytoin	11	0.046	0.053	0.122	0.090	0.419
Salicylic acid	6	0.007	0.067	0.129	0.106	0.258
Strychnine	7	0.038	0.054	0.119	0.055	0.508

TABLE 1.16 (continued)
The RMSE of Various Equations for Calculating Solute Solubility
in Water–Ethanol Mixtures

Solute	N^a	Equation 1.127	Equation 1.98 up to Power 3	Equation 1.98 up to Power 2	Equation 1.128	Equation 1.57
Tartaric acid	12	0.002	0.005	0.017	0.005	0.092
Theophylline	10	—	—	0.080	0.027	0.364
Triglycine	7	0.057	0.079	0.277	0.025	1.010
Tryptophan	8	0.042	0.019	0.097	0.050	0.372
DL-Valine	7	0.026	0.058	0.146	0.043	0.461
Zalcitabine	11	0.022	0.026	0.052	0.056	0.396
Ziduvudine	11	0.022	0.024	0.148	0.032	0.257
Overall		0.035	0.050	0.120	0.069	0.423

Sources: Machatha, S.G. and Yalkowsky S.H., *Int. J. Pharm.*, 286, 111, 2004; Machatha, S.G. and Yalkowsky, S.H., *J. Pharm. Sci.*, 94, 2731, 2005.

a N is the number of data points in each set.

where σ_A is the slope of the ascending part of the solubility profile and σ_B is the slope of the descending section of the profile. The RMSE values for the investigated data sets are reported in Table 1.16 along with the RMSE of Equation 1.127, the third-order polynomial, the second-order polynomial, and the bilinear (Equation 1.128) and log-linear (Equation 1.57) models collected from the literature [73,75]. It should be noted that the average RMSE of the bilinear model was more than that of Equation 1.127.

1.8 CONCLUDING REMARKS

There are various solubilization methods for increasing the solubility of low soluble drugs including the addition of permissible cosolvents, surface active agents, complexing compounds, changing pH of the solution for solutes with acidic/basic functional groups or salt formation, preparing soluble prodrugs, and, more recently, using ionic liquids [110]. These methods could solve the solubility problems of most of the drugs/drug candidates in practice; however, the problem for some candidates still remains and medicinal chemists replace these candidates with more optimized ones. The high rate of candidate rejection means more cost in drug discovery investigations and ultimately more expensive drugs in the market. In view of these points, practical and accurate solubility prediction methods are highly in demand in the pharmaceutical industry and efforts on aqueous solubility predictions should be continued until such an acceptable and practical prediction method is available. The usefulness of the prediction methods depends on a number of factors including (a) the fundamentals of the prediction method, (b) its ease of use, (c) its range of applicability, (d) the size of molecular descriptors and their quality, and (e) the accuracy of the predictions.

REFERENCES

1. Yalkowsky, S.H. and He, Y., *Handbook of Aqueous Solubility Data*. 2003, Boca Raton, FL: CRC Press.
2. US Pharmacopeial Convention, Rockville, MD, USP. 23rd Revision, 1998. pp. 220–221, 469, 480, 515, 659, 683, 1189, 1202, 1261, 1369, 1402, 1528, 1604, 2215, 2218.
3. Amidon, G.L., Lennernas, H., Shah, V.P., and Crison, J.R., A theoretical basis for a biopharmaceutic drug classification: The correlation of in vitro drug product dissolution and in vivo bioavailability. *Pharmaceutical Research*, 1995. 12: 413–420.

4. Pouton, C.W., Formulation of poorly water-soluble drugs for oral administration: Physicochemical and physiological issues and the lipid formulation classification system. *European Journal of Pharmaceutical Sciences*, 2006. 29: 278–287.

5. Lindenberg, M., Kopp, S., and Dressman, J.B., Classification of orally administered drugs on the World Health Organization model list of essential medicines according to the biopharmaceutics classification system. *European Journal of Pharmaceutics and Biopharmaceutics*, 2004. 58: 265–278.

6. Alsenz, J. and Kansy, M., High throughput solubility measurement in drug discovery and development. *Advanced Drug Delivery Reviews*, 2007. 59: 546–567.

7. Lipinski, C.A., Lombardo, F., Dominy, B.W., and Feeny, P.J., Experimental and computational approaches to estimate solubility and permeability in drug discovery and development settings. *Advanced Drug Delivery Reviews*, 2001. 46: 3–26.

8. Jorgensen, W.L. and Duffy, E.M, Prediction of drug solubility from structure. *Advanced Drug Delivery Reviews*, 2002. 54: 355–366.

9. Ren, S.S. and Anderson, B.D., What determines drugs solubility in lipid vehicles: Is it predictable? *Advanced Drug Delivery Reviews*, 2008. 60: 638–656.

10. Date, A.A. and Nagarsenker, M.S., Parenteral microemulsions: An overview. *International Journal of Pharmaceutics*, 2008. 355: 19–33.

11. Brewster, M.E. and Loftsson, T., Cyclodextrins as pharmaceutical solubilizers. *Advanced Drug Delivery Reviews*, 2007. 59: 645–666.

12. Higuchi, T. and Connors, K.A., Phase-solubility techniques. *Advances in Analytical Chemistry and Instrumentation*, 1965. 4: 117–212.

13. Glomme, A., Marz, J., and Dressman, J.B., Comparison of a miniaturized shake-flask solubility method with automated potentiometric acid/base titrations and calculated solubilities. *Journal of Pharmaceutical Sciences*, 2005. 94: 1–16.

14. Bevan, C.D. and Lloyd, R.S., A high-throughput screening method for the determination of aqueous drug solubility using laser nephelometry in microtiter plates. *Analytical Chemistry*, 2000. 72: 1781–1787.

15. Kibbey, C.E., Poole, S.K., Robinson, B., Jackson, J.D., and Durham, D., An integrated process for measuring the physicochemical properties of drug candidates in a preclinical discovery environment. *Journal of Pharmaceutical Sciences*, 2001. 90: 1164–1175.

16. Yalkowsky, S.H. and Valvani, S.C., Solubility and partitioning I: Solubility of nonelectrolytes in water. *Journal of Pharmaceutical Sciences*, 1980. 69: 912–922.

17. Jain, P., Sepassi, K., and Yalkowsky, S.H., Comparison of aqueous solubility estimation from AQUAFAC and the GSE. *International Journal of Pharmaceutics*, 2008. 360: 122–147.

18. Irmann, F., Eine einfache korrelation zwischen wasserloslichkeit und struktur von kohlenwasserstoffen und halogenkohlenwasserstoffen. *Chemical Engineering Technology*, 1965. 37: 789–798.

19. Hansch, C., Quinlan, J.E., and Lawerence, G.L., The linear free energy relationship between partition coefficients and the aqueous solubility of organic liquids. *Journal of Organic Chemistry*, 1968. 33: 347–350.

20. Isnard, P. and Lambert, S., Aqueous solubility and *n*-octanol/water partition coefficient correlations. *Chemosphere*, 1989. 18: 1837–1853.

21. Yang, G., Ran, Y., and Yalkwosky, S.H., Prediction of the aqueous solubility: Comparison of the general solubility equation and the method using an amended solvation energy relationship. *Journal of Pharmaceutical Sciences*, 2002. 91: 517–533.

22. Jain, N., Yang, G., Machatha, S.G., and Yalkowsky, S.H., Estimation of the aqueous solubility of weak electrolytes. *International Journal of Pharmaceutics*, 2006. 319: 169–171.

23. Abraham, M.H. and Le, J., The correlation and prediction of the solubility in water using an amended solvation energy relationship. *Journal of Pharmaceutical Sciences*, 1999. 88: 868–880.

24. Stovall, D.M., Givens, C., Keown, S., Hoover, K.R., Barnes, R., Harris, C., Lozano, J., Nguyen, M., Rodriguez, E., Acree, W.E., Jr., and Abraham, M.H., Solubility of crystalline nonelectrolyte solutes in organic solvents: Mathematical correlation of 4-chloro-3-nitrobenzoic acid and 2-chloro-5-nitrobenzoic acid solubilities with the Abraham solvation parameter model. *Physics and Chemistry of Liquids*, 2005. 43: 351–360.

25. Jouyban, A., Soltanpour, Sh., Soltani, S., and Acree, W.E., Jr., Solubility prediction of drugs in water–cosolvent mixtures using Abraham solvation parameters. *Journal of Pharmacy and Pharmaceutical Sciences*, 2007. 10: 263–277.

26. Hammett, L.P., The effect of structure upon the reactions of organic compounds. Benzene derivatives. *Journal of American Chemical Society*, 1937. 59: 96–103.

27. Taft, R.W., Jr., Polar and steric substituent constants for aliphatic and *o*-benzoate groups from rates of esterification and hydrolysis of esters. *Journal of American Chemical Society*, 1952. 74: 3120–3128.

28. Kamlet, M.J. and Taft, R.W., The solvatochromic comparison method. I. The beta-scale of solvent hydrogen-bond acceptor (HBA) basicities. *Journal of American Chemical Society*, 1976. 98: 377–383.

29. Abraham, M.R., Duce, P.P., Grellier, P.L., Prior, D.V., Morris, J.J., and Taylor, P.J., Hydrogen-bonding. Part 5. A thermodynamically-based scale of solute hydrogen-bond acidity. *Tetrahedron Letters*, 1988. 29: 1587–1590.

30. Blake-Taylor, B.H., Deleon, V.H., Acree, W.E., Jr., and Abraham, M.H., Mathematical correlation of salicylamide solubilities in organic solvents with the Abraham solvation parameter model. *Physics and Chemistry of Liquids*, 2007. 45: 289–298.

31. Jouyban, A., Review of the cosolvency models for predicting solubility of drugs in water–cosolvent mixtures. *Journal of Pharmacy and Pharmaceutical Sciences*, 2008. 11: 32–58.

32. Jouyban, A., Unpublished results. 2008.

33. Hou, T.J., Xia, K., Zhang, W., and Xu, X.J., ADME evaluation in drug discovery. 4. Prediction of aqueous solubility based on atom contribution approach. *Journal of Chemical Information and Computer Sciences*, 2004. 44: 266–275.

34. Faller, B. and Ertl, P., Computational approaches to determine drug solubility. *Advanced Drug Delivery Reviews*, 2007. 59: 533–545.

35. Duchowicz, P.R., Talevi, A., Bruno-Blanch, L.E., and Castro, E.A., New QSPR study for the prediction of aqueous solubility of drug-like compounds. *Bioorganic and Medicinal Chemistry*, 2008. 16: 7944–7955.

36. DRAGON 5.0, Evaluation version, http://www.disat.unimib.it/chm.

37. Huuskonen, J., Salo, M., and Takinen, J., Neural network modeling for estimation of the aqueous solubility of structurally related drugs. *Journal of Pharmaceutical Sciences*, 1997. 86: 450–454.

38. Huuskonen, J., Rantanen, J., and Livingstone, D., Prediction of aqueous solubility for a diverse set of organic compounds based on atom-type electrotopological indices. *European Journal of Medicinal Chemistry*, 2000. 35: 1081–1088.

39. Jouyban-Gharamaleki, A., Applications of mathematical modelling in pharmaceutical and analytical sciences, PhD dissertation. 2001, Bradford University, Bradford, U.K.

40. Huyskens, P.L. and Haulait-Pirson, M.C., Influence of H-bond chains in solvents on the solubility of inert substances. A new quantitative approach. *Journal of Molecular Liquids*, 1985. 31: 153–176.

41. Ruelle, P. and Kesselring, U.W., Solubility predictions for solid nitriles and tertiary amides based on the mobile order theory. *Pharmaceutical Research*, 1994. 11: 201–205.

42. Ruelle, P., Farinña-Cuendet, A., and Kesselring, U.W., Hydrophobic and solvation effects on the solubility of hydroxysteroids in various solvents: Quantitative and qualitative assessment by application of the mobile order and disorder theory. *Perspectives in Drug Discovery and Design*, 2000. 18: 61–112.

43. Ruelle, P., Farina-Cuendet, A., and Kesselring, U.W., The mobile order solubility equation applied to polyfunctional molecules: The non-hydroxysteroids in aqueous and non aqueous solvents. *International Journal of Pharmaceutics*, 1997. 157: 219–232.

44. Cheng, A. and Merz, K.M., Jr., Systems and method for aqueous solubility prediction, US patent. 2005.

45. Fillikov, A., Method of prediction of solubility of chemical compounds, US patent. 2006.

46. Jorgensen, W.L. and Duffy, E.M., Prediction of drug solubility from Monte Carlo simulations. *Bioorganic Medicinal Chemistry Letters*, 2000. 10: 1155–1158.

47. Jouyban, A., Fakhree, M.A.A., and Shayanfar, A., Solubility prediction methods for drug/drug like molecules. *Recent Patents on Chemical Engineering*, 2008. 1: 220–231.

48. Yalkowsky, S.H., Valvani, S.C., and Roseman, T.J., Solubility and partitioning. VI: Octanol solubility and octanol–water partition coefficients. *Journal of Pharmaceutical Sciences*, 1983. 72: 866–870.

49. Dearden, J.C. and O'Sullivan, J.G., Solubility of pharmaceuticals in cyclohexane. *Journal of Pharmacy and Pharmacology*, 1988. 40: 77P.

50. Sepassi, K. and Yalkowsky, S.H., Solubility prediction in octanol: A technical note. *AAPS PharmSciTech*, 2006. 7: E1–E8.

51. Martin, A., Wu, P.L., and Adjei, A., Extended Hansen solubility approach: Naphthalene in individual solvents. *Journal of Pharmaceutical Sciences*, 1981. 70: 1260–1264.

52. Beerbower, A., Wu, P.L., and Martin, A., Expanded solubility parameter approach 1. Naphthalene and benzoic acid in individual solvents. *Journal of Pharmaceutical Sciences*, 1984. 73: 179–188.

53. Gharagheizi, F., New procedure to calculate the Hansen solubility parameters of polymers. *Journal of Applied Polymer Science*, 2007. 103: 31–36.

54. Ho, D.L. and Glinka, C.J., New insights into Hansen's solubility parameters. *Journal of Polymer Science, Part B: Polymer Physics*, 2004. 42: 4337–4343.

55. Wu, P.L., Beerbower, A., and Martin, A., Extended Hansen approach: Calculating partial solubility parameters of solid solutes. *Journal of Pharmaceutical Sciences*, 1982. 71: 1285–1287.

56. Bustamante, P., Escalera, B., Martin, A., and Selles, E., Predicting the solubility of sulfamethoxy-pyridazine in individual solvents I: Calculating partial solubility parameters. *Journal of Pharmaceutical Sciences*, 1989. 78: 567–573.

57. Bustamante, P., Peña, M.A., and Barra, J., Partial-solubility parameters of naproxen and sodium diclofenac. *Journal of Pharmacy and Pharmacology*, 1998. 50: 975–982.

58. Martin, A., Wu, P.L., and Beerbower, A., Expanded solubility parameter approach II: *p*-Hydroxybenzoic acid and methyl *p*-hydroxybenzoate in individual solvents. *Journal of Pharmaceutical Sciences*, 1984. 73: 188–194.

59. Peña, M.A., Daali, Y., Barra, J., and Bustamante, P., Partial solubility parameters of lactose, mannitol and saccharose using the modified extended Hansen method and evaporation light scattering detection. *Chemical and Pharmaceutical Bulletin*, 2000. 48: 179–183.

60. Subrahmanyam, C.V.S. and Suresh, S., Solubility behaviour of haloperidol in individual solvents deter-mination of partial solubility parameters. *European Journal of Pharmaceutics and Biopharmaceutics*, 1999. 47: 289–294.

61. Verheyen, S., Augustijns, P., Kinget, R., and Van Den Mooter, G., Determination of partial solubility parameters of five benzodiazepines in individual solvents. *International Journal of Pharmaceutics*, 2001. 228: 199–207.

62. Bustamante, P., Peña, M.A., and Barra, J., Partial solubility parameters of piroxicam and niflumic acid. *International Journal of Pharmaceutics*, 1998. 174: 141–150.

63. Thimmasetty, J., Subrahmanyam, C.V.S., Sathesh Babu, P.R., Maulik, M.A., and Viswanath, B.A., Solubility behavior of pimozide in polar and nonpolar solvents: Partial solubility parameters approach. *Journal of Solution Chemistry*, 2008. 37: 1365–1378.

64. Bustamante, P., Peña, M.A., and Barra, J., The modified extended Hansen method to determine par-tial solubility parameters of drugs containing a single hydrogen bonding group and their sodium derivatives: Benzoic acid/Na and ibuprofen/Na. *International Journal of Pharmaceutics*, 2000. 194: 117–124.

65. Bustamante, P., Martin, A., and Gonzalez-Guisandez, M.A., Partial solubility parameters and solvato-chromic parameters for predicting the solubility of single and multiple drugs in individual solvents. *Journal of Pharmaceutical Sciences*, 1993. 82: 635–640.

66. Flanagan, K.B., Hoover, K.R., Garza, O., Hizon, A., Soto, T., Villegas, N., Acree, W.E., Jr., and Abraham, M.H., Mathematical correlation of 1-chloroanthraquinone solubilities in organic solvents with the Abraham solvation parameter model. *Physics and Chemistry of Liquids*, 2006. 44: 377–386.

67. Abraham, M.H. and Acree, W.E., Jr., Characterisation of the water–isopropyl myristate system. *International Journal of Pharmaceutics*, 2005. 294: 121–128.

68. Li, A., Predicting cosolvency. 3. Evaluation of the extended log-linear. *Industrial Engineering and Chemical Research*, 2001. 40: 5029–5035.

69. Jouyban, A. and Acree, W.E., Jr., In silico prediction of drug solubility in water–ethanol mixtures using Jouyban–Acree model. *Journal of Pharmacy and Pharmaceutical Sciences*, 2006. 9: 262–269.

70. Lee, Y.C., Myrdal, P.B., and Yalkowsky, S.H., Aqueous functional group activity coefficients (AQUAFAC) 4: Application to complex organic compounds. *Chemosphere*, 1996. 33: 2129–2144.

71. Millard, J.F., Alvarez-Nunez, F.A., and Yalkowsky, S.H., Solubilization by cosolvents. Establishment useful constants for the log-linear model. *International Journal of Pharmaceutics*, 2002. 245: 153–166.

72. Ikeda, H., Chiba, K., Kanou, A., and Hirayama, N., Prediction of solubility of drug by conductor-like screening model for real solvents. *Chemical and Pharmaceutical Bulletin*, 2005. 53: 253–255.

73. Machatha, S.G. and Yalkowsky S.H., Estimation of the ethanol/water solubility profile from the octanol/water partition coefficient. *International Journal of Pharmaceutics*, 2004. 286: 111–115.

74. Machatha, S.G., Bustamante, P., and Yalkowsky, S.H., Deviation from linearity of drug solubility in ethanol/water mixtures. *International Journal of Pharmaceutics*, 2004. 283: 83–88.

75. Machatha, S.G. and Yalkowsky, S.H., Bilinear model for the prediction of drug solubility in ethanol/water mixtures. *Journal of Pharmaceutical Sciences*, 2005. 94: 2731–2734.

76. Bodor, N. and Huang, M.J., A new method for the estimation of the aqueous solubility of organic com-pounds. *Journal of Pharmaceutical Sciences*, 1992. 81: 954–960.

77. Myrdal, P.B., Manka, A.M., and Yalkowsky, S.H., AQUAFAC 3: Aqueous functional, group activity coef-ficients: Application to the estimation of aqueous solubility. *Chemosphere*, 1995. 30: 1619–1637.

78. Sutter, J.M. and Jurs, P.C., Prediction of aqueous solubility for a diverse set of heteroatom-containing compounds using quantitative structure property relationship. *Journal of Chemical Information and Computer Sciences*, 1996. 36: 100–107.

79. Hansen, S.H., Rivevol, C., and Acree, W.E., Jr., Solubilities of anthracene, fluranthene and pyrene in organic solvents. Comparison of calculated values using UNIFAC and modified UNIFAC (Dortmund) with experimental data and values using mobile order theory. *Canadian Journal of Chemical Engineering*, 2000. 78: 1168–1174.

80. Jain, N. and Yalkowsky, S.H., Estimation of the aqueous solubility I: Applications to organic nonelectrolytes. *Journal of Pharmaceutical Sciences*, 2001. 90: 234–252.

81. Bustamante, P., Escalera, B., Martin, A., and Selles, E., A modification of the extended Hildebrand approach to predict the solubility of structurally related drugs in solvent mixtures. *Journal of Pharmacy and Pharmacology*, 1993. 45: 253–257.

82. Dickhut, R.M., Armstrong, D.E., and Andren A.W., The solubility of hydrophobic aromatic chemicals in organic solvent/water mixtures: Evaluation of four mixed solvent solubility estimation methods. *Environmental Toxicology and Chemistry*, 1991. 10: 881–889.

83. Jouyban, A., Majidi, M.R., Jalilzadeh, H., and Asadpour-Zeynali, K., Modeling drug solubility in water–cosolvent mixtures using an artificial neural network (ANN). *IL Farmaco*, 2004. 59: 505–512.

84. Jouyban-Gharamaleki, A., Barzegar-Jalali, M., and Acree, W.E., Jr., Solubility correlation of structurally related drugs in binary solvent mixtures. *International Journal of Pharmaceutics*, 1998. 166: 205–209.

85. Jouyban-Gharamaleki, A., Dastmalchi, S., Chan, H.K., Hanaee, J., Javanmard, A., and Barzegar-Jalali, M., Solubility prediction for furosemide in water–cosolvent mixtures using the minimum number of experiments. *Drug Development and Industrial Pharmacy*, 2001. 27: 577–583.

86. Jouyban-Gharamaleki, A., York, P., Hanna, M., and Clark, B.J., Solubility prediction of salmeterol xinafoate in water–dioxane mixtures. *International Journal of Pharmaceutics*, 2001. 216: 33–41.

87. Pinsuwan, S., Myrdal, P.B., Lee, Y.C., and Yalkowsky, S.H., Aqueous functional group activity coefficients: Applications to alcohols and acids. *Chemosphere*, 1997. 35: 2503–2513.

88. Ruelle, P. and Kesselring, U.W., Aqueous solubility prediction of environmental important chemicals from the mobile order thermodynamics. *Chemosphere*, 1997. 34: 275–298.

89. Subrahmanyam, C.V.S., Sreenivasa Reddy, M., Venkata Rao, J., and Gundu Rao, P., Irregular solution behaviour of paracetamol in binary solvents. *International Journal of Pharmaceutics*, 1992. 78: 17–24.

90. Bustamante, P., Romero, S., and Reillo, A., Thermodynamics of paracetamol in amphiprotic and amphiprotic–aprotic solvent mixtures. *Pharmaceutical Sciences*, 1995. 1: 505–507.

91. Regosz, A., Pelplinska, T., Kowalski, P., and Thiel, Z., Prediction of solubility of sulfonamides in water and organic solvents based on the extended regular solution theory. *International Journal of Pharmaceutics*, 1992. 88: 437–442.

92. Martin, A., Wu, P.L., Adjei, A., Lindstrom, R.E., and Elworthy, P.H., Extended Hildebrand solubility approach and the log linear solubility equation. *Journal of Pharmaceutical Sciences*, 1982. 71: 849–856.

93. Reillo, A., Cordoba, M., Escalera, B., Selles, E., and Cordoba, M., Jr., Prediction of sulfamethiazole solubility in dioxane—water mixtures. *Die Pharmazie*, 1995. 50: 472–475.

94. Dwivedi, S.K., Sattari, S., Jamali, F., and Mitchell, A.G., Ibuprofen racemate and enantiomers: Phase diagram, solubility and thermodynamic studies. *International Journal of Pharmaceutics*, 1992. 87: 95–104.

95. Khossravi, D. and Connors, K.A., Solvent effect on chemical processes. I: Solubility of aromatic and heterocyclic compounds in binary aqueous-organic solvents. *Journal of Pharmaceutical Sciences*, 1992. 81: 371–379.

96. Kishi, H. and Hashimoto Y., Evaluation of the procedure for the measurement of water solubility and *n*-octanol/water partition coefficient of chemicals. Results of a ring test in Japan. *Chemosphere*, 1989. 18: 1749–1759.

97. Bustamante, P., Romero, S., Pena, A., Escalera, B., and Reillo, A., Enthalpy–entropy compensations for the solubility of drugs in solvent mixtures; paracetamol, acetanilide and nalidixic acid in dioxane–water. *Journal of Pharmaceutical Sciences*, 1998. 87: 1590–1596.

98. Kulkarni, A.R., Soppimath, K.S., Dave, A.M., Mehta, M.H., and Aminabhavi, T.M., Solubility study of hazardous pesticide (chlorpyrifos) by gas chromatography. *Journal of Hazardous Materials A*, 2000. 80: 9–13.

99. Roy, D., Ducher, F., Laumain, A., and Legendre, J.Y., Determination of the aqueous solubility of drugs using a convenient 96-well plate-based assay. *Drug Development and Industrial Pharmacy*, 2001. 27: 107–109.

100. Romero, S., Escalera, B., and Bustamante, P., Solubility behaviour of polymorphs I and II of mefenamic acid in solvent mixtures. *International Journal of Pharmaceutics*, 1999. 178: 193–202.
101. Klopman, G., Wang, S., and Balthasar, D.M., Estimation of aqueous solubility of organic molecules by the group contribution approach. Application to the study of biodegradation. *Journal of Chemical Information and Computer Sciences*, 1992. 32: 474–482.
102. Kühne, R., Ebert, R.-U., Kleint, F., Schmidt, G., and Schüürmann, G., Group contribution methods to estimate water solubility of organic chemicals *Chemosphere*, 1995. 30: 2061–2077.
103. Yan, A. and Gasteiger, J., Prediction of aqueous solubility of organic compounds based on a 3D structure representation. *Journal of Chemical Information and Computer Sciences*, 2003. 43: 429–434.
104. Tetko, I.V., Tanchuk, V.Y., Kasheva, T.N., and Villa, A.E.P., Estimation of aqueous solubility of chemical compounds using E-state indices. *Journal of Chemical Information and Computer Sciences*, 2001. 41: 1488–1493.
105. Liu, R. and So, S.-S., Development of quantitative structure–property relationship models for early ADME evaluation in drug discovery. 1. Aqueous solubility. *Journal of Chemical Information and Computer Sciences*, 2001. 41: 1633–1639.
106. Wegner, J.K. and Zell, A., Prediction of aqueous solubility and partition coefficient optimized by a genetic algorithm based descriptor selection method. *Journal of Chemical Information and Computer Sciences*, 2003. 43: 1077–1084.
107. Myrdal, P.B. and Yalkowsky, S.H., Solubilization of drugs, in *Encyclopedia of Pharmaceutical Technology*. 1998. Dekker, New York, pp. 161–217.
108. Rubino, J.T., Cosolvents and cosolvency, in *Encyclopedia of Pharmaceutical Technology*. 1990. Dekker, New York, pp. 375–398.
109. Sanghvi, R., Narazaki, R., Machatha, S.G., and Yalkowsky, S.H., Solubility improvement of drugs using N-methyl pyrrolidone. *AAPS PharmSciTech*, 2008. 9: 366–376.
110. Mizuuchi, H., Jaitely, V., Murdan, S., and Florence, A.T., Room temperature ionic liquids and their mixtures: Potential pharmaceutical solvents. *European Journal of Pharmaceutical Sciences*, 2008. 33: 326–331.
111. Golightly, L.K., Smolinkse, S.S., Bennett, M.L., Sunderland III, E.W., and Rumack, B.H., Pharmaceutical excipients adverse effects associated with inactive ingredients in drug products (Part I). *Medical Toxicology and Adverse Drug Experience*, 1988. 3: 128–165.
112. Patel, D.M., Bernardo, P., Cooper, J., and Forrester, R.B., Glycerine, in *Handbook of Pharmaceutical Excipients*. 1986. America and Great Britain Pharmaceutical Societies, London, pp. 203–213.
113. Spiegel, A.J. and Noseworthy, M.M., Use of nonaqueous solvents in parenteral products. *Journal of Pharmaceutical Sciences*, 1963. 52: 917–927.
114. Tsai, P.S., Lipper, R.A., and Worthington, H.C, Propylene glycol, in *Handbook of Pharmaceutical Excipients*, 1986. America and Great Britain Pharmaceutical Societies, London, pp. 241–242.
115. Allen, M.J., Borody, T.J., Bugliosi, T.F., May, G.R., LaRusso, N.F., and Thistle, J.L., Rapid dissolution of gallstones by methyl *tert*-butyl ether—preliminary observations. *New England Journal of Medicine*, 1985. 312: 217–220.
116. Li, A. and Andren, A.W., Solubility of polychlorinated biphenyls in water/alcohol mixtures. 1. Experimental Data. *Environmental Science and Technology*, 1994. 28: 47–52.
117. Sunderland, V.B. and Watts, D.W., Alkaline ethanolysis of methyl 4-hydroxybenzoate and hydrolysis of methyl and ethyl 4-hydroxybenzoates in ethanol–water systems. *International Journal of Pharmaceutics*, 1985. 27: 1–15.
118. Acree, W.E., Jr., McCargar, J.W., Zvaigzne, A.I., and Teng, I.L., Mathematical representation of thermodynamic properties. Carbazole solubilities in binary alkane + dibutyl ether and alkane + tetrahydropyran solvent mixtures. *Physics and Chemistry of Liquids*, 1991. 23: 27–35.
119. Acree, W.E., Jr., Mathematical representation of thermodynamic properties. Part II. Derivation of the combined nearly ideal binary solvent (NIBS)/Redlich–Kister mathematical representation from a two-body and three-body interactional mixing model. *Thermochimica Acta*, 1992. 198: 71–79.
120. Adjei, A., Newburger, J., and Martin, A., Extended Hildebrand approach. Solubility of caffeine in dioxane–water mixtures. *Journal of Pharmaceutical Sciences*, 1980. 69: 659–661.
121. Barzegar-Jalali, M. and Jouyban-Gharamaleki, A., A general model from theoretical cosolvency models. *International Journal of Pharmaceutics*, 1997. 152: 247–250.
122. Jouyban-Gharamaleki, A., The modified Wilson model and predicting drug solubility in water–cosolvent mixtures. *Chemical and Pharmaceutical Bulletin*, 1998. 46: 1058–1061.

123. McHale, M.E.R., Kauppila, A.S.M.; Powell, J.C., and Acree, W.E, Jr., Solubility of anthracene in (binary alcohol + 2-butoxyethanol) solvent mixtures. *Journal of Chemical Thermodynamics*, 1996. 28: 209–214.

124. Ochsner, A.B., Belloto, R.J., Jr., and Sokoloski, T.D., Prediction of xanthine solubilities using statistical techniques. *Journal of Pharmaceutical Sciences*, 1985. 74: 132–135.

125. Williams, N.A. and Amidon, G.L., Excess free energy approach to the estimation of solubility in mixed solvent system. II. Ethanol–water mixtures. *Journal of Pharmaceutical Sciences*, 1984. 73: 14–18.

126. Yalkowsky, S.H. and Roseman, T.J., Solubilization of drugs by cosolvents, in *Techniques of Solubilization of Drugs*, S.H. Yalkowsky, ed. 1981. Dekker, New York, pp. 91–134.

127. Fan, C. and Jafvert, C.T., Margules equations applied to PAH solubilities in alcohol–water mixtures. *Environmental Science and Technology*, 1997. 31: 3516–3522.

128. Fredenslund, A., Jones, R.L, and Prausnitz, J.M., Group contribution estimation of activity coefficients in nonideal liquid mixtures. *AIChE Journal*, 1977. 21: 1086–1099.

129. Acree, W.E., Jr., and Rytting, J.H., Solubility in binary solvent systems: III. Predictive expressions based on molecular surface area. *Journal of Pharmaceutical Sciences*, 1983. 72: 292–296.

130. Jouyban-Gharamaleki, A. and Hanaee, J., A novel method for improvement of predictability of the CNIBS/R-K equation. *International Journal of Pharmaceutics*, 1997. 154: 245–247.

131. Chertkoff, M.J. and Martin, A.N., The solubility of benzoic acid in mixed solvents. *Journal of American Pharmaceutical Association*, 1960. 49: 444–447.

132. Paruta, A.N., Sciarrone, B.J., and Lordi, N.G., Solubility of salicylic acid as a function of dielectric constant. *Journal of Pharmaceutical Sciences*, 1964. 53: 1349–1353.

133. Martin, A., Wu, P.L., Adjei, A., Lindstrom, R.E., and Elworthy, P.H., Extended Hildebrand solubility approach and the log linear solubility equation. *Journal of Pharmaceutical Sciences*, 1982. 71: 849–856.

134. Yalkowsky, S.H., Amidon, G.L., Zografi, G., and Flynn, G.L., Solubility of nonelectrolytes in polar solvents III: Alkyl *p*-aminobenzoates in polar and mixed solvents. *Journal of Pharmaceutical Sciences*, 1975. 64: 48–52.

135. Jouyban, A., Soltanpour, Sh., and Chan, H.K., A simple relationship between dielectric constant of mixed solvents with solvent composition and temperature. *International Journal of Pharmaceutics*, 2004. 269: 353–360.

136. Martin, A., Paruta, A.N., and Adjei, A., Extended-Hildebrand solubility approach: Methylxanthines in mixed solvents. *Journal of Pharmaceutical Sciences*, 1981. 70: 1115–1120.

137. Li, A. and Yalkowsky, S.H., Solubility of organic solutes in ethanol–water mixtures. *Journal of Pharmaceutical Sciences*, 1994. 83: 1735–1740.

138. Valvani, S.C., Yalkowsky, S.H., and Roseman, T.J., Solubility and partitioning IV: Aqueous solubility and octanol–water partition coefficients of liquid nonelectrolytes. *Journal of Pharmaceutical Sciences*, 1981. 70: 502–507.

139. Li, A. and Yalkowsky, S.H., Predicting cosolvency. 1. Solubility ratio and solute log K_{ow}. *Industrial Engineering and Chemical Research*, 1998. 37: 4470–4475.

140. Yalkowsky, S.H. and Rubino, J.T., Solubilization by cosolvents I: Organic solutes in propylene glycol–water mixtures. *Journal of Pharmaceutical Sciences*, 1985. 74: 416–421.

141. Gould, P.L., Goodman, M., and Hanson, P.A., Investigation of the solubility relationships of polar, semi-polar and non-polar drugs in mixed co-solvent systems. *International Journal of Pharmaceutics*, 1984. 19: 149–159.

142. Chien, Y.W. and Lambert, H.J., Solubilization of steroids by multiple co-solvent systems. *Chemical and Pharmaceutical Bulletin*, 1975. 23: 1085–1090.

143. Rubino, J.T., Blanchard, J., and Yalkowsky, S.H., Solubilization by cosolvents II: Phenytoin in binary and ternary solvents. *Journal of Parenteral Science and Technology*, 1984. 38: 215–221.

144. Amidon, G.L. and Williams, N.A., A solubility equation for non-electrolytes in water. *International Journal of Pharmaceutics*, 1982. 11: 249–256.

145. Bustamante, P., Hinkley, D.V., Martin, A., and Shi, S., Statistical analysis of the extended Hansen method using bootstrap technique. *Journal of Pharmaceutical Sciences*, 1991. 80: 971–977.

146. Reillo, A., Bustamante, P., Escalera, B., Jimenez, M.M., and Selles, E., Solubility parameter-based methods for predicting the solubility of sulfapyridine in solvent mixtures. *Drug Development and Industrial Pharmacy*, 1995. 21: 2073–2084.

147. Jouyban-Gharamaleki, A., Romero, S., Bustamante, P., and Clark, B.J., Multiple solubility maxima of oxolinic acid in mixed solvents and a new extension of Hildebrand solubility approach. *Chemical and Pharmaceutical Bulletin*, 2000. 48: 175–178.

148. Jouyban, A. and Clark, B.J., Describing solubility of polymorphs in mixed solvents by CNIBS/R-K equation. *Die Pharmazie*, 2002. 57: 861–862.
149. Hwang, C.A., Holste, J.C., Hall, K.R., and Mansoori, G.A., A simple relation to predict or to correlate the excess functions of multicomponent mixtures. *Fluid Phase Equilibria*, 1991. 62: 173–189.
150. Acree, W.E., Jr., Comments concerning model for solubility estimation in mixed solvent systems. *International Journal of Pharmaceutics*, 1996. 127: 27–30.
151. Barzegar-Jalali, M. and Jouyban-Gharamaleki, A., Models for calculating solubility in binary solvent systems. *International Journal of Pharmaceutics*, 1996. 140: 237–248.
152. Zvaigzne, A.I. and Acree, W.E., Jr., Solubility of anthracene in binary alcohol + 2-methyl-1-propanol and alcohol + 3-methyl-1-butanol solvent mixtures. *Journal of Chemical Engineering Data*, 1995. 40: 917–919.
153. Jouyban-Gharamaleki, A. and Acree, W.E., Jr., Comparison of models for describing multiple peaks in solubility profiles. *International Journal of Pharmaceutics*, 1998. 167: 177–182.
154. Jouyban, A., Khoubnasabjafari, M., Chan, H.K., Altria, K.D., and Clark, B.J., Predicting electrophoretic mobility of beta-blockers in water–methanol mixed electrolyte system. *Chromatographia*, 2003. 57: 191–196.
155. Jouyban, A., Grosse, S.C., Chan, H.K., Coleman, M.W., and Clark, B.J., Mathematical representation of electrophoretic mobility of basic drugs in ternary solvent buffers in capillary zone electrophoresis. *Journal of Chromatography A*, 2003. 994: 191–198.
156. Jouyban-Gharamaleki, A., Khaledi, M.G., and Clark, B.J., Calculation of electrophoretic mobilities in water–organic modifier mixtures in capillary electrophoresis. *Journal of Chromatography A*, 2000. 868: 277–284.
157. Jouyban, A., Chan, H.K., Barzegar-Jalali, M., and Acree, W.E., Jr., A model to represent solvent effects on chemical stability of solutes in mixed solvent systems. *International Journal of Pharmaceutics*, 2002. 243: 167–172.
158. Jouyban, A., Chan, H.K., Clark, B.J., and Acree, W.E., Jr., Mathematical representation of apparent acid dissociation constants in aqueous-organic solvent mixtures. *International Journal of Pharmaceutics*, 2002. 246: 135–142.
159. Jouyban, A., Soltani, S., Chan, H.K., and Acree, W.E., Jr., Modeling acid dissociation constant of analytes in binary solvents at various temperatures using Jouyban–Acree model. *Thermochimica Acta*, 2005. 428: 119–123.
160. Jouyban, A., Rashidi, M.R., Vaez-Gharamaleki, Z., Matin, A.A., and Djozan, Dj., Mathematical representation of analyte's capacity factor in binary solvent mobile phases using Jouyban–Acree model. *Die Pharmazie*, 2005. 60: 827–829.
161. Jouyban, A., Fathi-Azarbayjani, A., Barzegar-Jalali, M., and Acree, W.E., Jr., Correlation of surface tension of mixed solvents with solvent composition. *Die Pharmazie*, 2004. 59: 937–941.
162. Jouyban, A., Khoubnasabjafari, M., Vaez-Gharamaleki, Z., Fekari, Z., and Acree, W.E., Jr., Calculation of the viscosity of binary liquids at various temperatures using Jouyban–Acree model. *Chemical and Pharmaceutical Bulletin*, 2005. 53: 519–523.
163. Jouyban, A., Fathi-Azarbayjani, A., Khoubnasabjafari, M., and Acree, W.E., Jr., Mathematical representation of the density of liquid mixtures at various temperatures using Jouyban–Acree model. *Indian Journal of Chemistry A*, 2005. 44: 1553–1560.
164. Jouyban, A., Khoubnasabjafari, M., and Acree, W.E., Jr., Modeling the solvatochromic parameter of mixed solvents with respect to solvent composition and temperature using Jouyban–Acree model. *Daru*, 2006. 14: 22–25.
165. Jouyban, A., Soltani, S., Khoubnasabjafari, M., and Acree, W.E., Jr., Refractive index correlation of solvent mixtures at various temperatures. *Asian Journal of Chemistry*, 2006. 18: 2037–2040.
166. Hasan, M., Shirude, D.F., Hiray, A.P., Sawant, A.B., and Kadam, U.B., Densities, viscosities and ultrasonic velocities of binary mixtures of methylbenzene with hexan-2-ol, heptan-2-ol and octan-2-ol at $T = 298.15$ and $308.15\,K$. *Fluid Phase Equilibria*, 2006. 252: 88–95.
167. Delaney, J.S., Predicting aqueous solubility from structure. *Drug Discovery Today*, 2005. 10: 289–295.
168. Strickley, R.G., Solubilizing excipients in oral and injectable formulations. *Pharmaceutical Research*, 2004. 21: 201–230.
169. Sweetana, S. and Akers, M.J., Solubility principles and practices for parenteral drug dosage form development. *PDA Journal of Pharmaceutical Science and Technology*, 1996. 50: 330–342.
170. Jouyban, A. and Acree, W.E., Jr., Comments concerning "solubility of anthracene in two binary solvents containing toluene". *Fluid Phase Equilibria*, 2003. 209: 155–159.

171. Jouyban, A., In silico prediction of drug solubility in water–dioxane mixtures using Jouyban–Acree model. *Die Pharmazie*, 2007. 62: 46–50.

172. Jouyban, A., Solubility prediction of drugs in water–PEG 400 mixtures. *Chemical and Pharmaceutical Bulletin*, 2006. 54: 1561–1566.

173. Jouyban, A., Prediction of drug solubility in water–propylene glycol mixtures using Jouyban–Acree model. *Die Pharmazie*, 2007. 62: 365–367.

174. Jouyban, A., Soltanpour, Sh., and Tamizi, E., Solubility prediction of solutes in aqueous mixtures of ethylene glycols. *Die Pharmazie*, 2008. 63: 548–550.

175. Jouyban, A., Fakhree, M.A.A., Hamzeh-Mivehroud, M., and Acree, W. E., Jr., Modeling the deviations of solubilities in water–dioxane mixtures from predicted solubilities by the Jouyban–Acree model. *Journal of Drug Delivery Science and Technology*, 2007. 17: 359–363.

176. Jouyban, A., Fakhree, M.A.A., Ghafourian, T., Saei, A.A., and Acree, W.E., Jr., Deviations of drug solubility in water–cosolvent mixtures from the Jouyban–Acree model. Effect of solute structure. *Die Pharmazie*, 2008. 63: 113–121.

177. PharmaAlgorithms, ADME Boxes, Version 3.0, PharmaAlgorithms Inc., 591 Indian Road, Toronto, ON M6P 2C4, Canada, 2006.

178. Jouyban, A., Soltanpour, Sh., Soltani, S., Tamizi, E., Fakhree, M.A.A., and Acree, W.E., Jr., Prediction of drug solubility in mixed solvents using computed Abraham parameters *Journal of Molecular Liquids*, 2009. 146: 82–88.

179. Jouyban-Gharamaleki, A., Clark, B.J., and Acree, W.E., Jr., Prediction of drug solubility in ternary solvent mixture. *Drug Development and Industrial Pharmacy*, 2000. 26: 971–973.

180. Jouyban, A., Chan, H.K., Chew, N.Y.K., Khoubnasabjafari, M., and Acree, W.E., Jr., Solubility prediction of paracetamol in binary and ternary solvent mixtures using Jouyban–Acree model. *Chemical and Pharmaceutical Bulletin*, 2006. 54: 428–431.

181. Jouyban, A., Chew, N.Y.K., Chan, H.K., Khoubnasabjafari, M., and Acree, W.E., Jr., Solubility prediction of salicylic acid in water–ethanol–propylene glycol mixtures using the Jouyban–Acree model. *Die Pharmazie*, 2006. 61: 417–419.

182. Jouyban, A. and Fakhree, M.A.A., A new definition of solubilization power of a cosolvent. *Die Pharmazie*, 2008. 63: 317–319.

183. Jouyban, A., Prediction of the optimized solvent composition for solubilization of drugs in water–cosolvent mixtures. *Die Pharmazie*, 2007. 62: 190–198.

184. Comor, J.J. and Kopecni, M.M., Prediction of gas chromatography solute activity coefficients in mixed stationary phases based on the Wilson equation. *Analytical Chemistry*, 1990. 62: 991–994.

185. Acree, W.E., Jr. and Zvaigzne, A.I., Thermodynamic properties of nonelectrolyte solution. Part 4. Estimation and mathematical representation of solute activity coefficients and solubilities in binary solvents using the NIBS and modified Wilson equations. *Thermochimica Acta*, 1991. 178: 151–167.

186. Tarantino, R., Bishop, E., Chen, F.-C., Iqbal, K., and Malick, A.W., *N*-methyl-2-pyrrolidone as a cosolvent: Relationship of cosolvent effect with solute polarity and the presence of proton-donating groups on model drug compounds. *Journal of Pharmaceutical Sciences*, 1994. 83: 1213–1216.

187. Wu, P.L. and Martin, A., Extended-Hildebrand solubility approach: *p*-hydroxybenzoic acid in mixtures of dioxane and water. *Journal of Pharmaceutical Sciences*, 1983. 72: 587–592.

188. Van den Mooter, G., Augustijns, P., and Kinget, R., Application of the thermodynamics of mobile order and disorder to explain the solubility of temazepam in aqueous solutions of polyethylene glycol 6000. *International Journal of Pharmaceutics*, 1998. 164: 81–89.

189. Rytting, E., Lentz, K.A., Chen, X.Q., Qian, F., and Venkatesh, S., A quantitative structure–property relationship for predicting drug solubility in PEG 400/water cosolvent systems. *Pharmaceutical Research*, 2004. 21: 237–244.

190. Barra, J., Lescure, F., Doelker, E., and Bustamante, P., The expanded Hansen approach to solubility parameters. Paracetamol and citric acid in individual solvents. *Journal of Pharmacy and Pharmacology*, 1997. 49: 644–651.

191. Richardson, P.J., McCafferty, D.F., and Woolfson, A.D., Determination of three-component partial solubility parameters for temazepam and the effects of change in partial molal volume on the thermodynamics of drug solubility. *International Journal of Pharmaceutics*, 1992. 78: 189–198.

192. Reulle, P. and Kesselring U.W., Aqueous solubility prediction of environmental important chemicals from the mobile order thermodynamics. *Chemosphere*, 1997. 34: 275–298.

2 Solubility Data in Organic Solvents

Solubility Data of Drugs in 1,2-Dichloroethane at Various Temperatures

Drug	T (°C)	Solubility	Solubility Unit	Reference
Enrofloxacin sodium	20.00	0.00248	Mole F.	[1]
Enrofloxacin sodium	22.00	0.00279	Mole F.	[1]
Enrofloxacin sodium	25.00	0.00358	Mole F.	[1]
Enrofloxacin sodium	27.00	0.00415	Mole F.	[1]
Enrofloxacin sodium	30.00	0.00656	Mole F.	[1]
Enrofloxacin sodium	32.00	0.00858	Mole F.	[1]
Enrofloxacin sodium	35.00	0.00996	Mole F.	[1]
Enrofloxacin sodium	37.00	0.01086	Mole F.	[1]
Imidacloprid	26.81	0.009866	Mole F.	[2]
Imidacloprid	29.72	0.01076	Mole F.	[2]
Imidacloprid	31.82	0.01163	Mole F.	[2]
Imidacloprid	36.93	0.01421	Mole F.	[2]
Imidacloprid	40.99	0.01606	Mole F.	[2]
Imidacloprid	44.28	0.01832	Mole F.	[2]
Imidacloprid	48.16	0.02132	Mole F.	[2]
Imidacloprid	52.52	0.02425	Mole F.	[2]
Imidacloprid	55.59	0.02709	Mole F.	[2]
Imidacloprid	58.19	0.03090	Mole F.	[2]
Imidacloprid	60.60	0.03452	Mole F.	[2]
Imidacloprid	63.76	0.03924	Mole F.	[2]
Imidacloprid	66.80	0.04415	Mole F.	[2]
Imidacloprid	69.66	0.04896	Mole F.	[2]
Imidacloprid	72.05	0.05463	Mole F.	[2]
Imidacloprid	73.95	0.06121	Mole F.	[2]
Xanthene	25.00	0.1549	Mole F.	[3]

Solubility Data of Drugs in Dichloromethane at Various Temperatures

Drug	T (°C)	Solubility	Solubility Unit	Reference
3,5-Di-*tert*-butyl-4-hydroxytoluene	N/A	0.45	Mole F.	[4]
Acephate	19.75	0.17580	Mole F.	[5]
Acephate	22.05	0.19002	Mole F.	[5]
Acephate	24.15	0.20333	Mole F.	[5]
Acephate	25.30	0.21045	Mole F.	[5]
Acephate	26.45	0.21741	Mole F.	[5]
Acephate	27.90	0.22608	Mole F.	[5]
Acephate	29.15	0.23333	Mole F.	[5]
Acephate	30.10	0.24053	Mole F.	[5]

(continued)

Solubility Data of Drugs in Dichloromethane at Various Temperatures   (continued)

Drug	T (°C)	Solubility	Solubility Unit	Reference
β-Carotene	Room	6000	mg/L	[6]
Betulin	5.05	0.000790	Mole F.	[7]
Betulin	10.05	0.000837	Mole F.	[7]
Betulin	15.05	0.000897	Mole F.	[7]
Betulin	25.05	0.000991	Mole F.	[7]
Carbamazepine	25.00	158	g/L	[8]
Cefazolin sodium pentahydrate	5.10	0.000001770	Mole F.	[9]
Cefazolin sodium pentahydrate	9.62	0.000002065	Mole F.	[9]
Cefazolin sodium pentahydrate	13.50	0.000002360	Mole F.	[9]
Cefazolin sodium pentahydrate	16.70	0.000002655	Mole F.	[9]
Cefazolin sodium pentahydrate	30.80	0.000002950	Mole F.	[9]
Cefazolin sodium pentahydrate	35.10	0.000003540	Mole F.	[9]
Cefazolin sodium pentahydrate	40.10	0.000004777	Mole F.	[9]
Cefazolin sodium pentahydrate	45.00	0.000005457	Mole F.	[9]
Enrofloxacin sodium	20.00	0.00809	Mole F.	[1]
Enrofloxacin sodium	22.00	0.00818	Mole F.	[1]
Enrofloxacin sodium	25.00	0.00835	Mole F.	[1]
Enrofloxacin sodium	27.00	0.00838	Mole F.	[1]
Enrofloxacin sodium	30.00	0.00844	Mole F.	[1]
Enrofloxacin sodium	32.00	0.00853	Mole F.	[1]
Enrofloxacin sodium	35.00	0.00868	Mole F.	[1]
Enrofloxacin sodium	37.00	0.00874	Mole F.	[1]
Imidacloprid	20.82	0.01318	Mole F.	[2]
Imidacloprid	22.81	0.01458	Mole F.	[2]
Imidacloprid	24.85	0.01567	Mole F.	[2]
Imidacloprid	26.90	0.01664	Mole F.	[2]
Imidacloprid	29.45	0.01768	Mole F.	[2]
Imidacloprid	30.84	0.01890	Mole F.	[2]
Imidacloprid	32.81	0.01991	Mole F.	[2]
Imidacloprid	34.65	0.02110	Mole F.	[2]
Imidacloprid	36.45	0.02231	Mole F.	[2]
Imidacloprid	38.51	0.02361	Mole F.	[2]
Imidacloprid	39.99	0.02475	Mole F.	[2]
Imidacloprid	40.78	0.02540	Mole F.	[2]
Lutein	Room	800	mg/L	[6]
Methyl elaidate	−26.10	6.01	% Weight	[10]
Methyl elaidate	−18.44	18.44	% Weight	[10]
Methyl elaidate	−7.40	50.67	% Weight	[10]
Methyl heptadecanoate	−1.90	13.31	% Weight	[10]
Methyl heptadecanoate	8.70	41.95	% Weight	[10]
Methyl palmitate	−9.90	4.92	% Weight	[10]
Methyl palmitate	0.20	18.86	% Weight	[10]
Methyl petroselaidate	−19.20	5.71	% Weight	[10]
Methyl petroselaidate	−10.80	18.24	% Weight	[10]
Methyl petroselaidate	0.70	50.60	% Weight	[10]
Methyl stearate	−17.10	0.30	% Weight	[10]
Methyl stearate	−4.80	2.28	% Weight	[10]
Methyl stearate	2.60	6.27	% Weight	[10]
Methyl stearate	13.40	27.56	% Weight	[10]
Methyl stearate	25.50	63.72	% Weight	[10]

Solubility Data of Drugs in Dichloromethane at Various Temperatures   (continued)

Drug	T (°C)	Solubility	Solubility Unit	Reference
Methyl stearate	34.40	87.52	% Weight	[10]
Methyl-3-(3,5-di-*tert*-butyl-4-hydroxyphenyl)-propionate	N/A	0.33	Mole F.	[4]
Octadecyl-3-(3,5-di-*tert*-butyl-4-hydroxyphenyl)-propionate	N/A	0.17	Mole F.	[4]
Paracetamol	30.00	0.48	g/L	[8]
Paracetamol	30.00	0.32	g/kg	[11]
Pentaerythrittol tetrakis(3-(3,5-di-*tert*-butyl-4-hydroxyphenyl)-propionate)	N/A	0.12	Mole F.	[4]
Pyoluteorin	5.05	0.01076	mol/L	[12]
Pyoluteorin	10.05	0.01152	mol/L	[12]
Pyoluteorin	15.05	0.01249	mol/L	[12]
Pyoluteorin	20.05	0.01382	mol/L	[12]
Pyoluteorin	25.05	0.01553	mol/L	[12]
Pyoluteorin	30.05	0.01765	mol/L	[12]
Pyoluteorin	35.05	0.02110	mol/L	[12]
Rifapentine	5.00	0.0159	Mole F.	[13]
Rifapentine	10.00	0.0165	Mole F.	[13]
Rifapentine	15.00	0.0175	Mole F.	[13]
Rifapentine	20.00	0.0184	Mole F.	[13]
Rifapentine	24.00	0.0196	Mole F.	[13]
Rifapentine	30.00	0.0209	Mole F.	[13]
Rifapentine	35.00	0.0224	Mole F.	[13]
Stearic acid	25.00	3.580	g/100 g	[14]
Stearic acid	30.00	8.850	g/100 g	[14]
Stearic acid	35.00	18.30	g/100 g	[14]
Temazepam	25.00	0.1322996	Mole F.	[15]
Testosterone propionate	25.00	0.27	Mole F.	[16]

Solubility Data of Drugs in 1,4-Dioxane at Various Temperatures

Drug	T (°C)	Solubility	Solubility Unit	Reference
16α,17α-Epoxyprogesterone	15.30	0.02265	Mole F.	[17]
16α,17α-Epoxyprogesterone	20.30	0.02658	Mole F.	[17]
16α,17α-Epoxyprogesterone	25.10	0.03097	Mole F.	[17]
16α,17α-Epoxyprogesterone	29.60	0.03525	Mole F.	[17]
16α,17α-Epoxyprogesterone	34.60	0.03996	Mole F.	[17]
16α,17α-Epoxyprogesterone	39.60	0.04519	Mole F.	[17]
16α,17α-Epoxyprogesterone	44.10	0.05562	Mole F.	[17]
16α,17α-Epoxyprogesterone	49.35	0.06303	Mole F.	[17]
16α,17α-Epoxyprogesterone	55.10	0.07104	Mole F.	[17]
16α,17α-Epoxyprogesterone	59.40	0.07822	Mole F.	[17]
2-(4-Ethylbenzoyl)benzoic acid	15.73	0.0042	Mole F.	[18]
2-(4-Ethylbenzoyl)benzoic acid	20.82	0.0066	Mole F.	[18]
2-(4-Ethylbenzoyl)benzoic acid	25.31	0.0091	Mole F.	[18]
2-(4-Ethylbenzoyl)benzoic acid	30.38	0.0121	Mole F.	[18]
2-(4-Ethylbenzoyl)benzoic acid	34.94	0.0153	Mole F.	[18]
2-(4-Ethylbenzoyl)benzoic acid	40.25	0.0195	Mole F.	[18]

(continued)

Solubility Data of Drugs in 1,4-Dioxane at Various Temperatures   (continued)

Drug	T (°C)	Solubility	Solubility Unit	Reference
2-(4-Ethylbenzoyl)benzoic acid	46.00	0.0240	Mole F.	[18]
2-(4-Ethylbenzoyl)benzoic acid	49.39	0.0278	Mole F.	[18]
2-(4-Ethylbenzoyl)benzoic acid	60.82	0.0353	Mole F.	[18]
2-Hydroxybenzoic acid	25.00	0.2945	Mole F.	[19]
3,5-Di-*tert*-butyl-4-hydroxytoluene	N/A	0.14	Mole F.	[4]
4-Aminobenzoic acid	25.00	0.06998	Mole F.	[20]
Acetylsalicylic acid	25.00	0.0516	Mole F.	[21]
Acetylsalicylic acid	25.00	0.12630	Mole F.	[22]
Anhydrous citric acid	25.00	0.14128	Mole F.	[23]
Benzoic acid	25.00	0.2853	Mole F.	[24]
Diclofenac	25.00	0.1055047	Mole F.	[25]
Flubiprofen	25.00	0.175	Mole F.	[26]
Haloperidol	25.00	0.00563	Mole F.	[27]
Ibuprofen	25.00	0.0371869	Mole F.	[28]
Irbesartan (form A)	14.75	0.000466	Mole F.	[29]
Irbesartan (form A)	24.85	0.000832	Mole F.	[29]
Irbesartan (form A)	30.50	0.001115	Mole F.	[29]
Irbesartan (form A)	39.75	0.001830	Mole F.	[29]
Irbesartan (form A)	44.90	0.002269	Mole F.	[29]
Irbesartan (form A)	49.70	0.002778	Mole F.	[29]
Irbesartan (form A)	35.05	0.001421	Mole F.	[29]
Irbesartan (form A)	20.30	0.000634	Mole F.	[29]
Lactose	25.00	0.0000293	Mole F.	[30]
Mannitol	25.00	0.0000615	Mole F.	[30]
Methyl-3-(3,5-di-*tert*-butyl-4-hydroxyphenyl)-propionate	N/A	0.35	Mole F.	[4]
Naproxen	25.00	0.1843167	Mole F.	[31]
Naproxen	25.00	0.10400	Mole F.	[32]
Niflumic acid	25.00	0.0483156	Mole F.	[33]
Octadecyl-3-(3,5-di-*tert*-butyl-4-hydroxyphenyl)-propionate	N/A	0.039	Mole F.	[4]
p-Aminobenzoic acid	25.00	0.0632285	Mole F.	[25]
Paracetamol	30.00	17.080000	g/kg	[11]
Paracetamol	25.00	0.03140	Mole F.	[23]
Pentaerythrittol tetrakis(3-(3,5-di-*tert*-butyl-4-hydroxyphenyl)-propionate)	N/A	0.0071	Mole F.	[4]
Phenothiazine	25.00	0.10260	Mole F.	[34]
p-Hydroxybenzoic acid	25.00	0.0844000	Mole F.	[35]
Pimozide	25.00	0.0113	Mole F.	[36]
Piroxicam	25.00	0.0049419	Mole F.	[33]
Saccharose	25.00	0.000000012	Mole F.	[30]
Salicylamide	25.00	0.1373	Mole F.	[37]
Salicylic acid	25.00	0.2978992	Mole F.	[25]
Salicylic acid	25.00	0.261	Mole F.	[38]
Sodium benzoate	25.00	0.0000005	Mole F.	[28]
Sodium diclofenac	25.00	0.0000535	Mole F.	[25]
Sodium diclofenac	25.00	0.0000535	Mole F.	[31]
Sodium ibuprofen	25.00	0.0001637	Mole F.	[28]

Solubility Data of Drugs in 1,4-Dioxane at Various Temperatures   (continued)

Drug	T (°C)	Solubility	Solubility Unit	Reference
Sodium p-aminobenzoate	25.00	0.0000117	Mole F.	[25]
Sodium salicylate	25.00	0.0008700	Mole F.	[25]
Sulfadiazine	25.00	0.0004948	Mole F.	[39]
Sulfamethoxypyridazine	25.00	0.0239000	Mole F.	[40]

Solubility Data of Drugs in 1-Butanol at Various Temperatures

Drug	T (°C)	Solubility	Solubility Unit	Reference
2-Hydroxybenzoic acid	25.00	0.1646	Mole F.	[19]
3,5-Di-tert-butyl-4-hydroxytoluene	N/A	0.13	Mole F.	[4]
4-Aminobenzoic acid	25.00	0.03139	Mole F.	[20]
5,5-Diethylbarbituric acid (Barbital)	25.00	41	g/L	[41]
5-Butyl-5-ethylbarbituric acid (Butetal)	25.00	278	g/L	[41]
5-Ethyl-5-(1-methylpropyl)-barbituric acid (Butabarbital)	25.00	28	g/L	[41]
5-Ethyl-5-(2-methylbutyl)-barbituric acid (Pentobarbital)	25.00	151	g/L	[41]
5-Ethyl-5-(2-methylpropyl)-barbituric acid	25.00	28	g/L	[41]
5-Ethyl-5-(3-methylbutyl)-barbituric acid (Amobarbital)	25.00	134	g/L	[41]
5-Ethyl-5-isopropylbarbituric acid (Probarbital)	25.00	13	g/L	[41]
5-Ethyl-5-pentylbarbituric acid	25.00	324	g/L	[41]
5-Ethyl-5-phenylbarbituric acid (Phenobarbital)	25.00	54	g/L	[41]
5-Ethyl-5-propylbarbiturate	25.00	78	g/L	[41]
Acetylsalicylic acid	25.00	0.0453	Mole F.	[21]
Acetylsalicylic acid	25.00	0.04616	Mole F.	[22]
Benzoic acid	25.00	0.2016	Mole F.	[24]
Benzoic acid	25.00	0.203	Mole F.	[43]
Berberine chloride	24.85	0.000054	Mole F.	[44]
Berberine chloride	29.85	0.000145	Mole F.	[44]
Berberine chloride	34.85	0.000179	Mole F.	[44]
Berberine chloride	39.85	0.000230	Mole F.	[44]
Betulin	5.05	0.001807	Mole F.	[7]
Betulin	15.05	0.002130	Mole F.	[7]
Betulin	25.05	0.002285	Mole F.	[7]
Betulin	35.05	0.002713	Mole F.	[7]
Butyl p-hydroxybenzoate	25.00	0.365	Mole F.	[43]
Butylparaben	25.00	0.364	Mole F.	[45]
Butylparaben	30.00	0.388	Mole F.	[45]
Butylparaben	35.00	0.439	Mole F.	[45]
Butylparaben	40.00	0.501	Mole F.	[45]
Carbamazepine (form III)	12.00	0.00302	Mole F.	[46]
Carbamazepine (form III)	18.40	0.00414	Mole F.	[46]
Carbamazepine (form III)	24.70	0.00536	Mole F.	[46]

(*continued*)

Solubility Data of Drugs in 1-Butanol at Various Temperatures   (continued)

Drug	T (°C)	Solubility	Solubility Unit	Reference
Carbamazepine (form III)	32.79	0.00714	Mole F.	[46]
Carbamazepine (form III)	38.30	0.00908	Mole F.	[46]
Carbamazepine (form III)	44.45	0.01176	Mole F.	[46]
Carbamazepine (form III)	50.75	0.01494	Mole F.	[46]
Carbamazepine (form III)	55.65	0.01826	Mole F.	[46]
Carbamazepine (form III)	70.49	0.02861	Mole F.	[46]
Cefazolin sodium pentahydrate	11.40	0.00001188	Mole F.	[9]
Cefazolin sodium pentahydrate	15.40	0.00001519	Mole F.	[9]
Cefazolin sodium pentahydrate	19.88	0.00002141	Mole F.	[9]
Cefazolin sodium pentahydrate	23.40	0.00002823	Mole F.	[9]
Cefazolin sodium pentahydrate	26.70	0.00003507	Mole F.	[9]
Cefazolin sodium pentahydrate	30.35	0.00004715	Mole F.	[9]
Cefazolin sodium pentahydrate	34.15	0.00006442	Mole F.	[9]
Cefazolin sodium pentahydrate	36.60	0.00008066	Mole F.	[9]
Celecoxib	25.00	29.030	g/L	[55]
Chlorpheniramine maleate	9.58	0.0010586	Mole F.	[47]
Chlorpheniramine maleate	15.07	0.0017418	Mole F.	[47]
Chlorpheniramine maleate	20.30	0.0023672	Mole F.	[47]
Chlorpheniramine maleate	24.58	0.0031028	Mole F.	[47]
Chlorpheniramine maleate	30.23	0.0044798	Mole F.	[47]
Chlorpheniramine maleate	35.50	0.0063824	Mole F.	[47]
Chlorpheniramine maleate	40.12	0.0091583	Mole F.	[47]
Chlorpheniramine maleate	45.00	0.0132260	Mole F.	[47]
Chlorpheniramine maleate	50.12	0.0192050	Mole F.	[47]
Chlorpheniramine maleate	55.00	0.0279800	Mole F.	[47]
Chlorpheniramine maleate	60.25	0.0411830	Mole F.	[47]
Clonazepam	25.00	2.7	g/L	[48]
D(−)-p-Hydroxyphenylglycine dane salt	20.00	0.0000997	Mole F.	[49]
D(−)-p-Hydroxyphenylglycine dane salt	25.16	0.0001101	Mole F.	[49]
D(−)-p-Hydroxyphenylglycine dane salt	29.80	0.0001223	Mole F.	[49]
D(−)-p-Hydroxyphenylglycine dane salt	35.04	0.0001393	Mole F.	[49]
D(−)-p-Hydroxyphenylglycine dane salt	39.80	0.0001578	Mole F.	[49]
D(−)-p-Hydroxyphenylglycine dane salt	45.01	0.0001812	Mole F.	[49]
D(−)-p-Hydroxyphenylglycine dane salt	50.10	0.0002074	Mole F.	[49]
D(−)-p-Hydroxyphenylglycine dane salt	55.14	0.0002365	Mole F.	[49]
D(−)-p-Hydroxyphenylglycine dane salt	60.02	0.0002678	Mole F.	[49]
D(−)-p-Hydroxyphenylglycine dane salt	65.17	0.0003040	Solubility F.	[49]
D(−)-p-Hydroxyphenylglycine dane salt	69.80	0.0003394	Mole F.	[49]
Diflunisal	25.00	0.0266	Mole F.	[26]
Ethylparaben	25.00	0.178	Mole F.	[45]
Ethylparaben	30.00	0.183	Mole F.	[45]
Ethylparaben	35.00	0.210	Mole F.	[45]
Ethylparaben	40.00	0.234	Mole F.	[45]
Ethyl p-hydroxybenzoate	25.00	0.187	Mole F.	[43]
Flubiprofen	25.00	0.0667	Mole F.	[26]
Glucose	40.00	0.59	g/L	[42]
Glucose	60.00	1.47	g/L	[42]
Haloperidol	25.00	0.00777	Mole F.	[27]
Hesperetin	15.05	0.0301	mol/L	[50]
Hesperetin	20.05	0.0320	mol/L	[50]

Solubility Data of Drugs in 1-Butanol at Various Temperatures (continued)

Drug	T (°C)	Solubility	Solubility Unit	Reference
Hesperetin	25.05	0.0370	mol/L	[50]
Hesperetin	30.05	0.0419	mol/L	[50]
Hesperetin	35.05	0.0468	mol/L	[50]
Hesperetin	40.05	0.0539	mol/L	[50]
Hesperetin	45.05	0.0641	mol/L	[50]
Hesperetin	50.05	0.0719	mol/L	[50]
Ketoprofen	25.00	0.0868	Mole F.	[51]
Lamivudine (form 2)	5.00	2.9	g/L	[52]
Lamivudine (form 2)	15.00	3.7	g/L	[52]
Lamivudine (form 2)	25.00	5.4	g/L	[52]
Lamivudine (form 2)	35.00	6.3	g/L	[52]
Lamivudine (form 2)	45.00	8.8	g/L	[52]
Lovastatin	12.55	0.0046046	Mole F.	[53]
Lovastatin	15.55	0.0049877	Mole F.	[53]
Lovastatin	17.50	0.0052554	Mole F.	[53]
Lovastatin	23.05	0.0061126	Mole F.	[53]
Lovastatin	28.05	0.0070227	Mole F.	[53]
Lovastatin	30.95	0.0076196	Mole F.	[53]
Lovastatin	32.05	0.0078606	Mole F.	[53]
Lovastatin	33.55	0.0082028	Mole F.	[53]
Lovastatin	35.55	0.0086850	Mole F.	[53]
Lovastatin	36.95	0.0090410	Mole F.	[53]
Luteolin	0.00	0.00138	Mole F.	[54]
Luteolin	10.00	0.00145	Mole F.	[54]
Luteolin	25.00	0.00179	Mole F.	[54]
Luteolin	40.00	0.00219	Mole F.	[54]
Luteolin	60.00	0.00290	Mole F.	[54]
Meloxicam	25.00	0.07	g/L	[8]
Meloxicam	25.00	0.285	g/L	[55]
Methylparaben	25.00	0.146	Mole F.	[45]
Methylparaben	30.00	0.154	Mole F.	[45]
Methylparaben	35.00	0.172	Mole F.	[45]
Methylparaben	40.00	0.191	Mole F.	[45]
Methyl p-hydroxybenzoate	25.00	0.147	Mole F.	[43]
Methyl p-hydroxybenzoate	25.00	0.1484	Mole F.	[35]
Methyl-3-(3,5-di-tert-butyl-4-hydroxyphenyl)-propionate	N/A	0.083	Mole F.	[4]
Naproxen	25.00	0.01416	Mole F.	[32]
Nimesulide	25.00	2.07	g/L	[8]
Nimesulide	25.00	2.12	g/L	[55]
Octadecyl-3-(3,5-di-tert-butyl-4-hydroxyphenyl)-propionate	N/A	0.011	Mole F.	[4]
Oleanolic acid	15.15	0.0320	mol/L	[56]
Oleanolic acid	20.15	0.0333	mol/L	[56]
Oleanolic acid	25.15	0.0382	mol/L	[56]
Oleanolic acid	30.15	0.0397	mol/L	[56]
Oleanolic acid	35.15	0.0441	mol/L	[56]
Oleanolic acid	40.15	0.0496	mol/L	[56]
Oleanolic acid	45.15	0.0590	mol/L	[56]

(continued)

Solubility Data of Drugs in 1-Butanol at Various Temperatures   (continued)

Drug	T (°C)	Solubility	Solubility Unit	Reference
Oleanolic acid	50.15	0.0647	mol/L	[56]
Oleanolic acid	55.15	0.0697	mol/L	[56]
Paracetamol	−5.00	47.55	g/kg	[11]
Paracetamol	0.00	51.96	g/kg	[11]
Paracetamol	5.00	57.21	g/kg	[11]
Paracetamol	10.00	63.31	g/kg	[11]
Paracetamol	15.00	69.29	g/kg	[11]
Paracetamol	20.00	77.07	g/kg	[11]
Paracetamol	25.00	83.27	g/kg	[11]
Paracetamol	30.00	93.64	g/kg	[11]
Paracetamol	30.00	73.6	g/L	[8]
Paroxetine HCl hemihydrate	22.10	0.0012670	Mole F.	[57]
Paroxetine HCl hemihydrate	26.00	0.0013487	Mole F.	[57]
Paroxetine HCl hemihydrate	30.25	0.0014356	Mole F.	[57]
Paroxetine HCl hemihydrate	34.10	0.0014998	Mole F.	[57]
Paroxetine HCl hemihydrate	37.85	0.0015780	Mole F.	[57]
Paroxetine HCl hemihydrate	42.00	0.0016795	Mole F.	[57]
Paroxetine HCl hemihydrate	46.00	0.0018095	Mole F.	[57]
Paroxetine HCl hemihydrate	49.85	0.0019007	Mole F.	[57]
Paroxetine HCl hemihydrate	54.05	0.0019604	Mole F.	[57]
Paroxetine HCl hemihydrate	58.00	0.0021602	Mole F.	[57]
Paroxetine HCl hemihydrate	62.10	0.0022272	Mole F.	[57]
Paroxetine HCl hemihydrate	66.10	0.0023580	Mole F.	[57]
Paroxetine HCl hemihydrate	70.00	0.0024481	Mole F.	[57]
Paroxetine HCl hemihydrate	74.25	0.0025409	Mole F.	[57]
Paroxetine HCl hemihydrate	78.00	0.0026034	Mole F.	[57]
Paroxetine HCl hemihydrate	81.95	0.0026632	Mole F.	[57]
Paroxetine HCl hemihydrate	85.75	0.0027042	Mole F.	[57]
Paroxetine HCl hemihydrate	90.00	0.0027618	Mole F.	[57]
Pentaerythrittol tetrakis(3-(3,5-di-*tert*-butyl-4-hydroxyphenyl)-propionate)	N/A	0.0015	Mole F.	[4]
Phenacetinum	10.22	0.016433	Mole F.	[58]
Phenacetinum	15.00	0.019652	Mole F.	[58]
Phenacetinum	20.58	0.024265	Mole F.	[58]
Phenacetinum	25.05	0.028897	Mole F.	[58]
Phenacetinum	30.03	0.035331	Mole F.	[58]
Phenacetinum	34.90	0.043203	Solubility F.	[58]
Phenacetinum	40.08	0.053612	Mole F.	[58]
Phenacetinum	45.28	0.066512	Mole F.	[58]
Phenacetinum	50.10	0.080964	Mole F.	[58]
Phenacetinum	55.00	0.098408	Mole F.	[58]
Phenacetinum	60.12	0.119910	Mole F.	[58]
Phenothiazine	25.00	0.01099	Mole F.	[34]
p-Hydroxybenzoic acid	25.00	0.1154	Mole F.	[35]
Pimozide	25.00	0.00249	Mole F.	[36]
Potassium clavulanate	0.80	0.0000606	Mole F.	[59]
Potassium clavulanate	3.90	0.0000656	Mole F.	[59]
Potassium clavulanate	7.90	0.0000706	Mole F.	[59]
Potassium clavulanate	11.90	0.0000756	Mole F.	[59]

Solubility Data of Drugs in 1-Butanol at Various Temperatures   (continued)

Drug	*T* (°C)	Solubility	Solubility Unit	Reference
Potassium clavulanate	16.10	0.0000812	Mole F.	[59]
Potassium clavulanate	20.00	0.0000918	Mole F.	[59]
Potassium clavulanate	23.90	0.0001053	Mole F.	[59]
Potassium clavulanate	28.05	0.0001187	Mole F.	[59]
Potassium clavulanate	31.95	0.0001359	Mole F.	[59]
Pravastatin sodium	4.95	0.00108	Mole F.	[60]
Pravastatin sodium	9.85	0.00118	Mole F.	[60]
Pravastatin sodium	14.85	0.00127	Mole F.	[60]
Pravastatin sodium	19.85	0.00144	Mole F.	[60]
Pravastatin sodium	24.85	0.00159	Mole F.	[60]
Pravastatin sodium	29.85	0.00178	Mole F.	[60]
Pravastatin sodium	34.85	0.00223	Mole F.	[60]
Pravastatin sodium	39.85	0.00290	Mole F.	[60]
Pravastatin sodium	45.05	0.00345	Mole F.	[60]
Pravastatin sodium	49.85	0.00408	Mole F.	[60]
Pravastatin sodium	54.85	0.00493	Mole F.	[60]
Propyl *p*-hydroxybenzoate	25.00	0.211	Mole F.	[43]
Propylparaben	25.00	0.206	Mole F.	[45]
Propylparaben	30.00	0.245	Mole F.	[45]
Propylparaben	35.00	0.278	Mole F.	[45]
Propylparaben	40.00	0.316	Mole F.	[45]
Rofecoxib	25.00	0.189	g/L	[55]
Rutin	10.00	0.000305	Mole F.	[61]
Rutin	25.00	0.000346	Mole F.	[61]
Rutin	40.00	0.0003969	Mole F.	[61]
Rutin	50.00	0.000446	Mole F.	[61]
Rutin	60.00	0.000508	Mole F.	[61]
Salicylamide	25.00	0.03037	Mole F.	[37]
Sulfadiazine	25.00	0.0000349	Mole F.	[39]
Sulfadiazine	25.00	0.000348	mol/L	[62]
Sulfadiazine	25.00	0.0000318	Mole F.	[63]
Sulfadiazine	30.00	0.0000409	Mole F.	[63]
Sulfadiazine	37.00	0.0000566	Mole F.	[63]
Sulfadimethoxine	25.00	0.0003890	Mole F.	[63]
Sulfadimethoxine	30.00	0.0005260	Mole F.	[63]
Sulfadimethoxine	37.00	0.0006700	Mole F.	[63]
Sulfamethoxypyridazine	25.00	0.0007300	Mole F.	[40]
Sulfanilamide	25.00	0.0189	mol/L	[62]
Sulfisoxazole	25.00	0.0161	mol/L	[62]
Sulfisomidine	25.00	0.0003440	Mole F.	[63]
Sulfisomidine	30.00	0.0004170	Mole F.	[63]
Sulfisomidine	37.00	0.0005560	Mole F.	[63]
Temazepam	24.00	0.0105	Volume F.	[64]
Temazepam	25.00	0.0049087	Mole F.	[15]
Thalidomide	32.00	0.07	g/L	[8]
Trimethoprim	3.20	0.0001570	Mole F.	[65]
Trimethoprim	9.48	0.0002536	Mole F.	[65]
Trimethoprim	14.81	0.0003589	Mole F.	[65]
Trimethoprim	19.69	0.0004727	Mole F.	[65]

(continued)

Solubility Data of Drugs in 1-Butanol at Various Temperatures   (continued)

Drug	T (°C)	Solubility	Solubility Unit	Reference
Trimethoprim	25.31	0.0006349	Mole F.	[65]
Trimethoprim	30.57	0.000844	Mole F.	[65]
Trimethoprim	35.69	0.001058	Mole F.	[65]
Trimethoprim	40.21	0.001325	Mole F.	[65]
Trimethoprim	45.30	0.001753	Mole F.	[65]
Trimethoprim	50.58	0.002347	Mole F.	[65]
Trimethoprim	56.06	0.003058	Mole F.	[65]
Trimethoprim	60.58	0.003862	Mole F.	[65]
Xanthene	25.00	0.01756	Mole F.	[3]
Xylitol	20.01	0.001128	Mole F.	[66]
Xylitol	25.06	0.001542	Mole F.	[66]
Xylitol	30.11	0.002117	Mole F.	[66]
Xylitol	35.12	0.002682	Mole F.	[66]
Xylitol	40.07	0.003422	Mole F.	[66]
Xylitol	45.01	0.004425	Mole F.	[66]
Xylitol	50.08	0.005883	Mole F.	[66]
Xylitol	55.17	0.007983	Mole F.	[66]
Xylitol	60.09	0.010860	Mole F.	[66]
Xylitol	65.07	0.014560	Mole F.	[66]
Xylitol	70.61	0.019420	Mole F.	[66]

Solubility Data of Drugs in 1-Decanol at Various Temperatures

Drug	T (°C)	Solubility	Solubility Unit	Reference
4-Aminobenzoic acid	25.00	0.01736	Mole F.	[20]
Acetylsalicylic acid	25.00	0.03652	Mole F.	[22]
Butylparaben	25.00	0.293	Mole F.	[45]
Butylparaben	30.00	0.371	Mole F.	[45]
Butylparaben	35.00	0.427	Mole F.	[45]
Butylparaben	40.00	0.491	Mole F.	[45]
Ethylparaben	25.00	0.115	Mole F.	[45]
Ethylparaben	30.00	0.175	Mole F.	[45]
Ethylparaben	35.00	0.192	Mole F.	[45]
Ethylparaben	40.00	0.215	Mole F.	[45]
Ibuprofen	25.00	0.2166	Mole F.	[67]
Ketoprofen	25.00	0.0831	Mole F.	[68]
Mestanolone	25.00	0.0138	Mole F.	[69]
Methandienone	25.00	0.0335	Mole F.	[69]
Methylparaben	25.00	0.088	Mole F.	[45]
Methylparaben	30.00	0.109	Mole F.	[45]
Methylparaben	35.00	0.128	Mole F.	[45]
Methylparaben	40.00	0.144	Mole F.	[45]
Methyltestosterone	25.00	0.0365	Mole F.	[69]
Nandrolone	25.00	0.0261	Mole F.	[69]
Naproxen	25.00	0.01630	Mole F.	[32]
Phenothiazine	25.00	0.01984	Mole F.	[34]
Propylparaben	25.00	0.171	Mole F.	[45]
Propylparaben	30.00	0.225	Mole F.	[45]
Propylparaben	35.00	0.264	Mole F.	[45]

Solubility Data of Drugs in 1-Decanol at Various Temperatures (continued)

Drug	T (°C)	Solubility	Solubility Unit	Reference
Propylparaben	40.00	0.285	Mole F.	[45]
Salicylamide	25.00	0.02082	Mole F.	[37]
Sulfadiazine	25.00	0.000388	mol/L	[62]
Sulfadiazine	25.00	0.0000740	Mole F.	[63]
Sulfadiazine	30.00	0.0000804	Mole F.	[63]
Sulfadiazine	37.00	0.0000947	Mole F.	[63]
Sulfadimethoxine	25.00	0.000224	Mole F.	[63]
Sulfadimethoxine	30.00	0.000269	Mole F.	[63]
Sulfadimethoxine	37.00	0.000337	Mole F.	[63]
Sulfisoxazole	25.00	0.00213	mol/L	[62]
Sulfisomidine	25.00	0.000180	Mole F.	[63]
Sulfisomidine	30.00	0.000204	Mole F.	[63]
Sulfisomidine	37.00	0.000253	Mole F.	[63]
Testosterone	25.00	0.0441	Mole F.	[69]
Xanthene	25.00	0.04528	Mole F.	[3]

Solubility Data of Drugs in 1-Heptanol at Various Temperatures

Drug	T (°C)	Solubility	Solubility Unit	Reference
4-Aminobenzoic acid	25.00	0.02277	Mole F.	[20]
Acetylsalicylic acid	25.00	0.03892	Mole F.	[22]
Benzoic acid	25.00	0.1946	Mole F.	[21]
Diflunisal	25.00	0.0383	Mole F.	[26]
Flubiprofen	25.00	0.0760	Mole F.	[26]
Ketoprofen	25.00	0.0674	Mole F.	[51]
Mestanolone	25.00	0.0137	Mole F.	[69]
Methandienone	25.00	0.0619	Mole F.	[69]
Methyl p-hydroxybenzoate	25.00	0.1483	Mole F.	[35]
Methyltestosterone	25.00	0.0425	Mole F.	[69]
Nandrolone	25.00	0.0852	Mole F.	[69]
Naproxen	25.00	0.01909	Mole F.	[32]
Paracetamol	30.00	37.43	g/kg	[11]
Paracetamol	30.00	28	g/L	[8]
Phenothiazine	25.00	0.01754	Mole F.	[34]
p-Hydroxybenzoic acid	25.00	0.1121	Mole F.	[35]
Pimozide	25.00	0.00412	Mole F.	[36]
Salicylamide	25.00	0.03026	Mole F.	[37]
Testosterone	25.00	0.0509	Mole F.	[69]
Thalidomide	32.00	0.26	g/L	[8]
Xanthene	25.00	0.03340	Mole F.	[3]

Solubility Data of Drugs in 1-Hexanol at Various Temperatures

Drug	T (°C)	Solubility	Solubility Unit	Reference
4-Aminobenzoic acid	25.00	0.02664	Mole F.	[20]
Acetylsalicylic acid	25.00	0.0393	Mole F.	[21]
Acetylsalicylic acid	25.00	0.03973	Mole F.	[22]
Benzoic acid	25.00	0.1905	Mole F.	[24]

(continued)

Solubility Data of Drugs in 1-Hexanol at Various Temperatures   (continued)

Drug	T (°C)	Solubility	Solubility Unit	Reference
Benzoic acid	25.00	0.201	Mole F.	[43]
Betulin	5.05	0.000744	Mole F.	[7]
Betulin	15.05	0.000846	Mole F.	[7]
Betulin	25.05	0.001020	Mole F.	[7]
Betulin	35.05	0.001273	Mole F.	[7]
Butyl p-hydroxybenzoate	25.00	0.371	Mole F.	[43]
Butylparaben	25.00	0.369	Mole F.	[45]
Butylparaben	30.00	0.395	Mole F.	[45]
Butylparaben	35.00	0.444	Mole F.	[45]
Butylparaben	40.00	0.507	Mole F.	[45]
Diflunisal	25.00	0.0331	Mole F.	[26]
Ethylparaben	25.00	0.189	Mole F.	[45]
Ethylparaben	30.00	0.195	Mole F.	[45]
Ethylparaben	35.00	0.212	Mole F.	[45]
Ethylparaben	40.00	0.228	Mole F.	[45]
Ethyl p-hydroxybenzoate	25.00	0.178	Mole F.	[43]
Flubiprofen	25.00	0.0716	Mole F.	[26]
Ketoprofen	25.00	0.0814	Mole F.	[51]
Lovastatin	20.70	0.0036267	Mole F.	[53]
Lovastatin	22.70	0.0039958	Mole F.	[53]
Lovastatin	25.80	0.0045493	Mole F.	[53]
Lovastatin	27.50	0.0049080	Mole F.	[53]
Lovastatin	30.75	0.0056535	Mole F.	[53]
Lovastatin	35.75	0.0070345	Mole F.	[53]
Lovastatin	38.85	0.0076794	Mole F.	[53]
Mestanolone	25.00	0.0108	Mole F.	[69]
Methandienone	25.00	0.0600	Mole F.	[69]
Methyl paraben	25.00	0.155	Mole F.	[45]
Methyl paraben	30.00	0.166	Mole F.	[45]
Methyl paraben	35.00	0.186	Mole F.	[45]
Methyl paraben	40.00	0.205	Mole F.	[45]
Methyl p-hydroxybenzoate	25.00	0.147	Mole F.	[43]
Methyl p-hydroxybenzoate	25.00	0.1477	Mole F.	[35]
Methyltestosterone	25.00	0.0424	Mole F.	[69]
Nandrolone	25.00	0.0889	Mole F.	[69]
Naproxen	25.00	0.01663	Mole F.	[32]
Paracetamol	30.00	49.71	g/kg	[11]
Paracetamol	30.00	38.6	g/L	[8]
Phenothiazine	25.00	0.01562	Mole F.	[34]
p-Hydroxybenzoic acid	25.00	0.1154	Mole F.	[35]
Pimozide	25.00	0.00331	Mole F.	[36]
Propyl p-hydroxybenzoate	25.00	0.216	Mole F.	[43]
Propylparaben	25.00	0.210	Mole F.	[45]
Propylparaben	30.00	0.256	Mole F.	[45]
Propylparaben	35.00	0.285	Mole F.	[45]
Propylparaben	40.00	0.303	Mole F.	[45]
Salicylamide	25.00	0.0327	Mole F.	[37]
Temazepam	25.00	0.0063267	Mole F.	[15]
Testosterone	25.00	0.0497	Mole F.	[69]
Thalidomide	32.00	0.24	g/L	[8]
Xanthene	25.00	0.02831	Mole F.	[3]

Solubility Data of Drugs in 1-Octanol at Various Temperatures

Drug	T (°C)	Solubility	Solubility Unit	Reference
15-s-15-Methyl prostaglandin F$_2$ α-methyl ester	30.00	1.32	Mole F.	[70]
2-Hydroxybenzoic acid	20.00	0.176	Mole F.	[71]
2-Hydroxybenzoic acid	25.00	0.186	Mole F.	[71]
2-Hydroxybenzoic acid	30.00	0.206	Mole F.	[71]
2-Hydroxybenzoic acid	37.00	0.235	Mole F.	[71]
2-Hydroxybenzoic acid	42.00	0.252	Mole F.	[71]
2-Hydroxybenzoic acid	25.00	0.2143	Mole F.	[19]
3-Hydroxybenzoic acid	20.00	0.0837	Mole F.	[71]
3-Hydroxybenzoic acid	25.00	0.0944	Mole F.	[71]
3-Hydroxybenzoic acid	30.00	0.0993	Mole F.	[71]
3-Hydroxybenzoic acid	37.00	0.113	Mole F.	[71]
3-Hydroxybenzoic acid	42.00	0.122	Mole F.	[71]
4-Aminobenzoic acid	25.00	0.02088	Mole F.	[20]
4-Hydroxybenzoic acid	20.00	0.129	Mole F.	[71]
4-Hydroxybenzoic acid	25.00	0.139	Mole F.	[71]
4-Hydroxybenzoic acid	30.00	0.154	Mole F.	[71]
4-Hydroxybenzoic acid	37.00	0.177	Mole F.	[71]
4-Hydroxybenzoic acid	42.00	0.186	Mole F.	[71]
Acetanilide	20.00	0.13	Mole F.	[72]
Acetanilide	25.00	0.147	Mole F.	[72]
Acetanilide	25.00	0.601	mol/L	[73]
Acetanilide	30.00	0.100	Mole F.	[70]
Acetanilide	30.00	0.172	Mole F.	[72]
Acetanilide	30.00	0.772	mol/L	[73]
Acetanilide	35.00	0.874	mol/L	[73]
Acetanilide	37.00	0.200	Mole F.	[72]
Acetanilide	40.00	1.145	mol/L	[73]
Acetanilide	42.00	0.230	Mole F.	[72]
Acetylsalicylic acid	25.00	0.0386	Mole F.	[21]
Acetylsalicylic acid	25.00	0.03581	Mole F.	[22]
Acetylsalicylic acid	30.00	0.0263	Mole F.	[70]
Alcofenac	5.00	0.303	mol/L	[74]
Alcofenac	25.00	0.610	mol/L	[74]
Alcofenac	37.00	1.303	mol/L	[74]
Aminopyrine	30.00	0.129	Mole F.	[70]
Antipyrine	30.00	0.0871	Mole F.	[70]
Atenolol	20.00	0.00338	Mole F.	[75]
Atenolol	25.00	0.00409	Mole F.	[75]
Atenolol	30.00	0.00530	Mole F.	[75]
Atenolol	37.00	0.00730	Mole F.	[75]
Atenolol	42.00	0.00874	Mole F.	[75]
Barbital	30.00	0.0158	Mole F.	[70]
Benzocaine	25.00	0.398	mol/L	[76]
Benzocaine	30.00	0.476	mol/L	[76]
Benzocaine	35.00	0.578	mol/L	[76]
Benzocaine	40.00	0.711	mol/L	[76]
Benzoic acid	25.00	0.1987	Mole F.	[24]
Benzoic acid	25.00	0.129	Mole F.	[43]

(continued)

Solubility Data of Drugs in 1-Octanol at Various Temperatures   (continued)

Drug	T (°C)	Solubility	Solubility Unit	Reference
Benzoic acid	30.00	0.112	Mole F.	[70]
Berberine chloride	24.85	0.000297	Mole F.	[44]
Berberine chloride	29.85	0.000313	Mole F.	[44]
Berberine chloride	34.85	0.000357	Mole F.	[44]
Berberine chloride	39.85	0.000435	Mole F.	[44]
Biphenyl	30.00	0.105	Mole F.	[70]
Butyl p-hydroxybenzoate	30.00	0.331	Mole F.	[70]
Butyl p-hydroxybenzoate	25.00	0.360	Mole F.	[43]
Butyl p-aminobenzoate	30.00	0.190	Mole F.	[70]
Butylparaben	25.00	0.331	Mole F.	[45]
Butylparaben	30.00	0.379	Mole F.	[45]
Butylparaben	35.00	0.440	Mole F.	[45]
Butylparaben	40.00	0.497	Mole F.	[45]
Caffeine	30.00	0.00245	Mole F.	[70]
Celecoxib	25.00	7.870	g/L	[55]
Cortisone	30.00	0.00263	Mole F.	[70]
Deoxycortisone	30.00	0.0263	Mole F.	[70]
Diclofenac	5.00	0.064	mol/L	[74]
Diclofenac	25.00	0.0150406	Mole F.	[25]
Diclofenac	25.00	0.078	mol/L	[74]
Diclofenac	37.00	0.089	mol/L	[74]
Diflunisal	20.00	0.0333	Mole F.	[77]
Diflunisal	25.00	0.0352	Mole F.	[26]
Diflunisal	25.00	0.0343	Mole F.	[77]
Diflunisal	30.00	0.0355	Mole F.	[77]
Diphenyl ethane	30.00	0.219	Mole F.	[70]
Ethyl p-aminobenzoate	30.00	0.0646	Mole F.	[70]
Ethylparaben	25.00	0.156	Mole F.	[45]
Ethylparaben	30.00	0.183	Mole F.	[45]
Ethylparaben	35.00	0.204	Mole F.	[45]
Ethylparaben	40.00	0.226	Mole F.	[45]
Ethyl p-hydroxybenzoate	25.00	0.170	Mole F.	[43]
Ethyl p-hydroxybenzoate	30.00	0.148	Mole F.	[70]
Fantiazac	5.00	0.113	mol/L	[74]
Fantiazac	25.00	0.170	mol/L	[74]
Fantiazac	37.00	0.223	mol/L	[74]
Fenbufen	5.00	0.005	mol/L	[74]
Fenbufen	20.00	0.00106	Mole F.	[77]
Fenbufen	25.00	0.00129	Mole F.	[77]
Fenbufen	25.00	0.012	mol/L	[74]
Fenbufen	30.00	0.00178	Mole F.	[77]
Fenbufen	37.00	0.019	mol/L	[74]
Flubiprofen	5.00	0.232	mol/L	[74]
Flubiprofen	18.00	0.0608	Mole F.	[77]
Flubiprofen	20.00	0.065	Mole F.	[77]
Flubiprofen	25.00	0.0817	Mole F.	[26]
Flubiprofen	25.00	0.0706	Mole F.	[77]
Flubiprofen	25.00	0.286	mol/L	[74]
Flubiprofen	30.00	0.0796	Mole F.	[77]

Solubility Data of Drugs in 1-Octanol at Various Temperatures (continued)

Drug	T (°C)	Solubility	Solubility Unit	Reference
Flubiprofen	30.00	0.0891	Mole F.	[70]
Flubiprofen	37.00	0.332	mol/L	[74]
Flufenamic acid	20.00	0.0772	Mole F.	[78]
Flufenamic acid	25.00	0.092	Mole F.	[78]
Flufenamic acid	30.00	0.103	Mole F.	[78]
Flufenamic acid	37.00	0.128	Mole F.	[78]
Flufenamic acid	42.00	0.149	Mole F.	[78]
Fluorene	30.00	0.0355	Mole F.	[70]
Fumaric acid	30.00	0.010	Mole F.	[70]
Gentisic acid	25.00	0.74131	mol/L	[79]
Hydrocortisone	25.00	0.00938	mol/L	[80]
Ibuprofen	25.00	1.912	mol/L	[81]
Ibuprofen	30.00	2.3737	mol/L	[81]
Ibuprofen	35.00	2.8274	mol/L	[81]
Ibuprofen	40.00	3.416	mol/L	[81]
Ibuprofen	5.00	0.059	mol/L	[74]
Ibuprofen	25.00	0.2201602	Mole F.	[28]
Ibuprofen	25.00	0.091	mol/L	[74]
Ibuprofen	25.00	0.1993	Mole F.	[67]
Ibuprofen	30.00	0.282	Mole F.	[70]
Ibuprofen	37.00	0.122	mol/L	[74]
Ketoprofen	5.00	0.227	mol/L	[74]
Ketoprofen	25.00	0.448	mol/L	[74]
Ketoprofen	25.00	0.0691	Mole F.	[51]
Ketoprofen	37.00	0.667	mol/L	[74]
Lactose	25.00	0.0001488	Mole F.	[30]
Lovastatin	12.55	0.0025518	Mole F.	[53]
Lovastatin	15.55	0.0027800	Mole F.	[53]
Lovastatin	17.50	0.0031255	Mole F.	[53]
Lovastatin	23.05	0.0039135	Mole F.	[53]
Lovastatin	28.05	0.0046741	Mole F.	[53]
Lovastatin	30.95	0.0049997	Mole F.	[53]
Lovastatin	32.05	0.0051493	Mole F.	[53]
Lovastatin	33.55	0.0056685	Mole F.	[53]
Lovastatin	35.55	0.0061930	Mole F.	[53]
Lovastatin	36.95	0.0065927	Mole F.	[53]
Meloxicam	25.00	0.187	g/L	[55]
Mestanolone	25.00	0.0158	Mole F.	[69]
Methandienone	25.00	0.0594	Mole F.	[69]
Methyl p-aminobenzoate	30.00	0.0380	Mole F.	[70]
Methylparaben	25.00	0.111	Mole F.	[45]
Methylparaben	30.00	0.151	Mole F.	[45]
Methylparaben	35.00	0.166	Mole F.	[45]
Methylparaben	40.00	0.172	Mole F.	[45]
Methyl p-hydroxybenzoate	25.00	0.129	Mole F.	[43]
Methyl p-hydroxybenzoate	25.00	0.1381	Mole F.	[35]
Methyl testosterone	30.00	0.0490	Mole F.	[70]
Methyl-p-hydroxybenzoate	30.00	0.101	Mole F.	[70]
Methyltestosterone	25.00	0.0421	Mole F.	[69]

(continued)

Solubility Data of Drugs in 1-Octanol at Various Temperatures   (continued)

Drug	T (°C)	Solubility	Solubility Unit	Reference
Nandrolone	25.00	0.0810	Mole F.	[69]
Naproxen	5.00	0.061	mol/L	[74]
Naproxen	20.00	0.0848	mol/L	[82]
Naproxen	25.00	0.0166441	Mole F.	[31]
Naproxen	25.00	0.116	mol/L	[74]
Naproxen	25.00	0.01604	Mole F.	[32]
Naproxen	25.00	0.1095	mol/L	[82]
Naproxen	30.00	0.1267	mol/L	[82]
Naproxen	35.00	0.1555	mol/L	[82]
Naproxen	37.00	0.157	mol/L	[74]
Naproxen	40.00	0.1826	mol/L	[82]
Niflumic acid	20.00	0.0259	Mole F.	[78]
Niflumic acid	25.00	0.0000457	Mole F.	[33]
Niflumic acid	25.00	0.0294	Mole F.	[78]
Niflumic acid	30.00	0.0336	Mole F.	[78]
Niflumic acid	37.00	0.0409	Mole F.	[78]
Niflumic acid	42.00	0.0468	Mole F.	[78]
Nimesulide	25.00	0.970	g/L	[55]
p-Aminobenzoic acid	30.00	0.0209	Mole F.	[70]
p-Aminobenzoic acid	25.00	0.0180610	Mole F.	[25]
Paracetamol	25.00	0.1351	mol/L	[73]
Paracetamol	30.00	0.150	mol/L	[73]
Paracetamol	35.00	0.172	mol/L	[73]
Paracetamol	40.00	0.201	mol/L	[73]
Paracetamol	20.00	0.0228	Mole F.	[72]
Paracetamol	25.00	0.0247	Mole F.	[72]
Paracetamol	30.00	0.0268	Mole F.	[72]
Paracetamol	30.00	27.47	g/kg	[11]
Paracetamol	37.00	0.0299	Mole F.	[72]
Paracetamol	42.00	0.0317	Mole F.	[72]
Phenacetin	20.00	0.0185	Mole F.	[72]
Phenacetin	25.00	0.0228	Mole F.	[72]
Phenacetin	25.00	0.1196	mol/L	[73]
Phenacetin	30.00	0.0186	Mole F.	[70]
Phenacetin	30.00	0.0273	Mole F.	[72]
Phenacetin	30.00	0.1427	mol/L	[73]
Phenacetin	35.00	0.1760	mol/L	[73]
Phenacetin	37.00	0.0340	Mole F.	[72]
Phenacetin	40.00	0.2045	mol/L	[73]
Phenacetin	42.00	0.0416	Mole F.	[72]
Phenobarbital	30.00	0.0107	Mole F.	[70]
Phenol	30.00	0.871	Mole F.	[70]
Phenothiazine	25.00	0.01855	Mole F.	[34]
p-Hydroxybenzoic acid	25.00	0.1032	Mole F.	[35]
p-Hydroxybenzoic acid	25.00	131.4	g/kg	[83]
Pimozide	25.00	0.00411	Mole F.	[36]
Pindolol	20.00	0.00161	Mole F.	[75]
Pindolol	25.00	0.00205	Mole F.	[75]
Pindolol	30.00	0.00236	Mole F.	[75]

Solubility Data of Drugs in 1-Octanol at Various Temperatures (continued)

Drug	T (°C)	Solubility	Solubility Unit	Reference
Pindolol	37.00	0.00309	Mole F.	[75]
Pindolol	42.00	0.00370	Mole F.	[75]
Piroxicam	25.00	0.0002996	Mole F.	[33]
Prednisolone	30.00	0.00316	Mole F.	[70]
Progestrone	30.00	0.0263	Mole F.	[70]
Propyl p-hydroxybenzoate	30.00	0.282	Mole F.	[70]
Propyl p-hydroxybenzoate	25.00	0.207	Mole F.	[43]
Propylparaben	25.00	0.200	Mole F.	[45]
Propylparaben	30.00	0.228	Mole F.	[45]
Propylparaben	35.00	0.271	Mole F.	[45]
Propylparaben	40.00	0.298	Mole F.	[45]
Prostaglandin E2	30.00	0.155	Mole F.	[70]
Prostaglandin F2 α	30.00	1.29	Mole F.	[70]
Rofecoxib	25.00	0.117	g/L	[55]
Salicylamide	25.00	0.02524	Mole F.	[37]
Salicylic acid	25.00	0.1548962	Mole F.	[25]
Salicylic acid	30.00	0.186	Mole F.	[70]
Sodium ibuprofen	25.00	0.0670847	Mole F.	[28]
Sodium benzoate	25.00	0.0000070	Mole F.	[28]
Sodium diclofenac	25.00	0.0000293	Mole F.	[31]
Sodium diclofenac	25.00	0.0000292	Mole F.	[25]
Sodium p-aminobenzoate	25.00	0.0001934	Mole F.	[25]
Sodium salicylate	25.00	0.0052528	Mole F.	[25]
Sulfacetamide	25.00	0.00933	mol/L	[84]
Sulfacetamide	30.00	0.0101	mol/L	[84]
Sulfacetamide	35.00	0.0112	mol/L	[84]
Sulfacetamide	40.00	0.0128	mol/L	[84]
Sulfadiazine	25.00	0.0000170	Mole F.	[39]
Sulfadiazine	25.00	0.0000141	Mole F.	[63]
Sulfadiazine	25.00	0.000088	mol/L	[84]
Sulfadiazine	30.00	0.0000176	Mole F.	[63]
Sulfadiazine	30.00	0.0001125	mol/L	[84]
Sulfadiazine	35.00	0.0001450	mol/L	[84]
Sulfadiazine	37.00	0.0000265	Mole F.	[63]
Sulfadiazine	40.00	0.0001920	mol/L	[84]
Sulfadimethoxine	25.00	0.0002040	Mole F.	[63]
Sulfadimethoxine	30.00	0.0002780	Mole F.	[63]
Sulfadimethoxine	37.00	0.0003590	Mole F.	[63]
Sulfamerazine	25.00	0.00044	mol/L	[84]
Sulfamerazine	30.00	0.00057	mol/L	[84]
Sulfamerazine	35.00	0.00067	mol/L	[84]
Sulfamerazine	40.00	0.000857	mol/L	[84]
Sulfamethazine	25.00	0.0016	mol/L	[84]
Sulfamethazine	30.00	0.00187	mol/L	[84]
Sulfamethazine	35.00	0.00211	mol/L	[84]
Sulfamethazine	40.00	0.00257	mol/L	[84]
Sulfamethoxazole	25.00	0.0061	mol/L	[84]
Sulfamethoxazole	30.00	0.0069	mol/L	[84]
Sulfamethoxazole	35.00	0.0083	mol/L	[84]

(continued)

Solubility Data of Drugs in 1-Octanol at Various Temperatures   (continued)

Drug	T (°C)	Solubility	Solubility Unit	Reference
Sulfamethoxazole	40.00	0.0092	mol/L	[84]
Sulfamethoxypyridazine	25.00	0.00031	Mole F.	[40]
Sulfanilamide	25.00	0.00321	mol/L	[84]
Sulfanilamide	30.00	0.00379	mol/L	[84]
Sulfanilamide	35.00	0.00471	mol/L	[84]
Sulfanilamide	40.00	0.00551	mol/L	[84]
Sulfapyridine	25.00	0.00050	mol/L	[84]
Sulfapyridine	30.00	0.00063	mol/L	[84]
Sulfapyridine	35.00	0.00074	mol/L	[84]
Sulfapyridine	40.00	0.00101	mol/L	[84]
Sulfisoxazole	25.00	0.00352	mol/L	[62]
Sulfathiazole	25.00	0.00060	mol/L	[84]
Sulfathiazole	30.00	0.00081	mol/L	[84]
Sulfathiazole	35.00	0.00105	mol/L	[84]
Sulfathiazole	40.00	0.00128	mol/L	[84]
Sulfisomidine	25.00	0.0001360	Mole F.	[63]
Sulfisomidine	30.00	0.0001830	Mole F.	[63]
Sulfisomidine	37.00	0.0002440	Mole F.	[63]
Testosterone	25.00	0.0569	Mole F.	[69]
Testosterone	30.00	0.0447	Mole F.	[70]
Thalidomide	32.00	0.10	g/L	[8]
Theophyline	30.00	0.00148	Mole F.	[70]
Triazolam	30.00	0.00117	Mole F.	[70]
Triclosan	20.00	0.0089	Mole F.	[85]
Triclosan	25.00	0.025	Mole F.	[85]
Triclosan	30.00	0.0785	Mole F.	[85]
Triclosan	35.00	0.157	Mole F.	[85]
Triclosan	40.00	0.285	Mole F.	[85]
Xanthene	25.00	0.03800	Mole F.	[3]

Solubility Data of Drugs in 1-Pentanol at Various Temperatures

Drug	T (°C)	Solubility	Solubility Unit	Reference
2-Hydroxybenzoic acid	25.00	0.1611	Mole F.	[19]
4-Aminobenzoic acid	25.00	0.02630	Mole F.	[20]
Acetylsalicylic acid	25.00	0.0395	Mole F.	[21]
Acetylsalicylic acid	25.00	0.03966	Mole F.	[22]
Anhydrous citric acid	25.00	0.06463	Mole F.	[23]
Benzoic acid	25.00	0.1839	Mole F.	[24]
Benzoic acid	25.00	0.226	Mole F.	[43]
Betulin	5.05	0.000950	Mole F.	[7]
Betulin	15.05	0.001204	Mole F.	[7]
Betulin	25.05	0.001432	Mole F.	[7]
Betulin	35.05	0.001713	Mole F.	[7]
Butyl p-hydroxybenzoate	25.00	0.364	Mole F.	[43]
Clonazepam	25.00	2.26	g/L	[48]
D(−)-p-Hydroxyphenylglycine dane salt	19.90	0.0000848	Mole F.	[49]
D(−)-p-Hydroxyphenylglycine dane salt	25.10	0.0001017	Mole F.	[49]
D(−)-p-Hydroxyphenylglycine dane salt	30.10	0.0001192	Mole F.	[49]

Solubility Data of Drugs in 1-Pentanol at Various Temperatures (continued)

Drug	T (°C)	Solubility	Solubility Unit	Reference
D(−)-p-Hydroxyphenylglycine dane salt	35.04	0.0001337	Mole F.	[49]
D(−)-p-Hydroxyphenylglycine dane salt	39.92	0.0001523	Mole F.	[49]
D(−)-p-Hydroxyphenylglycine dane salt	45.11	0.0001699	Mole F.	[49]
D(−)-p-Hydroxyphenylglycine dane salt	50.10	0.0001896	Mole F.	[49]
D(−)-p-Hydroxyphenylglycine dane salt	55.02	0.0002091	Mole F.	[49]
D(−)-p-Hydroxyphenylglycine dane salt	60.10	0.0002259	Mole F.	[49]
D(−)-p-Hydroxyphenylglycine dane salt	64.98	0.0002523	Mole F.	[49]
D(−)-p-Hydroxyphenylglycine dane salt	70.20	0.0002789	Mole F.	[49]
Diclofenac	25.00	0.0128682	Mole F.	[25]
Diflunisal	25.00	0.0326	Mole F.	[26]
Ethyl p-hydroxybenzoate	25.00	0.184	Mole F.	[43]
Flubiprofen	25.00	0.0716	Mole F.	[26]
Ibuprofen	25.00	0.2136107	Mole F.	[28]
Ibuprofen	25.00	0.1833	Mole F.	[67]
Ketoprofen	25.00	0.0778	Mole F.	[51]
Ketoprofen	25.00	0.0776	Mole F.	[68]
Lactose	25.00	0.0000517	Mole F.	[30]
Losartan potassium	20.00	0.04463	Mole F.	[86]
Losartan potassium	30.00	0.03109	Mole F.	[86]
Losartan potassium	40.00	0.02869	Mole F.	[86]
Losartan potassium	50.00	0.02342	Mole F.	[86]
Losartan potassium	60.00	0.02051	Mole F.	[86]
Losartan potassium	70.00	0.02184	Mole F.	[86]
Lovastatin	16.00	0.0031213	Mole F.	[53]
Lovastatin	17.95	0.0033974	Mole F.	[53]
Lovastatin	22.35	0.0040441	Mole F.	[53]
Lovastatin	26.05	0.0046819	Mole F.	[53]
Lovastatin	31.40	0.0057108	Mole F.	[53]
Lovastatin	32.50	0.0061291	Mole F.	[53]
Lovastatin	34.00	0.0064004	Mole F.	[53]
Lovastatin	36.00	0.0069139	Mole F.	[53]
Lovastatin	37.40	0.0073787	Mole F.	[53]
Mannitol	25.00	0.0001500	Mole F.	[30]
Mestanolone	25.00	0.0086	Mole F.	[69]
Methandienone	25.00	0.0578	Mole F.	[69]
Methyl p-hydroxybenzoate	25.00	0.159	Mole F.	[43]
Methyl p-hydroxybenzoate	25.00	0.1528	Mole F.	[35]
Methyltestosterone	25.00	0.0399	Mole F.	[69]
Nandrolone	25.00	0.1059	Mole F.	[69]
Naproxen	25.00	0.0317361	Mole F.	[31]
Naproxen	25.00	0.01561	Mole F.	[32]
Niflumic acid	25.00	0.0184074	Mole F.	[33]
p-Aminobenzoic acid	25.00	0.0233771	Mole F.	[25]
Paracetamol	30.00	67.82	g/kg	[11]
Paracetamol	30.00	50.8	g/L	[8]
Paracetamol	25.00	0.06990	Mole F.	[23]
Phenacetinum	9.80	0.011414	Mole F.	[58]
Phenacetinum	15.08	0.013629	Mole F.	[58]
Phenacetinum	20.12	0.016382	Mole F.	[58]

(continued)

Solubility Data of Drugs in 1-Pentanol at Various Temperatures (continued)

Drug	T (°C)	Solubility	Solubility Unit	Reference
Phenacetinum	25.05	0.019863	Mole F.	[58]
Phenacetinum	30.00	0.024323	Mole F.	[58]
Phenacetinum	34.80	0.029741	Mole F.	[58]
Phenacetinum	39.80	0.036708	Mole F.	[58]
Phenacetinum	44.72	0.045061	Mole F.	[58]
Phenacetinum	50.02	0.055923	Mole F.	[58]
Phenacetinum	55.15	0.068482	Mole F.	[58]
Phenacetinum	60.12	0.082759	Mole F.	[58]
Phenothiazine	25.00	0.01339	Mole F.	[34]
p-Hydroxybenzoic acid	25.00	0.0901	Mole F.	[35]
Pimozide	25.00	0.00324	Mole F.	[36]
Piroxicam	25.00	0.0002473	Mole F.	[33]
Propyl p-hydroxybenzoate	25.00	0.217	Mole F.	[43]
Saccharose	25.00	0.00000014	Mole F.	[30]
Salicylamide	25.00	0.03175	Mole F.	[37]
Salicylic acid	25.00	0.1547414	Mole F.	[25]
Sodium ibuprofen	25.00	0.0566876	Mole F.	[28]
Sodium benzoate	25.00	0.0008491	Mole F.	[28]
Sodium diclofenac	25.00	0.0072926	Mole F.	[31]
Sodium diclofenac	25.00	0.0072918	Mole F.	[25]
Sodium p-aminobenzoate	25.00	0.0005085	Mole F.	[25]
Sodium salicylate	25.00	0.0033965	Mole F.	[25]
Sulfadiazine	25.00	0.0000263	Mole F.	[39]
Sulfadiazine	25.00	0.000244	mol/L	[62]
Sulfadiazine	25.00	0.0000263	Mole F.	[63]
Sulfadiazine	30.00	0.0000331	Mole F.	[63]
Sulfadiazine	37.00	0.0000461	Mole F.	[63]
Sulfadimethoxine	25.00	0.000341	Mole F.	[63]
Sulfadimethoxine	30.00	0.000441	Mole F.	[63]
Sulfadimethoxine	37.00	0.000565	Mole F.	[63]
Sulfamethoxypyridazine	25.00	0.00057	Mole F.	[40]
Sulfisoxazole	25.00	0.00980	mol/L	[62]
Sulfisomidine	25.00	0.0002840	Mole F.	[63]
Sulfisomidine	30.00	0.0003430	Mole F.	[63]
Sulfisomidine	37.00	0.0004540	Mole F.	[63]
Testosterone	25.00	0.0450	Mole F.	[69]
Thalidomide	32.00	0.09	g/L	[8]
Xanthene	25.00	0.02212	Mole F.	[3]
Xylitol	20.05	0.0002980	Mole F.	[66]
Xylitol	25.03	0.0004830	Mole F.	[66]
Xylitol	30.11	0.0007339	Mole F.	[66]
Xylitol	35.14	0.001009	Mole F.	[66]
Xylitol	40.04	0.001364	Mole F.	[66]
Xylitol	45.09	0.001885	Mole F.	[66]
Xylitol	50.14	0.002629	Mole F.	[66]
Xylitol	55.08	0.003636	Mole F.	[66]
Xylitol	60.07	0.004998	Mole F.	[66]
Xylitol	65.18	0.006817	Mole F.	[66]
Xylitol	70.09	0.009034	Mole F.	[66]

Solubility Data of Drugs in 1-Propanol at Various Temperatures

Drug	T (°C)	Solubility	Solubility Unit	Reference
Ethyl p-hydroxybenzoate	25.00	0.171	Mole F.	[43]
Flubiprofen	25.00	0.0668	Mole F.	[26]
Glucose	60.00	3.72	g/L	[42]
Haloperidol	25.00	0.00326	Mole F.	[27]
Ibuprofen	25.00	0.1417	Mole F.	[67]
Ketoprofen	25.00	0.0845	Mole F.	[51]
Ketoprofen	25.00	0.0848	Mole F.	[68]
Lamivudine (form 2)	5.00	4.1	g/L	[52]
Lamivudine (form 2)	15.00	5.1	g/L	[52]
Lamivudine (form 2)	25.00	7.2	g/L	[52]
Lamivudine (form 2)	35.00	9.4	g/L	[52]
Lamivudine (form 2)	45.00	13.3	g/L	[52]
Losartan potassium	20.00	0.02601	Mole F.	[86]
Losartan potassium	30.00	0.02725	Mole F.	[86]
Losartan potassium	40.00	0.02985	Mole F.	[86]
Losartan potassium	50.00	0.03755	Mole F.	[86]
Losartan potassium	60.00	0.04519	Mole F.	[86]
Losartan potassium	70.00	0.04992	Mole F.	[86]
Lovastatin	13.00	0.0042104	Mole F.	[53]
Lovastatin	16.00	0.0045126	Mole F.	[53]
Lovastatin	17.95	0.0048848	Mole F.	[53]
Lovastatin	28.50	0.0064980	Mole F.	[53]
Lovastatin	31.40	0.0071873	Mole F.	[53]
Lovastatin	32.50	0.0073603	Mole F.	[53]
Lovastatin	34.00	0.0075945	Mole F.	[53]
Lovastatin	36.00	0.0080885	Mole F.	[53]
Lovastatin	37.40	0.0084415	Mole F.	[53]
Luteolin	0.00	0.00157	Mole F.	[54]
Luteolin	10.00	0.00203	Mole F.	[54]
Luteolin	25.00	0.00245	Mole F.	[54]
Luteolin	40.00	0.00271	Mole F.	[54]
Luteolin	60.00	0.00322	Mole F.	[54]
Methylparaben	25.00	0.138	Mole F.	[45]
Methylparaben	30.00	0.152	Mole F.	[45]
Methylparaben	35.00	0.180	Mole F.	[45]
Methylparaben	40.00	0.199	Mole F.	[45]
Methyl p-hydroxybenzoate	25.00	0.141	Mole F.	[43]
Methyl p-hydroxybenzoate	25.00	0.1486	Mole F.	[35]
Naproxen	25.00	0.01302	Mole F.	[32]
Paracetamol	−5.00	65.88	g/kg	[11]
Paracetamol	0.00	72.30	g/kg	[11]
Paracetamol	5.00	79.62	g/kg	[11]
Paracetamol	10.00	88.22	g/kg	[11]
Paracetamol	15.00	96.77	g/kg	[11]
Paracetamol	20.00	108.09	g/kg	[11]
Paracetamol	25.00	119.32	g/kg	[11]
Paracetamol	30.00	132.77	g/kg	[11]
Paracetamol	30.00	110	g/L	[8]

(continued)

Solubility Data of Drugs in 1-Propanol at Various Temperatures   (continued)

Drug	T (°C)	Solubility	Solubility Unit	Reference
Paroxetine HCl hemihydrate	26.00	0.0002709	Mole F.	[57]
Paroxetine HCl hemihydrate	30.30	0.0003130	Mole F.	[57]
Paroxetine HCl hemihydrate	34.10	0.0003372	Mole F.	[57]
Paroxetine HCl hemihydrate	38.10	0.0003695	Mole F.	[57]
Paroxetine HCl hemihydrate	42.00	0.0003957	Mole F.	[57]
Paroxetine HCl hemihydrate	46.00	0.0004205	Mole F.	[57]
Paroxetine HCl hemihydrate	50.00	0.0004425	Mole F.	[57]
Paroxetine HCl hemihydrate	54.10	0.0004674	Mole F.	[57]
Paroxetine HCl hemihydrate	58.00	0.0005031	Mole F.	[57]
Paroxetine HCl hemihydrate	62.00	0.0005234	Mole F.	[57]
Paroxetine HCl hemihydrate	66.10	0.0005585	Mole F.	[57]
Paroxetine HCl hemihydrate	70.00	0.0005817	Mole F.	[57]
Paroxetine HCl hemihydrate	74.20	0.0006293	Mole F.	[57]
Paroxetine HCl hemihydrate	78.00	0.0006669	Mole F.	[57]
Paroxetine HCl hemihydrate	81.95	0.0007345	Mole F.	[57]
Phenacetinum	9.50	0.0089512	Mole F.	[58]
Phenacetinum	16.10	0.012371	Mole F.	[58]
Phenacetinum	20.30	0.014756	Mole F.	[58]
Phenacetinum	25.10	0.017969	Mole F.	[58]
Phenacetinum	30.16	0.022224	Mole F.	[58]
Phenacetinum	34.70	0.027068	Mole F.	[58]
Phenacetinum	40.15	0.034520	Mole F.	[58]
Phenacetinum	45.00	0.042973	Mole F.	[58]
Phenacetinum	50.10	0.054049	Mole F.	[58]
Phenacetinum	55.30	0.067999	Mole F.	[58]
Phenacetinum	60.55	0.085175	Mole F.	[58]
Phenothiazine	25.00	0.00885	Mole F.	[34]
p-Hydroxybenzoic acid	25.00	0.1084	Mole F.	[35]
Pimozide	25.00	0.00151	Mole F.	[36]
Potassium clavulanate	0.55	0.0001049	Mole F.	[59]
Potassium clavulanate	4.00	0.0001259	Mole F.	[59]
Potassium clavulanate	8.00	0.0001421	Mole F.	[59]
Potassium clavulanate	11.95	0.0001555	Mole F.	[59]
Potassium clavulanate	16.00	0.0001717	Mole F.	[59]
Potassium clavulanate	20.00	0.0001872	Mole F.	[59]
Potassium clavulanate	24.00	0.0002057	Mole F.	[59]
Potassium clavulanate	28.05	0.0002284	Mole F.	[59]
Potassium clavulanate	32.20	0.0002722	Mole F.	[59]
Pravastatin sodium	4.85	0.00208	Mole F.	[60]
Pravastatin sodium	9.85	0.00216	Mole F.	[60]
Pravastatin sodium	14.85	0.00219	Mole F.	[60]
Pravastatin sodium	19.85	0.00227	Mole F.	[60]
Pravastatin sodium	24.85	0.00238	Mole F.	[60]
Pravastatin sodium	29.85	0.00274	Mole F.	[60]
Pravastatin sodium	34.85	0.00343	Mole F.	[60]
Pravastatin sodium	39.85	0.00449	Mole F.	[60]
Pravastatin sodium	44.85	0.00620	Mole F.	[60]
Pravastatin sodium	54.85	0.01216	Mole F.	[60]
Pravastatin sodium	59.85	0.01578	Mole F.	[60]

Solubility Data of Drugs in 1-Propanol at Various Temperatures   (continued)

Drug	T (°C)	Solubility	Solubility Unit	Reference
Propyl p-hydroxybenzoate	25.00	0.205	Mole F.	[43]
Propylparaben	25.00	0.198	Mole F.	[45]
Propylparaben	30.00	0.236	Mole F.	[45]
Propylparaben	35.00	0.273	Mole F.	[45]
Propylparaben	40.00	0.298	Mole F.	[45]
Ranitidine HCl (form 1)	21.30	0.179	g/100 g	[87]
Ranitidine HCl (form 1)	26.70	0.255	g/100 g	[87]
Ranitidine HCl (form 1)	33.20	0.417	g/100 g	[87]
Ranitidine HCl (form 1)	38.10	0.628	g/100 g	[87]
Ranitidine HCl (form 1)	42.80	0.788	g/100 g	[87]
Ranitidine HCl (form 2)	22.40	0.213	g/100 g	[87]
Ranitidine HCl (form 2)	25.60	0.279	g/100 g	[87]
Ranitidine HCl (form 2)	32.10	0.422	g/100 g	[87]
Ranitidine HCl (form 2)	39.20	0.731	g/100 g	[87]
Ranitidine HCl (form 2)	43.50	0.995	g/100 g	[87]
Rutin	10.00	0.001498	Mole F.	[61]
Rutin	25.00	0.001792	Mole F.	[61]
Rutin	40.00	0.002347	Mole F.	[61]
Rutin	50.00	0.002595	Mole F.	[61]
Rutin	60.00	0.002970	Mole F.	[61]
Salicylamide	25.00	0.03307	Mole F.	[37]
Sulfadiazine	25.00	0.0000596	Mole F.	[39]
Sulfadiazine	25.00	0.000559	mol/L	[62]
Sulfadiazine	25.00	0.0000432	Mole F.	[63]
Sulfadiazine	30.00	0.0000545	Mole F.	[63]
Sulfadiazine	37.00	0.0000744	Mole F.	[63]
Sulfadimethoxine	25.00	0.0004710	Mole F.	[63]
Sulfadimethoxine	30.00	0.0005630	Mole F.	[63]
Sulfadimethoxine	37.00	0.0007790	Mole F.	[63]
Sulfamethoxypyridazine	25.00	0.00083	Mole F.	[40]
Sulfisoxazole	25.00	0.0297	mol/L	[62]
Sulfisomidine	25.00	0.0004230	Mole F.	[63]
Sulfisomidine	30.00	0.0004890	Mole F.	[63]
Sulfisomidine	37.00	0.0006480	Mole F.	[63]
Temazepam	24.00	0.0083	Volume F.	[64]
Temazepam	25.00	0.0040287	Mole F.	[15]
Thalidomide	32.00	0.15	g/L	[8]
Trimethoprim	4.20	0.0002776	Mole F.	[65]
Trimethoprim	10.48	0.0003281	Mole F.	[65]
Trimethoprim	15.82	0.0004141	Mole F.	[65]
Trimethoprim	20.90	0.0005306	Mole F.	[65]
Trimethoprim	25.78	0.0006772	Mole F.	[65]
Trimethoprim	30.69	0.000866	Mole F.	[65]
Trimethoprim	35.29	0.001090	Mole F.	[65]
Trimethoprim	41.39	0.001479	Mole F.	[65]
Trimethoprim	46.02	0.001865	Mole F.	[65]
Trimethoprim	50.33	0.002312	Mole F.	[65]
Trimethoprim	55.42	0.002978	Mole F.	[65]
Trimethoprim	61.00	0.003915	Mole F.	[65]
Xanthene	25.00	0.01166	Mole F.	[3]

Solubility Data of Drugs in 2-Butanol at Various Temperatures

Drug	T (°C)	Solubility	Solubility Unit	Reference
2-Hydroxybenzoic acid	25.00	0.1869	Mole F.	[19]
4-Aminobenzoic acid	25.00	0.02808	Mole F.	[20]
Acetylsalicylic acid	25.00	0.05360	Mole F.	[22]
D(−)-p-Hydroxyphenylglycine dane salt	20.20	0.0000351	Mole F.	[49]
D(−)-p-Hydroxyphenylglycine dane salt	25.09	0.0000438	Mole F.	[49]
D(−)-p-Hydroxyphenylglycine dane salt	29.90	0.0000587	Mole F.	[49]
D(−)-p-Hydroxyphenylglycine dane salt	35.02	0.0000664	Mole F.	[49]
D(−)-p-Hydroxyphenylglycine dane salt	39.90	0.0000811	Mole F.	[49]
D(−)-p-Hydroxyphenylglycine dane salt	45.21	0.0000962	Mole F.	[49]
D(−)-p-Hydroxyphenylglycine dane salt	50.08	0.0001128	Mole F.	[49]
D(−)-p-Hydroxyphenylglycine dane salt	55.11	0.0001316	Mole F.	[49]
D(−)-p-Hydroxyphenylglycine dane salt	60.00	0.0001514	Mole F.	[49]
D(−)-p-Hydroxyphenylglycine dane salt	65.13	0.0001739	Mole F.	[49]
D(−)-p-Hydroxyphenylglycine dane salt	70.08	0.0001973	Mole F.	[49]
Ibuprofen	25.00	0.204	Mole F.	[67]
Ketoprofen	25.00	0.1480	Mole F.	[68]
Lamivudine (form 2)	5.00	2.1	g/L	[52]
Lamivudine (form 2)	15.00	3.0	g/L	[52]
Lamivudine (form 2)	25.00	4.0	g/L	[52]
Lamivudine (form 2)	35.00	5.6	g/L	[52]
Lamivudine (form 2)	45.00	7.8	g/L	[52]
Naproxen	25.00	0.01418	Mole F.	[32]
Paroxetine HCl hemihydrate	13.75	0.0003779	Mole F.	[57]
Paroxetine HCl hemihydrate	18.25	0.0003851	Mole F.	[57]
Paroxetine HCl hemihydrate	22.60	0.0004054	Mole F.	[57]
Paroxetine HCl hemihydrate	26.00	0.0004305	Mole F.	[57]
Paroxetine HCl hemihydrate	30.05	0.0004557	Mole F.	[57]
Paroxetine HCl hemihydrate	34.20	0.0005036	Mole F.	[57]
Paroxetine HCl hemihydrate	38.15	0.0005601	Mole F.	[57]
Paroxetine HCl hemihydrate	42.00	0.0006127	Mole F.	[57]
Paroxetine HCl hemihydrate	46.25	0.0006927	Mole F.	[57]
Paroxetine HCl hemihydrate	49.85	0.0007652	Mole F.	[57]
Paroxetine HCl hemihydrate	54.05	0.0008719	Mole F.	[57]
Paroxetine HCl hemihydrate	58.00	0.0009953	Mole F.	[57]
Paroxetine HCl hemihydrate	62.40	0.0011421	Mole F.	[57]
Paroxetine HCl hemihydrate	66.30	0.0012889	Mole F.	[57]
Paroxetine HCl hemihydrate	70.50	0.0014808	Mole F.	[57]
Paroxetine HCl hemihydrate	74.25	0.0016667	Mole F.	[57]
Paroxetine HCl hemihydrate	78.15	0.0018848	Mole F.	[57]
Phenothiazine	25.00	0.00732	Mole F.	[34]
Salicylamide	25.00	0.03533	Mole F.	[37]
Trimethoprim	3.70	0.0001532	Mole F.	[65]
Trimethoprim	10.51	0.0001790	Mole F.	[65]
Trimethoprim	16.93	0.0003002	Mole F.	[65]
Trimethoprim	21.52	0.0004220	Mole F.	[65]
Trimethoprim	25.30	0.0005623	Mole F.	[65]
Trimethoprim	30.59	0.0007581	Mole F.	[65]
Trimethoprim	35.69	0.001006	Mole F.	[65]
Trimethoprim	40.20	0.001315	Mole F.	[65]

Solubility Data of Drugs in 2-Butanol at Various Temperatures   (continued)

Drug	*T* (°C)	Solubility	Solubility Unit	Reference
Trimethoprim	45.31	0.001723	Mole F.	[65]
Trimethoprim	50.41	0.002309	Mole F.	[65]
Trimethoprim	55.51	0.002902	Mole F.	[65]
Trimethoprim	60.59	0.003773	Mole F.	[65]
Xanthene	25.00	0.01254	Mole F.	[3]

Solubility Data of Drugs in 2-Butanone at Various Temperatures

Drug	*T* (°C)	Solubility	Solubility Unit	Reference
2-Hydroxybenzoic acid	25.00	0.1852	Mole F.	[19]
3,5-Di-*tert*-butyl-4-hydroxytoluene	N/A	0.51	Mole F.	[4]
Carbamazepine	25.00	27.6	g/L	[8]
Imidacloprid	19.92	0.01588	Mole F.	[2]
Imidacloprid	26.29	0.01920	Mole F.	[2]
Imidacloprid	29.39	0.02128	Mole F.	[2]
Imidacloprid	33.15	0.02378	Mole F.	[2]
Imidacloprid	35.70	0.02603	Mole F.	[2]
Imidacloprid	38.09	0.02807	Mole F.	[2]
Imidacloprid	41.07	0.03076	Mole F.	[2]
Imidacloprid	45.04	0.03458	Mole F.	[2]
Imidacloprid	47.68	0.03794	Mole F.	[2]
Imidacloprid	50.72	0.04190	Mole F.	[2]
Imidacloprid	53.58	0.04598	Mole F.	[2]
Imidacloprid	56.66	0.05041	Mole F.	[2]
Imidacloprid	59.82	0.05730	Mole F.	[2]
Imidacloprid	62.06	0.06166	Mole F.	[2]
Imidacloprid	64.39	0.06779	Mole F.	[2]
Imidacloprid	67.50	0.07488	Mole F.	[2]
Imidacloprid	70.05	0.08231	Mole F.	[2]
Losartan (polymorph 1)	25.00	0.59	g/L	[88]
Losartan (polymorph 1)	35.00	0.78	g/L	[88]
Losartan (polymorph 1)	45.00	0.84	g/L	[88]
Losartan (polymorph 1)	55.00	0.94	g/L	[88]
Losartan (polymorph 1)	65.00	1.25	g/L	[88]
Losartan (polymorph 2)	25.00	1.95	g/L	[88]
Losartan (polymorph 2)	35.00	2.04	g/L	[88]
Losartan (polymorph 2)	45.00	2.13	g/L	[88]
Losartan (polymorph 2)	55.00	2.37	g/L	[88]
Losartan (polymorph 2)	65.00	2.63	g/L	[88]
Lovastatin	11.95	0.00752	Mole F.	[89]
Lovastatin	15.15	0.00878	Mole F.	[89]
Lovastatin	18.05	0.00918	Mole F.	[89]
Lovastatin	21.45	0.01105	Mole F.	[89]
Lovastatin	24.05	0.01151	Mole F.	[89]
Lovastatin	27.15	0.01324	Mole F.	[89]
Lovastatin	29.95	0.01470	Mole F.	[89]
Lovastatin	33.15	0.01751	Mole F.	[89]
Lovastatin	36.25	0.01972	Mole F.	[89]

(continued)

Solubility Data of Drugs in 2-Butanone at Various Temperatures   (continued)

Drug	T (°C)	Solubility	Solubility Unit	Reference
Lovastatin	39.05	0.02186	Mole F.	[89]
Methyl-3-(3,5-di-*tert*-butyl-4- hydroxyphenyl)-propionate	N/A	0.36	Mole F.	[4]
Octadecyl-3-(3,5-di-*tert*-butyl-4- hydroxyphenyl)-propionate	N/A	0.13	Mole F.	[4]
Paracetamol	30.00	69.99	g/kg	[11]
Paracetamol	30.00	54.6	g/L	[8]
Paroxetine HCl hemihydrate	27.80	0.0000091	Mole F.	[57]
Paroxetine HCl hemihydrate	30.20	0.0000108	Mole F.	[57]
Paroxetine HCl hemihydrate	33.45	0.0000142	Mole F.	[57]
Paroxetine HCl hemihydrate	37.50	0.0000188	Mole F.	[57]
Paroxetine HCl hemihydrate	39.50	0.0000235	Mole F.	[57]
Paroxetine HCl hemihydrate	43.55	0.0000314	Mole F.	[57]
Paroxetine HCl hemihydrate	50.60	0.0000587	Mole F.	[57]
Paroxetine HCl hemihydrate	55.75	0.0000863	Mole F.	[57]
Paroxetine HCl hemihydrate	58.20	0.0001112	Mole F.	[57]
Paroxetine HCl hemihydrate	60.20	0.0001393	Mole F.	[57]
Paroxetine HCl hemihydrate	63.45	0.0001799	Mole F.	[57]
Paroxetine HCl hemihydrate	65.90	0.0002131	Mole F.	[57]
Paroxetine HCl hemihydrate	68.55	0.0002536	Mole F.	[57]
Paroxetine HCl hemihydrate	69.85	0.0002939	Mole F.	[57]
Pentaerythrittol tetrakis(3-(3,5-di-*tert*- butyl-4-hydroxyphenyl)-propionate)	N/A	0.083	Mole F.	[4]
Stearic acid (polymorph B)	8.50	0.0018336	Mole F.	[90]
Stearic acid (polymorph B)	15.00	0.0039050	Mole F.	[90]
Stearic acid (polymorph B)	18.50	0.0057302	Mole F.	[90]
Stearic acid (polymorph B)	20.70	0.0069731	Mole F.	[90]
Stearic acid (polymorph B)	26.10	0.0129845	Mole F.	[90]
Stearic acid (polymorph B)	29.50	0.0185108	Mole F.	[90]
Stearic acid (polymorph C)	7.50	0.0018575	Mole F.	[90]
Stearic acid (polymorph C)	14.20	0.0040026	Mole F.	[90]
Stearic acid (polymorph C)	18.00	0.0059278	Mole F.	[90]
Stearic acid (polymorph C)	23.50	0.0101255	Mole F.	[90]
Stearic acid (polymorph C)	32.80	0.0251902	Mole F.	[90]

Solubility Data of Drugs in 2-Methyl-1-Propanol at Various Temperatures

Drug	T (°C)	Solubility	Solubility Unit	Reference
2-Hydroxybenzoic acid	25.00	0.1430	Mole F.	[19]
4-Aminobenzoic acid	25.00	0.01751	Mole F.	[20]
Acetylsalicylic acid	25.00	0.03186	Mole F.	[22]
Benzoic acid	25.00	0.1524	Mole F.	[24]
Ibuprofen	25.00	0.2011	Mole F.	[67]
Progesterone	25.00	35.6	g/L	[91]
Indomethacin	25.00	9.6	g/L	[91]

Solubility Data of Drugs in 2-Methyl-1-Propanol at Various Temperatures   (continued)

Drug	T (°C)	Solubility	Solubility Unit	Reference
Ketoprofen	25.00	0.1009	Mole F.	[68]
Naproxen	25.00	0.00864	Mole F.	[32]
Phenothiazine	25.00	0.00534	Mole F.	[34]
Potassium clavulanate	1.40	0.0000384	Mole F.	[59]
Potassium clavulanate	4.00	0.0000422	Mole F.	[59]
Potassium clavulanate	8.00	0.0000478	Mole F.	[59]
Potassium clavulanate	11.95	0.0000544	Mole F.	[59]
Potassium clavulanate	16.65	0.0000609	Mole F.	[59]
Potassium clavulanate	21.55	0.0000681	Mole F.	[59]
Potassium clavulanate	25.60	0.0000722	Mole F.	[59]
Potassium clavulanate	28.40	0.0000756	Mole F.	[59]
Potassium clavulanate	31.90	0.0000965	Mole F.	[59]
Salicylamide	25.00	0.02295	Mole F.	[37]
Xanthene	25.00	0.01077	Mole F.	[3]
p-Hydroxybenzoic acid	25.00	0.1297	Mole F.	[35]
Sulfadiazine	25.00	0.0000270	Mole F.	[39]
Sulfamethoxypyridazine	25.00	0.00043	Mole F.	[40]

Solubility Data of Drugs in 2-Methyl-2-Propanol at Various Temperatures

Drug	T (°C)	Solubility	Solubility Unit	Reference
2-Hydroxybenzoic acid	25.00	0.2193	Mole F.	[19]
Acetylsalicylic acid	25.00	0.06844	Mole F.	[22]
Glucose	40.00	1.24	g/L	[42]
Glucose	60.00	3.51	g/L	[42]
Phenothiazine	25.00	0.00583	Mole F.	[34]
Salicylamide	25.00	0.02939	Mole F.	[37]
Xanthene	25.00	0.01112	Mole F.	[3]

Solubility Data of Drugs in 2-Propanol at Various Temperatures

Drug	T (°C)	Solubility	Solubility Unit	Reference
1H-1,2,4-Triazole	10.02	0.1294	Mole F.	[92]
1H-1,2,4-Triazole	15.06	0.1441	Mole F.	[92]
1H-1,2,4-Triazole	20.01	0.1592	Mole F.	[92]
1H-1,2,4-Triazole	25.08	0.1756	Mole F.	[92]
1H-1,2,4-Triazole	30.12	0.1930	Mole F.	[92]
1H-1,2,4-Triazole	35.04	0.2114	Mole F.	[92]
1H-1,2,4-Triazole	40.09	0.2318	Mole F.	[92]
1H-1,2,4-Triazole	45.07	0.2537	Mole F.	[92]
1H-1,2,4-Triazole	50.12	0.2779	Mole F.	[92]

(*continued*)

Solubility Data of Drugs in 2-Propanol at Various Temperatures   (continued)

Drug	T (°C)	Solubility	Solubility Unit	Reference
1H-1,2,4-Triazole	55.05	0.3036	Mole F.	[92]
1H-1,2,4-Triazole	60.04	0.3320	Mole F.	[92]
1H-1,2,4-Triazole	65.03	0.3629	Mole F.	[92]
1H-1,2,4-Triazole	70.11	0.3972	Mole F.	[92]
2-(4-Ethylbenzoyl)benzoic acid	6.64	0.0023	Mole F.	[18]
2-(4-Ethylbenzoyl)benzoic acid	11.59	0.0037	Mole F.	[18]
2-(4-Ethylbenzoyl)benzoic acid	20.90	0.0085	Mole F.	[18]
2-(4-Ethylbenzoyl)benzoic acid	30.84	0.0174	Mole F.	[18]
2-(4-Ethylbenzoyl)benzoic acid	40.10	0.0305	Mole F.	[18]
2-(4-Ethylbenzoyl)benzoic acid	50.22	0.0464	Mole F.	[18]
2-(4-Ethylbenzoyl)benzoic acid	60.90	0.0631	Mole F.	[18]
2,6-Diaminopyridine	−6.90	0.02221	Mole F.	[93]
2,6-Diaminopyridine	−3.20	0.02796	Mole F.	[93]
2,6-Diaminopyridine	1.00	0.03024	Mole F.	[93]
2,6-Diaminopyridine	4.40	0.03793	Mole F.	[93]
2,6-Diaminopyridine	8.00	0.04009	Mole F.	[93]
2,6-Diaminopyridine	12.50	0.05116	Mole F.	[93]
2,6-Diaminopyridine	17.00	0.05648	Mole F.	[93]
2,6-Diaminopyridine	20.50	0.06799	Mole F.	[93]
2,6-Diaminopyridine	24.10	0.07325	Mole F.	[93]
2,6-Diaminopyridine	28.00	0.08868	Mole F.	[93]
2,6-Diaminopyridine	32.00	0.09760	Mole F.	[93]
2,6-Diaminopyridine	35.10	0.11392	Mole F.	[93]
2,6-Diaminopyridine	38.00	0.12011	Mole F.	[93]
2,6-Diaminopyridine	42.70	0.14734	Mole F.	[93]
2,6-Diaminopyridine	46.10	0.15935	Mole F.	[93]
2-Hydroxybenzoic acid	25.00	0.1789	Mole F.	[19]
4-Aminobenzoic acid	25.00	0.03218	Mole F.	[20]
4-Aminophenylacetic acid	20.00	0.74	g/L	[8]
Acephate	26.15	0.06062	Mole F.	[5]
Acephate	30.80	0.07730	Mole F.	[5]
Acephate	33.55	0.09255	Mole F.	[5]
Acephate	37.60	0.11816	Mole F.	[5]
Acephate	40.20	0.13717	Mole F.	[5]
Acephate	43.30	0.16263	Mole F.	[5]
Acephate	46.30	0.19257	Mole F.	[5]
Acephate	48.85	0.22410	Mole F.	[5]
Acetylsalicylic acid	8.45	0.013	Mole F.	[94]
Acetylsalicylic acid	18.65	0.032	Mole F.	[94]
Acetylsalicylic acid	25.00	0.05232	Mole F.	[22]
Acetylsalicylic acid	30.95	0.063	Mole F.	[94]
Acetylsalicylic acid	39.75	0.091	Mole F.	[94]
Acetylsalicylic acid	42.95	0.102	Mole F.	[94]
Acetylsalicylic acid	47.25	0.118	Mole F.	[94]
Acetylsalicylic acid	49.55	0.128	Mole F.	[94]
Acetylsalicylic acid	52.65	0.143	Mole F.	[94]
Acetylsalicylic acid	54.55	0.153	Mole F.	[94]
Acetylsalicylic acid	57.05	0.167	Mole F.	[94]

Solubility Data of Drugs in 2-Propanol at Various Temperatures (continued)

Drug	T (°C)	Solubility	Solubility Unit	Reference
Benzoic acid	25.00	0.1937	Mole F.	[24]
Berberine chloride	24.85	0.000171	Mole F.	[44]
Berberine chloride	29.85	0.000211	Mole F.	[44]
Berberine chloride	34.85	0.000252	Mole F.	[44]
Berberine chloride	39.85	0.000333	Mole F.	[44]
β-Carotene	Room	40	mg/L	[6]
Buspirone HCl (form 1)	25.00	0.50	g/100 g	[95]
Buspirone HCl (form 1)	30.00	1.11	g/100 g	[95]
Buspirone HCl (form 1)	35.00	1.66	g/100 g	[95]
Buspirone HCl (form 1)	40.00	2.55	g/100 g	[95]
Buspirone HCl (form 1)	45.00	3.43	g/100 g	[95]
Buspirone HCl (form 1)	55.00	4.77	g/100 g	[95]
Buspirone HCl (form 1)	65.30	10.72	g/100 g	[95]
Buspirone HCl (form 1)	70.5	15.76	g/100 g	[95]
Buspirone HCl (form 1)	81.00	37.62	g/100 g	[95]
Buspirone HCl (form 2)	25.00	0.87	g/100 g	[95]
Buspirone HCl (form 2)	30.00	1.89	g/100 g	[95]
Buspirone HCl (form 2)	35.00	1.91	g/100 g	[95]
Buspirone HCl (form 2)	40.00	2.60	g/100 g	[95]
Buspirone HCl (form 2)	45.00	6.83	g/100 g	[95]
Buspirone HCl (form 2)	55.00	12.49	g/100 g	[95]
Buspirone HCl (form 2)	65.30	22.20	g/100 g	[95]
Buspirone HCl (form 2)	70.5	28.56	g/100 g	[95]
Buspirone HCl (form 2)	81.00	42.14	g/100 g	[95]
Carbamazepine	25.00	9.95	g/L	[8]
Carbamazepine (form 3)	11.95	0.00183	Mole F.	[46]
Carbamazepine (form 3)	17.35	0.00265	Mole F.	[46]
Carbamazepine (form 3)	24.85	0.00360	Mole F.	[46]
Carbamazepine (form 3)	30.45	0.00425	Mole F.	[46]
Carbamazepine (form 3)	38.40	0.00609	Mole F.	[46]
Carbamazepine (form 3)	46.21	0.008100	Mole F.	[46]
Carbamazepine (form 3)	56.18	0.011400	Mole F.	[46]
Carbamazepine (form 3)	64.39	0.01692	Mole F.	[46]
Carbamazepine (form 1)	32.08	1.75	% Weight	[96]
Carbamazepine (form 1)	39.03	2.32	% Weight	[96]
Carbamazepine (form 1)	42.35	2.67	% Weight	[96]
Carbamazepine (form 1)	50.49	3.37	% Weight	[96]
Carbamazepine (form 1)	56.77	4.07	% Weight	[96]
Carbamazepine (form 3)	32.09	1.64	% Weight	[96]
Carbamazepine (form 3)	37.34	2.00	% Weight	[96]
Carbamazepine (form 3)	42.24	2.44	% Weight	[96]
Carbamazepine (form 3)	49.80	3.06	% Weight	[96]
Carbamazepine (form 3)	56.10	3.92	% Weight	[96]
Cholesteryl acetate	11.80	1.23	g/100 g	[97]
Cholesteryl acetate	16.70	5.07	g/100 g	[97]
Cholesteryl acetate	20.50	1.911	g/100 g	[97]
Cholesteryl acetate	30.00	2.827	g/100 g	[97]
Cholesteryl acetate	33.90	3.612	g/100 g	[97]

(continued)

Solubility Data of Drugs in 2-Propanol at Various Temperatures   (continued)

Drug	T (°C)	Solubility	Solubility Unit	Reference
Cholesteryl acetate	34.50	11.553	g/100 g	[97]
Cholesteryl acetate	36.50	5.101	g/100 g	[97]
Cholesteryl acetate	37.50	5.138	g/100 g	[97]
Cholesteryl acetate	44.00	19.900	g/100 g	[97]
Ciprofloxacin	20.00	0.0000130	Mole F.	[98]
Ciprofloxacin	30.00	0.0000215	Mole F.	[98]
Ciprofloxacin	40.00	0.0000328	Mole F.	[98]
Ciprofloxacin	50.00	0.0000420	Mole F.	[98]
D(−)-p-Hydroxyphenylglycine dane salt	20.10	0.0000574	Mole F.	[49]
D(−)-p-Hydroxyphenylglycine dane salt	25.08	0.0000638	Mole F.	[49]
D(−)-p-Hydroxyphenylglycine dane salt	29.90	0.0000722	Mole F.	[49]
D(−)-p-Hydroxyphenylglycine dane salt	35.26	0.0000840	Mole F.	[49]
D(−)-p-Hydroxyphenylglycine dane salt	40.00	0.0000966	Mole F.	[49]
D(−)-p-Hydroxyphenylglycine dane salt	45.11	0.0001125	Mole F.	[49]
D(−)-p-Hydroxyphenylglycine dane salt	50.00	0.0001299	Mole F.	[49]
D(−)-p-Hydroxyphenylglycine dane salt	55.04	0.0001502	Mole F.	[49]
D(−)-p-Hydroxyphenylglycine dane salt	60.20	0.0001733	Mole F.	[49]
D(−)-p-Hydroxyphenylglycine dane salt	65.12	0.0001977	Mole F.	[49]
D(−)-p-Hydroxyphenylglycine dane salt	70.10	0.0002245	Mole F.	[49]
Erythromycin A dihydrate	20.05	0.003000	Mole F.	[100]
Erythromycin A dihydrate	24.85	0.004012	Mole F.	[100]
Erythromycin A dihydrate	30.00	0.005560	Mole F.	[100]
Erythromycin A dihydrate	34.85	0.007542	Mole F.	[100]
Erythromycin A dihydrate	39.95	0.009953	Mole F.	[100]
Erythromycin A dihydrate	44.85	0.01430	Mole F.	[100]
Erythromycin A dihydrate	49.85	0.01940	Mole F.	[100]
Glucose	60.00	3.78	g/L	[42]
Haloperidol	25.00	0.0116	Mole F.	[27]
Hydroquinone	6.30	0.1064	Mole F.	[101]
Hydroquinone	15.15	0.1294	Mole F.	[101]
Hydroquinone	21.20	0.1469	Mole F.	[101]
Hydroquinone	28.60	0.1712	Mole F.	[101]
Hydroquinone	35.95	0.1998	Mole F.	[101]
Hydroquinone	41.00	0.2169	Mole F.	[101]
Hydroquinone	45.60	0.2325	Mole F.	[101]
Hydroquinone	54.45	0.2473	Mole F.	[101]
Hydroquinone	61.55	0.2623	Mole F.	[101]
Hydroquinone	69.40	0.2803	Mole F.	[101]
Ibuprofen	10.00	505.3	g/kg	[83]
Ibuprofen	15.00	658.8	g/kg	[83]
Ibuprofen	20.00	793.8	g/kg	[83]
Ibuprofen	25.00	0.2334	Mole F.	[67]
Ibuprofen	30.00	1297	g/kg	[83]
Ibuprofen	35.00	1540	g/kg	[83]
Ketoprofen	25.00	0.1269	Mole F.	[68]
L-(+)-Ascorbic acid	19.85	0.00051	Mole F.	[102]
L-(+)-Ascorbic acid	24.85	0.00063	Mole F.	[102]
L-(+)-Ascorbic acid	29.85	0.00078	Mole F.	[102]
L-(+)-Ascorbic acid	34.85	0.00100	Mole F.	[102]

Solubility Data of Drugs in 2-Propanol at Various Temperatures   (continued)

Drug	T (°C)	Solubility	Solubility Unit	Reference
L-(+)-Ascorbic acid	39.85	0.00124	Mole F.	[102]
L-(+)-Ascorbic acid	44.85	0.00165	Mole F.	[102]
L-(+)-Ascorbic acid	49.85	0.00205	Mole F.	[102]
Lamivudine (form 2)	5.00	2.7	g/L	[52]
Lamivudine (form 2)	15.00	3.8	g/L	[52]
Lamivudine (form 2)	25.00	4.9	g/L	[52]
Lamivudine (form 2)	35.00	6.4	g/L	[52]
Lamivudine (form 2)	45.00	8.0	g/L	[52]
Losartan potassium	20.00	0.002704	Mole F.	[86]
Losartan potassium	30.00	0.002792	Mole F.	[86]
Losartan potassium	40.00	0.004127	Mole F.	[86]
Losartan potassium	50.00	0.005469	Mole F.	[86]
Losartan potassium	60.00	0.006965	Mole F.	[86]
Losartan potassium	70.00	0.011690	Mole F.	[86]
Lutein	Room	400	mg/L	[6]
Luteolin	0.00	0.00107	Mole F.	[54]
Luteolin	10.00	0.00130	Mole F.	[54]
Luteolin	25.00	0.00194	Mole F.	[54]
Luteolin	40.00	0.00217	Mole F.	[54]
Luteolin	60.00	0.00267	Mole F.	[54]
Naproxen	25.00	0.01334	Mole F.	[32]
p-Aminophenylacetic acid	16.00	0.4	g/kg	[83]
p-Aminophenylacetic acid	20.00	0.6	g/kg	[83]
p-Aminophenylacetic acid	25.00	9.3	g/kg	[83]
p-Aminophenylacetic acid	30.00	1.0	g/kg	[83]
Paracetamol	−5.00	64.41	g/kg	[11]
Paracetamol	0.00	17.19	g/kg	[11]
Paracetamol	5.00	79.02	g/kg	[11]
Paracetamol	10.00	87.67	g/kg	[11]
Paracetamol	15.00	97.38	g/kg	[11]
Paracetamol	20.00	108.78	g/kg	[11]
Paracetamol	25.00	121.15	g/kg	[11]
Paracetamol	30.00	135.01	g/kg	[11]
Paracetamol	30.00	110	g/L	[8]
Phenothiazine	25.00	0.00600	Mole F.	[34]
Phenylacetic acid	20.00	1382	g/kg	[83]
p-Hydroxyphenylacetic acid	10.00	314.5	g/kg	[83]
p-Hydroxyphenylacetic acid	15.00	398.9	g/kg	[83]
p-Hydroxyphenylacetic acid	20.00	496.3	g/kg	[83]
p-Hydroxyphenylacetic acid	25.00	744.5	g/kg	[83]
p-Hydroxybenzoic acid	25.00	362	g/kg	[83]
Pimozide	25.00	0.00110	Mole F.	[36]
Potassium clavulanate	0.75	0.0000494	Mole F.	[59]
Potassium clavulanate	4.00	0.0000545	Mole F.	[59]
Potassium clavulanate	7.90	0.0000605	Mole F.	[59]
Potassium clavulanate	12.20	0.0000681	Mole F.	[59]
Potassium clavulanate	16.15	0.0000747	Mole F.	[59]
Potassium clavulanate	20.10	0.0000813	Mole F.	[59]
Potassium clavulanate	24.20	0.0000917	Mole F.	[59]
Potassium clavulanate	28.20	0.0001031	Mole F.	[59]

(continued)

Solubility Data of Drugs in 2-Propanol at Various Temperatures   (continued)

Drug	T (°C)	Solubility	Solubility Unit	Reference
Potassium clavulanate	32.25	0.0001231	Mole F.	[59]
Pravastatin sodium	4.85	0.000605	Mole F.	[60]
Pravastatin sodium	9.85	0.000684	Mole F.	[60]
Pravastatin sodium	14.85	0.000753	Mole F.	[60]
Pravastatin sodium	19.85	0.000818	Mole F.	[60]
Pravastatin sodium	24.85	0.000903	Mole F.	[60]
Pravastatin sodium	29.85	0.001030	Mole F.	[60]
Pravastatin sodium	34.85	0.0011	Mole F.	[60]
Pravastatin sodium	44.85	0.00141	Mole F.	[60]
Pravastatin sodium	49.75	0.00164	Mole F.	[60]
Pravastatin sodium	54.75	0.0019	Mole F.	[60]
Ranitidine HCl (form 1)	24.20	0.052	g/100 g	[87]
Ranitidine HCl (form 1)	31.20	0.081	g/100 g	[87]
Ranitidine HCl (form 1)	32.00	0.084	g/100 g	[87]
Ranitidine HCl (form 1)	35.00	0.104	g/100 g	[87]
Ranitidine HCl (form 1)	42.80	0.172	g/100 g	[87]
Ranitidine HCl (form 2)	25.20	0.061	g/100 g	[87]
Ranitidine HCl (form 2)	30.00	0.085	g/100 g	[87]
Ranitidine HCl (form 2)	34.00	0.114	g/100 g	[87]
Ranitidine HCl (form 2)	37.90	0.149	g/100 g	[87]
Ranitidine HCl (form 2)	42.00	0.180	g/100 g	[87]
Ricobendazole	25.00	0.91	g/L	[103]
Rutin	10.00	0.001303	Mole F.	[61]
Rutin	25.00	0.001570	Mole F.	[61]
Rutin	40.00	0.002077	Mole F.	[61]
Rutin	50.00	0.002361	Mole F.	[61]
Rutin	60.00	0.002669	Mole F.	[61]
Salicylamide	25.00	0.03309	Mole F.	[37]
Stearic acid	19.20	0.0099	Mole F.	[104]
Stearic acid	22.60	0.0124	Mole F.	[104]
Stearic acid	26.50	0.0172	Mole F.	[104]
Stearic acid	29.00	0.0231	Mole F.	[104]
Stearic acid	30.60	0.0282	Mole F.	[104]
Stearic acid	35.20	0.0452	Mole F.	[104]
Stearic acid	38.10	0.0625	Mole F.	[104]
Stearic acid	40.20	0.0814	Mole F.	[104]
Stearic acid	42.30	0.1027	Mole F.	[104]
Stearic acid	44.00	0.1223	Mole F.	[104]
Stearic acid	45.60	0.1435	Mole F.	[104]
Stearic acid	47.90	0.1789	Mole F.	[104]
Stearic acid	48.80	0.2101	Mole F.	[104]
Sulfadiazine	25.00	0.0000315	Mole F.	[39]
Sulfadiazine	25.00	0.00172	mol/L	[62]
Sulfaguanidine	25.00	0.00826	mol/L	[62]
Sulfamerazine	25.00	0.00658	mol/L	[62]
Sulfamethoxazole	25.00	0.0350	mol/L	[62]
Sulfamethoxypyridazine	25.00	0.00054	Mole F.	[40]
Sulfanilamide	25.00	0.0463	mol/L	[62]
Sulfapyridine	25.00	0.00702	mol/L	[62]
Sulfatiazole	25.00	0.0225	mol/L	[62]

Solubility Data of Drugs in 2-Propanol at Various Temperatures   (continued)

Drug	T (°C)	Solubility	Solubility Unit	Reference
Temazepam	25.00	0.0016571	Mole F.	[15]
Tetracycline	15.00	0.0003980	Mole F.	[98]
Tetracycline	20.00	0.0005620	Mole F.	[98]
Tetracycline	30.00	0.0007390	Mole F.	[98]
Trimethoprim	4.02	0.0001093	Mole F.	[65]
Trimethoprim	9.67	0.0001563	Mole F.	[65]
Trimethoprim	15.13	0.0002023	Mole F.	[65]
Trimethoprim	19.98	0.0002552	Mole F.	[65]
Trimethoprim	24.42	0.0003223	Mole F.	[65]
Trimethoprim	30.89	0.0004677	Mole F.	[65]
Trimethoprim	36.21	0.0006443	Mole F.	[65]
Trimethoprim	41.98	0.00091	Mole F.	[65]
Trimethoprim	46.38	0.00117	Mole F.	[65]
Trimethoprim	50.89	0.00151	Mole F.	[65]
Trimethoprim	55.22	0.00190	Mole F.	[65]
Trimethoprim	60.49	0.00247	Mole F.	[65]
Xanthene	25.00	0.008643	Mole F.	[3]
Xylitol	20.13	0.00113	Mole F.	[66]
Xylitol	25.05	0.00154	Mole F.	[66]
Xylitol	30.08	0.00212	Mole F.	[66]
Xylitol	35.04	0.00268	Mole F.	[66]
Xylitol	40.17	0.00342	Mole F.	[66]
Xylitol	45.06	0.00443	Mole F.	[66]
Xylitol	50.01	0.00588	Mole F.	[66]
Xylitol	55.06	0.00798	Mole F.	[66]
Xylitol	60.12	0.01086	Mole F.	[66]
Xylitol	65.04	0.01456	Mole F.	[66]
Xylitol	70.08	0.01942	Mole F.	[66]

Solubility Data of Drugs in 4-Methyl-2-Pentanone at Various Temperatures

Drug	T (°C)	Solubility	Solubility Unit	Reference
4-Aminophenylacetic	20.00	1.64	g/L	[8]
Ibuprofen	10.00	315.3	g/kg	[83]
Ibuprofen	15.00	390.9	g/kg	[83]
Ibuprofen	20.00	487.7	g/kg	[83]
Ibuprofen	30.00	763.7	g/kg	[83]
Ibuprofen	35.00	953.4	g/kg	[83]
p-Aminophenylacetic acid	16.00	1.1	g/kg	[83]
p-Aminophenylacetic acid	20.00	2.2	g/kg	[83]
p-Aminophenylacetic acid	25.00	1.4	g/kg	[83]
Paracetamol	30.00	17.81	g/kg	[11]
Paracetamol	30.00	14	g/L	[8]
p-Hydroxyphenylacetic acid	10.00	148.8	g/kg	[83]
p-Hydroxyphenylacetic acid	15.00	161.2	g/kg	[83]
p-Hydroxyphenylacetic acid	20.00	179.7	g/kg	[83]

Solubility Data of Drugs in Acetic Acid at Various Temperatures

Drug	T (°C)	Solubility	Solubility Unit	Reference
11α-Hydroxy-16α,17α-epoxyprogesterone	20.25	0.005377	Mole F.	[105]
11α-Hydroxy-16α,17α-epoxyprogesterone	22.75	0.005636	Mole F.	[105]
11α-Hydroxy-16α,17α-epoxyprogesterone	25.52	0.005974	Mole F.	[105]
11α-Hydroxy-16α,17α-epoxyprogesterone	28.52	0.006510	Mole F.	[105]
11α-Hydroxy-16α,17α-epoxyprogesterone	31.45	0.006989	Mole F.	[105]
11α-Hydroxy-16α,17α-epoxyprogesterone	34.75	0.007670	Mole F.	[105]
11α-Hydroxy-16α,17α-epoxyprogesterone	38.65	0.008420	Mole F.	[105]
11α-Hydroxy-16α,17α-epoxyprogesterone	41.70	0.009242	Mole F.	[105]
11α-Hydroxy-16α,17α-epoxyprogesterone	45.35	0.010340	Mole F.	[105]
11α-Hydroxy-16α,17α-epoxyprogesterone	49.75	0.011550	Mole F.	[105]
16α,17α-Epoxyprogesterone	20.45	0.01235	Mole F.	[17]
16α,17α-Epoxyprogesterone	22.60	0.01357	Mole F.	[17]
16α,17α-Epoxyprogesterone	25.10	0.01479	Mole F.	[17]
16α,17α-Epoxyprogesterone	28.15	0.01646	Mole F.	[17]
16α,17α-Epoxyprogesterone	30.80	0.01817	Mole F.	[17]
16α,17α-Epoxyprogesterone	33.70	0.02121	Mole F.	[17]
16α,17α-Epoxyprogesterone	37.60	0.02426	Mole F.	[17]
16α,17α-Epoxyprogesterone	40.40	0.02708	Mole F.	[17]
16α,17α-Epoxyprogesterone	43.50	0.03022	Mole F.	[17]
16α,17α-Epoxyprogesterone	46.10	0.03333	Mole F.	[17]
16α,17α-Epoxyprogesterone	49.00	0.03759	Mole F.	[17]
16α,17α-Epoxyprogesterone	52.90	0.04245	Mole F.	[17]
Anhydrous citric acid	25.00	0.02910	Mole F.	[23]
Benzoic acid	25.00	0.1675	Mole F.	[24]
Diclofenac	25.00	0.0056506	Mole F.	[25]
Dioxopromethazine HCl	5.02	0.0007593	Mole F.	[106]
Dioxopromethazine HCl	10.07	0.0009251	Mole F.	[106]
Dioxopromethazine HCl	15.04	0.001122	Mole F.	[106]
Dioxopromethazine HCl	20.11	0.001366	Mole F.	[106]
Dioxopromethazine HCl	25.02	0.001558	Mole F.	[106]
Dioxopromethazine HCl	30.04	0.001852	Mole F.	[106]
Dioxopromethazine HCl	35.06	0.002250	Mole F.	[106]
Dioxopromethazine HCl	40.06	0.002840	Mole F.	[106]
Dioxopromethazine HCl	45.03	0.003397	Mole F.	[106]
Hydroquinone	16.30	0.03210	Mole F.	[101]
Hydroquinone	21.50	0.03645	Mole F.	[101]
Hydroquinone	27.00	0.04193	Mole F.	[101]
Hydroquinone	33.20	0.04887	Mole F.	[101]
Hydroquinone	40.15	0.05791	Mole F.	[101]
Hydroquinone	45.00	0.06591	Mole F.	[101]
Hydroquinone	50.60	0.07522	Mole F.	[101]
Hydroquinone	56.50	0.08663	Mole F.	[101]
Hydroquinone	61.55	0.09831	Mole F.	[101]
Hydroquinone	68.10	0.11520	Mole F.	[101]
Ibuprofen	25.00	0.1136536	Mole F.	[28]
Lactose	25.00	0.0000578	Mole F.	[30]
Mannitol	25.00	0.0007205	Mole F.	[30]
Methyl p-hydroxybenzoate	25.00	0.0532	Mole F.	[35]

Solubility Data of Drugs in Acetic Acid at Various Temperatures   (continued)

Drug	T (°C)	Solubility	Solubility Unit	Reference
Naproxen	25.00	0.0152985	Mole F.	[31]
Niflumic acid	25.00	0.0078834	Mole F.	[33]
p-Aminobenzoic acid	25.00	0.0770732	Mole F.	[25]
Paracetamol	30.00	82.72	g/kg	[11]
Paracetamol	30.00	83.7	g/L	[8]
p-Hydroxybenzoic acid	25.00	0.0444	Mole F.	[35]
Piroxicam	25.00	0.0008805	Mole F.	[33]
Saccharose	25.00	0.000000039	Mole F.	[30]
Salicylamide	10.00	0.05854	Mole F.	[107]
Salicylamide	15.00	0.06525	Mole F.	[107]
Salicylamide	20.00	0.07295	Mole F.	[107]
Salicylamide	25.00	0.08239	Mole F.	[107]
Salicylamide	30.00	0.09278	Mole F.	[107]
Salicylamide	35.00	0.10359	Mole F.	[107]
Salicylamide	40.00	0.11715	Mole F.	[107]
Salicylamide	45.00	0.13138	Mole F.	[107]
Salicylamide	50.00	0.14516	Mole F.	[107]
Salicylic acid	10.00	0.03777	Mole F.	[108]
Salicylic acid	15.00	0.04293	Mole F.	[108]
Salicylic acid	20.00	0.04838	Mole F.	[108]
Salicylic acid	25.00	0.0597850	Mole F.	[25]
Salicylic acid	25.00	0.05493	Mole F.	[108]
Salicylic acid	30.00	0.06254	Mole F.	[108]
Salicylic acid	35.00	0.07111	Mole F.	[108]
Salicylic acid	40.00	0.08054	Mole F.	[108]
Salicylic acid	45.00	0.09182	Mole F.	[108]
Salicylic acid	50.00	0.10424	Mole F.	[108]
Sodium ibuprofen	25.00	0.0333766	Mole F.	[28]
Sodium benzoate	25.00	0.1517456	Mole F.	[28]
Sodium diclofenac	25.00	0.0044476	Mole F.	[31]
Sodium diclofenac	25.00	0.0044493	Mole F.	[25]
Sodium p-aminobenzoate	25.00	0.0903558	Mole F.	[25]
Sodium salicylate	25.00	0.0083458	Mole F.	[25]

Solubility Data of Drugs in Acetone at Various Temperatures

Drug	T (°C)	Solubility	Solubility Unit	Reference
11α-Hydroxy-16α,17α-epoxyprogesterone	10.22	0.001115	Mole F.	[105]
11α-Hydroxy-16α,17α-epoxyprogesterone	15.10	0.001254	Mole F.	[105]
11α-Hydroxy-16α,17α-epoxyprogesterone	18.90	0.001356	Mole F.	[105]
11α-Hydroxy-16α,17α-epoxyprogesterone	19.85	0.001421	Mole F.	[105]
11α-Hydroxy-16α,17α-epoxyprogesterone	25.20	0.001628	Mole F.	[105]
11α-Hydroxy-16α,17α-epoxyprogesterone	30.45	0.001898	Mole F.	[105]
11α-Hydroxy-16α,17α-epoxyprogesterone	35.35	0.002163	Mole F.	[105]
11α-Hydroxy-16α,17α-epoxyprogesterone	40.10	0.002463	Mole F.	[105]
11α-Hydroxy-16α,17α-epoxyprogesterone	45.60	0.002770	Mole F.	[105]

(continued)

Solubility Data of Drugs in Acetone at Various Temperatures   (continued)

Drug	T (°C)	Solubility	Solubility Unit	Reference
11α-Hydroxy-16α,17α-epoxyprogesterone	50.50	0.003108	Mole F.	[105]
16α,17α-Epoxyprogesterone	12.20	0.00895	Mole F.	[17]
16α,17α-Epoxyprogesterone	15.35	0.00985	Mole F.	[17]
16α,17α-Epoxyprogesterone	20.20	0.01150	Mole F.	[17]
16α,17α-Epoxyprogesterone	25.80	0.01372	Mole F.	[17]
16α,17α-Epoxyprogesterone	31.70	0.01624	Mole F.	[17]
16α,17α-Epoxyprogesterone	36.90	0.01838	Mole F.	[17]
16α,17α-Epoxyprogesterone	41.00	0.02035	Mole F.	[17]
16α,17α-Epoxyprogesterone	44.45	0.02225	Mole F.	[17]
16α,17α-Epoxyprogesterone	49.20	0.02507	Mole F.	[17]
16α,17α-Epoxyprogesterone	54.30	0.02853	Mole F.	[17]
2-(4-Ethylbenzoyl)benzoic acid	7.10	0.0015	Mole F.	[18]
2-(4-Ethylbenzoyl)benzoic acid	11.70	0.0022	Mole F.	[18]
2-(4-Ethylbenzoyl)benzoic acid	17.72	0.0032	Mole F.	[18]
2-(4-Ethylbenzoyl)benzoic acid	26.68	0.0054	Mole F.	[18]
2-(4-Ethylbenzoyl)benzoic acid	37.24	0.0087	Mole F.	[18]
2-(4-Ethylbenzoyl)benzoic acid	41.74	0.0107	Mole F.	[18]
2-(4-Ethylbenzoyl)benzoic acid	46.45	0.0132	Mole F.	[18]
2-(4-Ethylbenzoyl)benzoic acid	51.01	0.0156	Mole F.	[18]
2-Hydroxybenzoic acid	25.00	0.1817	Mole F.	[19]
3,5-Di-*tert*-butyl-4-hydroxytoluene	N/A	0.54	Mole F.	[4]
4-Aminophenylacetic acid	20.00	5.85	g/L	[8]
Acephate	26.30	0.06796	Mole F.	[5]
Acephate	29.45	0.08109	Mole F.	[5]
Acephate	32.05	0.09226	Mole F.	[5]
Acephate	33.25	0.09861	Mole F.	[5]
Acephate	36.65	0.11931	Mole F.	[5]
Acephate	38.25	0.13105	Mole F.	[5]
Acephate	40.60	0.14924	Mole F.	[5]
Acephate	42.65	0.16717	Mole F.	[5]
Acetic acid	−29.00	0.460	Mole F.	[109]
Acetic acid	−24.00	0.507	Mole F.	[109]
Acetic acid	−18.60	0.575	Mole F.	[109]
Acetic acid	−13.40	0.640	Mole F.	[109]
Acetic acid	−7.90	0.681	Mole F.	[109]
Acetic acid	−2.40	0.763	Mole F.	[109]
Acetic acid	3.70	0.821	Mole F.	[109]
Acetic acid	10.50	0.918	Mole F.	[109]
Acetylsalicylic acid	8.75	0.061	Mole F.	[94]
Acetylsalicylic acid	17.45	0.075	Mole F.	[94]
Acetylsalicylic acid	24.75	0.088	Mole F.	[94]
Acetylsalicylic acid	25.00	0.0828	Mole F.	[21]
Acetylsalicylic acid	31.25	0.101	Mole F.	[94]
Acetylsalicylic acid	37.45	0.114	Mole F.	[94]
Acetylsalicylic acid	42.15	0.127	Mole F.	[94]
Acetylsalicylic acid	46.65	0.139	Mole F.	[94]
Acetylsalicylic acid	50.15	0.151	Mole F.	[94]
Acetylsalicylic acid	53.15	0.162	Mole F.	[94]
Benzoic acid	25.00	0.1857	Mole F.	[24]

Solubility Data of Drugs in Acetone at Various Temperatures   (continued)

Drug	T (°C)	Solubility	Solubility Unit	Reference
β-Carotene	Room	200	mg/L	[6]
β-Sitosteryl maleate	16.60	0.0004899	Mole F.	[110]
β-Sitosteryl maleate	31.00	0.0008994	Mole F.	[110]
β-Sitosteryl maleate	37.50	0.001275	Mole F.	[110]
β-Sitosteryl maleate	40.20	0.001602	Mole F.	[110]
β-Sitosteryl maleate	44.00	0.001978	Mole F.	[110]
β-Sitosteryl maleate	47.80	0.002343	Mole F.	[110]
β-Sitosteryl maleate	49.00	0.002519	Mole F.	[110]
β-Sitosteryl maleate	52.00	0.003114	Mole F.	[110]
Betulin	10.05	0.000556	Mole F.	[7]
Betulin	15.05	0.000866	Mole F.	[7]
Betulin	25.05	0.001578	Mole F.	[7]
Betulin	35.05	0.002273	Mole F.	[7]
Carbamazepine	25.00	12	g/L	[8]
Cefotaxim sodium	5.00	0.881	g/100 g	[111]
Cefotaxim sodium	10.00	0.887	g/100 g	[111]
Cefotaxim sodium	15.00	0.905	g/100 g	[111]
Cefotaxim sodium	20.00	0.934	g/100 g	[111]
Cefotaxim sodium	30.00	0.987	g/100 g	[111]
Cefotaxim sodium	40.00	1.015	g/100 g	[111]
Ceftriaxone disodium	4.80	0.0000018	Mole F.	[112]
Ceftriaxone disodium	9.35	0.0000021	Mole F.	[112]
Ceftriaxone disodium	12.55	0.0000024	Mole F.	[112]
Ceftriaxone disodium	16.35	0.0000025	Mole F.	[112]
Ceftriaxone disodium	21.15	0.0000030	Mole F.	[112]
Ceftriaxone disodium	25.60	0.0000035	Mole F.	[112]
Ceftriaxone disodium	32.10	0.0000040	Mole F.	[112]
Ceftriaxone disodium	36.20	0.0000046	Mole F.	[112]
Ceftriaxone disodium	40.00	0.0000051	Mole F.	[112]
Ceftriaxone disodium	45.30	0.0000056	Mole F.	[112]
Cholestrole	11.00	1.219	g/100 g	[97]
Cholestrole	30.70	3.752	g/100 g	[97]
Cholestrole	33.90	4.575	g/100 g	[97]
Cholestrole	38.00	5.374	g/100 g	[97]
Ciprofloxacin	20.00	0.0000237	Mole F.	[98]
Ciprofloxacin	30.00	0.0000368	Mole F.	[98]
Ciprofloxacin	40.00	0.0000758	Mole F.	[98]
Ciprofloxacin	50.00	0.0001253	Mole F.	[98]
D(−)-p-Hydroxyphenylglycine dane salt	20.10	0.0004045	Mole F.	[49]
D(−)-p-Hydroxyphenylglycine dane salt	25.11	0.0004373	Mole F.	[49]
D(−)-p-Hydroxyphenylglycine dane salt	30.10	0.0004780	Mole F.	[49]
D(−)-p-Hydroxyphenylglycine dane salt	35.04	0.0005195	Mole F.	[49]
D(−)-p-Hydroxyphenylglycine dane salt	39.94	0.0005596	Mole F.	[49]
D(−)-p-Hydroxyphenylglycine dane salt	45.11	0.0006292	Mole F.	[49]
D(−)-p-Hydroxyphenylglycine dane salt	49.96	0.0006814	Mole F.	[49]
D(−)-p-Hydroxyphenylglycine dane salt	55.03	0.0007531	Mole F.	[49]
D(−)-p-Hydroxyphenylglycine dane salt	60.26	0.0008340	Mole F.	[49]
D(−)-p-Hydroxyphenylglycine dane salt	65.16	0.0009163	Mole F.	[49]

(continued)

Solubility Data of Drugs in Acetone at Various Temperatures   (continued)

Drug	*T* (°C)	Solubility	Solubility Unit	Reference
D(−)-*p*-Hydroxyphenylglycine dane salt	70.20	0.0010400	Mole F.	[49]
Diclofenac	25.00	0.0301974	Mole F.	[25]
Dioxopromethazine HCl	5.05	0.0001029	Mole F.	[106]
Dioxopromethazine HCl	10.01	0.0001236	Mole F.	[106]
Dioxopromethazine HCl	15.01	0.0001502	Mole F.	[106]
Dioxopromethazine HCl	20.02	0.0001770	Mole F.	[106]
Dioxopromethazine HCl	25.03	0.0001991	Mole F.	[106]
Dioxopromethazine HCl	30.02	0.0002163	Mole F.	[106]
Dioxopromethazine HCl	35.06	0.0002366	Mole F.	[106]
Dioxopromethazine HCl	40.05	0.0002585	Mole F.	[106]
Dioxopromethazine HCl	45.03	0.0002783	Mole F.	[106]
Dioxopromethazine HCl	50.02	0.0002984	Mole F.	[106]
Enrofloxacin sodium	20.00	0.00133	Mole F.	[1]
Enrofloxacin sodium	22.00	0.00147	Mole F.	[1]
Enrofloxacin sodium	25.00	0.00171	Mole F.	[1]
Enrofloxacin sodium	27.00	0.00198	Mole F.	[1]
Enrofloxacin sodium	30.00	0.00252	Mole F.	[1]
Enrofloxacin sodium	32.00	0.00273	Mole F.	[1]
Enrofloxacin sodium	35.00	0.00313	Mole F.	[1]
Enrofloxacin sodium	37.00	0.00347	Mole F.	[1]
Erythritol	13.20	0.0003098	Mass fraction	[113]
Erythritol	18.90	0.0003890	Mass fraction	[113]
Erythritol	23.40	0.0005487	Mass fraction	[113]
Erythritol	28.30	0.0008385	Mass fraction	[113]
Erythritol	31.90	0.0009286	Mass fraction	[113]
Erythritol	37.60	0.0013010	Mass fraction	[113]
Erythritol	41.60	0.0015262	Mass fraction	[113]
Erythritol	45.00	0.0016889	Mass fraction	[113]
Erythritol	48.30	0.0020689	Mass fraction	[113]
Erythritol	51.30	0.0023393	Mass fraction	[113]
Erythritol	55.20	0.0027474	Mass fraction	[113]
Erythromycin A dihydrate	20.05	0.009140	Mole F.	[100]
Erythromycin A dihydrate	24.85	0.01310	Mole F.	[100]
Erythromycin A dihydrate	30.00	0.02170	Mole F.	[100]
Erythromycin A dihydrate	34.85	0.02930	Mole F.	[100]
Erythromycin A dihydrate	39.95	0.04330	Mole F.	[100]
Erythromycin A dihydrate	44.85	0.05420	Mole F.	[100]
Erythromycin A dihydrate	49.85	0.08040	Mole F.	[100]
Flubiprofen	25.00	0.124	Mole F.	[26]
Haloperidol	25.00	0.00309	Mole F.	[27]
Hesperetin	15.05	0.285	mol/L	[50]
Hesperetin	20.05	0.312	mol/L	[50]
Hesperetin	25.05	0.340	mol/L	[50]
Hesperetin	30.05	0.366	mol/L	[50]
Hesperetin	35.05	0.383	mol/L	[50]
Hesperetin	40.05	0.385	mol/L	[50]
Hesperetin	45.05	0.394	mol/L	[50]
Hesperetin	50.05	0.429	mol/L	[50]

Solubility Data of Drugs in Acetone at Various Temperatures   (continued)

Drug	T (°C)	Solubility	Solubility Unit	Reference
Ibuprofen	10.00	587.6	g/kg	[83]
Ibuprofen	15.00	713.9	g/kg	[83]
Ibuprofen	20.00	883.3	g/kg	[83]
Ibuprofen	25.00	0.3508488	Mole F.	[28]
Ibuprofen	30.00	1357	g/kg	[83]
Ibuprofen	35.00	1679	g/kg	[83]
Imidacloprid	21.49	0.01198	Mole F.	[2]
Imidacloprid	26.90	0.01414	Mole F.	[2]
Imidacloprid	28.61	0.01484	Mole F.	[2]
Imidacloprid	31.90	0.01617	Mole F.	[2]
Imidacloprid	34.62	0.01745	Mole F.	[2]
Imidacloprid	37.02	0.01873	Mole F.	[2]
Imidacloprid	44.10	0.02323	Mole F.	[2]
Imidacloprid	45.99	0.02478	Mole F.	[2]
Imidacloprid	48.44	0.02645	Mole F.	[2]
Imidacloprid	54.74	0.03253	Mole F.	[2]
Imidacloprid	57.05	0.03526	Mole F.	[2]
Irbesartan (form A)	5.20	0.0002837	Mole F.	[29]
Irbesartan (form A)	10.00	0.0003640	Mole F.	[29]
Irbesartan (form A)	15.20	0.0004440	Mole F.	[29]
Irbesartan (form A)	20.45	0.000530	Mole F.	[29]
Irbesartan (form A)	25.10	0.000678	Mole F.	[29]
Irbesartan (form A)	29.80	0.000861	Mole F.	[29]
Irbesartan (form A)	35.25	0.001047	Mole F.	[29]
Irbesartan (form A)	39.60	0.001291	Mole F.	[29]
Irbesartan (form A)	45.40	0.001579	Mole F.	[29]
Irbesartan (form A)	49.95	0.001921	Mole F.	[29]
Isoniazid	27.85	0.00946	Mole F.	[114]
Isoniazid	29.85	0.01042	Mole F.	[114]
Isoniazid	34.85	0.01362	Mole F.	[114]
Isoniazid	37.85	0.01603	Mole F.	[114]
Isoniazid	39.85	0.01873	Mole F.	[114]
Isoquercitrin	50.00	0.00045	mol/L	[115]
L-(+)-Ascorbic acid	19.85	0.00026	Mole F.	[102]
L-(+)-Ascorbic acid	24.85	0.00032	Mole F.	[102]
L-(+)-Ascorbic acid	29.85	0.00037	Mole F.	[102]
L-(+)-Ascorbic acid	34.85	0.00043	Mole F.	[102]
L-(+)-Ascorbic acid	39.85	0.00051	Mole F.	[102]
L-(+)-Ascorbic acid	44.85	0.00058	Mole F.	[102]
L-(+)-Ascorbic acid	49.85	0.00064	Mole F.	[102]
Lactose	25.00	0.0000194	Mole F.	[30]
Lamivudine (form 2)	5.00	0.57	g/L	[52]
Lamivudine (form 2)	15.00	0.74	g/L	[52]
Lamivudine (form 2)	25.00	0.94	g/L	[52]
Lamivudine (form 2)	35.00	1.52	g/L	[52]
Lamivudine (form 2)	45.00	2.58	g/L	[52]
Lovastatin	5.50	0.006375	Mole F.	[116]
Lovastatin	10.05	0.007416	Mole F.	[116]
Lovastatin	11.95	0.00831	Mole F.	[89]
Lovastatin	15.05	0.008692	Mole F.	[116]

(continued)

Solubility Data of Drugs in Acetone at Various Temperatures   (continued)

Drug	T (°C)	Solubility	Solubility Unit	Reference
Lovastatin	15.15	0.00936	Mole F.	[89]
Lovastatin	18.05	0.01043	Mole F.	[89]
Lovastatin	20.10	0.010430	Mole F.	[116]
Lovastatin	21.45	0.01184	Mole F.	[89]
Lovastatin	24.05	0.01304	Mole F.	[89]
Lovastatin	25.60	0.01233	Mole F.	[116]
Lovastatin	27.15	0.01464	Mole F.	[89]
Lovastatin	29.95	0.01626	Mole F.	[89]
Lovastatin	30.80	0.01573	Mole F.	[116]
Lovastatin	33.15	0.01832	Mole F.	[89]
Lovastatin	35.10	0.01832	Mole F.	[116]
Lovastatin	36.25	0.02056	Mole F.	[89]
Lovastatin	39.05	0.02283	Mole F.	[89]
Lovastatin	40.10	0.02259	Mole F.	[116]
Lovastatin	45.00	0.02716	Mole F.	[116]
Lovastatin	49.50	0.03206	Mole F.	[116]
Lutein	Room	800	mg/L	[6]
Luteolin	0.00	0.00053	Mole F.	[54]
Luteolin	10.00	0.00091	Mole F.	[54]
Luteolin	25.00	0.00156	Mole F.	[54]
Luteolin	40.00	0.00234	Mole F.	[54]
Luteolin	60.00	0.00316	Mole F.	[54]
Methyl elaidate	−14.90	4.82	% Weight	[10]
Methyl elaidate	−6.00	21.67	% Weight	[10]
Methyl elaidate	−0.70	56.56	% Weight	[10]
Methyl heptadecanoate	4.10	4.95	% Weight	[10]
Methyl heptadecanoate	12.20	26.04	% Weight	[10]
Methyl heptadecanoate	17.70	55.8	% Weight	[10]
Methyl linoleate	−59.50	5.15	% Weight	[10]
Methyl linoleate	−48.50	36.44	% Weight	[10]
Methyl oleate	−40.70	4.94	% Weight	[10]
Methyl oleate	−31.40	36.22	% Weight	[10]
Methyl palmitate	0.70	4.53	% Weight	[10]
Methyl palmitate	10.90	20.11	% Weight	[10]
Methyl palmitate	17.50	56.12	% Weight	[10]
Methyl palmitoleate	−57.40	4.38	% Weight	[10]
Methyl palmitoleate	−47.60	19.43	% Weight	[10]
Methyl palmitoleate	−43.80	54.09	% Weight	[10]
Methyl petroselaidate	−7.60	5.04	% Weight	[10]
Methyl petroselaidate	1.60	21.7	% Weight	[10]
Methyl petroselaidate	7.10	56.68	% Weight	[10]
Methyl petroselinate	−23.90	4.04	% Weight	[10]
Methyl petroselinate	−15.60	21.99	% Weight	[10]
Methyl petroselinate	−11.00	56.91	% Weight	[10]
Methyl stearate	−7.70	0.06	% Weight	[10]
Methyl stearate	6.20	0.40	% Weight	[10]
Methyl stearate	18.60	2.77	% Weight	[10]
Methyl stearate	24.50	10.83	% Weight	[10]
Methyl stearate	31.90	47.37	% Weight	[10]
Methyl stearate	37.40	85.82	% Weight	[10]

Solubility Data of Drugs in Acetone at Various Temperatures   (continued)

Drug	T (°C)	Solubility	Solubility Unit	Reference
Methyl-3-(3,5-di-*tert*-butyl-4-hydroxyphenyl)-propionate	N/A	0.35	Mole F.	[4]
Naproxen	25.00	0.0692176	Mole F.	[31]
Niflumic acid	25.00	0.0009702	Mole F.	[33]
Octadecyl-3-(3,5-di-*tert*-butyl-4-hydroxyphenyl)-propionate	N/A	0.049	Mole F.	[4]
Oleanolic acid	15.15	0.0074	mol/L	[56]
Oleanolic acid	20.15	0.0112	mol/L	[56]
Oleanolic acid	25.15	0.0137	mol/L	[56]
Oleanolic acid	30.15	0.0155	mol/L	[56]
Oleanolic acid	35.15	0.0175	mol/L	[56]
Oleanolic acid	40.15	0.0196	mol/L	[56]
Oleanolic acid	45.15	0.0220	mol/L	[56]
Oleanolic acid	50.15	0.0254	mol/L	[56]
Oleanolic acid	55.15	0.0287	mol/L	[56]
p-Aminobenzoic acid	25.00	0.0526547	Mole F.	[25]
p-Aminophenylacetic acid	16.00	6.79	g/kg	[83]
p-Aminophenylacetic acid	20.00	8.8	g/kg	[83]
p-Aminophenylacetic acid	25.00	2.5	g/kg	[83]
p-Aminophenylacetic acid	30.00	12.6	g/kg	[83]
Paracetamol	−5.00	50.39	g/kg	[11]
Paracetamol	0.00	55.61	g/kg	[11]
Paracetamol	5.00	62.32	g/kg	[11]
Paracetamol	10.00	69.63	g/kg	[11]
Paracetamol	15.00	78.48	g/kg	[11]
Paracetamol	20.00	88.09	g/kg	[11]
Paracetamol	25.00	99.38	g/kg	[11]
Paracetamol	30.00	111.65	g/kg	[11]
Paracetamol	30.00	81.1	g/L	[8]
Pentaerythrittol tetrakis(3-(3,5-di-*tert*-butyl-4-hydroxyphenyl)-propionate)	N/A	0.073	Mole F.	[4]
Phenylacetic acid	20.00	1720	g/kg	[83]
p-Hydroxyphenylacetic acid	10.00	475.4	g/kg	[83]
p-Hydroxyphenylacetic acid	15.00	608.6	g/kg	[83]
p-Hydroxyphenylacetic acid	20.00	693	g/kg	[83]
p-Hydroxyphenylacetic acid	25.00	200.3	g/kg	[83]
p-Hydroxybenzoic acid	25.00	0.1185	Mole F.	[35]
p-Hydroxybenzoic acid	25.00	322.3	g/kg	[83]
Piroxicam	25.00	0.0028144	Mole F.	[33]
Puerarin	15.05	0.0325	Mole F.	[117]
Puerarin	20.05	0.0300	Mole F.	[117]
Puerarin	25.05	0.0280	Mole F.	[117]
Puerarin	30.05	0.0270	Mole F.	[117]
Puerarin	35.05	0.0250	Mole F.	[117]
Puerarin	40.05	0.0232	Mole F.	[117]
Puerarin	45.05	0.0211	Mole F.	[117]
Puerarin	50.05	0.0201	Mole F.	[117]
Puerarin	55.05	0.0188	Mole F.	[117]
Quercetin	50.00	0.08008	mol/L	[115]

(continued)

Solubility Data of Drugs in Acetone at Various Temperatures   (continued)

Drug	T (°C)	Solubility	Solubility Unit	Reference
Riboflavin	30.00	0.0356	g/L	[118]
Riboflavin	30.00	0.03	g/L	[8]
Rifapentine	5.00	0.000523	Mole F.	[13]
Rifapentine	10.00	0.000530	Mole F.	[13]
Rifapentine	15.00	0.000536	Mole F.	[13]
Rifapentine	20.00	0.000549	Mole F.	[13]
Rifapentine	24.00	0.000563	Mole F.	[13]
Rifapentine	30.00	0.000576	Mole F.	[13]
Rifapentine	35.00	0.000609	Mole F.	[13]
Rifapentine	40.00	0.000635	Mole F.	[13]
Rifapentine	45.00	0.000668	Mole F.	[13]
Rifapentine	50.00	0.000701	Mole F.	[13]
Rutin	10.00	0.000169	Mole F.	[61]
Rutin	25.00	0.000271	Mole F.	[61]
Rutin	40.00	0.000541	Mole F.	[61]
Rutin	50.00	0.000685	Mole F.	[61]
Rutin	50.00	0.01350	mol/L	[115]
Salicylamide	10.00	0.09930	Mole F.	[107]
Salicylamide	15.00	0.10785	Mole F.	[107]
Salicylamide	20.00	0.11783	Mole F.	[107]
Salicylamide	25.00	0.12943	Mole F.	[107]
Salicylamide	30.00	0.14062	Mole F.	[107]
Salicylamide	35.00	0.15352	Mole F.	[107]
Salicylamide	40.00	0.16819	Mole F.	[107]
Salicylamide	45.00	0.18161	Mole F.	[107]
Salicylamide	50.00	0.19842	Mole F.	[107]
Salicylic acid	10.00	0.14625	Mole F.	[108]
Salicylic acid	15.00	0.15685	Mole F.	[108]
Salicylic acid	20.00	0.16804	Mole F.	[108]
Salicylic acid	25.00	0.1319938	Mole F.	[25]
Salicylic acid	25.00	0.17924	Mole F.	[108]
Salicylic acid	30.00	0.19185	Mole F.	[108]
Salicylic acid	35.00	0.20238	Mole F.	[108]
Salicylic acid	40.00	0.21504	Mole F.	[108]
Salicylic acid	45.00	0.22801	Mole F.	[108]
Salicylic acid	50.00	0.24128	Mole F.	[108]
Sitosterol	30.70	4.583	g/100 g	[97]
Sitosterol	33.90	4.753	g/100 g	[97]
Sitosterol	38.00	5.267	g/100 g	[97]
Sitosterol	42.10	6.126	g/100 g	[97]
Sitosterol	45.30	7.443	g/100 g	[97]
Sodium ibuprofen	25.00	0.0003771	Mole F.	[28]
Sodium benzoate	25.00	0.0000207	Mole F.	[28]
Sodium diclofenac	20.00	4.49	mg/g	[119]
Sodium diclofenac	25.00	0.0000066	Mole F.	[31]
Sodium diclofenac	25.00	4.45	mg/g	[119]
Sodium diclofenac	30.00	4.67	mg/g	[119]
Sodium diclofenac	35.00	4.88	mg/g	[119]
Sodium diclofenac	40.00	5.25	mg/g	[119]

Solubility Data of Drugs in Acetone at Various Temperatures   (continued)

Drug	T (°C)	Solubility	Solubility Unit	Reference
Sodium diclofenac	25.00	0.0000066	Mole F.	[25]
Sodium p-aminobenzoate	25.00	0.0000210	Mole F.	[25]
Sodium salicylate	25.00	0.0123019	Mole F.	[25]
Stearic acid	18.80	0.0042	Mole F.	[104]
Stearic acid	21.60	0.0055	Mole F.	[104]
Stearic acid	24.90	0.0077	Mole F.	[104]
Stearic acid	28.00	0.0103	Mole F.	[104]
Stearic acid	31.70	0.0152	Mole F.	[104]
Stearic acid	34.50	0.0204	Mole F.	[104]
Stearic acid	37.90	0.0313	Mole F.	[104]
Stearic acid	40.80	0.0456	Mole F.	[104]
Stearic acid	43.30	0.0638	Mole F.	[104]
Stearic acid	45.90	0.0845	Mole F.	[104]
Stearic acid	47.20	0.1059	Mole F.	[104]
Stearic acid	48.40	0.1260	Mole F.	[104]
Stearic acid	49.40	0.1475	Mole F.	[104]
Stigmasteryl maleate	20.00	0.0003826	Mole F.	[110]
Stigmasteryl maleate	23.50	0.0004312	Mole F.	[110]
Stigmasteryl maleate	29.50	0.0002801	Mole F.	[110]
Stigmasteryl maleate	30.50	0.0005346	Mole F.	[110]
Stigmasteryl maleate	39.50	0.0004382	Mole F.	[110]
Stigmasteryl maleate	44.00	0.0007852	Mole F.	[110]
Stigmasteryl maleate	52.50	0.0010210	Mole F.	[110]
Sulfadiazine	25.00	0.0010421	Mole F.	[39]
Sulfaguanidine	25.00	0.0738	mol/L	[62]
Sulfamerazine (polymorph 1)	30.00	14.17	g/L	[120]
Sulfamerazine (polymorph 2)	30.00	12.45	g/L	[120]
Sulfamethoxypyridazine	25.00	0.00918	Mole F.	[40]
Sulfanilamide	25.00	1.42	mol/L	[62]
Sulfapyridine	25.00	0.0599	mol/L	[62]
Sulfatiazole	25.00	0.0724	mol/L	[62]
Tanshinone IIA	10.05	0.00960	mol/L	[121]
Tanshinone IIA	20.15	0.01211	mol/L	[121]
Tanshinone IIA	29.95	0.01372	mol/L	[121]
Tanshinone IIA	35.05	0.01553	mol/L	[121]
Tanshinone IIA	44.85	0.01965	mol/L	[121]
Tanshinone IIA	50.15	0.02392	mol/L	[121]
Temazepam	24.00	0.0587	Volume F.	[64]
Temazepam	25.00	0.0206991	Mole F.	[15]
Tetracycline	15.00	0.00499	Mole F.	[98]
Tetracycline	20.00	0.00520	Mole F.	[98]
Tetracycline	30.00	0.00779	Mole F.	[98]
Trimethoprim	3.79	0.0004556	Mole F.	[65]
Trimethoprim	8.97	0.0005405	Mole F.	[65]
Trimethoprim	15.47	0.0006638	Mole F.	[65]
Trimethoprim	19.79	0.0007599	Mole F.	[65]
Trimethoprim	24.81	0.0008896	Mole F.	[65]
Trimethoprim	31.02	0.001082	Mole F.	[65]

(continued)

Solubility Data of Drugs in Acetone at Various Temperatures   (continued)

Drug	T (°C)	Solubility	Solubility Unit	Reference
Trimethoprim	36.19	0.001275	Mole F.	[65]
Trimethoprim	41.19	0.001493	Mole F.	[65]
Trimethoprim	45.92	0.001731	Mole F.	[65]
Trimethoprim	50.36	0.001987	Mole F.	[65]
Xylitol	20.14	0.0002026	Mole F.	[66]
Xylitol	25.03	0.0002401	Mole F.	[66]
Xylitol	30.11	0.0003040	Mole F.	[66]
Xylitol	35.07	0.0004011	Mole F.	[66]
Xylitol	40.05	0.0005433	Mole F.	[66]
Xylitol	45.08	0.0007428	Mole F.	[66]
Xylitol	50.05	0.0010053	Mole F.	[66]

Solubility Data of Drugs Acetonitrile at Various Temperatures

Drug	T (°C)	Solubility	Solubility Unit	Reference
2,3-Dichlorophenol	25.00	2.86	Mole/100 g	[122]
2,3-Dimethylphenol	25.00	1.49	Mole/100 g	[122]
2,4,5-Trichlorophenol	25.00	2.56	Mole/100 g	[122]
2,4,6-Trichlorophenol	25.00	2.53	Mole/100 g	[122]
2,5-Dimethylphenol	25.00	1.23	Mole/100 g	[122]
2,6-Dichlorophenol	25.00	2.18	Mole/100 g	[122]
2-Iodophenol	25.00	7.44	Mole/100 g	[122]
2-Nitrophenol	25.00	3.39	Mole/100 g	[122]
3,4-Dichlorophenol	25.00	4.26	Mole/100 g	[122]
3,5-Dimethylphenol	25.00	2.46	Mole/100 g	[122]
3-Chlorophenol	25.00	4.63	Mole/100 g	[122]
3-Cyanophenol	25.00	1.51	Mole/100 g	[122]
3-Nitroaniline	25.00	0.295	Mole/100 g	[122]
3-Nitrophenol	25.00	1.63	Mole/100 g	[122]
4-Aminoacetophenone	25.00	0.0534	Mole/100 g	[122]
4-Bromophenol	25.00	2.74	Mole/100 g	[122]
4-Fluorophenol	25.00	3.82	Mole/100 g	[122]
4-Isopropylphenol	25.00	1.52	Mole/100 g	[122]
4-Methoxyphenol	25.00	3.1931248	Mole/100 g	[122]
4-Nitroaniline	25.00	0.176	Mole/100 g	[122]
4-Nitrophenol	25.00	1.49	Mole/100 g	[122]
4-Nitrotoluene	25.00	2.14	Mole/100 g	[122]
4-Phenylphenol	25.00	0.130	Mole/100 g	[122]
4-tert-Butylphenol	25.00	1.01	Mole/100 g	[122]
Acetylsalicylic acid	25.00	0.0185	Mole F.	[21]
Benzamide	25.00	0.0665	Mole/100 g	[122]
Benzoic acid	25.00	0.0539	Mole F.	[21]
β-Carotene	Room	10	mg/L	[6]
Butylparaben	25.00	0.571	Mole/100 g	[122]
Carbamazepine	25.00	40.4	g/L	[8]
Cholestrole	26.00	0.161	g/100 g	[97]
Cholestrole	30.70	0.187	g/100 g	[97]

Solubility Data of Drugs Acetonitrile at Various Temperatures   (continued)

Drug	T (°C)	Solubility	Solubility Unit	Reference
Cholestrole	35.00	0.240	g/100 g	[97]
Cholestrole	43.00	0.378	g/100 g	[97]
Cholestrole	47.50	0.492	g/100 g	[97]
Chrysin	50.00	0.00600	mol/L	[115]
Cortexolone	25.00	0.00209	Mole/100 g	[122]
Cortisone	25.00	0.00267	Mole/100 g	[122]
Diflunisal	25.00	0.00355	Mole F.	[26]
Eflucimibe (form A)	20.00	0.0001010	mol/L	[99]
Estradiol	25.00	0.00195	Mole/100 g	[122]
Estriol	25.00	11.6	Mole/100 g	[122]
Estrone	25.00	0.00172	Mole/100 g	[122]
Ethyl paraben	25.00	0.165	Mole/100 g	[122]
Flubiprofen	25.00	0.0308	Mole F.	[26]
Hesperetin	50.00	0.08500	mol/L	[115]
Hesperetin	60.00	0.09000	mol/L	[115]
Hesperetin	70.00	0.09300	mol/L	[115]
Hydrocortisone	25.00	0.00118	Mole/100 g	[122]
Isoquercitrin	50.00	0.00390	mol/L	[115]
Isoquercitrin	60.00	0.00370	mol/L	[115]
Isoquercitrin	70.00	0.00412	mol/L	[115]
L-(+)-Ascorbic acid	19.85	0.00020	Mole F.	[102]
L-(+)-Ascorbic acid	24.85	0.00024	Mole F.	[102]
L-(+)-Ascorbic acid	29.85	0.00028	Mole F.	[102]
L-(+)-Ascorbic acid	34.85	0.00034	Mole F.	[102]
L-(+)-Ascorbic acid	39.85	0.00040	Mole F.	[102]
L-(+)-Ascorbic acid	44.85	0.00045	Mole F.	[102]
L-(+)-Ascorbic acid	49.85	0.00060	Mole F.	[102]
Lamivudine (form 2)	5.00	0.47	g/L	[52]
Lamivudine (form 2)	15.00	0.64	g/L	[52]
Lamivudine (form 2)	25.00	0.91	g/L	[52]
Lamivudine (form 2)	35.00	1.45	g/L	[52]
Lamivudine (form 2)	45.00	2.15	g/L	[52]
Lidocaine	25.00	11.5	Mole/100 g	[122]
Lutein	Room	100	mg/L	[6]
Methyl 4-aminobenzoate	25.00	0.264	Mole/100 g	[122]
Naringenin	50.00	0.07700	mol/L	[115]
Naringenin	60.00	0.08000	mol/L	[115]
Naringenin	70.00	0.08800	mol/L	[115]
n-Butyl 4-aminobenzoate	25.00	0.864	Mole/100 g	[122]
Paracetamol	−5.00	9.44	g/kg	[11]
Paracetamol	0.00	11.18	g/kg	[11]
Paracetamol	5.00	13.44	g/kg	[11]
Paracetamol	10.00	15.98	g/kg	[11]
Paracetamol	15.00	19.34	g/kg	[11]
Paracetamol	20.00	23.10	g/kg	[11]
Paracetamol	25.00	27.54	g/kg	[11]
Paracetamol	30.00	32.83	g/kg	[11]
Paracetamol	30.00	22.4	g/L	[8]
Pentachlorophenol	25.00	0.0590	Mole/100 g	[122]

(continued)

Solubility Data of Drugs Acetonitrile at Various Temperatures   (continued)

Drug	T (°C)	Solubility	Solubility Unit	Reference
Phenol	25.00	8.52	Mole/100 g	[122]
Phenothiazine	25.00	0.01169	Mole F.	[34]
Phenyl benzoate	25.00	0.477	Mole/100 g	[122]
Prednisone	25.00	0.00121	Mole/100 g	[122]
Propyl paraben	25.00	0.176	Mole/100 g	[122]
Quercetin	50.00	0.00540	mol/L	[115]
Quercetin	60.00	0.00680	mol/L	[115]
Quercetin	70.00	0.00705	mol/L	[115]
Rutin	50.00	0.00050	mol/L	[115]
Salicylamide	10.00	0.02010	Mole F.	[107]
Salicylamide	15.00	0.02404	Mole F.	[107]
Salicylamide	20.00	0.02818	Mole F.	[107]
Salicylamide	25.00	0.03328	Mole F.	[107]
Salicylamide	30.00	0.03958	Mole F.	[107]
Salicylamide	35.00	0.04764	Mole F.	[107]
Salicylamide	40.00	0.05579	Mole F.	[107]
Salicylamide	45.00	0.06515	Mole F.	[107]
Salicylamide	50.00	0.07732	Mole F.	[107]
Salicylic acid	10.00	0.01917	Mole F.	[108]
Salicylic acid	15.00	0.02229	Mole F.	[108]
Salicylic acid	20.00	0.02549	Mole F.	[108]
Salicylic acid	25.00	0.02944	Mole F.	[108]
Salicylic acid	30.00	0.03443	Mole F.	[108]
Salicylic acid	35.00	0.03896	Mole F.	[108]
Salicylic acid	40.00	0.04530	Mole F.	[108]
Salicylic acid	45.00	0.05242	Mole F.	[108]
Salicylic acid	50.00	0.05937	Mole F.	[108]
Sulfamerazine (polymorph 1)	30.00	5.980	g/L	[120]
Sulfamerazine (polymorph 2)	30.00	5.207	g/L	[120]
Temazepam	25.00	0.0110091	Mole F.	[15]
Xanthene	25.00	0.01970	Mole F.	[3]

Solubility Data of Drugs in Acetophenone at Various Temperatures

Drug	T (°C)	Solubility	Solubility Unit	Reference
Benzoic acid	25.00	0.1878	Mole F.	[24]
Diclofenac	25.00	0.0447350	Mole F.	[25]
Ibuprofen	25.00	0.0030672	Mole F.	[28]
Naproxen	25.00	0.2952892	Mole F.	[31]
Niflumic acid	25.00	0.0007511	Mole F.	[33]
p-Aminobenzoic acid	25.00	0.0043351	Mole F.	[25]
p-Hydroxybenzoic acid	25.00	0.0223	Mole F.	[35]
Piroxicam	25.00	0.0002045	Mole F.	[33]
Salicylic acid	25.00	0.1527428	Mole F.	[25]
Sodium ibuprofen	25.00	0.0000428	Mole F.	[28]

Solubility Data of Drugs in Acetophenone at Various Temperatures   (continued)

Drug	T (°C)	Solubility	Solubility Unit	Reference
Sodium benzoate	25.00	0.0000331	Mole F.	[28]
Sodium diclofenac	25.00	0.0000508	Mole F.	[31]
Sodium diclofenac	25.00	0.0000508	Mole F.	[25]
Sodium p-aminobenzoate	25.00	0.0000271	Mole F.	[25]
Sodium salicylate	25.00	0.0000456	Mole F.	[25]
Temazepam	25.00	0.0651702	Mole F.	[15]

Solubility Data of Drugs in Benzene at Various Temperatures

Drug	T (°C)	Solubility	Solubility Unit	Reference
2-(4-Ethylbenzoyl)benzoic acid	9.90	0.0023	Mole F.	[18]
2-(4-Ethylbenzoyl)benzoic acid	15.53	0.0037	Mole F.	[18]
2-(4-Ethylbenzoyl)benzoic acid	20.37	0.0052	Mole F.	[18]
2-(4-Ethylbenzoyl)benzoic acid	29.63	0.0109	Mole F.	[18]
2-(4-Ethylbenzoyl)benzoic acid	40.00	0.0228	Mole F.	[18]
2-(4-Ethylbenzoyl)benzoic acid	49.50	0.0408	Mole F.	[18]
2-(4-Ethylbenzoyl)benzoic acid	55.23	0.0598	Mole F.	[18]
2-(4-Ethylbenzoyl)benzoic acid	59.55	0.0793	Mole F.	[18]
Acetylsalicylic acid	25.00	0.00101	Mole F.	[21]
Benzoic acid	25.00	0.0689	Mole F.	[24]
β-Carotene	Room	4000	mg/L	[6]
Chloramine	25.00	0.0224	Mole F.	[123]
Chlorpheniramine maleate	9.90	0.0000767	Mole F.	[47]
Chlorpheniramine maleate	14.90	0.0001228	Mole F.	[47]
Chlorpheniramine maleate	20.60	0.0001651	Mole F.	[47]
Chlorpheniramine maleate	25.63	0.0002011	Mole F.	[47]
Chlorpheniramine maleate	29.89	0.0002363	Mole F.	[47]
Chlorpheniramine maleate	35.10	0.0002882	Mole F.	[47]
Chlorpheniramine maleate	39.80	0.0003661	Mole F.	[47]
Chlorpheniramine maleate	45.12	0.0004557	Mole F.	[47]
Chlorpheniramine maleate	50.00	0.0006189	Mole F.	[47]
Chlorpheniramine maleate	54.38	0.0007808	Mole F.	[47]
Chlorpheniramine maleate	60.30	0.0011196	Mole F.	[47]
Diclofenac	25.00	0.0027203	Mole F.	[25]
Diflunisal	25.00	0.000471	Mole F.	[26]
Flubiprofen	25.00	0.0682	Mole F.	[26]
Haloperidol	25.00	0.01078	Mole F.	[27]
Hydrocortisone	25.00	0.000359	mol/L	[80]
Ibuprofen	25.00	0.0847203	Mole F.	[28]
Lactose	25.00	0.0000002	Mole F.	[30]
Lutein	Room	600	mg/L	[6]
Methyl p-aminobenzoate	25.00	0.0194482	Mole F.	[124]
Naproxen	25.00	0.0072431	Mole F.	[31]
n-Butyl p-aminobenzoate	25.00	0.3262798	Mole F.	[124]
Niflumic acid	25.00	0.0004704	Mole F.	[33]
n-Pentyl p-aminobenzoate	25.00	0.3753111	Mole F.	[124]

(continued)

Solubility Data of Drugs in Benzene at Various Temperatures   (continued)

Drug	T (°C)	Solubility	Solubility Unit	Reference
n-Propyl p-aminobenzoate	25.00	0.0658748	Mole F.	[124]
n-Propyl p-aminobenzoate	25.00	0.1939800	Mole F.	[124]
p-Aminobenzoic acid	25.00	0.0002101	Mole F.	[25]
Paracetamol	25.00	0.00011	Mole F.	[23]
Phenacetinum	9.95	0.0007884	Mole F.	[58]
Phenacetinum	15.00	0.0012485	Mole F.	[58]
Phenacetinum	19.95	0.0016052	Mole F.	[58]
Phenacetinum	24.75	0.0019537	Mole F.	[58]
Phenacetinum	30.10	0.0024537	Mole F.	[58]
Phenacetinum	35.08	0.0031283	Mole F.	[58]
Phenacetinum	40.00	0.0040896	Mole F.	[58]
Phenacetinum	45.00	0.0054662	Mole F.	[58]
Phenacetinum	49.70	0.0072153	Mole F.	[58]
Phenacetinum	55.20	0.009935	Mole F.	[58]
Phenacetinum	60.00	0.013028	Mole F.	[58]
p-Hydroxybenzoic acid	25.00	0.000033	Mole F.	[35]
Pimozide	25.00	0.000760	Mole F.	[36]
Piroxicam	25.00	0.0000722	Mole F.	[33]
Saccharose	25.00	0.00000002	Mole F.	[30]
Salicylic acid	25.00	0.0054508	Mole F.	[25]
Sodium ibuprofen	25.00	0.0000045	Mole F.	[28]
Sodium benzoate	25.00	0.0000004	Mole F.	[28]
Sodium diclofenac	25.00	0.0000045	Mole F.	[31]
Sodium diclofenac	25.00	0.0000045	Mole F.	[25]
Sodium p-aminobenzoate	25.00	0.0000002	Mole F.	[25]
Sodium salicylate	25.00	0.0000356	Mole F.	[25]
Sulfadiazine	25.00	0.0000029	Mole F.	[39]
Sulfamethoxazole	25.00	0.00200	mol/L	[62]
Sulfamethoxypyridazine	25.00	0.00006	Mole F.	[40]
Temazepam	25.00	0.0181321	Mole F.	[15]
Testosterone propionate	25.00	0.24	Mole F.	[16]
Vitamin K3	26.29	0.1088	Mole F.	[125]
Vitamin K3	31.68	0.1302	Mole F.	[125]
Vitamin K3	36.56	0.1523	Mole F.	[125]
Vitamin K3	40.77	0.1742	Mole F.	[125]
Vitamin K3	44.60	0.1970	Mole F.	[125]
Vitamin K3	48.71	0.2243	Mole F.	[125]
Vitamin K3	53.60	0.2616	Mole F.	[125]
Vitamin K3	57.60	0.2966	Mole F.	[125]
Vitamin K3	61.50	0.3343	Mole F.	[125]
Vitamin K3	65.81	0.3791	Mole F.	[125]

Solubility Data of Drugs in Benzyl Alcohol at Various Temperatures

Drug	T (°C)	Solubility	Solubility Unit	Reference
Benzoic acid	25.00	0.1441	Mole F.	[24]
Clonazepam	25.00	33.5	g/L	[126]
Clonazepam	25.00	33.5	g/L	[48]
Diazepam	25.00	369.3	g/L	[126]
Lorazepam	25.00	206.2	g/L	[126]
p-Hydroxybenzoic acid	25.00	0.0784	Mole F.	[35]
Prednisolone	25.00	118.0	g/L	[126]
Progesterone	25.00	423.2	g/L	[126]
Ricobendazole	25.00	17.90	g/L	[103]
Sulfadiazine	25.00	0.0003080	Mole F.	[39]
Sulfamethoxypyridazine	25.00	0.01347	Mole F.	[40]
Temazepam	25.00	0.1107518	Mole F.	[15]
Tetrazepam	25.00	332.7	g/L	[126]

Solubility Data of Drugs in Butyl Acetate at Various Temperatures

Drug	T (°C)	Solubility	Solubility Unit	Reference
1H-1,2,4-Triazole	10.05	0.0141	Mole F.	[92]
1H-1,2,4-Triazole	15.08	0.0166	Mole F.	[92]
1H-1,2,4-Triazole	20.11	0.0202	Mole F.	[92]
1H-1,2,4-Triazole	25.07	0.0240	Mole F.	[92]
1H-1,2,4-Triazole	30.09	0.0279	Mole F.	[92]
1H-1,2,4-Triazole	35.02	0.0320	Mole F.	[92]
1H-1,2,4-Triazole	40.13	0.0370	Mole F.	[92]
1H-1,2,4-Triazole	45.06	0.0428	Mole F.	[92]
1H-1,2,4-Triazole	50.01	0.0498	Mole F.	[92]
1H-1,2,4-Triazole	55.09	0.0586	Mole F.	[92]
1H-1,2,4-Triazole	60.04	0.0688	Mole F.	[92]
1H-1,2,4-Triazole	65.03	0.0809	Mole F.	[92]
1H-1,2,4-Triazole	70.08	0.0955	Mole F.	[92]
1H-1,2,4-Triazole	75.03	0.1130	Mole F.	[92]
1H-1,2,4-Triazole	80.04	0.1359	Mole F.	[92]
1H-1,2,4-Triazole	85.03	0.1667	Mole F.	[92]
1H-1,2,4-Triazole	90.11	0.2111	Mole F.	[92]
2-Hydroxybenzoic acid	25.00	0.1363	Mole F.	[19]
Acetylsalicylic acid	25.00	0.03345	Mole F.	[22]
Benzoic acid	25.00	0.1699	Mole F.	[24]
Chlorpheniramine maleate	10.00	0.0001650	Mole F.	[47]
Chlorpheniramine maleate	15.00	0.0002149	Mole F.	[47]
Chlorpheniramine maleate	21.80	0.0003681	Mole F.	[47]
Chlorpheniramine maleate	25.56	0.0004513	Mole F.	[47]
Chlorpheniramine maleate	30.16	0.0005493	Mole F.	[47]
Chlorpheniramine maleate	34.95	0.0006440	Mole F.	[47]
Chlorpheniramine maleate	40.00	0.0008805	Mole F.	[47]
Chlorpheniramine maleate	45.00	0.0010234	Mole F.	[47]
Chlorpheniramine maleate	49.90	0.0013170	Mole F.	[47]

(continued)

Solubility Data of Drugs in Butyl Acetate at Various Temperatures (continued)

Drug	T (°C)	Solubility	Solubility Unit	Reference
Chlorpheniramine maleate	54.50	0.0015496	Mole F.	[47]
Haloperidol	25.00	0.004946	Mole F.	[27]
Hydrocortisone	25.00	0.00408	mol/L	[80]
Hydroquinone	6.40	0.09877	Mole F.	[101]
Hydroquinone	10.80	0.1019	Mole F.	[101]
Hydroquinone	17.70	0.1100	Mole F.	[101]
Hydroquinone	24.05	0.1173	Mole F.	[101]
Hydroquinone	29.70	0.1247	Mole F.	[101]
Hydroquinone	37.85	0.1352	Mole F.	[101]
Hydroquinone	43.80	0.1450	Mole F.	[101]
Hydroquinone	49.90	0.1555	Mole F.	[101]
Hydroquinone	56.80	0.1677	Mole F.	[101]
Hydroquinone	62.95	0.1803	Mole F.	[101]
Hydroquinone	71.55	0.2049	Mole F.	[101]
Losartan potassium	20.00	0.002713	Mole F.	[86]
Losartan potassium	30.00	0.002380	Mole F.	[86]
Losartan potassium	40.00	0.000968	Mole F.	[86]
Losartan potassium	50.00	0.001206	Mole F.	[86]
Losartan potassium	60.00	0.001933	Mole F.	[86]
Losartan potassium	70.00	0.000907	Mole F.	[86]
Lovastatin	5.95	0.002727	Mole F.	[116]
Lovastatin	10.00	0.003190	Mole F.	[116]
Lovastatin	11.95	0.00407	Mole F.	[89]
Lovastatin	15.10	0.003763	Mole F.	[116]
Lovastatin	15.15	0.00452	Mole F.	[89]
Lovastatin	18.05	0.00494	Mole F.	[89]
Lovastatin	19.80	0.004410	Mole F.	[116]
Lovastatin	21.45	0.00557	Mole F.	[89]
Lovastatin	24.05	0.00611	Mole F.	[89]
Lovastatin	25.10	0.005279	Mole F.	[116]
Lovastatin	27.15	0.00679	Mole F.	[89]
Lovastatin	29.90	0.006403	Mole F.	[116]
Lovastatin	29.95	0.00748	Mole F.	[89]
Lovastatin	33.15	0.00836	Mole F.	[89]
Lovastatin	35.10	0.007541	Mole F.	[116]
Lovastatin	36.25	0.00931	Mole F.	[89]
Lovastatin	39.05	0.01032	Mole F.	[89]
Lovastatin	40.00	0.009004	Mole F.	[116]
Lovastatin	44.80	0.01069	Mole F.	[116]
Lovastatin	50.20	0.01312	Mole F.	[116]
Methyl p-hydroxybenzoate	25.00	0.1326	Mole F.	[35]
Naproxen	25.00	0.02342	Mole F.	[32]
p-Hydroxybenzoic acid	25.00	0.0574	Mole F.	[35]
Salicylamide	25.00	0.06075	Mole F.	[37]
Sulfadiazine	25.00	0.0000586	Mole F.	[39]
Sulfamethoxypyridazine	25.00	0.00088	Mole F.	[40]

Solubility Data of Drugs in Carbon Tetrachloride at Various Temperatures

Drug	T (°C)	Solubility	Solubility Unit	Reference
Acetic acid	−26.80	0.110	Mole F.	[109]
Acetic acid	−21.10	0.170	Mole F.	[109]
Acetic acid	−15.20	0.270	Mole F.	[109]
Acetic acid	−9.00	0.365	Mole F.	[109]
Acetic acid	−6.70	0.402	Mole F.	[109]
Acetic acid	−4.80	0.440	Mole F.	[109]
Acetic acid	−2.80	0.495	Mole F.	[109]
Acetic acid	−1.30	0.547	Mole F.	[109]
Acetic acid	0.70	0.595	Mole F.	[109]
Acetic acid	2.30	0.650	Mole F.	[109]
Acetic acid	3.80	0.689	Mole F.	[109]
Acetic acid	5.20	0.727	Mole F.	[109]
Acetic acid	7.50	0.795	Mole F.	[109]
Acetic acid	10.00	0.870	Mole F.	[109]
Acetic acid	12.00	0.925	Mole F.	[109]
Benzoic acid	25.00	0.0494	Mole F.	[24]
Chloramine	25.00	0.0131	Mole F.	[123]
D-p-Hydroxyphenylglycine	5.23	0.0002249	Mole F.	[127]
D-p-Hydroxyphenylglycine	10.08	0.0002273	Mole F.	[127]
D-p-Hydroxyphenylglycine	15.23	0.0002306	Mole F.	[127]
D-p-Hydroxyphenylglycine	20.21	0.0002340	Mole F.	[127]
D-p-Hydroxyphenylglycine	25.14	0.0002390	Mole F.	[127]
D-p-Hydroxyphenylglycine	30.06	0.0002436	Mole F.	[127]
D-p-Hydroxyphenylglycine	35.21	0.0002503	Mole F.	[127]
D-p-Hydroxyphenylglycine	40.15	0.0002580	Mole F.	[127]
D-p-Hydroxyphenylglycine	45.09	0.0002641	Mole F.	[127]
D-p-Hydroxyphenylglycine	50.04	0.0002723	Mole F.	[127]
Ethyl p-aminobenzoate	25.00	0.0092790	Mole F.	[124]
Hydrocortisone	25.00	0.0000284	mol/L	[80]
Methyl p-aminobenzoate	25.00	0.0031198	Mole F.	[124]
n-Butyl p-aminobenzoate	25.00	0.2231302	Mole F.	[124]
n-Pentyl p-aminobenzoate	25.00	0.2893842	Mole F.	[124]
n-Propyl p-aminobenzoate	25.00	0.0287246	Mole F.	[124]
Paracetamol	30.00	0.89	g/kg	[11]
Paracetamol	30.00	0.15	g/L	[8]
Pyoluteorin	5.05	0.00035	mol/L	[12]
Pyoluteorin	10.05	0.00046	mol/L	[12]
Pyoluteorin	15.05	0.00052	mol/L	[12]
Pyoluteorin	20.05	0.00088	mol/L	[12]
Pyoluteorin	25.05	0.00125	mol/L	[12]
Pyoluteorin	30.05	0.00163	mol/L	[12]
Pyoluteorin	35.05	0.00239	mol/L	[12]
Pyoluteorin	40.05	0.00297	mol/L	[12]
Pyoluteorin	45.05	0.00376	mol/L	[12]
Pyoluteorin	50.05	0.00452	mol/L	[12]
Pyoluteorin	55.05	0.00574	mol/L	[12]
Pyoluteorin	60.05	0.00674	mol/L	[12]
Salicylic acid	24.85	0.00299	Mass. F	[128]

(continued)

Solubility Data of Drugs in Carbon Tetrachloride at Various Temperatures　(continued)

Drug	T (°C)	Solubility	Solubility Unit	Reference
Salicylic acid	29.85	0.00377	Mass. F	[128]
Salicylic acid	34.85	0.00489	Mass. F	[128]
Salicylic acid	39.85	0.00631	Mass. F	[128]
Salicylic acid	44.85	0.00769	Mass. F	[128]
Salicylic acid	49.85	0.01019	Mass. F	[128]
Salicylic acid	54.85	0.01255	Mass. F	[128]
Salicylic acid	59.85	0.01512	Mass. F	[128]
Salicylic acid	64.85	0.01801	Mass. F	[128]
Salicylic acid	69.85	0.02127	Mass. F	[128]
Salicylic acid	74.85	0.02502	Mass. F	[128]
Stearic acid	25.00	5.47	g/100 g	[14]
Stearic acid	30.00	10.6	g/100 g	[14]
Stearic acid	35.00	18.8	g/100 g	[14]
Stearic acid	40.00	30.9	g/100 g	[14]
Testosterone propionate	25.00	0.16	Mole F.	[16]
Xanthene	25.00	0.1237	Mole F.	[3]

Solubility Data of Drugs in Chlorobenzene at Various Temperatures

Drug	T (°C)	Solubility	Solubility Unit	Reference
Benzoic acid	25.00	0.0862	Mole F.	[24]
Diclofenac	25.00	0.0036794	Mole F.	[25]
Ibuprofen	25.00	0.0201269	Mole F.	[28]
Lactose	25.00	0.0000049	Mole F.	[30]
Naproxen	25.00	0.0088911	Mole F.	[31]
Niflumic acid	25.00	0.0050216	Mole F.	[33]
p-Aminobenzoic acid	25.00	0.0004847	Mole F.	[25]
Piroxicam	25.00	0.0000704	Mole F.	[33]
Salicylic acid	25.00	0.0066976	Mole F.	[25]
Sodium ibuprofen	25.00	0.0000023	Mole F.	[28]
Sodium benzoate	25.00	0.0000058	Mole F.	[28]
Sodium diclofenac	25.00	0.0000220	Mole F.	[31]
Sodium diclofenac	25.00	0.0000220	Mole F.	[25]
Sodium p-aminobenzoate	25.00	0.0000018	Mole F.	[25]
Sodium salicylate	25.00	0.0000351	Mole F.	[25]
Testosterone propionate	25.00	0.27	Mole F.	[16]

Solubility Data of Drugs in Chloroform at Various Temperatures

Drug	T (°C)	Solubility	Solubility Unit	Reference
Acetanilide	25.00	1.86	mol/L	[73]
Acetanilide	30.00	2.08	mol/L	[73]
Acetanilide	35.00	2.45	mol/L	[73]
Acetanilide	40.00	2.63	mol/L	[73]
Acetic acid	−29.20	0.380	Mole F.	[109]

Solubility Data of Drugs in Chloroform at Various Temperatures   (continued)

Drug	T (°C)	Solubility	Solubility Unit	Reference
Acetic acid	−26.80	0.440	Mole F.	[109]
Acetic acid	−21.10	0.520	Mole F.	[109]
Acetic acid	−15.20	0.580	Mole F.	[109]
Acetic acid	−9.00	0.615	Mole F.	[109]
Acetic acid	−7.30	0.620	Mole F.	[109]
Acetic acid	−3.90	0.675	Mole F.	[109]
Acetic acid	−3.00	0.700	Mole F.	[109]
Acetic acid	0.70	0.740	Mole F.	[109]
Acetic acid	2.20	0.785	Mole F.	[109]
Acetic acid	4.30	0.805	Mole F.	[109]
Acetic acid	6.90	0.855	Mole F.	[109]
Acetic acid	10.00	0.900	Mole F.	[109]
Acetic acid	12.50	0.930	Mole F.	[109]
Acetylsalicylic acid	25.00	0.206	Mole F.	[21]
Benzoic acid	25.00	0.1283	Mole F.	[24]
β-Carotene	Room	2000	mg/L	[6]
Betulin	5.05	0.002602	Mole F.	[7]
Betulin	15.05	0.003377	Mole F.	[7]
Betulin	25.05	0.004370	Mole F.	[7]
Betulin	35.05	0.005111	Mole F.	[7]
Carbamazepine	25.00	169	g/L	[8]
Chloramine	25.00	0.0229	Mole F.	[123]
Diclofenac	25.00	0.0093910	Mole F.	[25]
Eflucimibe (form A)	20.00	0.0137	mol/L	[99]
Erythromycin A dihydrate	20.05	0.003914	Mole F.	[100]
Erythromycin A dihydrate	24.85	0.005399	Mole F.	[100]
Erythromycin A dihydrate	30.00	0.008749	Mole F.	[100]
Erythromycin A dihydrate	34.85	0.01341	Mole F.	[100]
Erythromycin A dihydrate	39.95	0.02159	Mole F.	[100]
Erythromycin A dihydrate	44.85	0.03129	Mole F.	[100]
Erythromycin A dihydrate	49.85	0.04734	Mole F.	[100]
Haloperidol	25.00	0.15988	Mole F.	[27]
Hydrocortisone	25.00	0.0112	mol/L	[80]
Ibuprofen	25.00	3.00	mol/L	[81]
Ibuprofen	30.00	3.4098	mol/L	[81]
Ibuprofen	35.00	3.81	mol/L	[81]
Ibuprofen	40.00	4.224	mol/L	[81]
Ibuprofen	10.00	457.5	g/kg	[83]
Ibuprofen	15.00	554.5	g/kg	[83]
Ibuprofen	20.00	644.8	g/kg	[83]
Ibuprofen	25.00	0.2538022	Mole F.	[28]
Irbesartan (form A)	4.55	0.001195	Mole F.	[29]
Irbesartan (form A)	10.15	0.001309	Mole F.	[29]
Irbesartan (form A)	14.55	0.001457	Mole F.	[29]
Irbesartan (form A)	19.55	0.001615	Mole F.	[29]
Irbesartan (form A)	24.60	0.001794	Mole F.	[29]
Irbesartan (form A)	29.70	0.002012	Mole F.	[29]
Irbesartan (form A)	34.65	0.002252	Mole F.	[29]

(continued)

Solubility Data of Drugs in Chloroform at Various Temperatures   (continued)

Drug	T (°C)	Solubility	Solubility Unit	Reference
Irbesartan (form A)	40.25	0.002592	Mole F.	[29]
Irbesartan (form A)	44.65	0.002935	Mole F.	[29]
Irbesartan (form A)	49.85	0.003351	Mole F.	[29]
Lutein	Room	6000	mg/L	[6]
Naproxen	20.00	0.238	mol/L	[82]
Naproxen	25.00	0.0302851	Mole F.	[31]
Naproxen	25.00	0.288	mol/L	[82]
Naproxen	30.00	0.321	mol/L	[82]
Naproxen	35.00	0.3827	mol/L	[82]
Naproxen	40.00	0.4392	mol/L	[82]
Niflumic acid	25.00	0.0012824	Mole F.	[33]
p-Aminobenzoic acid	25.00	0.0115162	Mole F.	[25]
p-Aminophenylacetic acid	16.00	0.9	g/kg	[83]
p-Aminophenylacetic acid	30.00	1.9	g/kg	[83]
Paracetamol	25.00	0.00318	mol/L	[73]
Paracetamol	30.00	0.00397	mol/L	[73]
Paracetamol	35.00	0.005486	mol/L	[73]
Paracetamol	40.00	0.00862	mol/L	[73]
Paracetamol	30.00	1.54	g/kg	[11]
Paracetamol	30.00	0.59	g/L	[8]
Paracetamol	25.00	0.00032	Mole F.	[23]
Phenacetin	25.00	0.434	mol/L	[73]
Phenacetin	30.00	0.552	mol/L	[73]
Phenacetin	35.00	0.618	mol/L	[73]
Phenacetin	40.00	0.8161	mol/L	[73]
Phenylacetic acid	20.00	637.1	g/kg	[83]
p-Hydroxyphenylacetic acid	10.00	1.3	g/kg	[83]
p-Hydroxyphenylacetic acid	15.00	2.5	g/kg	[83]
p-Hydroxyphenylacetic acid	25.00	183.6	g/kg	[83]
p-Hydroxybenzoic acid	25.00	0.00015	Mole F.	[35]
Piroxicam	25.00	0.0225956	Mole F.	[33]
p-Phenylphenol	25.00	0.10	mol/L	[129]
Pyoluteorin	5.05	0.0757	mol/L	[12]
Pyoluteorin	10.05	0.0788	mol/L	[12]
Pyoluteorin	15.05	0.0813	mol/L	[12]
Pyoluteorin	20.05	0.0901	mol/L	[12]
Pyoluteorin	25.05	0.0940	mol/L	[12]
Pyoluteorin	30.05	0.1007	mol/L	[12]
Pyoluteorin	35.05	0.1075	mol/L	[12]
Pyoluteorin	40.05	0.1220	mol/L	[12]
Pyoluteorin	45.05	0.1348	mol/L	[12]
Pyoluteorin	50.05	0.1567	mol/L	[12]
Pyoluteorin	55.05	0.1708	mol/L	[12]
Rifapentine	5.00	0.0254	Mole F.	[13]
Rifapentine	10.00	0.0265	Mole F.	[13]
Rifapentine	15.00	0.0281	Mole F.	[13]
Rifapentine	20.00	0.0287	Mole F.	[13]
Rifapentine	24.00	0.0293	Mole F.	[13]
Rifapentine	30.00	0.0299	Mole F.	[13]

Solubility Data of Drugs in Chloroform at Various Temperatures   (continued)

Drug	T (°C)	Solubility	Solubility Unit	Reference
Rifapentine	35.00	0.0308	Mole F.	[13]
Rifapentine	40.00	0.0316	Mole F.	[13]
Rifapentine	45.00	0.0326	Mole F.	[13]
Rifapentine	50.00	0.0334	Mole F.	[13]
Saccharose	25.00	0.000000047	Mole F.	[30]
Salicylic acid	25.00	0.0014575	Mole F.	[25]
Sodium ibuprofen	25.00	0.0004134	Mole F.	[28]
Sodium benzoate	25.00	0.0000011	Mole F.	[28]
Sodium diclofenac	25.00	0.0000049	Mole F.	[31]
Sodium diclofenac	25.00	0.0000050	Mole F.	[25]
Sodium p-aminobenzoate	25.00	0.0000001	Mole F.	[25]
Sodium salicylate	25.00	0.0000461	Mole F.	[25]
Sulfamethoxazole	25.00	0.00910	mol/L	[62]
Sulfamethoxypyridazine	25.00	0.00129	Mole F.	[40]
Testosterone propionate	25.00	0.35	Mole F.	[16]
Triclosan	20.00	0.33	Mole F.	[85]
Triclosan	25.00	0.412	Mole F.	[85]
Triclosan	30.00	0.501	Mole F.	[85]
Triclosan	35.00	0.561	Mole F.	[85]
Triclosan	40.00	0.633	Mole F.	[85]

Solubility Data of Drugs in Chloroform-Saturated Buffer at Various Temperatures

Drug	T (°C)	Solubility	Solubility Unit	Reference
Acetanilide	25.00	1.690	mol/L	[73]
Acetanilide	30.00	2.140	mol/L	[73]
Acetanilide	35.00	2.504	mol/L	[73]
Acetanilide	40.00	3.170	mol/L	[73]
Ibuprofen	25.00	0.0000619	mol/L	[81]
Ibuprofen	30.00	0.0000771	mol/L	[81]
Ibuprofen	35.00	0.0000940	mol/L	[81]
Ibuprofen	40.00	0.0001262	mol/L	[81]
Paracetamol	25.00	0.00268	mol/L	[73]
Paracetamol	30.00	0.00350	mol/L	[73]
Paracetamol	35.00	0.00452	mol/L	[73]
Paracetamol	40.00	0.00623	mol/L	[73]
Phenacetin	25.00	0.503	mol/L	[73]
Phenacetin	30.00	0.539	mol/L	[73]
Phenacetin	35.00	0.646	mol/L	[73]
Phenacetin	40.00	0.714	mol/L	[73]

Solubility Data of Drugs in Cyclohexane at Various Temperatures

Drug	T (°C)	Solubility	Solubility Unit	Reference
Acetanilide	25.00	0.00195	mol/L	[130]
Acetanilide	25.00	0.0015	mol/L	[73]
Acetanilide	30.00	0.00211	mol/L	[73]
Acetanilide	35.00	0.00345	mol/L	[73]
Acetanilide	40.00	0.00561	mol/L	[73]
Acetylsalicylic acid	25.00	0.000380	mol/L	[130]
Benzocaine	25.00	0.0236	mol/L	[130]
Benzocaine	25.00	0.0098	mol/L	[76]
Benzocaine	30.00	0.01225	mol/L	[76]
Benzocaine	35.00	0.0168	mol/L	[76]
Benzocaine	40.00	0.0203	mol/L	[76]
Benzoic acid	25.00	0.0102	Mole F.	[24]
Benzoic acid	25.00	0.560	mol/L	[130]
β-Carotene	Room	2000	mg/L	[6]
Betulin	10.05	0.0000168	Mole F.	[7]
Betulin	15.05	0.0000270	Mole F.	[7]
Betulin	25.05	0.0000634	Mole F.	[7]
Betulin	35.05	0.0001667	Mole F.	[7]
Biphenyl	25.00	4.55	mol/L	[130]
Chloramine	25.00	0.0060	Mole F.	[123]
Diclofenac	25.00	0.0000615	Mole F.	[25]
Ethyl p-aminobenzoate	25.00	0.0009978	Mole F.	[124]
Flubiprofen	25.00	0.0300	mol/L	[130]
Hydrocortisone	25.00	0.0000023	mol/L	[80]
Ibuprofen	25.00	0.937	mol/L	[81]
Ibuprofen	25.00	0.1520722	Mole F.	[28]
Ibuprofen	25.00	0.445	mol/L	[130]
Ibuprofen	30.00	1.2494	mol/L	[81]
Ibuprofen	35.00	1.504	mol/L	[81]
Ibuprofen	40.00	2.141	mol/L	[81]
Lactose	25.00	0.0000002	Mole F.	[30]
Losartan potassium	20.00	0.0000340	Mole F.	[86]
Losartan potassium	30.00	0.0000287	Mole F.	[86]
Losartan potassium	40.00	0.0000305	Mole F.	[86]
Losartan potassium	50.00	0.0000272	Mole F.	[86]
Losartan potassium	60.00	0.0000288	Mole F.	[86]
Losartan potassium	70.00	0.0000259	Mole F.	[86]
Lutein	Room	50	mg/L	[6]
Methyl p-aminobenzoate	25.00	0.0004222	Mole F.	[124]
Methyl p-hydroxybenzoate	25.00	0.00118	mol/L	[130]
Naproxen	20.00	0.0003280	mol/L	[82]
Naproxen	25.00	0.0001197	Mole F.	[31]
Naproxen	25.00	0.0005394	mol/L	[82]
Naproxen	30.00	0.0007690	mol/L	[82]
Naproxen	35.00	0.001075	mol/L	[82]

Solubility Data of Drugs in Cyclohexane at Various Temperatures (continued)

Drug	T (°C)	Solubility	Solubility Unit	Reference
Naproxen	40.00	0.001479	mol/L	[82]
n-Butyl p-aminobenzoate	25.00	0.0036979	Mole F.	[124]
Niflumic acid	25.00	0.0000162	Mole F.	[33]
n-Pentyl p-aminobenzoate	25.00	0.0048441	Mole F.	[124]
n-Propyl p-aminobenzoate	25.00	0.0020294	Mole F.	[124]
p-Aminobenzoic acid	25.00	0.0000009	Mole F.	[25]
Paracetamol	25.00	0.000278	mol/L	[73]
Paracetamol	25.00	0.0000200	Mole F.	[23]
Paracetamol	30.00	0.000350	mol/L	[73]
Paracetamol	35.00	0.000470	mol/L	[73]
Paracetamol	40.00	0.000566	mol/L	[73]
Phenacetin	25.00	0.000945	mol/L	[130]
Phenacetin	25.00	0.0003207	mol/L	[73]
Phenacetin	30.00	0.0005360	mol/L	[73]
Phenacetin	35.00	0.0009289	mol/L	[73]
Phenacetin	40.00	0.00148	mol/L	[73]
Phenol	25.00	1.57	mol/L	[130]
Phenothiazine	25.00	0.000979	Mole F.	[34]
Piroxicam	25.00	0.0000179	Mole F.	[33]
Propyl p-hydroxybenzoate	25.00	0.00303	mol/L	[130]
Salicylic acid	25.00	0.0005509	Mole F.	[25]
Salicylic acid	25.00	0.00715	mol/L	[130]
Sodium benzoate	25.00	0.0000007	Mole F.	[28]
Sodium diclofenac	25.00	0.0000032	Mole F.	[31]
Sodium diclofenac	25.00	0.0000032	Mole F.	[25]
Sodium ibuprofen	25.00	0.0000085	Mole F.	[28]
Sodium p-aminobenzoate	25.00	0.0000001	Mole F.	[25]
Sodium salicylate	25.00	0.0000330	Mole F.	[25]
Temazepam	24.00	0.000343	Volume F.	[64]
Temazepam	25.00	0.0015765	Mole F.	[15]
Androstanolone acetate	25.00	0.007	Mole F.	[131]
Nandrolone acetate	25.00	0.037	Mole F.	[131]
Testosterone acetate	25.00	0.004	Mole F.	[131]
Androstanolone butyrate	25.00	0.022	Mole F.	[131]
Nandrolone propionate	25.00	0.084	Mole F.	[131]
Testosterone butyrate	25.00	0.014	Mole F.	[131]
Androstanolone formate	25.00	0.003	Mole F.	[131]
Nandrolone formate	25.00	0.009	Mole F.	[131]
Testosterone formate	25.00	0.004	Mole F.	[131]
Testosterone propionate	25.00	0.012	Mole F.	[16]
Testosterone valerate	25.00	0.018	Mole F.	[131]
Xanthene	25.00	0.04203	Mole F.	[3]

Solubility Data of Drugs in Cyclohexanone at Various Temperatures

Drug	T (°C)	Solubility	Solubility Unit	Reference
2-Hydroxybenzoic acid	25.00	0.2301000	Mole F.	[19]
3,5-Di-*tert*-butyl-4-hydroxytoluene	N/A	0.60	Mole F.	[4]
β-Carotene	Room	2000	mg/L	[6]
Lutein	Room	4000	mg/L	[6]
Methyl-3-(3,5-di-*tert*-butyl-4-hydroxyphenyl)-propionate	N/A	0.38	Mole F.	[4]
Octadecyl-3-(3,5-di-*tert*-butyl-4-hydroxyphenyl)-propionate	N/A	0.15	Mole F.	[4]
Pentaerythrittol tetrakis(3-(3,5-di-*tert*-butyl-4-hydroxyphenyl)-propionate)	N/A	0.10	Mole F.	[4]
Temazepam	25.00	0.0679788	Mole F.	[15]

Solubility Data of Drugs in Cyclopentane at Various Temperatures

Drug	T (°C)	Solubility	Solubility Unit	Reference
Acetic acid	−21.00	0.055	Mole F.	[109]
Acetic acid	−15.20	0.065	Mole F.	[109]
Acetic acid	−9.00	0.095	Mole F.	[109]
Acetic acid	−3.90	0.100	Mole F.	[109]
Acetic acid	−2.40	0.115	Mole F.	[109]
Acetic acid	−0.10	0.155	Mole F.	[109]
Acetic acid	3.20	0.230	Mole F.	[109]
Acetic acid	4.30	0.275	Mole F.	[109]
Acetic acid	5.60	0.345	Mole F.	[109]
Acetic acid	6.50	0.490	Mole F.	[109]
Acetic acid	8.70	0.775	Mole F.	[109]
Acetic acid	10.10	0.780	Mole F.	[109]
Acetic acid	11.30	0.845	Mole F.	[109]
Acetic acid	11.90	0.870	Mole F.	[109]
Acetic acid	12.50	0.905	Mole F.	[109]
Testosterone propionate	25.00	0.007	Mole F.	[16]

Solubility Data of Drugs in Decane at Various Temperatures

Drug	T (°C)	Solubility	Solubility Unit	Reference
3,5-Di-*tert*-butyl-4-hydroxytoluene	N/A	0.32	Mole F.	[4]
Benzoic acid	25.00	0.0154	Mole F.	[24]
Mestanolone	25.00	0.0002740	Mole F.	[69]
Methandienone	25.00	0.0003580	Mole F.	[69]
Methyl *p*-hydroxybenzoate	25.00	0.000121	Mole F.	[35]
Methyl-3-(3,5-di-*tert*-butyl-4-hydroxyphenyl)-propionate	N/A	0.064	Mole F.	[4]
Methyltestosterone	25.00	0.0004020	Mole F.	[69]
Nandrolone	25.00	0.0003820	Mole F.	[69]

Solubility Data of Drugs in Decane at Various Temperatures   (continued)

Drug	T (°C)	Solubility	Solubility Unit	Reference
Octadecyl-3-(3,5-di*tert*-butyl-4-hydroxyphenyl)-propionate	N/A	0.073	Mole F.	[4]
Pentaerythrittol tetrakis(3-(3,5-di-*tert*-butyl-4-hydroxyphenyl)-propionate)	N/A	0.00040	Mole F.	[4]
Phenothiazine	25.00	0.001056	Mole F.	[34]
Stearic acid (polymorph A)	18.60	0.0011250	Mole F.	[90]
Stearic acid (polymorph A)	25.60	0.0038143	Mole F.	[90]
Stearic acid (polymorph A)	29.50	0.0076122	Mole F.	[90]
Stearic acid (polymorph A)	33.70	0.0161473	Mole F.	[90]
Stearic acid (polymorph B)	5.70	0.0000692	Mole F.	[90]
Stearic acid (polymorph B)	17.30	0.0168231	Mole F.	[90]
Stearic acid (polymorph B)	20.00	0.0015338	Mole F.	[90]
Stearic acid (polymorph B)	25.00	0.0034825	Mole F.	[90]
Stearic acid (polymorph B)	30.40	0.4690715	Mole F.	[90]
Stearic acid (polymorph B)	34.50	0.0191204	Mole F.	[90]
Stearic acid (polymorph B)	35.80	0.0233537	Mole F.	[90]
Stearic acid (bulk)	17.90	0.0010200	Mole F.	[90]
Stearic acid (bulk)	22.40	0.0020498	Mole F.	[90]
Stearic acid (bulk)	26.00	0.0041114	Mole F.	[90]
Stearic acid (bulk)	27.90	0.0060000	Mole F.	[90]
Stearic acid (bulk)	29.80	0.0082627	Mole F.	[90]
Stearic acid (bulk)	32.10	0.0118551	Mole F.	[90]
Stearic acid (bulk)	33.10	0.0142357	Mole F.	[90]
Stearic acid (bulk)	33.80	0.0164078	Mole F.	[90]
Stearic acid (bulk)	35.00	0.0196240	Mole F.	[90]
Stearic acid (bulk)	37.10	0.0243798	Mole F.	[90]
Stearic acid (bulk)	37.60	0.0291296	Mole F.	[90]
Stearic acid (polymorph C)	19.50	0.0013740	Mole F.	[90]
Stearic acid (polymorph C)	24.70	0.0034721	Mole F.	[90]
Stearic acid (polymorph C)	35.50	0.0209420	Mole F.	[90]
Testosterone	25.00	0.0002320	Mole F.	[69]
Testosterone propionate	25.00	0.0095	Mole F.	[131]
Xanthene	25.00	0.04610	Mole F.	[3]

Solubility Data of Drugs in Diethyl Ether at Various Temperatures

Drug	T (°C)	Solubility	Solubility Unit	Reference
Acetylsalicylic acid	25.00	0.03529	Mole F.	[22]
Atractylenolide III	10.05	0.002770	Mole F.	[132]
Atractylenolide III	15.05	0.003749	Mole F.	[132]
Atractylenolide III	20.05	0.005624	Mole F.	[132]
Atractylenolide III	25.05	0.007383	Mole F.	[132]
Atractylenolide III	30.05	0.009387	Mole F.	[132]
Benzoic acid	25.00	0.1837	Mole F.	[24]
Diclofenac	25.00	0.0021059	Mole F.	[25]
Ibuprofen	25.00	0.0222258	Mole F.	[28]
Lactose	25.00	0.0000024	Mole F.	[30]

(*continued*)

Solubility Data of Drugs in Diethyl Ether at Various Temperatures   (continued)

Drug	T (°C)	Solubility	Solubility Unit	Reference
Mannitol	25.00	0.0000095	Mole F.	[30]
Methyl *p*-hydroxybenzoate	25.00	0.0840	Mole F.	[35]
Naproxen	25.00	0.0029908	Mole F.	[31]
Naproxen	25.00	0.01984	Mole F.	[32]
Niflumic acid	25.00	0.0278478	Mole F.	[33]
p-Aminobenzoic acid	25.00	0.0135686	Mole F.	[25]
Phenothiazine	25.00	0.02581	Mole F.	[34]
p-Hydroxybenzoic acid	25.00	0.0521	Mole F.	[35]
Piroxicam	25.00	0.0000505	Mole F.	[33]
Salicylic acid	25.00	0.1521330	Mole F.	[25]
Sodium ibuprofen	25.00	0.0006050	Mole F.	[28]
Sodium benzoate	25.00	0.0000019	Mole F.	[28]
Sodium diclofenac	25.00	0.0005274	Mole F.	[31]
Sodium diclofenac	25.00	0.0005272	Mole F.	[25]
Sodium *p*-aminobenzoate	25.00	0.0001303	Mole F.	[25]
Sodium salicylate	25.00	0.0008709	Mole F.	[25]
Temazepam	24.00	0.0044	Volume F.	[64]

Solubility Data of Drugs in Dimethyl Formamide at Various Temperatures

Drug	T (°C)	Solubility	Solubility Unit	Reference
2-(4-Ethylbenzoyl)benzoic acid	8.05	0.3433	Mole F.	[18]
2-(4-Ethylbenzoyl)benzoic acid	13.34	0.3674	Mole F.	[18]
2-(4-Ethylbenzoyl)benzoic acid	17.48	0.3822	Mole F.	[18]
2-(4-Ethylbenzoyl)benzoic acid	23.27	0.4200	Mole F.	[18]
2-(4-Ethylbenzoyl)benzoic acid	32.50	0.4513	Mole F.	[18]
2-(4-Ethylbenzoyl)benzoic acid	41.19	0.4704	Mole F.	[18]
2-(4-Ethylbenzoyl)benzoic acid	52.40	0.5073	Mole F.	[18]
2-(4-Ethylbenzoyl)benzoic acid	62.60	0.5318	Mole F.	[18]
Anhydrous citric acid	25.00	0.1872	Mole F.	[23]
Benzoic acid	25.00	0.4909	Mole F.	[24]
β-Carotene	Room	200	mg/L	[6]
Cefazolin sodium pentahydrate	5.00	0.02399	Mole F.	[9]
Cefazolin sodium pentahydrate	10.00	0.02637	Mole F.	[9]
Cefazolin sodium pentahydrate	15.00	0.02961	Mole F.	[9]
Cefazolin sodium pentahydrate	21.00	0.04032	Mole F.	[9]
Ceftriaxone disodium	5.05	0.0000568	Mole F.	[112]
Ceftriaxone disodium	9.50	0.0000670	Mole F.	[112]
Ceftriaxone disodium	16.40	0.0000908	Mole F.	[112]
Ceftriaxone disodium	21.30	0.0001190	Mole F.	[112]
Ceftriaxone disodium	27.20	0.0001510	Mole F.	[112]
Ceftriaxone disodium	30.05	0.0001764	Mole F.	[112]
Ceftriaxone disodium	32.50	0.0001898	Mole F.	[112]
Ceftriaxone disodium	37.50	0.0002229	Mole F.	[112]
Ceftriaxone disodium	40.65	0.0002675	Mole F.	[112]
Ceftriaxone disodium	44.45	0.0002863	Mole F.	[112]
Diclofenac	25.00	0.2204686	Mole F.	[25]

Solubility Data of Drugs in Dimethyl Formamide at Various Temperatures   (continued)

Drug	T (°C)	Solubility	Solubility Unit	Reference
Dioxopromethazine HCl	5.05	0.003806	Mole F.	[106]
Dioxopromethazine HCl	10.05	0.004184	Mole F.	[106]
Dioxopromethazine HCl	15.12	0.004480	Mole F.	[106]
Dioxopromethazine HCl	20.03	0.004744	Mole F.	[106]
Dioxopromethazine HCl	25.02	0.005139	Mole F.	[106]
Dioxopromethazine HCl	30.04	0.005537	Mole F.	[106]
Dioxopromethazine HCl	35.08	0.006014	Mole F.	[106]
Dioxopromethazine HCl	40.02	0.006580	Mole F.	[106]
Dioxopromethazine HCl	45.08	0.007163	Mole F.	[106]
Dioxopromethazine HCl	50.03	0.007827	Mole F.	[106]
D-p-Hydroxyphenylglycine	5.12	0.0000199	Mole F.	[127]
D-p-Hydroxyphenylglycine	10.08	0.0000239	Mole F.	[127]
D-p-Hydroxyphenylglycine	15.03	0.0000267	Mole F.	[127]
D-p-Hydroxyphenylglycine	20.05	0.0000338	Mole F.	[127]
D-p-Hydroxyphenylglycine	25.24	0.0000403	Mole F.	[127]
D-p-Hydroxyphenylglycine	30.03	0.0000493	Mole F.	[127]
D-p-Hydroxyphenylglycine	35.11	0.0000600	Mole F.	[127]
D-p-Hydroxyphenylglycine	40.05	0.0000704	Mole F.	[127]
D-p-Hydroxyphenylglycine	45.24	0.0000868	Mole F.	[127]
D-p-Hydroxyphenylglycine	50.00	0.0001017	Mole F.	[127]
Eflucimibe (form A)	20.00	0.0323	mol/L	[99]
Haloperidol	25.00	0.04047	Mole F.	[27]
Ibuprofen	25.00	0.12771	Mole F.	[28]
Lactose	25.00	0.0016680	Mole F.	[30]
Lutein	Room	1000	mg/L	[6]
Mannitol	25.00	0.0016894	Mole F.	[30]
Methyl p-hydroxybenzoate	25.00	0.4605	Mole F.	[35]
Naproxen	25.00	0.1383180	Mole F.	[31]
Niflumic acid	25.00	0.4671990	Mole F.	[33]
p-Aminobenzoic acid	25.00	0.0476439	Mole F.	[25]
Paracetamol	30.00	1012.02	g/kg	[11]
Paracetamol	25.00	0.5174400	Mole F.	[23]
Paroxetine HCl hemihydrate	22.35	0.0012354	Mole F.	[57]
Paroxetine HCl hemihydrate	25.40	0.0013659	Mole F.	[57]
Paroxetine HCl hemihydrate	28.90	0.0015244	Mole F.	[57]
Paroxetine HCl hemihydrate	33.00	0.0016974	Mole F.	[57]
Paroxetine HCl hemihydrate	37.00	0.0020238	Mole F.	[57]
Paroxetine HCl hemihydrate	40.90	0.0023307	Mole F.	[57]
Paroxetine HCl hemihydrate	44.70	0.0025497	Mole F.	[57]
Paroxetine HCl hemihydrate	49.10	0.0028470	Mole F.	[57]
Paroxetine HCl hemihydrate	53.00	0.0031549	Mole F.	[57]
Paroxetine HCl hemihydrate	57.00	0.0034685	Mole F.	[57]
Paroxetine HCl hemihydrate	61.05	0.0037763	Mole F.	[57]
Paroxetine HCl hemihydrate	65.20	0.0040947	Mole F.	[57]
Paroxetine HCl hemihydrate	69.00	0.0044107	Mole F.	[57]
Paroxetine HCl hemihydrate	73.05	0.0047220	Mole F.	[57]
Paroxetine HCl hemihydrate	77.30	0.0050506	Mole F.	[57]
Paroxetine HCl hemihydrate	82.00	0.0053707	Mole F.	[57]

(continued)

Solubility Data of Drugs in Dimethyl Formamide at Various Temperatures   (continued)

Drug	T (°C)	Solubility	Solubility Unit	Reference
p-Hydroxybenzoic acid	25.00	0.2137000	Mole F.	[35]
Piroxicam	25.00	0.0167560	Mole F.	[33]
Ricobendazole	25.00	5.88	g/L	[103]
Saccharose	25.00	0.0000462	Mole F.	[30]
Salicylic acid	25.00	0.3840433	Mole F.	[25]
Sodium ibuprofen	25.00	0.0011453	Mole F.	[28]
Sodium benzoate	25.00	0.0002731	Mole F.	[28]
Sodium cefotaxim	5.00	0.040727	Mole F.	[133]
Sodium cefotaxim	7.00	0.041300	Mole F.	[133]
Sodium cefotaxim	10.00	0.043607	Mole F.	[133]
Sodium cefotaxim	12.00	0.046185	Mole F.	[133]
Sodium cefotaxim	15.00	0.047428	Mole F.	[133]
Sodium cefotaxim	17.00	0.053743	Mole F.	[133]
Sodium cefotaxim	20.00	0.056962	Mole F.	[133]
Sodium cefotaxim	22.00	0.060021	Mole F.	[133]
Sodium cefotaxim	25.00	0.066202	Mole F.	[133]
Sodium cefotaxim	27.00	0.069022	Mole F.	[133]
Sodium cefotaxim	30.00	0.075678	Mole F.	[133]
Sodium cefotaxim	32.00	0.078517	Mole F.	[133]
Sodium cefotaxim	35.00	0.084511	Mole F.	[133]
Sodium diclofenac	25.00	0.0826947	Mole F.	[31]
Sodium diclofenac	25.00	0.0826616	Mole F.	[25]
Sodium p-aminobenzoate	25.00	0.0001498	Mole F.	[25]
Sodium salicylate	25.00	0.3902374	Mole F.	[25]
Sulfadiazine	25.00	0.0710068	Mole F.	[39]
Sulfamethoxypyridazine	25.00	0.08807	Mole F.	[40]
Temazepam	25.00	0.1137762	Mole F.	[15]
Xylitol	20.06	0.04111	Mole F.	[66]
Xylitol	25.14	0.05042	Mole F.	[66]
Xylitol	30.11	0.06088	Mole F.	[66]
Xylitol	35.09	0.07359	Mole F.	[66]
Xylitol	40.07	0.08946	Mole F.	[66]
Xylitol	45.17	0.1099	Mole F.	[66]
Xylitol	50.05	0.1344	Mole F.	[66]
Xylitol	55.09	0.1656	Mole F.	[66]
Xylitol	60.12	0.2037	Mole F.	[66]

Solubility Data of Drugs in Dimethyl Sulfoxide at Various Temperatures

Drug	T (°C)	Solubility	Solubility Unit	Reference
Benzoic acid	25.00	0.5102	Mole F.	[24]
β-Carotene	Room	30	mg/L	[6]
Clindamycin phosphate	25.00	0.000422	Mole F.	[134]
Clindamycin phosphate	30.00	0.000519	Mole F.	[134]
Clindamycin phosphate	35.00	0.000702	Mole F.	[134]
Clindamycin phosphate	40.00	0.000893	Mole F.	[134]
Clindamycin phosphate	45.00	0.001197	Mole F.	[134]

Solubility Data of Drugs in Dimethyl Sulfoxide at Various Temperatures   (continued)

Drug	*T* (°C)	Solubility	Solubility Unit	Reference
Clindamycin phosphate	50.00	0.001506	Mole F.	[134]
Clindamycin phosphate	55.00	0.002169	Mole F.	[134]
Clindamycin phosphate	60.00	0.002859	Mole F.	[134]
Eflucimibe (form A)	20.00	0.00548	mol/L	[99]
Haloperidol	25.00	0.010576	Mole F.	[27]
Lutein	Room	1000	mg/L	[6]
Luteolin	25.00	0.05130	Mole F.	[54]
Luteolin	40.00	0.06625	Mole F.	[54]
Luteolin	60.00	0.08498	Mole F.	[54]
Methyl *p*-hydroxybenzoate	25.00	0.5839	Mole F.	[35]
Paracetamol	30.00	1132.56	g/kg	[11]
p-Hydroxybenzoic acid	25.00	0.3674	Mole F.	[35]
Riboflavin	30.00	39.7	g/L	[118]
Riboflavin	30.00	39.6	g/L	[8]
Ricobendazole	25.00	16.50	g/L	[103]
Sodium cefotaxim	20.00	0.073768	Mole F.	[133]
Sodium cefotaxim	22.00	0.074391	Mole F.	[133]
Sodium cefotaxim	24.00	0.075085	Mole F.	[133]
Sodium cefotaxim	25.00	0.075490	Mole F.	[133]
Sodium cefotaxim	27.00	0.076973	Mole F.	[133]
Sodium cefotaxim	29.00	0.077655	Mole F.	[133]
Sodium cefotaxim	30.00	0.077911	Mole F.	[133]
Sodium cefotaxim	32.00	0.078328	Mole F.	[133]
Sodium cefotaxim	34.00	0.078537	Mole F.	[133]
Sodium cefotaxim	35.00	0.078797	Mole F.	[133]
Sodium diclofenac	20.00	112	mg/g	[119]
Sodium diclofenac	25.00	135	mg/g	[119]
Sodium diclofenac	30.00	151	mg/g	[119]
Sodium diclofenac	35.00	182	mg/g	[119]
Sodium diclofenac	40.00	212	mg/g	[119]
Sulfadiazine	25.00	0.2097974	Mole F.	[39]
Sulfamethoxypyridazine	25.00	0.20980	Mole F.	[40]

Solubility Data of Drugs in Ethanol at Various Temperatures

Drug	*T* (°C)	Solubility	Solubility Unit	Reference
11α-Hydroxy-16α,17α-epoxyprogesterone	10.23	0.0004817	Mole F.	[105]
11α-Hydroxy-16α,17α-epoxyprogesterone	14.42	0.0005563	Mole F.	[105]
11α-Hydroxy-16α,17α-epoxyprogesterone	20.41	0.0006723	Mole F.	[105]
11α-Hydroxy-16α,17α-epoxyprogesterone	25.05	0.0007946	Mole F.	[105]
11α-Hydroxy-16α,17α-epoxyprogesterone	30.45	0.0009534	Mole F.	[105]
11α-Hydroxy-16α,17α-epoxyprogesterone	34.91	0.001106	Mole F.	[105]
11α-Hydroxy-16α,17α-epoxyprogesterone	35.40	0.001133	Mole F.	[105]
11α-Hydroxy-16α,17α-epoxyprogesterone	40.15	0.001323	Mole F.	[105]
11α-Hydroxy-16α,17α-epoxyprogesterone	45.63	0.001601	Mole F.	[105]
11α-Hydroxy-16α,17α-epoxyprogesterone	51.45	0.001882	Mole F.	[105]

(continued)

Solubility Data of Drugs in Ethanol at Various Temperatures (continued)

Drug	T (°C)	Solubility	Solubility Unit	Reference
16α,17α-epoxyprogesterone	12.95	0.00134	Mole F.	[17]
16α,17α-epoxyprogesterone	15.00	0.00145	Mole F.	[17]
16α,17α-epoxyprogesterone	20.20	0.00171	Mole F.	[17]
16α,17α-epoxyprogesterone	24.55	0.00211	Mole F.	[17]
16α,17α-epoxyprogesterone	29.90	0.00258	Mole F.	[17]
16α,17α-epoxyprogesterone	35.70	0.00340	Mole F.	[17]
16α,17α-epoxyprogesterone	39.50	0.00400	Mole F.	[17]
16α,17α-epoxyprogesterone	44.40	0.00487	Mole F.	[17]
16α,17α-epoxyprogesterone	49.55	0.00596	Mole F.	[17]
16α,17α-epoxyprogesterone	55.30	0.00729	Mole F.	[17]
1H-1,2,4-Triazole	10.03	0.1314	Mole F.	[92]
1H-1,2,4-Triazole	15.10	0.1471	Mole F.	[92]
1H-1,2,4-Triazole	20.06	0.1643	Mole F.	[92]
1H-1,2,4-Triazole	25.03	0.1836	Mole F.	[92]
1H-1,2,4-Triazole	30.00	0.2047	Mole F.	[92]
1H-1,2,4-Triazole	34.99	0.2278	Mole F.	[92]
1H-1,2,4-Triazole	40.06	0.2531	Mole F.	[92]
1H-1,2,4-Triazole	45.08	0.2799	Mole F.	[92]
1H-1,2,4-Triazole	50.07	0.3084	Mole F.	[92]
1H-1,2,4-Triazole	55.12	0.3389	Mole F.	[92]
1H-1,2,4-Triazole	60.13	0.3709	Mole F.	[92]
2-(4-Ethylbenzoyl)benzoic acid	9.14	0.0733	Mole F.	[18]
2-(4-Ethylbenzoyl)benzoic acid	16.53	0.1001	Mole F.	[18]
2-(4-Ethylbenzoyl)benzoic acid	21.76	0.1227	Mole F.	[18]
2-(4-Ethylbenzoyl)benzoic acid	31.45	0.1644	Mole F.	[18]
2-(4-Ethylbenzoyl)benzoic acid	41.15	0.2141	Mole F.	[18]
2-(4-Ethylbenzoyl)benzoic acid	49.76	0.2556	Mole F.	[18]
2-(4-Ethylbenzoyl)benzoic acid	55.50	0.2895	Mole F.	[18]
2-(4-Ethylbenzoyl)benzoic acid	65.04	0.3539	Mole F.	[18]
2-(4-Ethylbenzoyl)benzoic acid	69.67	0.3876	Mole F.	[18]
2,6-Diaminopyridine	−6.20	0.07515	Mole F.	[93]
2,6-Diaminopyridine	−2.80	0.08533	Mole F.	[93]
2,6-Diaminopyridine	1.00	0.09924	Mole F.	[93]
2,6-Diaminopyridine	4.30	0.10832	Mole F.	[93]
2,6-Diaminopyridine	8.00	0.12261	Mole F.	[93]
2,6-Diaminopyridine	13.20	0.14793	Mole F.	[93]
2,6-Diaminopyridine	17.00	0.16553	Mole F.	[93]
2,6-Diaminopyridine	20.40	0.18231	Mole F.	[93]
2,6-Diaminopyridine	24.00	0.20355	Mole F.	[93]
2,6-Diaminopyridine	28.00	0.22557	Mole F.	[93]
2,6-Diaminopyridine	31.40	0.23644	Mole F.	[93]
2,6-Diaminopyridine	34.80	0.26611	Mole F.	[93]
2,6-Diaminopyridine	38.90	0.28172	Mole F.	[93]
2,6-Diaminopyridine	43.10	0.31865	Mole F.	[93]
2,6-Diaminopyridine	48.10	0.34208	Mole F.	[93]
4-Aminobenzoic acid	25.00	0.05062	Mole F.	[20]
4-Aminophenylacetic acid	20.00	1.76	g/L	[8]
5,5-Diethylbarbituric acid (Barbital)	25.00	95	g/L	[41]

Solubility Data of Drugs in Ethanol at Various Temperatures (continued)

Drug	T (°C)	Solubility	Solubility Unit	Reference
5-Butyl-5-ethylbarbituric acid (Butetal)	25.00	392	g/L	[41]
5-Ethyl-5-(1-methylpropyl)-barbituric acid (Butabarbital)	25.00	80	g/L	[41]
5-Ethyl-5-(2-methylbutyl)-barbituric acid (Pentobarbital)	25.00	235	g/L	[41]
5-Ethyl-5-(2-methylpropyl)-barbituric acid	25.00	72	g/L	[41]
5-Ethyl-5-(3-methylbutyl)-barbituric acid (Amobarbital)	25.00	215	g/L	[41]
5-Ethyl-5-isopropylbarbituric acid (Probarbital)	25.00	65	g/L	[41]
5-Ethyl-5-pentylbarbituric acid	25.00	444	g/L	[41]
5-Ethyl-5-phenylbarbituric acid (Phenobarbital)	25.00	127	g/L	[41]
5-Ethyl-5-propylbarbiturate	25.00	195	g/L	[41]
Acetic acid	−29.40	0.457	Mole F.	[109]
Acetic acid	−24.90	0.503	Mole F.	[109]
Acetic acid	−20.90	0.540	Mole F.	[109]
Acetic acid	−20.30	0.548	Mole F.	[109]
Acetic acid	−16.90	0.573	Mole F.	[109]
Acetic acid	−14.90	0.595	Mole F.	[109]
Acetic acid	−12.30	0.623	Mole F.	[109]
Acetic acid	−8.00	0.657	Mole F.	[109]
Acetic acid	−6.70	0.700	Mole F.	[109]
Acetic acid	−4.10	0.718	Mole F.	[109]
Acetic acid	1.90	0.816	Mole F.	[109]
Acetic acid	7.20	0.861	Mole F.	[109]
Acetic acid	11.20	0.922	Mole F.	[109]
Acetylsalicylic acid	3.15	0.0250	Mole F.	[94]
Acetylsalicylic acid	18.75	0.049	Mole F.	[94]
Acetylsalicylic acid	25.00	0.0855	Mole F.	[21]
Acetylsalicylic acid	29.25	0.071	Mole F.	[94]
Acetylsalicylic acid	36.85	0.093	Mole F.	[94]
Acetylsalicylic acid	43.35	0.113	Mole F.	[94]
Acetylsalicylic acid	48.35	0.133	Mole F.	[94]
Acetylsalicylic acid	52.75	0.152	Mole F.	[94]
Acetylsalicylic acid	57.25	0.170	Mole F.	[94]
Acetylsalicylic acid	60.25	0.187	Mole F.	[94]
Acetylsalicylic acid	63.45	0.204	Mole F.	[94]
Anhydrous citric acid	25.00	0.14363	Mole F.	[23]
Atractylenolide III	10.05	0.004426	Mole F.	[132]
Atractylenolide III	15.05	0.005475	Mole F.	[132]
Atractylenolide III	20.05	0.006733	Mole F.	[132]
Atractylenolide III	25.05	0.008085	Mole F.	[132]
Atractylenolide III	30.05	0.009657	Mole F.	[132]
Atractylenolide III	35.05	0.01131	Mole F.	[132]
Atractylenolide III	40.05	0.01309	Mole F.	[132]
Atractylenolide III	45.05	0.01507	Mole F.	[132]
Atractylenolide III	50.05	0.01713	Mole F.	[132]
Benzoic acid	25.00	0.1789	Mole F.	[24]

(continued)

Solubility Data of Drugs in Ethanol at Various Temperatures   (continued)

Drug	T (°C)	Solubility	Solubility Unit	Reference
Benzoyl peroxide	25.00	17.8	mg/g	[135]
Berberine chloride	24.85	0.000444	Mole F.	[44]
Berberine chloride	29.85	0.000528	Mole F.	[44]
Berberine chloride	34.85	0.000637	Mole F.	[44]
Berberine chloride	39.85	0.000778	Mole F.	[44]
β-Carotene	Room	30	mg/L	[6]
Betulin	5.05	0.000579	Mole F.	[7]
Betulin	15.05	0.000658	Mole F.	[7]
Betulin	25.05	0.000898	Mole F.	[7]
Betulin	35.05	0.001133	Mole F.	[7]
Butylparaben	25.00	0.360	Mole F.	[45]
Butylparaben	30.00	0.395	Mole F.	[45]
Butylparaben	35.00	0.437	Mole F.	[45]
Butylparaben	40.00	0.500	Mole F.	[45]
Caffeine (form 1)	25.00	5.66	g/L	[136]
Caffeine (form 2)	25.00	5.57	g/L	[136]
Caffeine hydrate	25.00	7.32	g/L	[136]
Carbamazepine	25.00	24.4	g/L	[8]
Carbamazepine (form III)	5.65	0.003090	Mole F.	[46]
Carbamazepine (form III)	10.20	0.003820	Mole F.	[46]
Carbamazepine (form III)	14.55	0.004200	Mole F.	[46]
Carbamazepine (form III)	19.05	0.004800	Mole F.	[46]
Carbamazepine (form III)	22.65	0.005390	Mole F.	[46]
Carbamazepine (form III)	27.95	0.006490	Mole F.	[46]
Carbamazepine (form III)	32.55	0.007190	Mole F.	[46]
Carbamazepine (form III)	39.55	0.008260	Mole F.	[46]
Carbamazepine (form III)	45.85	0.009870	Mole F.	[46]
Carbamazepine (form III)	56.60	0.01647	Mole F.	[46]
Carbamazepine (form III)	65.01	0.02578	Mole F.	[46]
Carbamazepine (form 1)	25.00	23.50	g/L	[136]
Carbamazepine (form 2)	25.00	27.04	g/L	[136]
Carbamazepine hydrate	25.00	23.83	g/L	[136]
Cefazolin sodium pentahydrate	2.30	0.00022	Mole F.	[9]
Cefazolin sodium pentahydrate	10.50	0.0002357	Mole F.	[9]
Cefazolin sodium pentahydrate	15.00	0.0002738	Mole F.	[9]
Cefazolin sodium pentahydrate	18.60	0.0002988	Mole F.	[9]
Cefazolin sodium pentahydrate	20.00	0.0003176	Mole F.	[9]
Cefazolin sodium pentahydrate	21.80	0.0003406	Mole F.	[9]
Cefazolin sodium pentahydrate	25.10	0.0003929	Mole F.	[9]
Cefazolin sodium pentahydrate	27.20	0.0004352	Mole F.	[9]
Cefazolin sodium pentahydrate	31.70	0.0005205	Mole F.	[9]
Cefazolin sodium pentahydrate	32.50	0.0005533	Mole F.	[9]
Cefazolin sodium pentahydrate	35.90	0.0006419	Mole F.	[9]
Cefotaxim sodium	5.00	0.195	g/100 g	[111]
Cefotaxim sodium	10.00	0.216	g/100 g	[111]
Cefotaxim sodium	15.00	0.264	g/100 g	[111]
Cefotaxim sodium	20.00	0.307	g/100 g	[111]

Solubility Data of Drugs in Ethanol at Various Temperatures   (continued)

Drug	T (°C)	Solubility	Solubility Unit	Reference
Cefotaxim sodium	30.00	0.379	g/100 g	[111]
Cefotaxim sodium	40.00	0.429	g/100 g	[111]
Ceftriaxone disodium	4.70	0.0000130	Mole F.	[112]
Ceftriaxone disodium	8.90	0.0000165	Mole F.	[112]
Ceftriaxone disodium	13.25	0.0000190	Mole F.	[112]
Ceftriaxone disodium	18.55	0.0000212	Mole F.	[112]
Ceftriaxone disodium	23.95	0.0000224	Mole F.	[112]
Ceftriaxone disodium	26.10	0.0000231	Mole F.	[112]
Ceftriaxone disodium	31.00	0.0000251	Mole F.	[112]
Ceftriaxone disodium	36.25	0.0000331	Mole F.	[112]
Ceftriaxone disodium	41.70	0.0000387	Mole F.	[112]
Ceftriaxone disodium	43.20	0.0000437	Mole F.	[112]
Celecoxib	25.00	63.346	g/L	[55]
Chlorpheniramine maleate	9.50	0.0026158	Mole F.	[47]
Chlorpheniramine maleate	16.00	0.0037752	Mole F.	[47]
Chlorpheniramine maleate	20.90	0.0050769	Mole F.	[47]
Chlorpheniramine maleate	25.00	0.0065584	Mole F.	[47]
Chlorpheniramine maleate	30.30	0.0091899	Mole F.	[47]
Chlorpheniramine maleate	35.29	0.0128680	Mole F.	[47]
Chlorpheniramine maleate	39.78	0.0177940	Mole F.	[47]
Chlorpheniramine maleate	45.18	0.0261900	Mole F.	[47]
Chlorpheniramine maleate	50.05	0.0376680	Mole F.	[47]
Chlorpheniramine maleate	54.70	0.0531800	Mole F.	[47]
Chlorpheniramine maleate	57.92	0.0673130	Mole F.	[47]
Cholesteryl acetate	11.50	0.373	g/100 g	[97]
Cholesteryl acetate	20.50	0.655	g/100 g	[97]
Cholesteryl acetate	30.00	0.970	g/100 g	[97]
Cholesteryl acetate	33.00	1.092	g/100 g	[97]
Cholesteryl acetate	33.90	1.192	g/100 g	[97]
Cholesteryl acetate	37.50	1.446	g/100 g	[97]
Cholestrole	27.50	3.456	g/100 g	[97]
Cholestrole	30.70	3.962	g/100 g	[97]
Cholestrole	33.90	4.445	g/100 g	[97]
Cholestrole	38.00	5.258	g/100 g	[97]
Cholestrole	44.20	0.811	g/100 g	[97]
Ciprofloxacin	20.00	0.0000081	Mole F.	[98]
Ciprofloxacin	30.00	0.0000140	Mole F.	[98]
Ciprofloxacin	40.00	0.0000201	Mole F.	[98]
Ciprofloxacin	50.00	0.0000336	Mole F.	[98]
Clindamycin phosphate	10.00	0.00000759	Mole F.	[134]
Clindamycin phosphate	15.00	0.00000843	Mole F.	[134]
Clindamycin phosphate	20.00	0.0000101	Mole F.	[134]
Clindamycin phosphate	25.00	0.0000126	Mole F.	[134]
Clindamycin phosphate	30.00	0.0000139	Mole F.	[134]
Clindamycin phosphate	35.00	0.0000156	Mole F.	[134]
Clindamycin phosphate	40.00	0.0000176	Mole F.	[134]
Clindamycin phosphate	45.00	0.0000190	Mole F.	[134]
Danazol	25.00	22.80	g/L	[137]
Dexamethasone sodium phosphate	5.00	0.0001316	Mole F.	[138]

(continued)

Solubility Data of Drugs in Ethanol at Various Temperatures   (continued)

Drug	T (°C)	Solubility	Solubility Unit	Reference
Dexamethasone sodium phosphate	10.00	0.0001368	Mole F.	[138]
Dexamethasone sodium phosphate	15.00	0.0001548	Mole F.	[138]
Dexamethasone sodium phosphate	20.00	0.0001605	Mole F.	[138]
Dexamethasone sodium phosphate	25.00	0.0001751	Mole F.	[138]
Dexamethasone sodium phosphate	30.00	0.0002032	Mole F.	[138]
Dexamethasone sodium phosphate	35.00	0.0002373	Mole F.	[138]
Dexamethasone sodium phosphate	40.00	0.0002563	Mole F.	[138]
Dexamethasone sodium phosphate	45.00	0.0002849	Mole F.	[138]
Diclofenac	25.00	0.0087299	Mole F.	[25]
Diflunisal	25.00	0.0191	Mole F.	[26]
Digoxine	25.00	2.241	g/L	[137]
Dioxopromethazine HCl	5.01	0.0002686	Mole F.	[106]
Dioxopromethazine HCl	10.06	0.0002961	Mole F.	[106]
Dioxopromethazine HCl	15.07	0.0003225	Mole F.	[106]
Dioxopromethazine HCl	20.03	0.0003650	Mole F.	[106]
Dioxopromethazine HCl	25.06	0.0004190	Mole F.	[106]
Dioxopromethazine HCl	30.07	0.0004999	Mole F.	[106]
Dioxopromethazine HCl	35.03	0.0005887	Mole F.	[106]
Dioxopromethazine HCl	40.02	0.0006860	Mole F.	[106]
Dioxopromethazine HCl	45.05	0.0008286	Mole F.	[106]
D-p-Hydroxyphenylglycine	5.13	0.0000926	Mole F.	[127]
D-p-Hydroxyphenylglycine	10.27	0.0000930	Mole F.	[127]
D-p-Hydroxyphenylglycine	15.03	0.0000934	Mole F.	[127]
D-p-Hydroxyphenylglycine	20.09	0.0000937	Mole F.	[127]
D-p-Hydroxyphenylglycine	25.23	0.0000940	Mole F.	[127]
D-p-Hydroxyphenylglycine	30.02	0.0000942	Mole F.	[127]
D-p-Hydroxyphenylglycine	35.24	0.0000943	Mole F.	[127]
D-p-Hydroxyphenylglycine	40.04	0.0000944	Mole F.	[127]
D-p-Hydroxyphenylglycine	45.22	0.0000945	Mole F.	[127]
D-p-Hydroxyphenylglycine	50.11	0.0000945	Mole F.	[127]
Eflucimibe (form A)	19.85	0.000554	Mole F.	[99]
Eflucimibe (form A)	20.00	0.000654	mol/L	[99]
Eflucimibe (form A)	24.85	0.000731	Mole F.	[99]
Eflucimibe (form A)	29.85	0.00104	Mole F.	[99]
Eflucimibe (form A)	34.85	0.00147	Mole F.	[99]
Eflucimibe (form A)	36.85	0.00169	Mole F.	[99]
Eflucimibe (form A)	39.85	0.00207	Mole F.	[99]
Eflucimibe (form A)	44.85	0.00300	Mole F.	[99]
Eflucimibe (form A)	49.85	0.00445	Mole F.	[99]
Eflucimibe (form A)	53.85	0.005650	Mole F.	[99]
Eflucimibe (form A)	54.85	0.005970	Mole F.	[99]
Eflucimibe (form B)	19.85	0.000577	Mole F.	[99]
Eflucimibe (form B)	24.85	0.000791	Mole F.	[99]
Eflucimibe (form B)	29.85	0.001130	Mole F.	[99]
Eflucimibe (form B)	34.85	0.001620	Mole F.	[99]
Eflucimibe (form B)	36.85	0.001860	Mole F.	[99]
Eflucimibe (form B)	39.85	0.002300	Mole F.	[99]
Eflucimibe (form B)	44.85	0.003300	Mole F.	[99]
Eflucimibe (form B)	49.85	0.004750	Mole F.	[99]

Solubility Data of Drugs in Ethanol at Various Temperatures (continued)

Drug	T (°C)	Solubility	Solubility Unit	Reference
Eflucimibe (form B)	53.85	0.006380	Mole F.	[99]
Eflucimibe (form B)	54.85	0.006870	Mole F.	[99]
Erythritol	14.40	0.003464	Mass fraction	[113]
Erythritol	16.60	0.003719	Mass fraction	[113]
Erythritol	20.60	0.004213	Mass fraction	[113]
Erythritol	26.30	0.005547	Mass fraction	[113]
Erythritol	30.30	0.008320	Mass fraction	[113]
Erythritol	34.00	0.009519	Mass fraction	[113]
Erythritol	39.50	0.013456	Mass fraction	[113]
Erythritol	44.80	0.018095	Mass fraction	[113]
Erythritol	49.60	0.023567	Mass fraction	[113]
Erythritol	52.20	0.025738	Mass fraction	[113]
Erythritol	62.80	0.041785	Mass fraction	[113]
Erythritol	66.00	0.048504	Mass fraction	[113]
Erythromycin A dihydrate	20.05	0.01483	Mole F.	[100]
Erythromycin A dihydrate	24.85	0.01919	Mole F.	[100]
Erythromycin A dihydrate	30.00	0.02544	Mole F.	[100]
Erythromycin A dihydrate	34.85	0.03383	Mole F.	[100]
Erythromycin A dihydrate	39.95	0.04600	Mole F.	[100]
Erythromycin A dihydrate	44.85	0.06200	Mole F.	[100]
Erythromycin A dihydrate	49.85	0.08100	Mole F.	[100]
Ethylparaben	25.00	0.165	Mole F.	[45]
Ethylparaben	30.00	0.184	Mole F.	[45]
Ethylparaben	35.00	0.218	Mole F.	[45]
Ethylparaben	40.00	0.220	Mole F.	[45]
Flubiprofen	25.00	0.0612	Mole F.	[26]
Gallic acid	25.05	0.1890	Mass fraction	
Gallic acid	30.05	0.1894	Mass fraction	[139]
Gallic acid	35.05	0.1900	Mass fraction	[139]
Gallic acid	40.05	0.1917	Mass fraction	[139]
Gallic acid	45.05	0.1955	Mass fraction	[139]
Gallic acid	50.05	0.2002	Mass fraction	[139]
Gallic acid	55.05	0.2045	Mass fraction	[139]
Gallic acid	60.05	0.2093	Mass fraction	[139]
Glucose	40.00	3.81	g/L	[42]
Glucose	60.00	9.02	g/L	[42]
Haloperidol	25.00	0.00570	Mole F.	[27]
Hesperetin	15.05	0.069	mol/L	[50]
Hesperetin	20.05	0.073	mol/L	[50]
Hesperetin	25.05	0.087	mol/L	[50]
Hesperetin	30.05	0.093	mol/L	[50]
Hesperetin	35.05	0.107	mol/L	[50]
Hesperetin	40.05	0. 106	mol/L	[50]
Hesperetin	45.05	0.131	mol/L	[50]
Hesperetin	50.05	0.132	mol/L	[50]
Hydroquinone	3.50	0.1630	Mole F.	[101]
Hydroquinone	7.70	0.1685	Mole F.	[101]
Hydroquinone	12.15	0.1748	Mole F.	[101]
Hydroquinone	18.25	0.1828	Mole F.	[101]

(continued)

Solubility Data of Drugs in Ethanol at Various Temperatures   (continued)

Drug	T (°C)	Solubility	Solubility Unit	Reference
Hydroquinone	24.50	0.1914	Mole F.	[101]
Hydroquinone	31.40	0.2012	Mole F.	[101]
Hydroquinone	38.35	0.2123	Mole F.	[101]
Hydroquinone	46.40	0.2261	Mole F.	[101]
Hydroquinone	53.80	0.2392	Mole F.	[101]
Hydroquinone	58.85	0.2468	Mole F.	[101]
Hydroquinone	63.60	0.2610	Mole F.	[101]
Hydroquinone	69.00	0.2800	Mole F.	[101]
Ibuprofen	10.00	592.6	g/kg	[83]
Ibuprofen	15.00	765.8	g/kg	[83]
Ibuprofen	20.00	886.5	g/kg	[83]
Ibuprofen	25.00	0.1422314	Mole F.	[28]
Ibuprofen	25.00	0.08392	Mole F.	[67]
Ibuprofen	30.00	1420	g/kg	[83]
Ibuprofen	35.00	2230	g/kg	[83]
Imidacloprid	25.54	0.0005024	Mole F.	[2]
Imidacloprid	27.97	0.0005693	Mole F.	[2]
Imidacloprid	32.51	0.0007937	Mole F.	[2]
Imidacloprid	35.72	0.001002	Mole F.	[2]
Imidacloprid	39.19	0.001176	Mole F.	[2]
Imidacloprid	42.81	0.001362	Mole F.	[2]
Imidacloprid	44.92	0.001513	Mole F.	[2]
Imidacloprid	47.59	0.001703	Mole F.	[2]
Imidacloprid	50.81	0.001967	Mole F.	[2]
Imidacloprid	54.39	0.002252	Mole F.	[2]
Imidacloprid	56.91	0.002514	Mole F.	[2]
Imidacloprid	60.42	0.002886	Mole F.	[2]
Irbesartan (form A)	5.45	0.0003328	Mole F.	[29]
Irbesartan (form A)	10.40	0.000489	Mole F.	[29]
Irbesartan (form A)	15.35	0.000627	Mole F.	[29]
Irbesartan (form A)	19.55	0.000809	Mole F.	[29]
Irbesartan (form A)	24.65	0.001059	Mole F.	[29]
Irbesartan (form A)	29.90	0.001220	Mole F.	[29]
Irbesartan (form A)	35.35	0.001593	Mole F.	[29]
Irbesartan (form A)	40.15	0.002098	Mole F.	[29]
Irbesartan (form A)	45.35	0.002630	Mole F.	[29]
Irbesartan (form A)	50.45	0.003411	Solubility Unit	[29]
Isoniazid	27.85	0.00545	Mole F.	[114]
Isoniazid	29.85	0.00579	Mole F.	[114]
Isoniazid	34.85	0.00717	Mole F.	[114]
Isoniazid	37.85	0.00816	Mole F.	[114]
Isoniazid	39.85	0.00887	Mole F.	[114]
Ketoconazol	25.00	22.46	g/L	[137]
Ketoprofen	25.00	0.0640	Mole F.	[51]
Ketoprofen	25.00	0.0701	Mole F.	[68]
L-(+)-Ascorbic acid	19.85	0.00229	Mole F.	[102]
L-(+)-Ascorbic acid	24.85	0.00274	Mole F.	[102]
L-(+)-Ascorbic acid	29.85	0.00321	Mole F.	[102]
L-(+)-Ascorbic acid	34.85	0.00385	Mole F.	[102]

Solubility Data of Drugs in Ethanol at Various Temperatures   (continued)

Drug	T (°C)	Solubility	Solubility Unit	Reference
L-(+)-Ascorbic acid	39.85	0.00463	Mole F.	[102]
L-(+)-Ascorbic acid	44.85	0.00535	Mole F.	[102]
L-(+)-Ascorbic acid	49.85	0.0059	Mole F.	[102]
Lactose	25.00	0.0000090	Mole F.	[30]
Lamivudine (form 2)	5.00	7.0	g/L	[52]
Lamivudine (form 2)	15.00	8.8	g/L	[52]
Lamivudine (form 2)	25.00	11.4	g/L	[52]
Lamivudine (form 2)	35.00	15.2	g/L	[52]
Lamivudine (form 2)	45.00	19.6	g/L	[52]
Losartan potassium	20.00	0.03807	Mole F.	[86]
Losartan potassium	30.00	0.04729	Mole F.	[86]
Losartan potassium	40.00	0.05471	Mole F.	[86]
Losartan potassium	50.00	0.07022	Mole F.	[86]
Losartan potassium	60.00	0.07394	Mole F.	[86]
Losartan potassium	70.00	0.07846	Mole F.	[86]
Lovastatin	5.10	0.001402	Mole F.	[116]
Lovastatin	10.15	0.001797	Mole F.	[116]
Lovastatin	13.00	0.0020574	Mole F.	[53]
Lovastatin	15.10	0.002256	Mole F.	[116]
Lovastatin	16.00	0.0023045	Mole F.	[53]
Lovastatin	17.95	0.0025077	Mole F.	[53]
Lovastatin	21.80	0.002981	Mole F.	[116]
Lovastatin	23.50	0.0032270	Mole F.	[53]
Lovastatin	25.00	0.003529	Mole F.	[116]
Lovastatin	28.50	0.0040839	Mole F.	[53]
Lovastatin	29.90	0.004460	Mole F.	[116]
Lovastatin	31.40	0.0046648	Mole F.	[53]
Lovastatin	34.00	0.0054305	Mole F.	[53]
Lovastatin	34.65	0.005536	Mole F.	[116]
Lovastatin	36.00	0.0058507	Mole F.	[53]
Lovastatin	37.40	0.0062596	Mole F.	[53]
Lovastatin	39.90	0.007094	Mole F.	[116]
Lovastatin	45.50	0.009326	Mole F.	[116]
Lovastatin	49.50	0.01119	Mole F.	[116]
Lutein	Room	300	mg/L	[6]
Luteolin	0.00	0.00076	Mole F.	[54]
Luteolin	10.00	0.00114	Mole F.	[54]
Luteolin	25.00	0.00185	Mole F.	[54]
Luteolin	40.00	0.00268	Mole F.	[54]
Luteolin	60.00	0.00378	Mole F.	[54]
Mannitol	25.00	0.0001129	Mole F.	[30]
Mefenamic acid	25.00	5.96	g/L	[137]
Meloxicam	25.00	0.21	g/L	[8]
Meloxicam	25.00	0.354	g/L	[55]
Methyl paraben	25.00	0.147	Mole F.	[45]
Methyl paraben	30.00	0.156	Mole F.	[45]
Methyl paraben	35.00	0.174	Mole F.	[45]
Methyl paraben	40.00	0.201	Mole F.	[45]

(continued)

Solubility Data of Drugs in Ethanol at Various Temperatures (continued)

Drug	T (°C)	Solubility	Solubility Unit	Reference
Methyl p-hydroxybenzoate	25.00	0.1495	Mole F.	[35]
Naproxen	25.00	0.0201148	Mole F.	[31]
N-Epoxymethyl-1,8-naphthalimide	N/A	1.25	g/L	[140]
Niflumic acid	25.00	0.0144943	Mole F.	[33]
Nimesulide	25.00	3.18	g/L	[8]
Nimesulide	25.00	3.32	g/L	[55]
Oleanolic acid	15.15	0.0089	mol/L	[56]
Oleanolic acid	20.15	0.0111	mol/L	[56]
Oleanolic acid	25.15	0.0135	mol/L	[56]
Oleanolic acid	30.15	0.0149	mol/L	[56]
Oleanolic acid	35.15	0.0157	mol/L	[56]
Oleanolic acid	40.15	0.0171	mol/L	[56]
Oleanolic acid	45.15	0.0211	mol/L	[56]
Oleanolic acid	50.15	0.0221	mol/L	[56]
Oleanolic acid	55.15	0.0261	mol/L	[56]
p-Aminobenzoic acid	25.00	0.0465141	Mole F.	[25]
p-Aminophenylacetic acid	16.00	1.1	g/kg	[83]
p-Aminophenylacetic acid	20.00	1.4	g/kg	[83]
p-Aminophenylacetic acid	25.00	1.8	g/kg	[83]
p-Aminophenylacetic acid	30.00	2.3	g/kg	[83]
Paracetamol	−5.00	118.56	g/kg	[11]
Paracetamol	0.00	129.65	g/kg	[11]
Paracetamol	5.00	141.82	g/kg	[11]
Paracetamol	10.00	156.14	g/kg	[11]
Paracetamol	15.00	171.400	g/kg	[11]
Paracetamol	20.00	190.61	g/kg	[11]
Paracetamol	25.00	209.91	g/kg	[11]
Paracetamol	30.00	232.75	g/kg	[11]
Paracetamol	30.00	152	g/L	[8]
Paracetamol	25.00	0.19664	Mole F.	[23]
Paroxetine HCl hemihydrate	27.00	0.0032598	Mole F.	[57]
Paroxetine HCl hemihydrate	30.05	0.0036201	Mole F.	[57]
Paroxetine HCl hemihydrate	32.95	0.0039162	Mole F.	[57]
Paroxetine HCl hemihydrate	36.85	0.0041637	Mole F.	[57]
Paroxetine HCl hemihydrate	39.95	0.0043103	Mole F.	[57]
Paroxetine HCl hemihydrate	43.00	0.0044358	Mole F.	[57]
Paroxetine HCl hemihydrate	45.90	0.0045115	Mole F.	[57]
Paroxetine HCl hemihydrate	48.90	0.0046440	Mole F.	[57]
Paroxetine HCl hemihydrate	52.10	0.0047787	Mole F.	[57]
Paroxetine HCl hemihydrate	54.85	0.0049146	Mole F.	[57]
Paroxetine HCl hemihydrate	58.15	0.0049409	Mole F.	[57]
Paroxetine HCl hemihydrate	61.15	0.0050889	Mole F.	[57]
Paroxetine HCl hemihydrate	63.95	0.0052088	Mole F.	[57]
Paroxetine HCl hemihydrate	67.05	0.0053509	Mole F.	[57]
Paroxetine HCl hemihydrate	70.05	0.0054782	Mole F.	[57]
Phenacetinum	9.50	0.0096408	Mole F.	[58]
Phenacetinum	16.10	0.012861	Mole F.	[58]
Phenacetinum	20.26	0.015170	Mole F.	[58]

Solubility Data of Drugs in Ethanol at Various Temperatures   (continued)

Drug	T (°C)	Solubility	Solubility Unit	Reference
Phenacetinum	25.10	0.018376	Mole F.	[58]
Phenacetinum	30.16	0.022580	Mole F.	[58]
Phenacetinum	34.70	0.027324	Mole F.	[58]
Phenacetinum	40.15	0.034532	Mole F.	[58]
Phenacetinum	45.10	0.042789	Mole F.	[58]
Phenacetinum	50.10	0.053061	Mole F.	[58]
Phenacetinum	55.50	0.066650	Mole F.	[58]
Phenacetinum	60.55	0.082236	Mole F.	[58]
Phenazine-1-carboxylic acid	5.05	0.00370	mol/L	[141]
Phenazine-1-carboxylic acid	10.05	0.00430	mol/L	[141]
Phenazine-1-carboxylic acid	15.05	0.00490	mol/L	[141]
Phenazine-1-carboxylic acid	20.05	0.00560	mol/L	[141]
Phenazine-1-carboxylic acid	25.05	0.00640	mol/L	[141]
Phenazine-1-carboxylic acid	30.05	0.00740	mol/L	[141]
Phenazine-1-carboxylic acid	35.05	0.00840	mol/L	[141]
Phenazine-1-carboxylic acid	40.05	0.00960	mol/L	[141]
Phenazine-1-carboxylic acid	45.05	0.01090	mol/L	[141]
Phenazine-1-carboxylic acid	50.05	0.01230	mol/L	[141]
Phenazine-1-carboxylic acid	55.05	0.01390	mol/L	[141]
Phenothiazine	25.00	0.00890	Mole F.	[34]
Phenylacetic acid	20.00	2064	g/kg	[83]
p-Hydroxyphenylacetic acid	10.00	567.8	g/kg	[83]
p-Hydroxyphenylacetic acid	15.00	702.3	g/kg	[83]
p-Hydroxyphenylacetic acid	20.00	937.5	g/kg	[83]
p-Hydroxyphenylacetic acid	25.00	601.4	g/kg	[83]
p-Hydroxybenzoic acid	25.00	0.1213	Mole F.	[35]
p-Hydroxybenzoic acid	25.00	432.7	g/kg	[83]
Pimozide	25.00	0.0012	Mole F.	[36]
Piroxicam	25.00	0.0001412	Mole F.	[33]
Potassium clavulanate	0.80	0.0004627	Mole F.	[59]
Potassium clavulanate	3.95	0.0004912	Mole F.	[59]
Potassium clavulanate	7.85	0.0005211	Mole F.	[59]
Potassium clavulanate	12.25	0.0005653	Mole F.	[59]
Potassium clavulanate	15.95	0.0006134	Mole F.	[59]
Potassium clavulanate	20.20	0.0006639	Mole F.	[59]
Potassium clavulanate	23.90	0.0007120	Mole F.	[59]
Potassium clavulanate	28.30	0.0007767	Mole F.	[59]
Potassium clavulanate	32.10	0.0008636	Mole F.	[59]
Pravastatin sodium	4.85	0.0031	Mole F.	[60]
Pravastatin sodium	9.85	0.00325	Mole F.	[60]
Pravastatin sodium	14.85	0.00346	Mole F.	[60]
Pravastatin sodium	19.85	0.00366	Mole F.	[60]
Pravastatin sodium	24.85	0.00406	Mole F.	[60]
Pravastatin sodium	29.85	0.00429	Mole F.	[60]
Pravastatin sodium	34.85	0.00463	Mole F.	[60]
Pravastatin sodium	39.85	0.00493	Mole F.	[60]
Pravastatin sodium	44.85	0.00569	Mole F.	[60]
Pravastatin sodium	49.85	0.00692	Mole F.	[60]
Pravastatin sodium	54.85	0.00793	Mole F.	[60]

(continued)

Solubility Data of Drugs in Ethanol at Various Temperatures   (continued)

Drug	T (°C)	Solubility	Solubility Unit	Reference
Pravastatin sodium	59.85	0.00942	Mole F.	[60]
Propylparaben	25.00	0.197	Mole F.	[45]
Propylparaben	30.00	0.228	Mole F.	[45]
Propylparaben	35.00	0.296	Mole F.	[45]
Propylparaben	40.00	0.298	Mole F.	[45]
Puerarin	15.05	0.2954	Mole F.	[117]
Puerarin	20.05	0.2601	Mole F.	[117]
Puerarin	25.05	0.2302	Mole F.	[117]
Puerarin	30.05	0.2081	Mole F.	[117]
Puerarin	35.05	0.1860	Mole F.	[117]
Puerarin	40.05	0.1660	Mole F.	[117]
Puerarin	45.05	0.1422	Mole F.	[117]
Puerarin	50.05	0.1298	Mole F.	[117]
Puerarin	55.05	0.1200	Mole F.	[117]
Ricobendazole	25.00	1.16	g/L	[103]
Rifapentine	5.00	0.000515	Mole F.	[13]
Rifapentine	10.00	0.000525	Mole F.	[13]
Rifapentine	15.00	0.000541	Mole F.	[13]
Rifapentine	20.00	0.000562	Mole F.	[13]
Rifapentine	24.00	0.000583	Mole F.	[13]
Rifapentine	30.00	0.000604	Mole F.	[13]
Rifapentine	35.00	0.000625	Mole F.	[13]
Rifapentine	40.00	0.000651	Mole F.	[13]
Rifapentine	45.00	0.000677	Mole F.	[13]
Rifapentine	50.00	0.000703	Mole F.	[13]
Rofecoxib	25.00	0.683	g/L	[55]
Rutin	10.00	0.003721	Mole F.	[61]
Rutin	25.00	0.004365	Mole F.	[61]
Rutin	40.00	0.005918	Mole F.	[61]
Rutin	50.00	0.006532	Mole F.	[61]
Rutin	60.00	0.007453	Mole F.	[61]
Saccharose	25.00	0.000000071	Mole F.	[30]
Salicylamide	25.00	0.03847	Mole F.	[37]
Salicylic acid	24.85	0.32538	Mole F.	[128]
Salicylic acid	25.00	0.1100302	Mole F.	[25]
Salicylic acid	25.00	0.1450	Mole F.	[38]
Salicylic acid	29.85	0.35696	Mass. F	[128]
Salicylic acid	34.85	0.38513	Mass. F	[128]
Salicylic acid	39.85	0.40553	Mass. F	[128]
Salicylic acid	44.85	0.42740	Mass. F	[128]
Salicylic acid	49.85	0.44890	Mass. F	[128]
Salicylic acid	54.85	0.47786	Mass. F	[128]
Salicylic acid	59.85	0.49958	Mass. F	[128]
Salicylic acid	64.85	0.52519	Mass. F	[128]
Salicylic acid	69.85	0.54340	Mass. F	[128]
Salicylic acid	74.85	0.56977	Mass. F	[128]
Sitosterol	30.40	2.901	g/100 g	[97]
Sitosterol	33.90	3.504	g/100 g	[97]
Sitosterol	38.00	3.901	g/100 g	[97]

Solubility Data of Drugs in Ethanol at Various Temperatures   (continued)

Drug	T (°C)	Solubility	Solubility Unit	Reference
Sitosterol	42.10	4.628	g/100 g	[97]
Sodium ibuprofen	25.00	0.0775137	Mole F.	[28]
Sodium benzoate	25.00	0.0014653	Mole F.	[28]
Sodium diclofenac	25.00	0.0079825	Mole F.	[31]
Sodium diclofenac	25.00	0.0079785	Mole F.	[25]
Sodium p-aminobenzoate	25.00	0.0015869	Mole F.	[25]
Sodium salicylate	25.00	0.0124629	Mole F.	[25]
Stearic acid	18.30	0.0025	Mole F.	[104]
Stearic acid	21.80	0.0031	Mole F.	[104]
Stearic acid	26.70	0.0057	Mole F.	[104]
Stearic acid	31.70	0.0103	Mole F.	[104]
Stearic acid	36.30	0.0180	Mole F.	[104]
Stearic acid	39.20	0.0280	Mole F.	[104]
Stearic acid	41.60	0.0406	Mole F.	[104]
Stearic acid	43.10	0.0498	Mole F.	[104]
Stearic acid	43.70	0.0560	Mole F.	[104]
Stearic acid	45.40	0.0731	Mole F.	[104]
Stearic acid	46.80	0.0926	Mole F.	[104]
Stearic acid	48.80	0.1241	Mole F.	[104]
Stearic acid	50.50	0.1546	Mole F.	[104]
Stearic acid	15.00	0.79	g/100 g	[14]
Stearic acid	20.00	1.27	g/100 g	[14]
Stearic acid	25.00	2.23	g/100 g	[14]
Stearic acid	30.00	4.35	g/100 g	[14]
Stearic acid	35.00	9.86	g/100 g	[14]
Stearic acid	40.00	27.3	g/100 g	[14]
Sulfadiazine	25.00	0.0000840	Mole F.	[39]
Sulfadiazine	25.00	0.00132	mol/L	[62]
Sulfadiazine	25.00	0.0000768	Mole F.	[63]
Sulfadiazine	30.00	0.0000936	Mole F.	[63]
Sulfadiazine	37.00	0.0001240	Mole F.	[63]
Sulfadimethoxine	25.00	0.0007140	Mole F.	[63]
Sulfadimethoxine	30.00	0.0008580	Mole F.	[63]
Sulfadimethoxine	37.00	0.0011000	Mole F.	[63]
Sulfamethoxypyridazine	25.00	0.00126	Mole F.	[40]
Sulfisoxazole	25.00	0.0714	mol/L	[62]
Sulfisomidine	25.00	0.0005530	Mole F.	[63]
Sulfisomidine	30.00	0.0006380	Mole F.	[63]
Sulfisomidine	37.00	0.0008200	Mole F.	[63]
Tanshinone IIA	10.05	0.00249	mol/L	[121]
Tanshinone IIA	20.15	0.00295	mol/L	[121]
Tanshinone IIA	29.95	0.00389	mol/L	[121]
Tanshinone IIA	35.05	0.00441	mol/L	[121]
Tanshinone IIA	44.85	0.00661	mol/L	[121]
Tanshinone IIA	50.15	0.00582	mol/L	[121]
Temazepam	24.00	0.0156	Volume F.	[64]
Temazepam	25.00	0.0029997	Mole F.	[15]
Tetracycline	15.00	0.00613	Mole F.	[98]
Tetracycline	20.00	0.00732	Mole F.	[98]

(continued)

Solubility Data of Drugs in Ethanol at Various Temperatures   (continued)

Drug	T (°C)	Solubility	Solubility Unit	Reference
Tetracycline	30.00	0.01150	Mole F.	[98]
Thalidomide	32.00	0.49	g/L	[8]
Triclosan	20.00	0.332	Mole F.	[85]
Triclosan	25.00	0.449	Mole F.	[85]
Triclosan	30.00	0.570	Mole F.	[85]
Triclosan	35.00	0.654	Mole F.	[85]
Triclosan	40.00	0.751	Mole F.	[85]
Trimethoprim	8.00	0.0003559	Mole F.	[65]
Trimethoprim	14.10	0.0004605	Mole F.	[65]
Trimethoprim	19.49	0.0005455	Mole F.	[65]
Trimethoprim	24.80	0.0006808	Mole F.	[65]
Trimethoprim	28.90	0.0008332	Mole F.	[65]
Trimethoprim	34.70	0.001049	Mole F.	[65]
Trimethoprim	38.20	0.001238	Mole F.	[65]
Trimethoprim	42.09	0.001514	Mole F.	[65]
Trimethoprim	45.70	0.001772	Mole F.	[65]
Trimethoprim	50.80	0.002226	Mole F.	[65]
Trimethoprim	53.50	0.002553	Mole F.	[65]
Trimethoprim	57.51	0.003085	Mole F.	[65]
Trimethoprim	62.10	0.003877	Mole F.	[65]
Xanthene	25.00	0.006231	Mole F.	[3]
Xylitol	20.08	0.001728	Mole F.	[66]
Xylitol	25.13	0.002419	Mole F.	[66]
Xylitol	30.02	0.003091	Mole F.	[66]
Xylitol	35.06	0.003974	Mole F.	[66]
Xylitol	40.04	0.005224	Mole F.	[66]
Xylitol	45.14	0.007090	Mole F.	[66]
Xylitol	50.09	0.009657	Mole F.	[66]
Xylitol	55.04	0.013150	Mole F.	[66]
Xylitol	60.07	0.017850	Mole F.	[66]

Solubility Data of Drugs in Ethyl Acetate at Various Temperatures

Drug	T (°C)	Solubility	Solubility Unit	Reference
11α-Hydroxy-16α,17α-epoxyprogesterone	8.62	0.0006591	Mole F.	[105]
11α-Hydroxy-16α,17α-epoxyprogesterone	12.55	0.0007348	Mole F.	[105]
11α-Hydroxy-16α,17α-epoxyprogesterone	17.30	0.0008238	Mole F.	[105]
11α-Hydroxy-16α,17α-epoxyprogesterone	21.40	0.0009128	Mole F.	[105]
11α-Hydroxy-16α,17α-epoxyprogesterone	25.38	0.0010366	Mole F.	[105]
11α-Hydroxy-16α,17α-epoxyprogesterone	30.62	0.001193	Mole F.	[105]
11α-Hydroxy-16α,17α-epoxyprogesterone	35.24	0.001354	Mole F.	[105]
11α-Hydroxy-16α,17α-epoxyprogesterone	39.45	0.001540	Mole F.	[105]
11α-Hydroxy-16α,17α-epoxyprogesterone	44.52	0.001741	Mole F.	[105]
11α-Hydroxy-16α,17α-epoxyprogesterone	49.22	0.001957	Mole F.	[105]
16α,17α-Epoxyprogesterone	12.30	0.00752	Mole F.	[17]
16α,17α-Epoxyprogesterone	15.55	0.00819	Mole F.	[17]

Solubility Data of Drugs in Ethyl Acetate at Various Temperatures (continued)

Drug	*T* (°C)	Solubility	Solubility Unit	Reference
16α,17α-Epoxyprogesterone	20.70	0.00968	Mole F.	[17]
16α,17α-Epoxyprogesterone	24.80	0.01102	Mole F.	[17]
16α,17α-Epoxyprogesterone	29.60	0.01257	Mole F.	[17]
16α,17α-Epoxyprogesterone	35.00	0.01422	Mole F.	[17]
16α,17α-Epoxyprogesterone	39.60	0.01580	Mole F.	[17]
16α,17α-Epoxyprogesterone	44.70	0.01765	Mole F.	[17]
16α,17α-Epoxyprogesterone	49.65	0.01985	Mole F.	[17]
16α,17α-Epoxyprogesterone	55.30	0.02230	Mole F.	[17]
1H-1,2,4-Triazole	10.06	0.0268	Mole F.	[92]
1H-1,2,4-Triazole	15.14	0.0320	Mole F.	[92]
1H-1,2,4-Triazole	20.06	0.0365	Mole F.	[92]
1H-1,2,4-Triazole	25.04	0.0410	Mole F.	[92]
1H-1,2,4-Triazole	30.03	0.0457	Mole F.	[92]
1H-1,2,4-Triazole	35.12	0.0513	Mole F.	[92]
1H-1,2,4-Triazole	40.05	0.0578	Mole F.	[92]
1H-1,2,4-Triazole	45.06	0.0660	Mole F.	[92]
1H-1,2,4-Triazole	50.01	0.0759	Mole F.	[92]
1H-1,2,4-Triazole	55.02	0.0884	Mole F.	[92]
1H-1,2,4-Triazole	60.12	0.1040	Mole F.	[92]
2-Hydroxybenzoic acid	25.00	0.1425	Mole F.	[19]
4-Aminophenylacetic acid	20.00	0.73	g/L	[8]
Acephate	34.75	0.03496	Mole F.	[5]
Acephate	37.60	0.04082	Mole F.	[5]
Acephate	41.05	0.04901	Mole F.	[5]
Acephate	45.25	0.06475	Mole F.	[5]
Acephate	48.40	0.07792	Mole F.	[5]
Acephate	51.80	0.09648	Mole F.	[5]
Acephate	54.45	0.11850	Mole F.	[5]
Acetylsalicylic acid	25.00	0.0448	Mole F.	[21]
Acetylsalicylic acid	25.00	0.04583	Mole F.	[22]
Anhydrous citric acid	25.00	0.00618	Mole F.	[23]
Atractylenolide III	10.05	0.02243	Mole F.	[132]
Atractylenolide III	15.05	0.02285	Mole F.	[132]
Atractylenolide III	20.05	0.02362	Mole F.	[132]
Atractylenolide III	25.05	0.02420	Mole F.	[132]
Atractylenolide III	30.05	0.02521	Mole F.	[132]
Atractylenolide III	35.05	0.02610	Mole F.	[132]
Atractylenolide III	40.05	0.02711	Mole F.	[132]
Atractylenolide III	45.05	0.02815	Mole F.	[132]
Atractylenolide III	50.05	0.02928	Mole F.	[132]
Benzoic acid	25.00	0.1649	Mole F.	[24]
β-Carotene	Room	500	mg/L	[6]
β-Sitosteryl maleate	15.00	0.003501	Mole F.	[110]
β-Sitosteryl maleate	20.00	0.004773	Mole F.	[110]
β-Sitosteryl maleate	25.00	0.005864	Mole F.	[110]
β-Sitosteryl maleate	30.00	0.007755	Mole F.	[110]
β-Sitosteryl maleate	35.00	0.009282	Mole F.	[110]
β-Sitosteryl maleate	40.00	0.01218	Mole F.	[110]

(continued)

Solubility Data of Drugs in Ethyl Acetate at Various Temperatures   (continued)

Drug	T (°C)	Solubility	Solubility Unit	Reference
β-Sitosteryl maleate	44.70	0.01800	Mole F.	[110]
Betulin	5.05	0.000963	Mole F.	[7]
Betulin	15.05	0.001550	Mole F.	[7]
Betulin	25.05	0.002066	Mole F.	[7]
Betulin	35.05	0.003047	Mole F.	[7]
Carbamazepine	25.00	10.7	g/L	[8]
Chlorpheniramine maleate	9.71	0.0005947	Mole F.	[47]
Chlorpheniramine maleate	16.30	0.0008069	Mole F.	[47]
Chlorpheniramine maleate	20.50	0.0009306	Mole F.	[47]
Chlorpheniramine maleate	25.45	0.0011364	Mole F.	[47]
Chlorpheniramine maleate	29.10	0.0013634	Mole F.	[47]
Chlorpheniramine maleate	35.00	0.0017435	Mole F.	[47]
Chlorpheniramine maleate	40.00	0.0021522	Mole F.	[47]
Chlorpheniramine maleate	45.00	0.0027072	Mole F.	[47]
Chlorpheniramine maleate	50.00	0.0034322	Mole F.	[47]
Chlorpheniramine maleate	54.95	0.0042694	Mole F.	[47]
Chlorpheniramine maleate	60.00	0.0054172	Mole F.	[47]
Diclofenac	25.00	0.0228913	Mole F.	[25]
Flubiprofen	25.00	0.111	Mole F.	[26]
Gallic acid	25.05	0.01276	Mass fraction	[139]
Gallic acid	30.05	0.01290	Mass fraction	[139]
Gallic acid	35.05	0.01303	Mass fraction	[139]
Gallic acid	40.05	0.01335	Mass fraction	[139]
Gallic acid	45.05	0.01438	Mass fraction	[139]
Gallic acid	50.05	0.01544	Mass fraction	[139]
Gallic acid	55.05	0.01598	Mass fraction	[139]
Gallic acid	60.05	0.01689	Mass fraction	[139]
Haloperidol	25.00	0.01007	Mole F.	[27]
Hydrocortisone	25.00	0.00776	mol/L	[80]
Hydroquinone	5.55	0.07333	Mole F.	[101]
Hydroquinone	17.00	0.08800	Mole F.	[101]
Hydroquinone	25.65	0.10170	Mole F.	[101]
Hydroquinone	31.45	0.11030	Mole F.	[101]
Hydroquinone	37.65	0.12050	Mole F.	[101]
Hydroquinone	43.55	0.13120	Mole F.	[101]
Hydroquinone	50.10	0.14440	Mole F.	[101]
Hydroquinone	57.45	0.15950	Mole F.	[101]
Hydroquinone	65.20	0.17700	Mole F.	[101]
Hydroquinone	71.95	0.19210	Mole F.	[101]
Ibuprofen	10.00	327.2	g/kg	[83]
Ibuprofen	15.00	414.8	g/kg	[83]
Ibuprofen	20.00	531.6	g/kg	[83]
Ibuprofen	25.00	0.3348073	Mole F.	[28]
Ibuprofen	30.00	856.4	g/kg	[83]
Ibuprofen	35.00	1084	g/kg	[83]
Isoniazid	27.85	0.00159	Mole F.	[114]
Isoniazid	29.85	0.00172	Mole F.	[114]
Isoniazid	34.85	0.00206	Mole F.	[114]
Isoniazid	37.85	0.00225	Mole F.	[114]

Solubility Data of Drugs in Ethyl Acetate at Various Temperatures   (continued)

Drug	T (°C)	Solubility	Solubility Unit	Reference
Isoniazid	39.85	0.00251	Mole F.	[114]
Ketoprofen	25.00	0.1530	Mole F.	[68]
L-(+)-Ascorbic acid	19.85	0.00010	Mole F.	[102]
L-(+)-Ascorbic acid	24.85	0.00013	Mole F.	[102]
L-(+)-Ascorbic acid	29.85	0.00017	Mole F.	[102]
L-(+)-Ascorbic acid	34.85	0.00023	Mole F.	[102]
L-(+)-Ascorbic acid	39.85	0.00030	Mole F.	[102]
L-(+)-Ascorbic acid	44.85	0.00038	Mole F.	[102]
L-(+)-Ascorbic acid	49.85	0.00048	Mole F.	[102]
Lactose	25.00	0.0000006	Mole F.	[30]
Lamivudine (form 2)	5.00	0.019	g/L	[52]
Lamivudine (form 2)	15.00	0.037	g/L	[52]
Lamivudine (form 2)	25.00	0.057	g/L	[52]
Lamivudine (form 2)	35.00	0.086	g/L	[52]
Lamivudine (form 2)	45.00	0.128	g/L	[52]
Losartan potassium	20.00	0.002509	Mole F.	[86]
Losartan potassium	30.00	0.001518	Mole F.	[86]
Losartan potassium	40.00	0.002522	Mole F.	[86]
Losartan potassium	50.00	0.001381	Mole F.	[86]
Losartan potassium	60.00	0.002883	Mole F.	[86]
Losartan potassium	70.00	0.002788	Mole F.	[86]
Lovastatin	5.45	0.002911	Mole F.	[116]
Lovastatin	9.10	0.003397	Mole F.	[116]
Lovastatin	11.95	0.00417	Mole F.	[89]
Lovastatin	15.10	0.004078	Mole F.	[116]
Lovastatin	15.15	0.00470	Mole F.	[89]
Lovastatin	18.05	0.00517	Mole F.	[89]
Lovastatin	18.50	0.004713	Mole F.	[116]
Lovastatin	21.45	0.00586	Mole F.	[89]
Lovastatin	24.05	0.00656	Mole F.	[89]
Lovastatin	24.70	0.005684	Mole F.	[116]
Lovastatin	27.15	0.00706	Mole F.	[89]
Lovastatin	29.95	0.00778	Mole F.	[89]
Lovastatin	30.00	0.006881	Mole F.	[116]
Lovastatin	33.15	0.00870	Mole F.	[89]
Lovastatin	36.25	0.00967	Mole F.	[89]
Lovastatin	36.70	0.008492	Mole F.	[116]
Lovastatin	39.05	0.01078	Mole F.	[89]
Lovastatin	39.80	0.009872	Mole F.	[116]
Lovastatin	45.10	0.011970	Mole F.	[116]
Lovastatin	50.50	0.01481	Mole F.	[116]
Lutein	Room	800	mg/L	[6]
Mannitol	25.00	0.0000595	Mole F.	[30]
Mefenamic acid (form 1)	26.17	1.04	% Weight	[96]
Mefenamic acid (form 1)	37.93	1.59	% Weight	[96]
Mefenamic acid (form 1)	43.45	2.13	% Weight	[96]
Mefenamic acid (form 1)	50.62	2.74	% Weight	[96]
Mefenamic acid (form 1)	57.74	3.46	% Weight	[96]
Mefenamic acid (form 2)	33.26	1.97	% Weight	[96]

(continued)

Solubility Data of Drugs in Ethyl Acetate at Various Temperatures   (continued)

Drug	T (°C)	Solubility	Solubility Unit	Reference
Mefenamic acid (form 2)	39.66	2.50	% Weight	[96]
Mefenamic acid (form 2)	44.95	2.91	% Weight	[96]
Mefenamic acid (form 2)	52.20	3.59	% Weight	[96]
Mefenamic acid (form 2)	58.06	4.18	% Weight	[96]
Methyl p-hydroxybenzoate	25.00	0.1270	Mole F.	[35]
Naproxen	25.00	0.0354760	Mole F.	[31]
Niflumic acid	25.00	0.0000910	Mole F.	[33]
p-Aminobenzoic acid	25.00	0.0575558	Mole F.	[25]
p-Aminophenylacetic acid	16.00	0.6	g/kg	[83]
p-Aminophenylacetic acid	20.00	1.2	g/kg	[83]
p-Aminophenylacetic acid	25.00	18.0	g/kg	[83]
Paracetamol	−5.00	4.46	g/kg	[11]
Paracetamol	0.00	5.27	g/kg	[11]
Paracetamol	5.00	5.78	g/kg	[11]
Paracetamol	10.00	6.42	g/kg	[11]
Paracetamol	15.00	7.37	g/kg	[11]
Paracetamol	20.00	8.52	g/kg	[11]
Paracetamol	25.00	9.45	g/kg	[11]
Paracetamol	25.00	0.01236	Mole F.	[23]
Paracetamol	30.00	10.73	g/kg	[11]
Paracetamol	30.00	9.84	g/L	[8]
Paroxetine HCl hemihydrate	20.50	0.0000044	Mole F.	[57]
Paroxetine HCl hemihydrate	23.45	0.0000066	Mole F.	[57]
Paroxetine HCl hemihydrate	26.90	0.0000101	Mole F.	[57]
Paroxetine HCl hemihydrate	30.40	0.0000154	Mole F.	[57]
Paroxetine HCl hemihydrate	34.00	0.0000211	Mole F.	[57]
Paroxetine HCl hemihydrate	37.70	0.0000303	Mole F.	[57]
Paroxetine HCl hemihydrate	40.90	0.0000415	Mole F.	[57]
Paroxetine HCl hemihydrate	43.85	0.0000544	Mole F.	[57]
Paroxetine HCl hemihydrate	47.05	0.0000694	Mole F.	[57]
Paroxetine HCl hemihydrate	50.00	0.0000887	Mole F.	[57]
Paroxetine HCl hemihydrate	52.85	0.0001078	Mole F.	[57]
Paroxetine HCl hemihydrate	56.65	0.0001293	Mole F.	[57]
Paroxetine HCl hemihydrate	58.95	0.0001506	Mole F.	[57]
Paroxetine HCl hemihydrate	62.05	0.0001760	Mole F.	[57]
Paroxetine HCl hemihydrate	65.05	0.0002039	Mole F.	[57]
Paroxetine HCl hemihydrate	68.00	0.0002334	Mole F.	[57]
Paroxetine HCl hemihydrate	71.05	0.0002642	Mole F.	[57]
Phenacetinum	9.50	0.0059313	Mole F.	[58]
Phenacetinum	16.30	0.0082669	Mole F.	[58]
Phenacetinum	20.15	0.0095839	Mole F.	[58]
Phenacetinum	25.20	0.011494	Mole F.	[58]
Phenacetinum	30.02	0.013683	Mole F.	[58]
Phenacetinum	34.88	0.016438	Mole F.	[58]
Phenacetinum	39.90	0.020055	Mole F.	[58]
Phenacetinum	45.10	0.024842	Mole F.	[58]
Phenacetinum	50.10	0.030650	Mole F.	[58]
Phenacetinum	55.00	0.037683	Mole F.	[58]

Solubility Data of Drugs in Ethyl Acetate at Various Temperatures (continued)

Drug	T (°C)	Solubility	Solubility Unit	Reference
Phenacetinum	60.00	0.0464240	Mole F.	[58]
Phenylacetic acid	20.00	825.4	g/kg	[83]
p-Hydroxyphenylacetic acid	10.00	134.2	g/kg	[83]
p-Hydroxyphenylacetic acid	15.00	147.7	g/kg	[83]
p-Hydroxyphenylacetic acid	20.00	164.3	g/kg	[83]
p-Hydroxybenzoic acid	25.00	0.0737	Mole F.	[35]
p-Hydroxybenzoic acid	25.00	117.7	g/kg	[83]
Pimozide	25.00	0.0021	Mole F.	[36]
Piroxicam	25.00	0.0024443	Mole F.	[33]
Ricobendazole	25.00	0.21	g/L	[103]
Rutin	10.00	0.001144	Mole F.	[61]
Rutin	25.00	0.001085	Mole F.	[61]
Rutin	40.00	0.001003	Mole F.	[61]
Rutin	50.00	0.000953	Mole F.	[61]
Rutin	60.00	0.000894	Mole F.	[61]
Saccharose	25.00	0.00000004	Mole F.	[30]
Salicylamide	10.00	0.05528	Mole F.	[107]
Salicylamide	15.00	0.06070	Mole F.	[107]
Salicylamide	20.00	0.06703	Mole F.	[107]
Salicylamide	25.00	0.07448	Mole F.	[37]
Salicylamide	25.00	0.07549	Mole F.	[107]
Salicylamide	30.00	0.08345	Mole F.	[107]
Salicylamide	35.00	0.09281	Mole F.	[107]
Salicylamide	40.00	0.10276	Mole F.	[107]
Salicylamide	45.00	0.11432	Mole F.	[107]
Salicylamide	50.00	0.12716	Mole F.	[107]
Salicylic acid	10.00	0.10908	Mole F.	[108]
Salicylic acid	15.00	0.11647	Mole F.	[108]
Salicylic acid	20.00	0.12514	Mole F.	[108]
Salicylic acid	24.85	0.20076	Mole F.	[128]
Salicylic acid	25.00	0.1223340	Mole F.	[25]
Salicylic acid	25.00	0.13571	Mole F.	[108]
Salicylic acid	25.00	0.1383	Mass. F	[38]
Salicylic acid	29.85	0.21788	Mole F.	[128]
Salicylic acid	30.00	0.14517	Mass. F	[108]
Salicylic acid	34.85	0.23582	Mole F.	[128]
Salicylic acid	35.00	0.15656	Mass. F	[108]
Salicylic acid	39.85	0.25246	Mole F.	[128]
Salicylic acid	40.00	0.16752	Mass. F	[108]
Salicylic acid	44.85	0.26924	Mole F.	[128]
Salicylic acid	45.00	0.17988	Mass. F	[108]
Salicylic acid	49.85	0.28596	Mole F.	[128]
Salicylic acid	50.00	0.19215	Mass. F	[108]
Salicylic acid	54.85	0.30102	Mass. F	[128]
Salicylic acid	59.85	0.31802	Mass. F	[128]
Salicylic acid	64.85	0.33476	Mass. F	[128]
Salicylic acid	69.85	0.35151	Mass. F	[128]
Salicylic acid	74.85	0.36921	Mole F.	[128]
Sodium benzoate	25.00	0.0000022	Mole F.	[28]

(continued)

Solubility Data of Drugs in Ethyl Acetate at Various Temperatures   (continued)

Drug	T (°C)	Solubility	Solubility Unit	Reference
Sodium diclofenac	20.00	1.19	mg/g	[119]
Sodium diclofenac	25.00	0.0156874	Mole F.	[31]
Sodium diclofenac	25.00	1.26	mg/g	[119]
Sodium diclofenac	25.00	0.0156858	Mole F.	[25]
Sodium diclofenac	30.00	1.40	mg/g	[119]
Sodium diclofenac	35.00	1.60	mg/g	[119]
Sodium diclofenac	40.00	0.60	mg/g	[119]
Sodium ibuprofen	25.00	0.0006204	Mole F.	[28]
Sodium p-aminobenzoate	25.00	0.0004905	Mole F.	[25]
Sodium salicylate	25.00	0.0014159	Mole F.	[25]
Stigmasteryl maleate	15.00	0.001460	Mole F.	[110]
Stigmasteryl maleate	20.10	0.001789	Mole F.	[110]
Stigmasteryl maleate	25.00	0.002025	Mole F.	[110]
Stigmasteryl maleate	29.50	0.002801	Mole F.	[110]
Stigmasteryl maleate	35.50	0.003951	Mole F.	[110]
Stigmasteryl maleate	39.50	0.004382	Mole F.	[110]
Stigmasteryl maleate	44.20	0.005301	Mole F.	[110]
Sulfadiazine	25.00	0.0001189	Mole F.	[39]
Sulfamethoxypyridazine	25.00	0.00205	Mole F.	[40]
Tanshinone IIA	10.05	0.00809	mol/L	[121]
Tanshinone IIA	20.15	0.01128	mol/L	[121]
Tanshinone IIA	29.95	0.01281	mol/L	[121]
Tanshinone IIA	35.05	0.01482	mol/L	[121]
Tanshinone IIA	44.85	0.01833	mol/L	[121]
Tanshinone IIA	50.15	0.02091	mol/L	[121]
Temazepam	24.00	0.0344	Volume F.	[64]
Temazepam	25.00	0.0144756	Mole F.	[15]

Solubility Data of Drugs in Ethyl Acetoacetate at Various Temperatures

Drug	T (°C)	Solubility	Solubility Unit	Reference
Acetic acid	−29.00	0.365	Mole F.	[109]
Acetic acid	−24.50	0.420	Mole F.	[109]
Acetic acid	−20.80	0.442	Mole F.	[109]
Acetic acid	−14.90	0.510	Mole F.	[109]
Acetic acid	−12.30	0.551	Mole F.	[109]
Acetic acid	−9.00	0.610	Mole F.	[109]
Acetic acid	−4.80	0.660	Mole F.	[109]
Acetic acid	−2.00	0.730	Mole F.	[109]
Acetic acid	2.70	0.782	Mole F.	[109]
Acetic acid	7.10	0.855	Mole F.	[109]
Acetic acid	10.40	0.915	Mole F.	[109]

Solubility Data of Drugs in Ethylbenzene at Various Temperatures

Drug	T (°C)	Solubility	Solubility Unit	Reference
2,6-Diaminopyridine	0.10	0.00022	Mole F.	[93]
2,6-Diaminopyridine	4.00	0.00026	Mole F.	[93]
2,6-Diaminopyridine	9.10	0.00033	Mole F.	[93]
2,6-Diaminopyridine	12.40	0.00035	Mole F.	[93]
2,6-Diaminopyridine	16.30	0.00046	Mole F.	[93]
2,6-Diaminopyridine	19.90	0.00049	Mole F.	[93]
2,6-Diaminopyridine	25.50	0.00066	Mole F.	[93]
2,6-Diaminopyridine	29.10	0.00069	Mole F.	[93]
2,6-Diaminopyridine	32.20	0.00086	Mole F.	[93]
2,6-Diaminopyridine	36.40	0.00090	Mole F.	[93]
2,6-Diaminopyridine	40.00	0.00113	Mole F.	[93]
2,6-Diaminopyridine	44.20	0.00120	Mole F.	[93]
2,6-Diaminopyridine	48.10	0.00150	Mole F.	[93]
2,6-Diaminopyridine	53.00	0.00164	Mole F.	[93]
2,6-Diaminopyridine	56.30	0.00197	Mole F.	[93]
2,6-Diaminopyridine	60.20	0.00212	Mole F.	[93]
2,6-Diaminopyridine	63.80	0.00252	Mole F.	[93]
2,6-Diaminopyridine	67.90	0.00288	Mole F.	[93]
2,6-Diaminopyridine	73.10	0.00330	Mole F.	[93]
2,6-Diaminopyridine	76.40	0.00385	Mole F.	[93]
2,6-Diaminopyridine	79.90	0.00430	Mole F.	[93]
2,6-Diaminopyridine	83.90	0.00474	Mole F.	[93]
2,6-Diaminopyridine	88.00	0.00577	Mole F.	[93]
Vitamin K3	29.70	0.09763	Mole F.	[125]
Vitamin K3	36.29	0.1218	Mole F.	[125]
Vitamin K3	41.62	0.1464	Mole F.	[125]
Vitamin K3	46.38	0.1726	Mole F.	[125]
Vitamin K3	50.60	0.1994	Mole F.	[125]
Vitamin K3	54.78	0.2283	Mole F.	[125]
Vitamin K3	58.60	0.2602	Mole F.	[125]
Vitamin K3	62.29	0.2967	Mole F.	[125]
Vitamin K3	66.27	0.3423	Mole F.	[125]
Vitamin K3	70.71	0.3957	Mole F.	[125]

Solubility Data of Drugs in Ethyl Formate at Various Temperatures

Drug	T (°C)	Solubility	Solubility Unit	Reference
1H-1,2,4-Triazole	10.06	0.02083	Mole F.	[92]
1H-1,2,4-Triazole	15.08	0.02460	Mole F.	[92]
1H-1,2,4-Triazole	20.14	0.02864	Mole F.	[92]
1H-1,2,4-Triazole	25.09	0.03309	Mole F.	[92]
1H-1,2,4-Triazole	30.11	0.03838	Mole F.	[92]
1H-1,2,4-Triazole	35.02	0.04459	Mole F.	[92]
1H-1,2,4-Triazole	40.09	0.05235	Mole F.	[92]
1H-1,2,4-Triazole	45.12	0.06168	Mole F.	[92]
1H-1,2,4-Triazole	50.16	0.07295	Mole F.	[92]

(continued)

Solubility Data of Drugs in Ethyl Formate at Various Temperatures   (continued)

Drug	T (°C)	Solubility	Solubility Unit	Reference
2-(4-Ethylbenzoyl)benzoic acid	7.45	0.0014	Mole F.	[18]
2-(4-Ethylbenzoyl)benzoic acid	16.53	0.0027	Mole F.	[18]
2-(4-Ethylbenzoyl)benzoic acid	21.76	0.0038	Mole F.	[18]
2-(4-Ethylbenzoyl)benzoic acid	26.50	0.0050	Mole F.	[18]
2-(4-Ethylbenzoyl)benzoic acid	31.45	0.0060	Mole F.	[18]
2-(4-Ethylbenzoyl)benzoic acid	36.44	0.0075	Mole F.	[18]
2-(4-Ethylbenzoyl)benzoic acid	41.15	0.0087	Mole F.	[18]
2-(4-Ethylbenzoyl)benzoic acid	45.52	0.0100	Mole F.	[18]
2-(4-Ethylbenzoyl)benzoic acid	49.76	0.0113	Mole F.	[18]
Betulin	5.05	0.000301	Mole F.	[7]
Betulin	15.05	0.000333	Mole F.	[7]
Betulin	25.05	0.000316	Mole F.	[7]
Betulin	35.05	0.000395	Mole F.	[7]
Chlorpheniramine maleate	9.75	0.0010428	Mole F.	[47]
Chlorpheniramine maleate	15.50	0.0015435	Mole F.	[47]
Chlorpheniramine maleate	20.50	0.0022614	Mole F.	[47]
Chlorpheniramine maleate	25.50	0.0030746	Mole F.	[47]
Chlorpheniramine maleate	30.00	0.0040683	Mole F.	[47]
Chlorpheniramine maleate	34.95	0.0053387	Mole F.	[47]
Chlorpheniramine maleate	40.00	0.0070663	Mole F.	[47]
Chlorpheniramine maleate	45.00	0.0091291	Mole F.	[47]
Chlorpheniramine maleate	49.95	0.0117980	Mole F.	[47]
Chlorpheniramine maleate	54.95	0.0149870	Mole F.	[47]
Diclofenac	25.00	0.0040300	Mole F.	[25]
Ibuprofen	25.00	0.0858718	Mole F.	[28]
Naproxen	25.00	0.0132932	Mole F.	[31]
Niflumic acid	25.00	0.0183156	Mole F.	[33]
p-Aminobenzoic acid	25.00	0.0029234	Mole F.	[25]
Piroxicam	25.00	0.0088441	Mole F.	[33]
Salicylic acid	25.00	0.0075818	Mole F.	[25]
Sodium ibuprofen	25.00	0.0000374	Mole F.	[28]
Sodium benzoate	25.00	0.0000250	Mole F.	[28]
Sodium diclofenac	25.00	0.0003070	Mole F.	[31]
Sodium diclofenac	25.00	0.0003069	Mole F.	[25]
Sodium p-aminobenzoate	25.00	0.0000257	Mole F.	[25]
Sodium salicylate	25.00	0.0000547	Mole F.	[25]

Solubility Data of Drugs in Ethylene Dichloride at Various Temperatures

Drug	T (°C)	Solubility	Solubility Unit	Reference
Benzoic acid	25.00	0.0695	Mole F.	[24]
Diclofenac	25.00	0.00403	Mole F.	[25]
Ibuprofen	25.00	0.0858718	Mole F.	[28]
Naproxen	25.00	0.0132932	Mole F.	[31]
Niflumic acid	25.00	0.0183156	Mole F.	[33]
p-Aminobenzoic acid	25.00	0.0029234	Mole F.	[25]
p-Hydroxybenzoic acid	25.00	0.00011	Mole F.	[35]
Piroxicam	25.00	0.0088441	Mole F.	[33]
Salicylic acid	25.00	0.0075818	Mole F.	[25]
Sodium ibuprofen	25.00	0.0000374	Mole F.	[28]
Sodium benzoate	25.00	0.0000250	Mole F.	[28]
Sodium diclofenac	25.00	0.0003070	Mole F.	[31]
Sodium diclofenac	25.00	0.0003069	Mole F.	[25]
Sodium p-aminobenzoate	25.00	0.0000257	Mole F.	[25]
Sodium salicylate	25.00	0.0000547	Mole F.	[25]

Solubility Data of Drugs in Ethylene Glycol at Various Temperatures

Drug	T (°C)	Solubility	Solubility Unit	Reference
Anhydrous citric acid	25.00	0.21509	Mole F.	[23]
Benzoic acid	25.00	0.0884	Mole F.	[24]
Celecoxib	25.00	3.856	g/L	[55]
Diclofenac	25.00	0.0016968	Mole F.	[25]
Ibuprofen	25.00	0.0190403	Mole F.	[28]
Mannitol	25.00	0.0025084	Mole F.	[30]
Meloxicam	25.00	0.094	g/L	[55]
Methyl p-hydroxybenzoate	25.00	0.0480	Mole F.	[35]
Naproxen	25.00	0.0038492	Mole F.	[31]
Niflumic acid	25.00	0.0010584	Mole F.	[33]
Nimesulide	25.00	0.510	g/L	[55]
p-Aminobenzoic acid	25.00	0.0693215	Mole F.	[25]
Paracetamol	30.00	144.30	g/kg	[11]
Paracetamol	30.00	141	g/L	[8]
Paracetamol	25.00	0. 12199	Mole F.	[23]
Phenothiazine	25.00	0.00191	Mole F.	[34]
p-Hydroxybenzoic acid	25.00	0.1132	Mole F.	[35]
Piroxicam	25.00	0.0001400	Mole F.	[33]
Rofecoxib	25.00	0.126	g/L	[55]
Saccharose	25.00	0.0000090	Mole F.	[30]
Salicylic acid	25.00	0.0850939	Mole F.	[25]
Sodium ibuprofen	25.00	0.5085458	Mole F.	[28]
Sodium benzoate	25.00	0.0802988	Mole F.	[28]
Sodium diclofenac	25.00	0.0371980	Mole F.	[31]

(continued)

Solubility Data of Drugs in Ethylene Glycol at Various Temperatures   (continued)

Drug	T (°C)	Solubility	Solubility Unit	Reference
Sodium diclofenac	25.00	0.0371794	Mole F.	[25]
Sodium p-aminobenzoate	25.00	0.0656774	Mole F.	[25]
Sodium salicylate	25.00	0.2642129	Mole F.	[25]
Sulfadiazine	25.00	0.0003206	Mole F.	[39]
Sulfamethoxypyridazine	25.00	0.00496	Mole F.	[40]
Temazepam	25.00	0.0024212	Mole F.	[15]

Solubility Data of Drugs in Formamide at Various Temperatures

Drug	T (°C)	Solubility	Solubility Unit	Reference
Anhydrous citric acid	25.00	0.14467	Mole F.	[23]
Benzoic acid	25.00	0.1525	Mole F.	[24]
Ceftriaxone disodium	5.25	0.006641	Mole F.	[112]
Ceftriaxone disodium	9.10	0.008193	Mole F.	[112]
Ceftriaxone disodium	14.60	0.01083	Mole F.	[112]
Ceftriaxone disodium	20.90	0.01488	Mole F.	[112]
Ceftriaxone disodium	26.50	0.01945	Mole F.	[112]
Ceftriaxone disodium	31.95	0.02474	Mole F.	[112]
Ceftriaxone disodium	34.55	0.02787	Mole F.	[112]
Ceftriaxone disodium	38.65	0.03330	Mole F.	[112]
Ceftriaxone disodium	42.10	0.03852	Mole F.	[112]
Diclofenac	25.00	0.0017590	Mole F.	[25]
Ibuprofen	25.00	0.0014323	Mole F.	[28]
Lactose	25.00	0.0254485	Mole F.	[30]
Mannitol	25.00	0.0060337	Mole F.	[30]
Methyl p-hydroxybenzoate	25.00	0.0765	Mole F.	[35]
Naproxen	25.00	0.0065401	Mole F.	[31]
Niflumic acid	25.00	0.0027367	Mole F.	[33]
p-Aminobenzoic acid	25.00	0.0122283	Mole F.	[25]
Paracetamol	25.00	0.15947	Mole F.	[23]
p-Hydroxybenzoic acid	25.00	0.0341	Mole F.	[35]
Piroxicam	25.00	0.0004995	Mole F.	[33]
Saccharose	25.00	0.0000151	Mole F.	[30]
Salicylic acid	25.00	0.0426384	Mole F.	[25]
Sodium ibuprofen	25.00	0.0215819	Mole F.	[28]
Sodium benzoate	25.00	0.0425192	Mole F.	[28]
Sodium cefotaxim	5.00	0.048303	Mole F.	[133]
Sodium cefotaxim	7.00	0.050265	Mole F.	[133]
Sodium cefotaxim	10.00	0.052955	Mole F.	[133]
Sodium cefotaxim	12.00	0.056184	Mole F.	[133]
Sodium cefotaxim	15.00	0.059646	Mole F.	[133]
Sodium cefotaxim	17.00	0.061517	Mole F.	[133]
Sodium cefotaxim	20.00	0.064359	Mole F.	[133]
Sodium cefotaxim	22.00	0.065158	Mole F.	[133]
Sodium cefotaxim	25.00	0.067434	Mole F.	[133]

Solubility Data of Drugs in Formamide at Various Temperatures   (continued)

Drug	T (°C)	Solubility	Solubility Unit	Reference
Sodium cefotaxim	27.00	0.068772	Mole F.	[133]
Sodium cefotaxim	30.00	0.070191	Mole F.	[133]
Sodium cefotaxim	32.00	0.071863	Mole F.	[133]
Sodium cefotaxim	35.00	0.073271	Mole F.	[133]
Sodium diclofenac	25.00	0.0565856	Mole F.	[31]
Sodium diclofenac	25.00	0.0565856	Mole F.	[25]
Sodium p-aminobenzoate	25.00	0.1764002	Mole F.	[25]
Sodium salicylate	25.00	0.1990897	Mole F.	[25]
Temazepam	25.00	0.0050084	Mole F.	[15]

Solubility Data of Drugs in Glycerol at Various Temperatures

Drug	T (°C)	Solubility	Solubility Unit	Reference
Anhydrous citric acid	25.00	0.03151	Mole F.	[23]
Benzoic acid	25.00	0.0164	Mole F.	[24]
Diclofenac	25.00	0.0003081	Mole F.	[25]
Griseofulvin	130.00	0.00037	Mole F.	[142]
Haloperidol	25.00	0.000424	Mole F.	[27]
Ibuprofen	25.00	0.0027581	Mole F.	[28]
Lactose	25.00	0.0042476	Mole F.	[30]
Meloxicam	25.00	0.138	g/L	[55]
Methyl p-hydroxybenzoate	25.00	0.0064	Mole F.	[35]
Naproxen	25.00	0.0005578	Mole F.	[31]
Niflumic acid	25.00	0.0006523	Mole F.	[33]
Nimesulide	25.00	0.218	g/L	[55]
p-Aminobenzoic acid	25.00	0.0717907	Mole F.	[25]
Paracetamol	25.00	0.03375	Mole F.	[23]
p-Hydroxybenzoic acid	25.00	0.0301	Mole F.	[35]
Piroxicam	25.00	0.0000283	Mole F.	[33]
Rofecoxib	25.00	0.108	g/L	[55]
Saccharose	25.00	0.0000025	Mole F.	[30]
Salicylic acid	25.00	0.0746459	Mole F.	[25]
Sodium ibuprofen	25.00	0.2822111	Mole F.	[28]
Sodium benzoate	25.00	0.0538582	Mole F.	[28]
Sodium diclofenac	25.00	0.1763473	Mole F.	[31]
Sodium diclofenac	25.00	0.1755204	Mole F.	[25]
Sodium p-aminobenzoate	25.00	0.0751702	Mole F.	[25]
Sodium salicylate	25.00	0.1168341	Mole F.	[25]
Sulfadiazine	25.00	0.0000759	Mole F.	[39]
Sulfamethoxypyridazine	25.00	0.00128	Mole F.	[40]

Solubility Data of Drugs in Heptane at Various Temperatures

Drug	*T* (°C)	Solubility	Solubility Unit	Reference
Acetic acid	−29.20	0.020	Mole F.	[109]
Acetic acid	−14.10	0.040	Mole F.	[109]
Acetic acid	−4.90	0.075	Mole F.	[109]
Acetic acid	−3.50	0.075	Mole F.	[109]
Acetic acid	−2.60	0.080	Mole F.	[109]
Acetic acid	−0.90	0.115	Mole F.	[109]
Acetic acid	1.70	0.145	Mole F.	[109]
Acetic acid	3.50	0.155	Mole F.	[109]
Acetic acid	5.40	0.205	Mole F.	[109]
Acetic acid	7.50	0.240	Mole F.	[109]
Acetic acid	9.50	0.305	Mole F.	[109]
Acetic acid	11.90	0.550	Mole F.	[109]
Acetic acid	12.40	0.680	Mole F.	[109]
Acetic acid	13.70	0.885	Mole F.	[109]
Acetic acid	14.80	0.935	Mole F.	[109]
Benzoic acid	25.00	0.0117000	Mole F.	[24]
Chloramine	25.00	0.0040	Mole F.	[123]
Diclofenac	25.00	0.0000599	Mole F.	[25]
Eflucimibe (form A)	19.85	0.0000094	Mole F.	[99]
Eflucimibe (form A)	20.00	0.0000032	mol/L	[99]
Eflucimibe (form A)	24.85	0.0000183	Mole F.	[99]
Eflucimibe (form A)	29.85	0.0000272	Mole F.	[99]
Eflucimibe (form A)	34.85	0.0000412	Mole F.	[99]
Eflucimibe (form A)	39.85	0.0000801	Mole F.	[99]
Eflucimibe (form A)	44.85	0.0001290	Mole F.	[99]
Eflucimibe (form A)	49.85	0.0002170	Mole F.	[99]
Eflucimibe (form A)	53.85	0.0003640	Mole F.	[99]
Eflucimibe (form A)	54.85	0.0003910	Mole F.	[99]
Eflucimibe (form B)	19.85	0.0000095	Mole F.	[99]
Eflucimibe (form B)	24.85	0.0000191	Mole F.	[99]
Eflucimibe (form B)	29.85	0.0000350	Mole F.	[99]
Eflucimibe (form B)	34.85	0.0000452	Mole F.	[99]
Eflucimibe (form B)	39.85	0.0000876	Mole F.	[99]
Eflucimibe (form B)	44.85	0.0001330	Mole F.	[99]
Eflucimibe (form B)	49.85	0.0002320	Mole F.	[99]
Eflucimibe (form B)	53.85	0.0003830	Mole F.	[99]
Eflucimibe (form B)	54.85	0.0004120	Mole F.	[99]
Ethyl *p*-aminobenzoate	25.00	0.0009210	Mole F.	[124]
Flubiprofen	25.00	0.000631	Mole F.	[26]
Ibuprofen	25.00	0.0559834	Mole F.	[28]
Mestanolone	25.00	0.0001690	Mole F.	[69]
Methandienone	25.00	0.0002680	Mole F.	[69]
Methyl *p*-aminobenzoate	25.00	0.0004057	Mole F.	[124]
Methyl *p*-hydroxybenzoate	25.00	0.000083	Mole F.	[35]
Methyltestosterone	25.00	0.000299	Mole F.	[69]
Nandrolone	25.00	0.0002870	Mole F.	[69]
Naproxen	25.00	0.0018008	Mole F.	[31]
n-Butyl *p*-aminobenzoate	25.00	0.0027122	Mole F.	[124]
Niflumic acid	25.00	0.0071761	Mole F.	[33]

Solubility Data of Drugs in Heptane at Various Temperatures   (continued)

Drug	T (°C)	Solubility	Solubility Unit	Reference
n-Pentyl p-aminobenzoate	25.00	0.0032471	Mole F.	[124]
n-Propyl p-aminobenzoate	25.00	0.0016287	Mole F.	[124]
p-Aminobenzoic acid	25.00	0.0000061	Mole F.	[25]
Phenothiazine	25.00	0.000696	Mole F.	[34]
Piroxicam	25.00	0.0000135	Mole F.	[33]
Salicylic acid	25.00	0.0011160	Mole F.	[25]
Sodium ibuprofen	25.00	0.0000124	Mole F.	[28]
Sodium benzoate	25.00	0.0000044	Mole F.	[28]
Sodium diclofenac	25.00	0.0185034	Mole F.	[31]
Sodium diclofenac	25.00	0.0184997	Mole F.	[25]
Sodium p-aminobenzoate	25.00	0.0000012	Mole F.	[25]
Sodium salicylate	25.00	0.0000246	Mole F.	[25]
Stearic acid	20.30	0.0023	Mole F.	[104]
Stearic acid	24.20	0.0053	Mole F.	[104]
Stearic acid	27.50	0.0089	Mole F.	[104]
Stearic acid	31.50	0.0163	Mole F.	[104]
Stearic acid	34.20	0.0227	Mole F.	[104]
Stearic acid	36.70	0.0366	Mole F.	[104]
Stearic acid	40.30	0.0562	Mole F.	[104]
Stearic acid	41.40	0.0661	Mole F.	[104]
Stearic acid	42.70	0.0779	Mole F.	[104]
Stearic acid	44.40	0.1005	Mole F.	[104]
Stearic acid	46.20	0.1279	Mole F.	[104]
Stearic acid	48.90	0.1786	Mole F.	[104]
Stearic acid	51.70	0.2401	Mole F.	[104]
Temazepam	24.00	0.000118	Volume F.	[64]
Temazepam	25.00	0.0006161	Mole F.	[15]
Testosterone	25.00	0.0001550	Mole F.	[69]
Testosterone propionate	25.00	0.0058	Mole F.	[16]
Triclosan	20.00	0.0156	Mole F.	[85]
Triclosan	25.00	0.0286	Mole F.	[85]
Triclosan	30.00	0.0421	Mole F.	[85]
Triclosan	35.00	0.0605	Mole F.	[85]
Triclosan	40.00	0.0806	Mole F.	[85]
Xanthene	25.00	0.03543	Mole F.	[3]

Solubility Data of Drugs in Hexane at Various Temperatures

Drug	T (°C)	Solubility	Solubility Unit	Reference
2-Hydroxybenzoic acid	20.00	0.000390	Mole F.	[71]
2-Hydroxybenzoic acid	25.00	0.000491	Mole F.	[71]
2-Hydroxybenzoic acid	30.00	0.000643	Mole F.	[71]
2-Hydroxybenzoic acid	37.00	0.000959	Mole F.	[71]
2-Hydroxybenzoic acid	42.00	0.001230	Mole F.	[71]
3-Hydroxybenzoic acid	20.00	0.0000061	Mole F.	[71]
3-Hydroxybenzoic acid	25.00	0.0000092	Mole F.	[71]
3-Hydroxybenzoic acid	30.00	0.0000112	Mole F.	[71]

(continued)

Solubility Data of Drugs in Hexane at Various Temperatures   (continued)

Drug	T (°C)	Solubility	Solubility Unit	Reference
3-Hydroxybenzoic acid	37.00	0.0000174	Mole F.	[71]
3-Hydroxybenzoic acid	42.00	0.0000262	Mole F.	[71]
4-Hydroxybenzoic acid	20.00	0.0000021	Mole F.	[71]
4-Hydroxybenzoic acid	21.00	0.0000025	Mole F.	[71]
4-Hydroxybenzoic acid	25.00	0.0000031	Mole F.	[71]
4-Hydroxybenzoic acid	30.00	0.0000038	Mole F.	[71]
4-Hydroxybenzoic acid	37.00	0.0000060	Mole F.	[71]
4-Hydroxybenzoic acid	42.00	0.0000071	Mole F.	[71]
Acetanilide	20.00	0.0000098	Mole F.	[72]
Acetanilide	25.00	0.0000129	Mole F.	[72]
Acetanilide	30.00	0.0000177	Mole F.	[72]
Acetanilide	33.00	0.0000208	Mole F.	[72]
Acetanilide	37.00	0.0000275	Mole F.	[72]
Acetanilide	42.00	0.0000372	Mole F.	[72]
Acetylsalicylic acid	25.00	0.0341	Mole F.	[21]
Atenolol	15.00	0.000000263	Mole F.	[75]
Atenolol	20.00	0.000000305	Mole F.	[75]
Atenolol	25.00	0.000000377	Mole F.	[75]
Atenolol	30.00	0.000000381	Mole F.	[75]
Atenolol	37.00	0.000000305	Mole F.	[75]
Atenolol	42.00	0.000000504	Mole F.	[75]
Atractylenolide III	10.05	0.0000244	Mole F.	[132]
Atractylenolide III	15.05	0.0000412	Mole F.	[132]
Atractylenolide III	20.05	0.0000639	Mole F.	[132]
Atractylenolide III	25.05	0.0000832	Mole F.	[132]
Atractylenolide III	30.05	0.0001128	Mole F.	[132]
Atractylenolide III	35.05	0.0001534	Mole F.	[132]
Atractylenolide III	40.05	0.0002017	Mole F.	[132]
Atractylenolide III	45.05	0.0002716	Mole F.	[132]
Atractylenolide III	50.05	0.0003362	Mole F.	[132]
Benzoic acid	25.00	0.0095	Mole F.	[24]
β-Carotene	Room	600	mg/L	[6]
Diflunisal	20.00	0.0000090	Mole F.	[77]
Diflunisal	25.00	0.0000112	Mole F.	[77]
Diflunisal	30.00	0.0000143	Mole F.	[77]
Ethyl p-aminobenzoate	25.00	0.0008502	Mole F.	[124]
Fenbufen	30.00	0.00000104	Mole F.	[77]
Flubiprofen	20.00	0.000328	Mole F.	[77]
Flubiprofen	25.00	0.000494	Mole F.	[26]
Flubiprofen	25.00	0.000443	Mole F.	[77]
Flubiprofen	30.00	0.000666	Mole F.	[77]
Flufenamic acid	20.00	0.0000145	Mole F.	[78]
Flufenamic acid	25.00	0.0000198	Mole F.	[78]
Flufenamic acid	30.00	0.0000259	Mole F.	[78]
Flufenamic acid	37.00	0.0000399	Mole F.	[78]
Flufenamic acid	42.00	0.0000512	Mole F.	[78]
Hydrocortisone	25.00	0.0000095	mol/L	[80]
Lutein	Room	20	mg/L	[6]
Luteolin	0.00	0.00110	Mole F.	[54]

Solubility Data of Drugs in Hexane at Various Temperatures (continued)

Drug	T (°C)	Solubility	Solubility Unit	Reference
Luteolin	10.00	0.00123	Mole F.	[54]
Luteolin	25.00	0.00154	Mole F.	[54]
Luteolin	40.00	0.00173	Mole F.	[54]
Luteolin	60.00	0.00201	Mole F.	[54]
Mestanolone	25.00	0.000180	Mole F.	[69]
Methyl elaidate	−29.50	3.39	% Weight	[10]
Methyl elaidate	−17.00	14.83	% Weight	[10]
Methyl elaidate	−5.50	47.25	% Weight	[10]
Methyl heptadecanoate	0.70	14.33	% Weight	[10]
Methyl heptadecanoate	12.40	46.06	% Weight	[10]
Methyl linoleate	−53.50	28.42	% Weight	[10]
Methyl linoleate	−53.30	28.46	% Weight	[10]
Methyl linoleate	−47.00	59.56	% Weight	[10]
Methyl oleate	−51.10	6.63	% Weight	[10]
Methyl oleate	−34.50	47.50	% Weight	[10]
Methyl palmitate	−14.6	3.13	% Weight	[10]
Methyl palmitate	−0.1	14.51	% Weight	[10]
Methyl palmitate	12.3	44.62	% Weight	[10]
Methyl palmitolate	−54.10	22.10	% Weight	[10]
Methyl palmitolate	45.30	57.55	% Weight	[10]
Methyl p-aminobenzoate	25.00	0.0003598	Mole F.	[124]
Methyl petroselaidate	−22.20	2.76	% Weight	[10]
Methyl petroselaidate	−8.90	15.69	% Weight	[10]
Methyl petroselaidate	2.60	47.44	% Weight	[10]
Methyl petroselinate	−38.30	3.55	% Weight	[10]
Methyl petroselinate	−26.70	15.7	% Weight	[10]
Methyl petroselinate	−15.90	46.86	% Weight	[10]
Methyl p-hydroxybenzoate	25.00	0.000077	Mole F.	[35]
Methyl stearate	−13.50	0.33	% Weight	[10]
Methyl stearate	−1.10	1.45	% Weight	[10]
Methyl stearate	10.70	5.84	% Weight	[10]
Methyl stearate	21.00	18.87	% Weight	[10]
Methyl stearate	33.80	65.08	% Weight	[10]
Methyl stearate	36.20	86.16	% Weight	[10]
Methyltestosterone	25.00	0.000244	Mole F.	[69]
n-Butyl p-aminobenzoate	25.00	0.0024055	Mole F.	[124]
Niflumic acid	20.00	0.0000140	Mole F.	[78]
Niflumic acid	25.00	0.0000165	Mole F.	[78]
Niflumic acid	30.00	0.0000203	Mole F.	[78]
Niflumic acid	37.00	0.0000247	Mole F.	[78]
Niflumic acid	42.00	0.0000291	Mole F.	[78]
N-Methyl thalidomide	25.00	0.090	g/L	[143]
n-Pentyl p-aminobenzoate	25.00	0.0030276	Mole F.	[124]
N-Pentyl thalidomide	25.00	0.530	g/L	[143]
n-Propyl p-aminobenzoate	25.00	0.0015338	Mole F.	[124]
N-Propyl thalidomide	25.00	0.220	g/L	[143]
Paracetamol	25.00	0.0000041	Mole F.	[72]
Paracetamol	28.00	0.0000048	Mole F.	[72]

(continued)

Solubility Data of Drugs in Hexane at Various Temperatures   (continued)

Drug	T (°C)	Solubility	Solubility Unit	Reference
Paracetamol	30.00	0.0000053	Mole F.	[72]
Paracetamol	33.00	0.0000063	Mole F.	[72]
Paracetamol	37.00	0.0000077	Mole F.	[72]
Paracetamol	42.00	0.0000102	Mole F.	[72]
Phenacetin	20.00	0.0000145	Mole F.	[72]
Phenacetin	25.00	0.0000198	Mole F.	[72]
Phenacetin	30.00	0.0000259	Mole F.	[72]
Phenacetin	37.00	0.0000399	Mole F.	[72]
Phenacetin	42.00	0.0000512	Mole F.	[72]
Phenothiazine	25.00	0.000585	Mole F.	[34]
Pimozide	25.00	0.0000074	Mole F.	[36]
Pindolol	15.00	0.000000302	Mole F.	[75]
Pindolol	20.00	0.000000381	Mole F.	[75]
Pindolol	25.00	0.000000369	Mole F.	[75]
Pindolol	30.00	0.000000408	Mole F.	[75]
Pindolol	37.00	0.000000400	Mole F.	[75]
Pindolol	42.00	0.000000484	Mole F.	[75]
Stearic acid	20.60	0.0034	Mole F.	[104]
Stearic acid	21.80	0.0045	Mole F.	[104]
Stearic acid	25.30	0.0073	Mole F.	[104]
Stearic acid	27.50	0.0103	Mole F.	[104]
Stearic acid	30.40	0.0161	Mole F.	[104]
Stearic acid	34.10	0.0295	Mole F.	[104]
Stearic acid	36.90	0.0460	Mole F.	[104]
Stearic acid	40.30	0.0710	Mole F.	[104]
Stearic acid	41.90	0.0856	Mole F.	[104]
Stearic acid	43.60	0.1076	Mole F.	[104]
Stearic acid	45.90	0.1423	Mole F.	[104]
Stearic acid	48.40	0.1873	Mole F.	[104]
Stearic acid	50.40	0.2397	Mole F.	[104]
Temazepam	24.00	0.000107	Volume F.	[64]
Temazepam	25.00	0.0005855	Mole F.	[15]
Testosterone propionate	25.00	0.005	Mole F.	[16]
Thalidomide	25.00	0.0001	g/L	[143]
Xanthene	25.00	0.02949	Mole F.	[3]

Solubility Data of Drugs in Isopropyl Acetate at Various Temperatures

Drug	T (°C)	Solubility	Solubility Unit	Reference
Hydrocortisone	25.00	0.00386	mol/L	[80]
Lovastatin	11.95	0.00335	Mole F.	[89]
Lovastatin	15.15	0.00372	Mole F.	[89]
Lovastatin	18.05	0.00412	Mole F.	[89]
Lovastatin	21.45	0.00459	Mole F.	[89]
Lovastatin	24.05	0.00497	Mole F.	[89]
Lovastatin	27.15	0.00553	Mole F.	[89]

Solubility Data of Drugs in Isopropyl Acetate at Various Temperatures (continued)

Drug	T (°C)	Solubility	Solubility Unit	Reference
Lovastatin	29.95	0.00606	Mole F.	[89]
Lovastatin	33.15	0.00677	Mole F.	[89]
Lovastatin	36.25	0.00756	Mole F.	[89]
Lovastatin	39.05	0.00832	Mole F.	[89]

Solubility Data of Drugs in Isobutyl Acetate at Various Temperatures

Drug	T (°C)	Solubility	Solubility Unit	Reference
Lovastatin	11.95	0.00345	Mole F.	[89]
Lovastatin	15.15	0.00383	Mole F.	[89]
Lovastatin	18.05	0.00422	Mole F.	[89]
Lovastatin	21.45	0.00470	Mole F.	[89]
Lovastatin	24.05	0.00504	Mole F.	[89]
Lovastatin	27.15	0.00570	Mole F.	[89]
Lovastatin	29.95	0.00627	Mole F.	[89]
Lovastatin	33.15	0.00702	Mole F.	[89]
Lovastatin	36.25	0.00777	Mole F.	[89]
Lovastatin	39.05	0.00864	Mole F.	[89]

Solubility Data of Drugs in Isopropyl Myristate at Various Temperatures

Drug	T (°C)	Solubility	Solubility Unit	Reference
1-Alkylaminocarbonyl-5-flurouracil (C1)	32.00	0.0002992	mol/L	[144]
1-Alkylaminocarbonyl-5-flurouracil (C2)	32.00	0.0027925	mol/L	[144]
1-Alkylaminocarbonyl-5-flurouracil (C3)	32.00	0.0123880	mol/L	[144]
1-Alkylaminocarbonyl-5-flurouracil (C4)	32.00	0.0246037	mol/L	[144]
1-Alkylaminocarbonyl-5-flurouracil (C8)	32.00	0.0467735	mol/L	[144]
1-Alkylcarbonyl-5-flurouracil (C1)	32.00	0.0220800	mol/L	[144]
1-Alkylcarbonyl-5-flurouracil (C2)	32.00	0.0363915	mol/L	[144]
1-Alkylcarbonyl-5-flurouracil (C3)	32.00	0.0174181	mol/L	[144]
1-Alkylcarbonyl-5-flurouracil (C4)	32.00	0.0391742	mol/L	[144]
1-Alkylcarbonyl-5-flurouracil (C5)	32.00	0.1127197	mol/L	[144]
1-Alkylcarbonyl-5-flurouracil (C7)	32.00	0.1106624	mol/L	[144]
1-Pivaloyloxymethyl-5-flurouracil	32.00	0.0077804	mol/L	[144]
5-Flurouracil	32.00	0.0000492	mol/L	[144]
6,9-bis(Alkylcarbonyloxymethyl)-6-mercaptopurin (C1)	32.00	0.0052723	mol/L	[144]
6,9-bis(Alkylcarbonyloxymethyl)-6-mercaptopurin (C2)	32.00	0.0336512	mol/L	[144]
6,9-bis(Alkylcarbonyloxymethyl)-6-mercaptopurin (C3)	32.00	0.0909913	mol/L	[144]
6,9-bis(Alkylcarbonyloxymethyl)-6-mercaptopurin (C4)	32.00	0.1741807	mol/L	[144]

(continued)

Solubility Data of Drugs in Isopropyl Myristate at Various Temperatures   (continued)

Drug	T (°C)	Solubility	Solubility Unit	Reference
6-Alkylcarbonyloxymethyl- 6-mercaptopurin (C1)	32.00	0.0010520	mol/L	[144]
6-Alkylcarbonyloxymethyl- 6-mercaptopurin (C2)	32.00	0.0023014	mol/L	[144]
6-Alkylcarbonyloxymethyl- 6-mercaptopurin (C3)	32.00	0.0032885	mol/L	[144]
6-Alkylcarbonyloxymethyl- 6-mercaptopurin (C4)	32.00	0.0042073	mol/L	[144]
6-Alkylcarbonyloxymethyl- 6-mercaptopurin(C5)	32.00	0.0036813	mol/L	[144]
6-Alkylcarbonyloxymethyl- 6-mercaptopurin (C7)	32.00	0.0041495	mol/L	[144]
6-Mercaptopurin	32.00	0.0000224	mol/L	[144]
7-Alkylcarbonyloxymethyltheophyline (C1)	32.00	0.0027479	mol/L	[144]
7-Alkylcarbonyloxymethyltheophyline (C2)	32.00	0.0029309	mol/L	[144]
7-Alkylcarbonyloxymethyltheophyline (C3)	32.00	0.0254097	mol/L	[144]
7-Alkylcarbonyloxymethyltheophyline (C4)	32.00	0.0439542	mol/L	[144]
7-Alkylcarbonyloxymethyltheophyline (C5)	32.00	0.0778037	mol/L	[144]
Acetanilide	25.00	0.0722	mol/L	[73]
Acetanilide	30.00	0.0820	mol/L	[73]
Acetanilide	35.00	0.0952	mol/L	[73]
Acetanilide	40.00	0.1111	mol/L	[73]
Alkylcarbonyloxymethyl-5-flurouracil (C1)	32.00	0.0032885	mol/L	[144]
Alkylcarbonyloxymethyl-5-flurouracil (C2)	32.00	0.0098401	mol/L	[144]
Alkylcarbonyloxymethyl-5-flurouracil (C3)	32.00	0.0143880	mol/L	[144]
Alkylcarbonyloxymethyl-5-flurouracil (C4)	32.00	0.0147911	mol/L	[144]
Alkylcarbonyloxymethyl-5-flurouracil (C5)	32.00	0.0146893	mol/L	[144]
Alkylcarbonyloxymethyl-5-flurouracil (C7)	32.00	0.0100000	mol/L	[144]
Alkylcarbonyloxymethyl-5-flurouracil (C9)	32.00	0.0042756	mol/L	[144]
Alkyloxycarbonyl-5-flurouracil (C1)	32.00	0.0021281	mol/L	[144]
Alkyloxycarbonyl-5-flurouracil (C2)	32.00	0.0130918	mol/L	[144]
Alkyloxycarbonyl-5-flurouracil (C3)	32.00	0.0152055	mol/L	[144]
Alkyloxycarbonyl-5-flurouracil (C4)	32.00	0.0338065	mol/L	[144]
Alkyloxycarbonyl-5-flurouracil (C6)	32.00	0.1534617	mol/L	[144]
Alkyloxycarbonyl-5-flurouracil (C8)	32.00	0.0363915	mol/L	[144]
Benzocaine	25.00	0.207	mol/L	[76]
Benzocaine	30.00	0.249	mol/L	[76]
Benzocaine	35.00	0.336	mol/L	[76]
Benzocaine	40.00	0.401	mol/L	[76]
Benztropine	37.00	782.50	g/L	[145]
Ibuprofen	25.00	0.5847	mol/L	[81]
Ibuprofen	30.00	0.775	mol/L	[81]
Ibuprofen	35.00	0.980	mol/L	[81]
Ibuprofen	40.00	1.203	mol/L	[81]
Naproxen	20.00	0.01801	mol/L	[82]
Naproxen	25.00	0.0214	mol/L	[82]
Naproxen	30.00	0.02404	mol/L	[82]

Solubility Data of Drugs in Isopropyl Myristate at Various Temperatures (continued)

Drug	T (°C)	Solubility	Solubility Unit	Reference
Naproxen	35.00	0.0288	mol/L	[82]
Naproxen	40.00	0.03273	mol/L	[82]
Paracetamol	25.00	0.002723	mol/L	[73]
Paracetamol	30.00	0.003169	mol/L	[73]
Paracetamol	35.00	0.00371	mol/L	[73]
Paracetamol	40.00	0.00453	mol/L	[73]
Phenacetin	25.00	0.00996	mol/L	[73]
Phenacetin	30.00	0.01155	mol/L	[73]
Phenacetin	35.00	0.01460	mol/L	[73]
Phenacetin	40.00	0.01624	mol/L	[73]
Progestrone	25.00	17.0	g/L	[91]
Indomethacine	25.00	1.7	g/L	[91]
Theophylline	32.00	0.0003396	mol/L	[144]
Triclosan	20.00	0.0422	Mole F.	[85]
Triclosan	25.00	0.0475	Mole F.	[85]
Triclosan	30.00	0.0527	Mole F.	[85]
Triclosan	35.00	0.0563	Mole F.	[85]
Triclosan	40.00	0.0594	Mole F.	[85]

Solubility Data of Drugs in Isopropyl Myristate-Saturated Buffer at Various Temperatures

Drug	T (°C)	Solubility	Solubility Unit	Reference
Paracetamol	25.00	0.00332	mol/L	[73]
Paracetamol	30.00	0.0034920	mol/L	[73]
Paracetamol	35.00	0.0037380	mol/L	[73]
Paracetamol	40.00	0.00388	mol/L	[73]
Acetanilide	25.00	0.07910	mol/L	[73]
Acetanilide	30.00	0.09190	mol/L	[73]
Acetanilide	35.00	0.10590	mol/L	[73]
Acetanilide	40.00	0.11970	mol/L	[73]
Benzocaine	25.00	0.0059	mol/L	[76]
Benzocaine	30.00	0.0069	mol/L	[76]
Benzocaine	35.00	0.0088	mol/L	[76]
Benzocaine	40.00	0.01025	mol/L	[76]
Ibuprofen	25.00	0.0000548	mol/L	[81]
Ibuprofen	30.00	0.0000736	mol/L	[81]
Ibuprofen	35.00	0.0000949	mol/L	[81]
Ibuprofen	40.00	0.0001222	mol/L	[81]
Phenacetin	25.00	0.01055	mol/L	[73]
Phenacetin	30.00	0.01353	mol/L	[73]
Phenacetin	35.00	0.01558	mol/L	[73]
Phenacetin	40.00	0.01930	mol/L	[73]

Solubility Data of Drugs in Methanol at Various Temperatures

Drug	T (°C)	Solubility	Solubility Unit	Reference
11α-Hydroxy-16α,17α-epoxyprogesterone	11.42	0.0005982	Mole F.	[105]
11α-Hydroxy-16α,17α-epoxyprogesterone	15.95	0.0006772	Mole F.	[105]
11α-Hydroxy-16α,17α-epoxyprogesterone	20.20	0.0007822	Mole F.	[105]
11α-Hydroxy-16α,17α-epoxyprogesterone	24.92	0.0009242	Mole F.	[105]
11α-Hydroxy-16α,17α-epoxyprogesterone	29.70	0.001083	Mole F.	[105]
11α-Hydroxy-16α,17α-epoxyprogesterone	34.90	0.001268	Mole F.	[105]
11α-Hydroxy-16α,17α-epoxyprogesterone	39.42	0.001459	Mole F.	[105]
11α-Hydroxy-16α,17α-epoxyprogesterone	44.86	0.001713	Mole F.	[105]
11α-Hydroxy-16α,17α-epoxyprogesterone	49.41	0.002007	Mole F.	[105]
16α,17α-Epoxyprogesterone	9.35	0.00100	Mole F.	[17]
16α,17α-Epoxyprogesterone	15.55	0.00133	Mole F.	[17]
16α,17α-Epoxyprogesterone	19.95	0.00159	Mole F.	[17]
16α,17α-Epoxyprogesterone	24.75	0.00192	Mole F.	[17]
16α,17α-Epoxyprogesterone	29.80	0.00240	Mole F.	[17]
16α,17α-Epoxyprogesterone	34.60	0.00290	Mole F.	[17]
16α,17α-Epoxyprogesterone	39.65	0.00349	Mole F.	[17]
16α,17α-Epoxyprogesterone	44.50	0.00424	Mole F.	[17]
16α,17α-Epoxyprogesterone	49.30	0.00516	Mole F.	[17]
16α,17α-Epoxyprogesterone	54.45	0.00627	Mole F.	[17]
2-(4-Ethylbenzoyl)benzoic acid	6.75	0.0627	Mole F.	[18]
2-(4-Ethylbenzoyl)benzoic acid	12.45	0.0827	Mole F.	[18]
2-(4-Ethylbenzoyl)benzoic acid	17.66	0.1014	Mole F.	[18]
2-(4-Ethylbenzoyl)benzoic acid	22.30	0.1184	Mole F.	[18]
2-(4-Ethylbenzoyl)benzoic acid	32.41	0.1644	Mole F.	[18]
2-(4-Ethylbenzoyl)benzoic acid	41.25	0.2016	Mole F.	[18]
2-(4-Ethylbenzoyl)benzoic acid	50.65	0.2482	Mole F.	[18]
2-(4-Ethylbenzoyl)benzoic acid	55.54	0.2727	Mole F.	[18]
2-(4-Ethylbenzoyl)benzoic acid	60.38	0.3005	Mole F.	[18]
2,6-Diaminopyridine	−7.00	0.05984	Mole F.	[93]
2,6-Diaminopyridine	−3.20	0.06600	Mole F.	[93]
2,6-Diaminopyridine	1.00	0.07786	Mole F.	[93]
2,6-Diaminopyridine	4.90	0.08883	Mole F.	[93]
2,6-Diaminopyridine	8.00	0.09331	Mole F.	[93]
2,6-Diaminopyridine	12.80	0.11146	Mole F.	[93]
2,6-Diaminopyridine	17.00	0.12519	Mole F.	[93]
2,6-Diaminopyridine	20.20	0.13135	Mole F.	[93]
2,6-Diaminopyridine	24.00	0.14964	Mole F.	[93]
2,6-Diaminopyridine	28.00	0.16484	Mole F.	[93]
2,6-Diaminopyridine	31.40	0.17823	Mole F.	[93]
2,6-Diaminopyridine	34.80	0.18758	Mole F.	[93]
2,6-Diaminopyridine	38.00	0.20739	Mole F.	[93]
2,6-Diaminopyridine	44.70	0.23324	Mole F.	[93]
4-Aminophenylacetic acid	20.00	3.950	g/L	[8]
5,5-Diethylbarbituric acid (Barbital)	25.00	166	g/L	[41]
5-Butyl-5-ethylbarbituric acid (Butetal)	25.00	530	g/L	[41]
5-Ethyl-5-(1-methylpropyl)-barbituric acid (Butabarbital)	25.00	132	g/L	[41]

Solubility Data of Drugs in Methanol at Various Temperatures   (continued)

Drug	T (°C)	Solubility	Solubility Unit	Reference
5-Ethyl-5-(2-methylbutyl)-barbituric acid (Pentobarbital)	25.00	313	g/L	[41]
5-Ethyl-5-(2-methylpropyl)-barbituric acid	25.00	123	g/L	[41]
5-Ethyl-5-(3-methylbutyl)-barbituric acid (Amobarbital)	25.00	293	g/L	[41]
5-Ethyl-5-isopropylbarbituric acid (Probarbital)	25.00	129	g/L	[41]
5-Ethyl-5-pentylbarbituric acid	25.00	553	g/L	[41]
5-Ethyl-5-phenylbarbituric acid (Phenobarbital)	25.00	252	g/L	[41]
5-Ethyl-5-propylbarbiturate	25.00	325	g/L	[41]
Acetic acid	−30.30	0.464	Mole F.	[109]
Acetic acid	−24.70	0.552	Mole F.	[109]
Acetic acid	−17.10	0.602	Mole F.	[109]
Acetic acid	−12.30	0.663	Mole F.	[109]
Acetic acid	−7.70	0.712	Mole F.	[109]
Acetic acid	−2.00	0.778	Mole F.	[109]
Acetic acid	2.70	0.820	Mole F.	[109]
Acetic acid	7.10	0.880	Mole F.	[109]
Acetic acid	10.40	0.934	Mole F.	[109]
Acetylsalicylic acid	25.00	0.0719	Mole F.	[21]
Anhydrous citric acid	25.00	0.15753	Mole F.	[23]
Benzoic acid	25.00	0.1632	Mole F.	[24]
β-Carotene	Room	10	mg/L	[6]
β-Glucose	40.00	14.24	g/L	[42]
Betulin	5.05	0.000179	Mole F.	[7]
Betulin	15.05	0.000248	Mole F.	[7]
Betulin	25.05	0.000328	Mole F.	[7]
Betulin	35.05	0.000397	Mole F.	[7]
Butylparaben	25.00	0.336	Mole F.	[45]
Butylparaben	30.00	0.369	Mole F.	[45]
Butylparaben	35.00	0.457	Mole F.	[45]
Butylparaben	40.00	0.491	Mole F.	[45]
Carbamazepine	25.00	76.8	g/L	[8]
Carbamazepine (form III)	3.65	0.006400	Mole F.	[46]
Carbamazepine (form III)	11.90	0.008790	Mole F.	[46]
Carbamazepine (form III)	17.15	0.01030	Mole F.	[46]
Carbamazepine (form III)	28.35	0.01414	Mole F.	[46]
Carbamazepine (form III)	38.25	0.01858	Mole F.	[46]
Carbamazepine (form III)	47.55	0.02529	Mole F.	[46]
Carbamazepine (form III)	53.65	0.02990	Mole F.	[46]
Cefazolin sodium pentahydrate	15.70	0.0008872	Mole F.	[9]
Cefazolin sodium pentahydrate	23.40	0.001011	Mole F.	[9]
Cefazolin sodium pentahydrate	30.00	0.001355	Mole F.	[9]
Cefazolin sodium pentahydrate	33.70	0.001778	Mole F.	[9]
Cefazolin sodium pentahydrate	37.40	0.002692	Mole F.	[9]
Cefotaxim sodium	5.00	1.385	g/100 g	[111]
Cefotaxim sodium	10.00	1.409	g/100 g	[111]

(continued)

Solubility Data of Drugs in Methanol at Various Temperatures   (continued)

Drug	T (°C)	Solubility	Solubility Unit	Reference
Cefotaxim sodium	15.00	1.417	g/100 g	[111]
Cefotaxim sodium	20.00	1.453	g/100 g	[111]
Cefotaxim sodium	30.00	1.531	g/100 g	[111]
Cefotaxim sodium	40.00	1.573	g/100 g	[111]
Ceftriaxone disodium	5.15	0.0004184	Mole F.	[112]
Ceftriaxone disodium	10.05	0.0005129	Mole F.	[112]
Ceftriaxone disodium	14.85	0.0006491	Mole F.	[112]
Ceftriaxone disodium	20.10	0.0008052	Mole F.	[112]
Ceftriaxone disodium	24.75	0.0009813	Mole F.	[112]
Ceftriaxone disodium	29.95	0.001216	Mole F.	[112]
Ceftriaxone disodium	32.45	0.001345	Mole F.	[112]
Ceftriaxone disodium	36.50	0.001540	Mole F.	[112]
Ceftriaxone disodium	40.65	0.001855	Mole F.	[112]
Ceftriaxone disodium	44.75	0.002166	Mole F.	[112]
Celecoxib	25.00	113.94	g/L	[55]
Cholesteryl acetate	31.50	0.390	g/100 g	[97]
Cholesteryl acetate	32.30	0.420	g/100 g	[97]
Cholesteryl acetate	35.10	0.490	g/100 g	[97]
Cholesteryl acetate	39.00	0.633	g/100 g	[97]
Cholesteryl acetate	41.20	0.689	g/100 g	[97]
Cholesteryl acetate	44.30	0.832	g/100 g	[97]
Cholestrole	20.00	0.648	g/100 g	[97]
Cholestrole	30.70	0.987	g/100 g	[97]
Cholestrole	33.90	1.143	g/100 g	[97]
Cholestrole	40.00	1.510	g/100 g	[97]
Cholestrole	45.00	2.008	g/100 g	[97]
Clindamycin phosphate	13.00	0.000337	Mole F.	[134]
Clindamycin phosphate	18.00	0.000359	Mole F.	[134]
Clindamycin phosphate	23.00	0.000394	Mole F.	[134]
Clindamycin phosphate	28.00	0.000441	Mole F.	[134]
Clindamycin phosphate	33.00	0.000491	Mole F.	[134]
Clindamycin phosphate	38.00	0.000541	Mole F.	[134]
Clindamycin phosphate	43.00	0.000591	Mole F.	[134]
Clindamycin phosphate	48.00	0.000648	Mole F.	[134]
D(−)-p-Hydroxyphenylglycine dane salt	20.00	0.0034600	Mole F.	[49]
D(−)-p-Hydroxyphenylglycine dane salt	25.07	0.0037030	Mole F.	[49]
D(−)-p-Hydroxyphenylglycine dane salt	30.00	0.0039580	Mole F.	[49]
D(−)-p-Hydroxyphenylglycine dane salt	35.04	0.0042370	Mole F.	[49]
D(−)-p-Hydroxyphenylglycine dane salt	40.20	0.0045420	Mole F.	[49]
D(−)-p-Hydroxyphenylglycine dane salt	45.12	0.0048500	Mole F.	[49]
D(−)-p-Hydroxyphenylglycine dane salt	50.20	0.0051870	Mole F.	[49]
D(−)-p-Hydroxyphenylglycine dane salt	55.01	0.0055240	Mole F.	[49]
D(−)-p-Hydroxyphenylglycine dane salt	60.20	0.0059960	Mole F.	[49]
Dexamethasone sodium phosphate	5.00	0.01105	Mole F.	[138]
Dexamethasone sodium phosphate	10.00	0.01072	Mole F.	[138]
Dexamethasone sodium phosphate	15.00	0.01049	Mole F.	[138]
Dexamethasone sodium phosphate	20.00	0.009823	Mole F.	[138]
Dexamethasone sodium phosphate	25.00	0.009523	Mole F.	[138]
Dexamethasone sodium phosphate	30.00	0.009036	Mole F.	[138]

Solubility Data of Drugs in Methanol at Various Temperatures   (continued)

Drug	T (°C)	Solubility	Solubility Unit	Reference
Dexamethasone sodium phosphate	35.00	0.008862	Mole F.	[138]
Dexamethasone sodium phosphate	40.00	0.008301	Mole F.	[138]
Dexamethasone sodium phosphate	45.00	0.007817	Mole F.	[138]
Diclofenac	25.00	0.0058694	Mole F.	[25]
Diflunisal	25.00	0.0151	Mole F.	[26]
Dioxopromethazine HCl	20.01	0.002080	Mole F.	[106]
Dioxopromethazine HCl	25.03	0.002769	Mole F.	[106]
Dioxopromethazine HCl	30.03	0.003475	Mole F.	[106]
Dioxopromethazine HCl	35.05	0.004334	Mole F.	[106]
Dioxopromethazine HCl	40.02	0.005479	Mole F.	[106]
Dioxopromethazine HCl	45.07	0.006886	Mole F.	[106]
Dioxopromethazine HCl	55.08	0.010020	Mole F.	[106]
Dioxopromethazine HCl	60.04	0.012280	Mole F.	[106]
Dioxopromethazine HCl	65.05	0.014860	Mole F.	[106]
D-*p*-Hydroxyphenylglycine	5.02	0.0004199	Mole F.	[127]
D-*p*-Hydroxyphenylglycine	10.07	0.0004207	Mole F.	[127]
D-*p*-Hydroxyphenylglycine	15.06	0.0004215	Mole F.	[127]
D-*p*-Hydroxyphenylglycine	20.13	0.0004225	Mole F.	[127]
D-*p*-Hydroxyphenylglycine	25.09	0.0004236	Mole F.	[127]
D-*p*-Hydroxyphenylglycine	30.17	0.0004247	Mole F.	[127]
D-*p*-Hydroxyphenylglycine	35.05	0.0004260	Mole F.	[127]
D-*p*-Hydroxyphenylglycine	40.11	0.0004273	Mole F.	[127]
D-*p*-Hydroxyphenylglycine	45.12	0.0004288	Mole F.	[127]
D-*p*-Hydroxyphenylglycine	50.10	0.0004303	Mole F.	[127]
Eflucimibe (form A)	20.00	0.0001420	mol/L	[99]
Ellagic acid	37.00	0.6717	g/L	[146]
Enrofloxacin sodium	20.00	0.00096	Mole F.	[1]
Enrofloxacin sodium	22.00	0.00102	Mole F.	[1]
Enrofloxacin sodium	25.00	0.00115	Mole F.	[1]
Enrofloxacin sodium	27.00	0.00120	Mole F.	[1]
Enrofloxacin sodium	30.00	0.00130	Mole F.	[1]
Enrofloxacin sodium	32.00	0.00135	Mole F.	[1]
Enrofloxacin sodium	35.00	0.00144	Mole F.	[1]
Enrofloxacin sodium	37.00	0.00149	Mole F.	[1]
Erythritol	14.20	0.01639	Mass fraction	[113]
Erythritol	18.00	0.01687	Mass fraction	[113]
Erythritol	21.50	0.01825	Mass fraction	[113]
Erythritol	24.80	0.02218	Mass fraction	[113]
Erythritol	29.10	0.02806	Mass fraction	[113]
Erythritol	34.50	0.03594	Mass fraction	[113]
Erythritol	37.10	0.03876	Mass fraction	[113]
Erythritol	40.35	0.04615	Mass fraction	[113]
Erythritol	44.85	0.05760	Mass fraction	[113]
Erythritol	46.85	0.06254	Mass fraction	[113]
Erythritol	49.60	0.07231	Mass fraction	[113]
Erythritol	54.20	0.08621	Mass fraction	[113]
Erythromycin A dihydrate	20.05	0.01162	Mole F.	[100]
Erythromycin A dihydrate	24.85	0.01432	Mole F.	[100]

(continued)

Solubility Data of Drugs in Methanol at Various Temperatures   (continued)

Drug	T (°C)	Solubility	Solubility Unit	Reference
Erythromycin A dihydrate	30.00	0.01789	Mole F.	[100]
Erythromycin A dihydrate	34.85	0.02154	Mole F.	[100]
Erythromycin A dihydrate	39.95	0.02534	Mole F.	[100]
Erythromycin A dihydrate	44.85	0.03050	Mole F.	[100]
Erythromycin A dihydrate	49.85	0.03730	Mole F.	[100]
Ethylparaben	25.00	0.135	Mole F.	[45]
Ethylparaben	30.00	0.157	Mole F.	[45]
Ethylparaben	35.00	0.185	Mole F.	[45]
Ethylparaben	40.00	0.208	Mole F.	[45]
Flubiprofen	25.00	0.0478	Mole F.	[26]
Gallic acid	25.05	0.2793	Mass fraction	[139]
Gallic acid	30.05	0.2883	Mass fraction	[139]
Gallic acid	35.05	0.2930	Mass fraction	[139]
Gallic acid	40.05	0.2959	Mass fraction	[139]
Gallic acid	45.05	0.3013	Mass fraction	[139]
Gallic acid	50.05	0.3048	Mass fraction	[139]
Gallic acid	55.05	0.3107	Mass fraction	[139]
Gallic acid	60.05	0.3174	Mass fraction	[139]
Glucose	40.00	28.76	g/L	[42]
Haloperidol	25.00	0.00153	Mole F.	[27]
Hesperetin	15.05	0.085	mol/L	[50]
Hesperetin	20.05	0.092	mol/L	[50]
Hesperetin	25.05	0.106	mol/L	[50]
Hesperetin	30.05	0.111	mol/L	[50]
Hesperetin	35.05	0.127	mol/L	[50]
Hesperetin	40.05	0.142	mol/L	[50]
Hesperetin	45.05	0.162	mol/L	[50]
Hesperetin	50.05	0.187	mol/L	[50]
Hydroquinone	8.50	0.1047	Mole F.	[101]
Hydroquinone	15.55	0.1121	Mole F.	[101]
Hydroquinone	20.65	0.1177	Mole F.	[101]
Hydroquinone	28.00	0.1270	Mole F.	[101]
Hydroquinone	33.70	0.1335	Mole F.	[101]
Hydroquinone	41.45	0.1437	Mole F.	[101]
Hydroquinone	46.70	0.1521	Mole F.	[101]
Hydroquinone	53.50	0.1608	Mole F.	[101]
Hydroquinone	62.05	0.1766	Mole F.	[101]
Hydroquinone	70.25	0.1949	Mole F.	[101]
Ibuprofen	10.00	725.8	g/kg	[83]
Ibuprofen	15.00	922.3	g/kg	[83]
Ibuprofen	20.00	1035	g/kg	[83]
Ibuprofen	25.00	0.0246815	Mole F.	[28]
Ibuprofen	25.00	0.06053	Mole F.	[67]
Ibuprofen	30.00	1752	g/kg	[83]
Ibuprofen	35.00	2730	g/kg	[83]
Imidacloprid	20.64	0.001021	Mole F.	[2]
Imidacloprid	22.23	0.001112	Mole F.	[2]
Imidacloprid	26.66	0.001361	Mole F.	[2]
Imidacloprid	29.00	0.001557	Mole F.	[2]

Solubility Data of Drugs in Methanol at Various Temperatures   (continued)

Drug	T (°C)	Solubility	Solubility Unit	Reference
Imidacloprid	32.71	0.001793	Mole F.	[2]
Imidacloprid	36.17	0.002107	Mole F.	[2]
Imidacloprid	39.47	0.002467	Mole F.	[2]
Imidacloprid	42.23	0.002828	Mole F.	[2]
Imidacloprid	47.44	0.003609	Mole F.	[2]
Imidacloprid	50.22	0.004131	Mole F.	[2]
Imidacloprid	54.96	0.005296	Mole F.	[2]
Imidacloprid	57.60	0.005984	Mole F.	[2]
Imidacloprid	60.19	0.006760	Mole F.	[2]
Isoniazid	27.85	0.01070	Mole F.	[114]
Isoniazid	29.85	0.01139	Mole F.	[114]
Isoniazid	34.85	0.01371	Mole F.	[114]
Isoniazid	37.85	0.01541	Mole F.	[114]
Isoniazid	39.85	0.01634	Mole F.	[114]
Ketoprofen	25.00	0.0396	Mole F.	[51]
Ketoprofen	25.00	0.0428	Mole F.	[68]
L-(+)-Ascorbic acid	19.85	0.01006	Mole F.	[102]
L-(+)-Ascorbic acid	24.85	0.01083	Mole F.	[102]
L-(+)-Ascorbic acid	29.85	0.01214	Mole F.	[102]
L-(+)-Ascorbic acid	34.85	0.01326	Mole F.	[102]
L-(+)-Ascorbic acid	39.85	0.01512	Mole F.	[102]
L-(+)-Ascorbic acid	44.85	0.01676	Mole F.	[102]
L-(+)-Ascorbic acid	49.85	0.01832	Mole F.	[102]
Lactose	25.00	0.0000765	Mole F.	[30]
Lamivudine (form 1)	5.00	18.7	g/L	[52]
Lamivudine (form 1)	15.00	22.8	g/L	[52]
Lamivudine (form 1)	25.00	28.3	g/L	[52]
Lamivudine (form 1)	35.00	35.8	g/L	[52]
Lamivudine (form 1)	45.00	48.8	g/L	[52]
Losartan potassium	20.00	0.02771	Mole F.	[86]
Losartan potassium	30.00	0.03746	Mole F.	[86]
Losartan potassium	40.00	0.04449	Mole F.	[86]
Losartan potassium	50.00	0.04828	Mole F.	[86]
Losartan potassium	60.00	0.05079	Mole F.	[86]
Losartan potassium	70.00	0.05329	Mole F.	[86]
Lovastatin	5.40	0.001277	Mole F.	[116]
Lovastatin	10.25	0.001595	Mole F.	[116]
Lovastatin	15.05	0.001971	Mole F.	[116]
Lovastatin	21.00	0.002514	Mole F.	[116]
Lovastatin	25.20	0.003147	Mole F.	[116]
Lovastatin	30.15	0.003863	Mole F.	[116]
Lovastatin	34.90	0.004959	Mole F.	[116]
Lovastatin	40.80	0.006293	Mole F.	[116]
Lovastatin	44.70	0.007770	Mole F.	[116]
Lovastatin	48.30	0.009292	Mole F.	[116]
Lutein	Room	200	mg/L	[6]
Luteolin	0.00	0.00046	Mole F.	[54]
Luteolin	10.00	0.00048	Mole F.	[54]

(continued)

Solubility Data of Drugs in Methanol at Various Temperatures   (continued)

Drug	T (°C)	Solubility	Solubility Unit	Reference
Luteolin	25.00	0.00054	Mole F.	[54]
Luteolin	40.00	0.00071	Mole F.	[54]
Luteolin	60.00	0.00086	Mole F.	[54]
Mannitol	25.00	0.0003632	Mole F.	[30]
Meloxicam	25.00	0.31	g/L	[8]
Meloxicam	25.00	0.382	g/L	[55]
Methylparaben	25.00	0.121	Mole F.	[45]
Methylparaben	30.00	0.139	Mole F.	[45]
Methylparaben	35.00	0.162	Mole F.	[45]
Methylparaben	40.00	0.178	Mole F.	[45]
Methyl p-hydroxybenzoate	25.00	0.1254	Mole F.	[35]
Naproxen	25.00	0.0145801	Mole F.	[31]
Niflumic acid	25.00	0.0075516	Mole F.	[33]
Nimesulide	25.00	8.58	g/L	[8]
Nimesulide	25.00	8.812	g/L	[55]
p-Aminobenzoic acid	25.00	0.0539337	Mole F.	[25]
p-Aminophenylacetic acid	16.00	3.7	g/kg	[83]
p-Aminophenylacetic acid	20.00	4.3	g/kg	[83]
p-Aminophenylacetic acid	25.00	4.7	g/kg	[83]
Paracetamol	−5.00	174.48	g/kg	[11]
Paracetamol	0.00	191.48	g/kg	[11]
Paracetamol	5.00	215.09	g/kg	[11]
Paracetamol	10.00	239.60	g/kg	[11]
Paracetamol	15.00	265.43	g/kg	[11]
Paracetamol	20.00	297.81	g/kg	[11]
Paracetamol	25.00	332.11	g/kg	[11]
Paracetamol	30.00	371.61	g/kg	[11]
Paracetamol	30.00	274	g/L	[8]
Paracetamol	25.00	0.24732	Mole F.	[23]
Paroxetine HCl hemihydrate	14.35	0.0006913	Mole F.	[57]
Paroxetine HCl hemihydrate	17.45	0.0008111	Mole F.	[57]
Paroxetine HCl hemihydrate	20.45	0.0009581	Mole F.	[57]
Paroxetine HCl hemihydrate	23.40	0.0010868	Mole F.	[57]
Paroxetine HCl hemihydrate	27.65	0.0012807	Mole F.	[57]
Paroxetine HCl hemihydrate	30.55	0.0013832	Mole F.	[57]
Paroxetine HCl hemihydrate	33.25	0.0014726	Mole F.	[57]
Paroxetine HCl hemihydrate	37.20	0.0016865	Mole F.	[57]
Paroxetine HCl hemihydrate	40.10	0.0018015	Mole F.	[57]
Paroxetine HCl hemihydrate	42.90	0.0018952	Mole F.	[57]
Paroxetine HCl hemihydrate	44.95	0.0019915	Mole F.	[57]
Paroxetine HCl hemihydrate	47.85	0.0020977	Mole F.	[57]
Paroxetine HCl hemihydrate	50.10	0.0021706	Mole F.	[57]
Paroxetine HCl hemihydrate	54.55	0.0023536	Mole F.	[57]
Paroxetine HCl hemihydrate	58.05	0.0024873	Mole F.	[57]
Paroxetine HCl hemihydrate	62.15	0.0026353	Mole F.	[57]
Phenacetinum	10.59	0.010460	Mole F.	[58]
Phenacetinum	15.00	0.012790	Mole F.	[58]
Phenacetinum	20.68	0.016571	Mole F.	[58]
Phenacetinum	25.03	0.020259	Mole F.	[58]

Solubility Data of Drugs in Methanol at Various Temperatures   (continued)

Drug	T (°C)	Solubility	Solubility Unit	Reference
Phenacetinum	30.03	0.025574	Mole F.	[58]
Phenacetinum	34.90	0.032072	Mole F.	[58]
Phenacetinum	40.10	0.040701	Mole F.	[58]
Phenacetinum	44.95	0.050556	Mole F.	[58]
Phenacetinum	50.10	0.063176	Mole F.	[58]
Phenacetinum	55.00	0.077480	Mole F.	[58]
Phenacetinum	60.30	0.095735	Mole F.	[58]
Phenazine-1-carboxylic acid	5.05	0.00340	mol/L	[141]
Phenazine-1-carboxylic acid	10.05	0.00370	mol/L	[141]
Phenazine-1-carboxylic acid	15.05	0.00410	mol/L	[141]
Phenazine-1-carboxylic acid	20.05	0.00460	mol/L	[141]
Phenazine-1-carboxylic acid	25.05	0.00540	mol/L	[141]
Phenazine-1-carboxylic acid	30.05	0.00630	mol/L	[141]
Phenazine-1-carboxylic acid	35.05	0.00760	mol/L	[141]
Phenazine-1-carboxylic acid	40.05	0.00930	mol/L	[141]
Phenazine-1-carboxylic acid	45.05	0.01160	mol/L	[141]
Phenazine-1-carboxylic acid	50.05	0.01460	mol/L	[141]
Phenazine-1-carboxylic acid	55.05	0.01880	mol/L	[141]
Phenothiazine	25.00	0.00512	Mole F.	[34]
Phenylacetic acid	20.00	2915	g/kg	[83]
p-Hydroxyphenylacetic acid	10.00	934.4	g/kg	[83]
p-Hydroxyphenylacetic acid	15.00	1061	g/kg	[83]
p-Hydroxyphenylacetic acid	20.00	1608	g/kg	[83]
p-Hydroxyphenylacetic acid	25.00	1466	g/kg	[83]
p-Hydroxybenzoic acid	25.00	0.1142	Mole F.	[35]
p-Hydroxybenzoic acid	25.00	555.8	g/kg	[83]
Pimozide	25.00	0.000821	Mole F.	[36]
Piroxicam	25.00	0.0000206	Mole F.	[33]
Pravastatin sodium	4.65	0.02246	Mole F.	[60]
Pravastatin sodium	9.85	0.02448	Mole F.	[60]
Pravastatin sodium	14.75	0.02695	Mole F.	[60]
Pravastatin sodium	19.85	0.03009	Mole F.	[60]
Pravastatin sodium	24.65	0.03262	Mole F.	[60]
Pravastatin sodium	29.75	0.03732	Mole F.	[60]
Pravastatin sodium	34.85	0.04175	Mole F.	[60]
Pravastatin sodium	39.85	0.04921	Mole F.	[60]
Pravastatin sodium	44.85	0.05880	Mole F.	[60]
Pravastatin sodium	50.85	0.06867	Mole F.	[60]
Propylparaben	25.00	0.172	Mole F.	[45]
Propylparaben	30.00	0.207	Mole F.	[45]
Propylparaben	35.00	0.240	Mole F.	[45]
Propylparaben	40.00	0.276	Mole F.	[45]
Ranitidine HCl (form 1)	12.40	12.965	g/100 g	[87]
Ranitidine HCl (form 1)	17.10	17.873	g/100 g	[87]
Ranitidine HCl (form 1)	26.70	36.034	g/100 g	[87]
Ranitidine HCl (form 1)	28.50	40.026	g/100 g	[87]
Ranitidine HCl (form 1)	31.00	46.943	g/100 g	[87]
Ranitidine HCl (form 1)	43.00	92.868	g/100 g	[87]

(continued)

Solubility Data of Drugs in Methanol at Various Temperatures   (continued)

Drug	T (°C)	Solubility	Solubility Unit	Reference
Ranitidine HCl (form 2)	7.90	10.466	g/100 g	[87]
Ranitidine HCl (form 2)	14.80	17.101	g/100 g	[87]
Ranitidine HCl (form 2)	29.90	47.834	g/100 g	[87]
Ranitidine HCl (form 2)	32.10	56.406	g/100 g	[87]
Ranitidine HCl (form 2)	34.90	68.133	g/100 g	[87]
Ranitidine HCl (form 2)	41.50	98.104	g/100 g	[87]
Riboflavin	30.00	0.0328	g/L	[118]
Riboflavin	30.00	0.11	g/L	[8]
Rifapentine	5.00	0.0220	Mole F.	[13]
Rifapentine	10.00	0.0224	Mole F.	[13]
Rifapentine	15.00	0.0229	Mole F.	[13]
Rifapentine	20.00	0.0235	Mole F.	[13]
Rifapentine	24.00	0.0242	Mole F.	[13]
Rifapentine	30.00	0.0249	Mole F.	[13]
Rifapentine	35.00	0.0257	Mole F.	[13]
Rifapentine	40.00	0.0267	Mole F.	[13]
Rifapentine	45.00	0.0278	Mole F.	[13]
Rifapentine	50.00	0.0290	Mole F.	[13]
Rofecoxib	25.00	0.835	g/L	[55]
Rutin	10.00	0.002289	Mole F.	[61]
Rutin	25.00	0.003034	Mole F.	[61]
Rutin	40.00	0.003989	Mole F.	[61]
Rutin	50.00	0.004660	Mole F.	[61]
Rutin	60.00	0.005238	Mole F.	[61]
Saccharose	25.00	0.000000662	Mole F.	[30]
Salicylamide	10.00	0.02672	Mole F.	[107]
Salicylamide	15.00	0.03069	Mole F.	[107]
Salicylamide	20.00	0.03512	Mole F.	[107]
Salicylamide	25.00	0.04083	Mole F.	[37]
Salicylamide	25.00	0.04060	Mole F.	[107]
Salicylamide	30.00	0.04666	Mole F.	[107]
Salicylamide	35.00	0.05358	Mole F.	[107]
Salicylamide	40.00	0.06251	Mole F.	[107]
Salicylamide	45.00	0.07191	Mole F.	[107]
Salicylamide	50.00	0.08359	Mole F.	[107]
Salicylic acid	10.00	0.09935	Mole F.	[108]
Salicylic acid	15.00	0.10781	Mole F.	[108]
Salicylic acid	20.00	0.11772	Mole F.	[108]
Salicylic acid	25.00	0.1321259	Mole F.	[25]
Salicylic acid	25.00	0.12802	Mole F.	[108]
Salicylic acid	25.00	0.1223	Mole F.	[38]
Salicylic acid	30.00	0.13937	Mole F.	[108]
Salicylic acid	35.00	0.15053	Mole F.	[108]
Salicylic acid	40.00	0.16359	Mole F.	[108]
Salicylic acid	45.00	0.17681	Mole F.	[108]
Salicylic acid	50.00	0.19168	Mole F.	[108]
Sitosterol	30.40	0.492	g/100 g	[97]
Sitosterol	33.90	0.571	g/100 g	[97]
Sitosterol	38.00	0.680	g/100 g	[97]

Solubility Data of Drugs in Methanol at Various Temperatures   (continued)

Drug	*T* (°C)	Solubility	Solubility Unit	Reference
Sitosterol	45.30	0.824	g/100 g	[97]
Sodium ibuprofen	25.00	0.2693887	Mole F.	[28]
Sodium benzoate	25.00	0.0141860	Mole F.	[28]
Sodium diclofenac	25.00	0.0598808	Mole F.	[31]
Sodium diclofenac	25.00	0.0599047	Mole F.	[25]
Sodium *p*-aminobenzoate	25.00	0.0261214	Mole F.	[25]
Sodium salicylate	25.00	0.0539876	Mole F.	[25]
Stearic acid (polymorph B)	2.20	0.0001009	Mole F.	[90]
Stearic acid (polymorph B)	6.30	0.0002402	Mole F.	[90]
Stearic acid (polymorph B)	11.75	0.0004321	Mole F.	[90]
Stearic acid (polymorph B)	13.40	0.0005009	Mole F.	[90]
Stearic acid (polymorph B)	16.50	0.0006801	Mole F.	[90]
Stearic acid (polymorph B)	18.80	0.0007563	Mole F.	[90]
Stearic acid (polymorph B)	21.20	0.0011646	Mole F.	[90]
Stearic acid (polymorph B)	27.20	0.0022397	Mole F.	[90]
Stearic acid (polymorph B)	31.00	0.0036479	Mole F.	[90]
Stearic acid (polymorph B)	34.40	0.0057148	Mole F.	[90]
Stearic acid (bulk)	13.50	0.0005030	Mole F.	[90]
Stearic acid (bulk)	15.20	0.0004028	Mole F.	[90]
Stearic acid (bulk)	17.60	0.0004479	Mole F.	[90]
Stearic acid (bulk)	19.70	0.0006749	Mole F.	[90]
Stearic acid (bulk)	23.50	0.0011182	Mole F.	[90]
Stearic acid (bulk)	24.50	0.0014315	Mole F.	[90]
Stearic acid (bulk)	27.80	0.0020153	Mole F.	[90]
Stearic acid (bulk)	32.80	0.0039859	Mole F.	[90]
Stearic acid (polymorph C)	11.00	0.0004379	Mole F.	[90]
Stearic acid (polymorph C)	15.70	0.0006832	Mole F.	[90]
Stearic acid (polymorph C)	20.40	0.0011715	Mole F.	[90]
Stearic acid (polymorph C)	27.00	0.0022584	Mole F.	[90]
Stearic acid (polymorph C)	29.20	0.0027345	Mole F.	[90]
Stearic acid (polymorph C)	35.10	0.0056937	Mole F.	[90]
Stearic acid (polymorph C)	35.50	0.0065388	Mole F.	[90]
Stearic acid (polymorph C)	39.00	0.0102334	Mole F.	[90]
Sulfadiazine	25.00	0.0002340	Mole F.	[39]
Sulfadiazine	25.00	0.00471	mol/L	[62]
Sulfadiazine	25.00	0.0001930	Mole F.	[63]
Sulfadiazine	30.00	0.0002290	Mole F.	[63]
Sulfadiazine	37.00	0.0002990	Mole F.	[63]
Sulfadimethoxine	25.00	0.0011600	Mole F.	[63]
Sulfadimethoxine	30.00	0.0013900	Mole F.	[63]
Sulfadimethoxine	37.00	0.0017700	Mole F.	[63]
Sulfamerazine (polymorph 1)	30.00	5.669	g/L	[120]
Sulfamerazine (polymorph 2)	30.00	5.003	g/L	[120]
Sulfamethoxazole	25.00	0.356	mol/L	[62]
Sulfamethoxypyridazine	25.00	0.00294	Mole F.	[40]
Sulfisoxazole	25.00	0.184	mol/L	[62]
Sulfisomidine	25.00	0.0011200	Mole F.	[63]
Sulfisomidine	30.00	0.0012700	Mole F.	[63]

(continued)

Solubility Data of Drugs in Methanol at Various Temperatures   (continued)

Drug	T (°C)	Solubility	Solubility Unit	Reference
Sulfisomidine	37.00	0.0016500	Mole F.	[63]
Tanshinone IIA	10.05	0.00207	mol/L	[121]
Tanshinone IIA	20.15	0.00258	mol/L	[121]
Tanshinone IIA	29.95	0.00363	mol/L	[121]
Tanshinone IIA	35.05	0.00422	mol/L	[121]
Tanshinone IIA	44.85	0.00630	mol/L	[121]
Tanshinone IIA	50.15	0.00801	mol/L	[121]
Temazepam	25.00	0.0054740	Mole F.	[15]
Thalidomide	32.00	1.18	g/L	[8]
Trimethoprim	3.87	0.0008282	Mole F.	[65]
Trimethoprim	9.70	0.001002	Mole F.	[65]
Trimethoprim	14.94	0.001201	Mole F.	[65]
Trimethoprim	19.88	0.001432	Mole F.	[65]
Trimethoprim	25.02	0.001728	Mole F.	[65]
Trimethoprim	31.09	0.002171	Mole F.	[65]
Trimethoprim	36.21	0.002649	Mole F.	[65]
Trimethoprim	40.97	0.003203	Mole F.	[65]
Trimethoprim	45.91	0.003917	Mole F.	[65]
Trimethoprim	50.38	0.004710	Mole F.	[65]
Trimethoprim	55.00	0.005707	Mole F.	[65]
Trimethoprim	60.58	0.007198	Mole F.	[65]
Xanthene	25.00	0.004455	Mole F.	[3]

Solubility Data of Drugs in Methyl Acetate at Various Temperatures

Drug	T (°C)	Solubility	Solubility Unit	Reference
1H-1,2,4-Triazole	10.12	0.0968	Mole F.	[92]
1H-1,2,4-Triazole	15.06	0.1064	Mole F.	[92]
1H-1,2,4-Triazole	20.04	0.1168	Mole F.	[92]
1H-1,2,4-Triazole	25.06	0.1282	Mole F.	[92]
1H-1,2,4-Triazole	30.05	0.1408	Mole F.	[92]
1H-1,2,4-Triazole	35.14	0.1552	Mole F.	[92]
1H-1,2,4-Triazole	40.03	0.1710	Mole F.	[92]
1H-1,2,4-Triazole	45.01	0.1892	Mole F.	[92]
1H-1,2,4-Triazole	50.09	0.2104	Mole F.	[92]
Acetylsalicylic acid	25.00	0.05287	Mole F.	[22]
Betulin	5.05	0.000598	Mole F.	[7]
Betulin	15.05	0.000705	Mole F.	[7]
Betulin	25.05	0.000797	Mole F.	[7]
Betulin	35.05	0.000917	Mole F.	[7]
Hydrocortisone	25.00	0.0162	mol/L	[80]
Lovastatin	11.95	0.00309	Mole F.	[89]
Lovastatin	15.15	0.00340	Mole F.	[89]
Lovastatin	18.05	0.00376	Mole F.	[89]
Lovastatin	21.45	0.00417	Mole F.	[89]
Lovastatin	24.05	0.00447	Mole F.	[89]

Solubility Data of Drugs in Methyl Acetate at Various Temperatures   (continued)

Drug	T (°C)	Solubility	Solubility Unit	Reference
Lovastatin	27.15	0.00515	Mole F.	[89]
Lovastatin	29.95	0.00543	Mole F.	[89]
Lovastatin	33.15	0.00619	Mole F.	[89]
Lovastatin	36.25	0.00674	Mole F.	[89]
Lovastatin	39.05	0.00746	Mole F.	[89]
Naproxen	25.00	0.02746	Mole F.	[32]
Salicylamide	25.00	0.08377	Mole F.	[37]
Temazepam	25.00	0.0199276	Mole F.	[15]

Solubility Data of Drugs in *m*-Xylene at Various Temperatures

Drug	T (°C)	Solubility	Solubility Unit	Reference
Benzamide	57.00	0.5	% Weight	[147]
Benzamide	68.00	1.0	% Weight	[147]
Benzamide	95.00	5.0	% Weight	[147]
Benzamide	105.00	10.0	% Weight	[147]
Benzamide	108.00	20.0	% Weight	[147]
Benzamide	110.00	40.0	% Weight	[147]
Benzamide	112.00	50.0	% Weight	[147]
Vitamin K3	28.50	0.1270	Mole F.	[125]
Vitamin K3	34.59	0.1515	Mole F.	[125]
Vitamin K3	39.61	0.1761	Mole F.	[125]
Vitamin K3	44.21	0.2021	Mole F.	[125]
Vitamin K3	48.49	0.2295	Mole F.	[125]
Vitamin K3	52.79	0.2581	Mole F.	[125]
Vitamin K3	56.58	0.2881	Mole F.	[125]
Vitamin K3	60.49	0.3230	Mole F.	[125]
Vitamin K3	64.46	0.3645	Mole F.	[125]
Vitamin K3	68.40	0.4109	Mole F.	[125]

Solubility Data of Drugs in *N*-Methyl Pyrrolidone at Various Temperatures

Drug	T (°C)	Solubility	Solubility Unit	Reference
2-(4-Ethylbenzoyl)benzoic acid	6.95	0.1626	Mole F.	[18]
2-(4-Ethylbenzoyl)benzoic acid	11.00	0.1755	Mole F.	[18]
2-(4-Ethylbenzoyl)benzoic acid	20.29	0.2095	Mole F.	[18]
2-(4-Ethylbenzoyl)benzoic acid	30.61	0.2604	Mole F.	[18]
2-(4-Ethylbenzoyl)benzoic acid	40.00	0.3147	Mole F.	[18]
2-(4-Ethylbenzoyl)benzoic acid	49.70	0.3958	Mole F.	[18]
2-(4-Ethylbenzoyl)benzoic acid	61.15	0.4843	Mole F.	[18]
Benzoic acid	23.20	0.5162	Mole F.	[148]
Benzoic acid	28.00	0.5235	Mole F.	[148]
Benzoic acid	31.90	0.5302	Mole F.	[148]

(continued)

Solubility Data of Drugs in *N*-Methyl Pyrrolidone at Various Temperatures   (continued)

Drug	T (°C)	Solubility	Solubility Unit	Reference
Benzoic acid	34.80	0.5353	Mole F.	[148]
Benzoic acid	37.90	0.5409	Mole F.	[148]
Benzoic acid	41.10	0.5470	Mole F.	[148]
Benzoic acid	45.20	0.5535	Mole F.	[148]
Benzoic acid	48.80	0.5607	Mole F.	[148]
Benzoic acid	54.20	0.5744	Mole F.	[148]
Benzoic acid	60.40	0.5913	Mole F.	[148]
Benzoic acid	65.70	0.6082	Mole F.	[148]
Benzoic acid	70.10	0.6228	Mole F.	[148]
Benzoic acid	74.60	0.6395	Mole F.	[148]
Benzoic acid	78.00	0.6535	Mole F.	[148]
Benzoic acid	81.90	0.6691	Mole F.	[148]
Benzoic acid	86.70	0.6933	Mole F.	[148]
Benzoic acid	91.40	0.7181	Mole F.	[148]
Benzoic acid	95.90	0.7447	Mole F.	[148]
Benzoic acid	98.20	0.7592	Mole F.	[148]
Dioxopromethazine HCl	5.01	0.006413	Mole F.	[106]
Dioxopromethazine HCl	10.05	0.006924	Mole F.	[106]
Dioxopromethazine HCl	15.09	0.007468	Mole F.	[106]
Dioxopromethazine HCl	20.01	0.007905	Mole F.	[106]
Dioxopromethazine HCl	25.02	0.008426	Mole F.	[106]
Dioxopromethazine HCl	35.02	0.009755	Mole F.	[106]
Dioxopromethazine HCl	40.06	0.010410	Mole F.	[106]
Dioxopromethazine HCl	45.08	0.011110	Mole F.	[106]
Dioxopromethazine HCl	50.12	0.011920	Mole F.	[106]
Sodium cefotaxim	5.00	0.000254	Mole F.	[133]
Sodium cefotaxim	7.00	0.000935	Mole F.	[133]
Sodium cefotaxim	10.00	0.001523	Mole F.	[133]
Sodium cefotaxim	12.00	0.001915	Mole F.	[133]
Sodium cefotaxim	15.00	0.002788	Mole F.	[133]
Sodium cefotaxim	17.00	0.003829	Mole F.	[133]
Sodium cefotaxim	20.00	0.005241	Mole F.	[133]
Sodium cefotaxim	22.00	0.008686	Mole F.	[133]
Sodium cefotaxim	25.00	0.011798	Mole F.	[133]
Sodium cefotaxim	27.00	0.038016	Mole F.	[133]
Sodium cefotaxim	30.00	0.073655	Mole F.	[133]
Sodium cefotaxim	32.00	0.082957	Mole F.	[133]
Sodium cefotaxim	35.00	0.102970	Mole F.	[133]

Solubility Data of Drugs in Octane at Various Temperatures

Drug	T (°C)	Solubility	Solubility Unit	Reference
Benzoic acid	25.00	0.0129	Mole F.	[21]
Flubiprofen	25.00	0.000616	Mole F.	[26]
Mestanolone	25.00	0.0001780	Mole F.	[69]
Methandienone	25.00	0.0002980	Mole F.	[69]

Solubility Data of Drugs in Octane at Various Temperatures   (continued)

Drug	T (°C)	Solubility	Solubility Unit	Reference
Methyltestosterone	25.00	0.000347	Mole F.	[69]
Nandrolone	25.00	0.000320	Mole F.	[69]
Phenothiazine	25.00	0.000858	Mole F.	[34]
Testosterone	25.00	0.000179	Mole F.	[69]
Testosterone propionate	25.00	0.0081	Mole F.	[131]
Xanthene	25.00	0.03976	Mole F.	[3]

Solubility Data of Drugs in Octanol-Saturated Buffer at Various Temperatures

Drug	T (°C)	Solubility	Solubility Unit	Reference
Paracetamol	25.00	0.222	mol/L	[73]
Paracetamol	30.00	0.234	mol/L	[73]
Paracetamol	35.00	0.2555	mol/L	[73]
Paracetamol	40.00	0.2750	mol/L	[73]
Acetanilide	25.00	1.0280	mol/L	[73]
Acetanilide	30.00	1.11	mol/L	[73]
Acetanilide	35.00	1.232	mol/L	[73]
Acetanilide	40.00	1.34	mol/L	[73]
Benzocaine	25.00	0.00593	mol/L	[76]
Benzocaine	30.00	0.00698	mol/L	[76]
Benzocaine	35.00	0.00855	mol/L	[76]
Benzocaine	40.00	0.01026	mol/L	[76]
Ibuprofen	25.00	0.0000557	mol/L	[81]
Ibuprofen	30.00	0.0000757	mol/L	[81]
Ibuprofen	35.00	0.0000992	mol/L	[81]
Ibuprofen	40.00	0.0001285	mol/L	[81]
Phenacetin	25.00	0.161	mol/L	[73]
Phenacetin	30.00	0.217	mol/L	[73]
Phenacetin	35.00	0.262	mol/L	[73]
Phenacetin	40.00	0.376	mol/L	[73]

Solubility Data of Drugs in Octanol-Saturated Water at Various Temperatures

Drug	T (°C)	Solubility	Solubility Unit	Reference
Sulfacetamide	25.00	0.0362	mol/L	[84]
Sulfacetamide	30.00	0.0442	mol/L	[84]
Sulfacetamide	35.00	0.0527	mol/L	[84]
Sulfacetamide	40.00	0.0632	mol/L	[84]
Sulfadiazine	25.00	0.000246	mol/L	[84]
Sulfadiazine	30.00	0.000325	mol/L	[84]
Sulfadiazine	35.00	0.000424	mol/L	[84]
Sulfadiazine	40.00	0.000541	mol/L	[84]
Sulfamerazine	25.00	0.000750	mol/L	[84]
Sulfamerazine	30.00	0.000901	mol/L	[84]

(continued)

Solubility Data of Drugs in Octanol-Saturated Water at Various Temperatures   (continued)

Drug	T (°C)	Solubility	Solubility Unit	Reference
Sulfamerazine	35.00	0.001150	mol/L	[84]
Sulfamerazine	40.00	0.001400	mol/L	[84]
Sulfamethazine	25.00	0.001570	mol/L	[84]
Sulfamethazine	30.00	0.001860	mol/L	[84]
Sulfamethazine	35.00	0.002290	mol/L	[84]
Sulfamethazine	40.00	0.002750	mol/L	[84]
Sulfamethoxazole	25.00	0.001520	mol/L	[84]
Sulfamethoxazole	30.00	0.001930	mol/L	[84]
Sulfamethoxazole	35.00	0.002440	mol/L	[84]
Sulfamethoxazole	40.00	0.003110	mol/L	[84]
Sulfanilamide	25.00	0.0405	mol/L	[84]
Sulfanilamide	30.00	0.0527	mol/L	[84]
Sulfanilamide	35.00	0.0767	mol/L	[84]
Sulfanilamide	40.00	0.0935	mol/L	[84]
Sulfapyridine	25.00	0.000976	mol/L	[84]
Sulfapyridine	30.00	0.001280	mol/L	[84]
Sulfapyridine	35.00	0.001600	mol/L	[84]
Sulfapyridine	40.00	0.002070	mol/L	[84]
Sulfathiazole	25.00	0.00174	mol/L	[84]
Sulfathiazole	30.00	0.00233	mol/L	[84]
Sulfathiazole	35.00	0.00320	mol/L	[84]
Sulfathiazole	40.00	0.00403	mol/L	[84]

Solubility Data of Drugs in o-Xylene at Various Temperatures

Drug	T (°C)	Solubility	Solubility Unit	Reference
2,6-Diaminopyridine	0.20	0.00128	Mole F.	[93]
2,6-Diaminopyridine	4.00	0.00140	Mole F.	[93]
2,6-Diaminopyridine	9.10	0.00185	Mole F.	[93]
2,6-Diaminopyridine	12.40	0.00201	Mole F.	[93]
2,6-Diaminopyridine	16.30	0.00252	Mole F.	[93]
2,6-Diaminopyridine	20.50	0.00283	Mole F.	[93]
2,6-Diaminopyridine	24.40	0.00351	Mole F.	[93]
2,6-Diaminopyridine	29.10	0.00395	Mole F.	[93]
2,6-Diaminopyridine	32.20	0.00469	Mole F.	[93]
2,6-Diaminopyridine	36.40	0.00517	Mole F.	[93]
2,6-Diaminopyridine	41.00	0.00640	Mole F.	[93]
2,6-Diaminopyridine	44.20	0.00695	Mole F.	[93]
2,6-Diaminopyridine	48.10	0.00838	Mole F.	[93]
2,6-Diaminopyridine	52.80	0.00937	Mole F.	[93]
2,6-Diaminopyridine	56.30	0.01104	Mole F.	[93]
2,6-Diaminopyridine	60.20	0.01208	Mole F.	[93]
2,6-Diaminopyridine	63.80	0.01439	Mole F.	[93]
2,6-Diaminopyridine	68.00	0.01534	Mole F.	[93]
2,6-Diaminopyridine	73.10	0.01942	Mole F.	[93]
2,6-Diaminopyridine	76.40	0.02264	Mole F.	[93]

Solubility Data of Drugs in *o*-Xylene at Various Temperatures   (continued)

Drug	*T* (°C)	Solubility	Solubility Unit	Reference
2,6-Diaminopyridine	80.30	0.02401	Mole F.	[93]
2,6-Diaminopyridine	83.90	0.02922	Mole F.	[93]
2,6-Diaminopyridine	87.90	0.03403	Mole F.	[93]
Vitamin K3	29.80	0.1404	Mole F.	[125]
Vitamin K3	36.09	0.1687	Mole F.	[125]
Vitamin K3	41.10	0.1935	Mole F.	[125]
Vitamin K3	45.19	0.2186	Mole F.	[125]
Vitamin K3	49.30	0.2453	Mole F.	[125]
Vitamin K3	53.62	0.2744	Mole F.	[125]
Vitamin K3	57.39	0.3050	Mole F.	[125]
Vitamin K3	61.08	0.3396	Mole F.	[125]
Vitamin K3	65.17	0.3829	Mole F.	[125]
Vitamin K3	69.28	0.4321	Mole F.	[125]

Solubility Data of Drugs in Polyethylene Glycol 400 (PEG 400) at Various Temperatures

Drug	*T* (°C)	Solubility	Solubility Unit	Reference
Benzoyl peroxide	25.00	39.8	mg/g	[135]
Caffeine (form 1)	25.00	12.98	g/L	[136]
Caffeine (form 2)	25.00	12.76	g/L	[136]
Caffeine hydrate	25.00	12.85	g/L	[136]
Carbamazepine hydrate	25.00	82.56	g/L	[136]
Celecoxib	25.00	414.804	g/L	[55]
Danazol	25.00	33.87	g/L	[137]
Digoxine	25.00	0.347	g/L	[137]
Ketoconazol	25.00	14.57	g/L	[137]
Mefenamic acid	25.00	24.46	g/L	[137]
Meloxicam	25.00	3.763	g/L	[55]
Nimesulide	25.00	63.12	g/L	[55]
Rofecoxib	25.00	11.234	g/L	[55]

Solubility Data of Drugs in Propionic Acid at Various Temperatures

Drug	*T* (°C)	Solubility	Solubility Unit	Reference
Benzoic acid	25.00	0.1887	Mole F.	[24]
Diclofenac	25.00	0.0098135	Mole F.	[25]
Ibuprofen	25.00	0.2099050	Mole F.	[28]
Lactose	25.00	0.0000423	Mole F.	[30]
Methyl *p*-hydroxybenzoate	25.00	0.0386	Mole F.	[35]
Naproxen	25.00	0.0287132	Mole F.	[31]
Niflumic acid	25.00	0.0207129	Mole F.	[33]
p-Aminobenzoic acid	25.00	0.0489968	Mole F.	[25]
Paracetamol	25.00	0.01111	Mole F.	[23]

(continued)

Solubility Data of Drugs in Propionic Acid at Various Temperatures   (continued)

Drug	T (°C)	Solubility	Solubility Unit	Reference
p-Hydroxybenzoic acid	25.00	0.0347	Mole F.	[35]
Piroxicam	25.00	0.0016028	Mole F.	[33]
Salicylic acid	25.00	0.0661388	Mole F.	[25]
Sodium ibuprofen	25.00	0.0305436	Mole F.	[28]
Sodium benzoate	25.00	0.1507398	Mole F.	[28]
Sodium diclofenac	25.00	0.0396446	Mole F.	[31]
Sodium diclofenac	25.00	0.0396367	Mole F.	[25]
Sodium p-aminobenzoate	25.00	0.1235635	Mole F.	[25]
Sodium salicylate	25.00	0.0237779	Mole F.	[25]

Solubility Data of Drugs in Propyl Acetate at Various Temperatures

Drug	T (°C)	Solubility	Solubility Unit	Reference
Lovastatin	11.95	0.00395	Mole F.	[89]
Lovastatin	15.15	0.00438	Mole F.	[89]
Lovastatin	18.05	0.00481	Mole F.	[89]
Lovastatin	21.45	0.00544	Mole F.	[89]
Lovastatin	24.05	0.00587	Mole F.	[89]
Lovastatin	27.15	0.00658	Mole F.	[89]
Lovastatin	29.95	0.00724	Mole F.	[89]
Lovastatin	33.15	0.00803	Mole F.	[89]
Lovastatin	36.25	0.00901	Mole F.	[89]
Lovastatin	39.05	0.00997	Mole F.	[89]
Methyl p-hydroxybenzoate	25.00	0.1366	Mole F.	[35]
Salicylamide	25.00	0.06546	Mole F.	[37]

Solubility Data of Drugs in Propylene Glycol at Various Temperatures

Drug	T (°C)	Solubility	Solubility Unit	Reference
1H-1,2,4-Triazole	10.01	0.1745	Mole F.	[92]
1H-1,2,4-Triazole	15.12	0.2125	Mole F.	[92]
1H-1,2,4-Triazole	20.11	0.2488	Mole F.	[92]
1H-1,2,4-Triazole	25.01	0.2837	Mole F.	[92]
1H-1,2,4-Triazole	30.04	0.3187	Mole F.	[92]
1H-1,2,4-Triazole	35.03	0.3527	Mole F.	[92]
1H-1,2,4-Triazole	40.05	0.3862	Mole F.	[92]
1H-1,2,4-Triazole	45.13	0.4194	Mole F.	[92]
1H-1,2,4-Triazole	50.06	0.4511	Mole F.	[92]
1H-1,2,4-Triazole	55.07	0.4827	Mole F.	[92]
1H-1,2,4-Triazole	60.08	0.5137	Mole F.	[92]
1H-1,2,4-Triazole	65.14	0.5446	Mole F.	[92]
1H-1,2,4-Triazole	70.06	0.5742	Mole F.	[92]
1H-1,2,4-Triazole	75.08	0.6039	Mole F.	[92]

Solubility Data of Drugs in Propylene Glycol at Various Temperatures   (continued)

Drug	T (°C)	Solubility	Solubility Unit	Reference
1H-1,2,4-Triazole	80.09	0.6333	Mole F.	[92]
Acetylsalicylic acid	22.55	0.017	Mole F.	[94]
Acetylsalicylic acid	27.95	0.025	Mole F.	[94]
Acetylsalicylic acid	31.95	0.033	Mole F.	[94]
Acetylsalicylic acid	35.05	0.041	Mole F.	[94]
Acetylsalicylic acid	40.45	0.056	Mole F.	[94]
Acetylsalicylic acid	45.15	0.071	Mole F.	[94]
Acetylsalicylic acid	50.55	0.092	Mole F.	[94]
Acetylsalicylic acid	54.75	0.112	Mole F.	[94]
Acetylsalicylic acid	57.55	0.126	Mole F.	[94]
Acetylsalicylic acid	60.75	0.145	Mole F.	[94]
Anhydrous citric acid	25.00	0.04491	Mole F.	[23]
Benzoyl peroxide	25.00	2.95	mg/g	[135]
Benztropine	37.00	435.80	g/L	[145]
Celecoxib	25.00	30.023	g/L	[55]
Danazol	25.00	9.05	g/L	[137]
Diclofenac	25.00	0.0049124	Mole F.	[25]
Digoxine	25.00	1.070	g/L	[137]
Haloperidol	25.00	0.0009	Mole F.	[27]
Hydrocortisone	25.00	0.0460	mol/L	[80]
Ibuprofen	25.00	0.0858890	Mole F.	[28]
Ketoconazol	25.00	22.66	g/L	[137]
Lactose	25.00	0.0012121	Mole F.	[30]
Mannitol	25.00	0.0014402	Mole F.	[30]
Mefenamic acid	25.00	1.23	g/L	[137]
Meloxicam	25.00	0.307	g/L	[55]
Methyl p-hydroxybenzoate	25.00	0.0941	Mole F.	[35]
Naproxen	25.00	0.0076703	Mole F.	[31]
Niflumic acid	25.00	0.0006255	Mole F.	[33]
Nimesulide	25.00	1.76	g/L	[55]
p-Aminobenzoic acid	25.00	0.0826616	Mole F.	[25]
Paracetamol	25.00	0.11505	Mole F.	[23]
p-Hydroxybenzoic acid	25.00	0.1308	Mole F.	[35]
Pimozide	25.00	0.000252	Mole F.	[36]
Piroxicam	25.00	0.0002706	Mole F.	[33]
Ricobendazole	25.00	2.59	g/L	[103]
Rofecoxib	25.00	1.152	g/L	[55]
Saccharose	25.00	0.0000044	Mole F.	[30]
Salicylic acid	25.00	0.1136082	Mole F.	[25]
Sodium ibuprofen	25.00	0.2430228	Mole F.	[28]
Sodium benzoate	25.00	0.0188000	Mole F.	[28]
Sodium diclofenac	25.00	0.0000126	Mole F.	[31]
Sodium diclofenac	25.00	0.0000126	Mole F.	[25]
Sodium p-aminobenzoate	25.00	0.1277091	Mole F.	[25]
Sodium salicylate	25.00	0.2029085	Mole F.	[25]
Sulfadiazine	25.00	0.0003589	Mole F.	[39]
Sulfamethoxypyridazine	25.00	0.00531	Mole F.	[40]
Temazepam	25.00	0.0050604	Mole F.	[15]

Solubility Data of Drugs in *p*-Xylene at Various Temperatures

Drug	T (°C)	Solubility	Solubility Unit	Reference
Vitamin K3	29.35	0.1289	Mole F.	[125]
Vitamin K3	36.38	0.1579	Mole F.	[125]
Vitamin K3	41.49	0.1841	Mole F.	[125]
Vitamin K3	45.81	0.2103	Mole F.	[125]
Vitamin K3	50.20	0.2386	Mole F.	[125]
Vitamin K3	54.53	0.2687	Mole F.	[125]
Vitamin K3	58.87	0.3039	Mole F.	[125]
Vitamin K3	62.74	0.3445	Mole F.	[125]
Vitamin K3	66.95	0.3899	Mole F.	[125]
Vitamin K3	71.09	0.4403	Mole F.	[125]

Solubility Data of Drugs in *sec*-Butyl Acetate at Various Temperatures

Drug	T (°C)	Solubility	Solubility Unit	Reference
Lovastatin	11.95	0.00365	Mole F.	[89]
Lovastatin	15.15	0.00401	Mole F.	[89]
Lovastatin	18.05	0.00446	Mole F.	[89]
Lovastatin	21.45	0.00500	Mole F.	[89]
Lovastatin	24.05	0.00545	Mole F.	[89]
Lovastatin	27.15	0.00606	Mole F.	[89]
Lovastatin	29.95	0.00666	Mole F.	[89]
Lovastatin	33.15	0.00745	Mole F.	[89]
Lovastatin	36.25	0.00830	Mole F.	[89]
Lovastatin	39.05	0.00915	Mole F.	[89]

Solubility Data of Drugs in *tert*-Butyl Acetate at Various Temperatures

Drug	T (°C)	Solubility	Solubility Unit	Reference
Lovastatin	11.95	0.00275	Mole F.	[89]
Lovastatin	15.15	0.00314	Mole F.	[89]
Lovastatin	18.05	0.00346	Mole F.	[89]
Lovastatin	21.45	0.00386	Mole F.	[89]
Lovastatin	24.05	0.00420	Mole F.	[89]
Lovastatin	27.15	0.00479	Mole F.	[89]
Lovastatin	29.95	0.00510	Mole F.	[89]
Lovastatin	33.15	0.00569	Mole F.	[89]
Lovastatin	36.25	0.00638	Mole F.	[89]
Lovastatin	39.05	0.00693	Mole F.	[89]

Solubility Data of Drugs in Tetrahydrofurane at Various Temperatures

Drug	T (°C)	Solubility	Solubility Unit	Reference
2-(4-Ethylbenzoyl)benzoic acid	8.05	0.0064	Mole F.	[18]
2-(4-Ethylbenzoyl)benzoic acid	13.34	0.0085	Mole F.	[18]
2-(4-Ethylbenzoyl)benzoic acid	23.27	0.0125	Mole F.	[18]
2-(4-Ethylbenzoyl)benzoic acid	32.50	0.0178	Mole F.	[18]
2-(4-Ethylbenzoyl)benzoic acid	41.19	0.0246	Mole F.	[18]
2-(4-Ethylbenzoyl)benzoic acid	52.40	0.0377	Mole F.	[18]
2-(4-Ethylbenzoyl)benzoic acid	57.19	0.0442	Mole F.	[18]
2-(4-Ethylbenzoyl)benzoic acid	62.60	0.0541	Mole F.	[18]
2-Hydroxybenzoic acid	25.00	0.3642	Mole F.	[19]
3,5-Di-*tert*-butyl-4-hydroxytoluene	N/A	0.77	Mole F.	[4]
Acetylsalicylic acid	25.00	0.1904	Mole F.	[22]
β-Carotene	Room	10000	mg/L	[6]
Carbamazepine (form III)	5.20	0.01066	Mole F.	[46]
Carbamazepine (form III)	13.49	0.01253	Mole F.	[46]
Carbamazepine (form III)	16.40	0.01350	Mole F.	[46]
Carbamazepine (form III)	22.40	0.01526	Mole F.	[46]
Carbamazepine (form III)	30.60	0.01763	Mole F.	[46]
Carbamazepine (form III)	38.30	0.02063	Mole F.	[46]
Carbamazepine (form III)	53.28	0.02689	Mole F.	[46]
Carbamazepine (form III)	57.31	0.03063	Mole F.	[46]
Eflucimibe (form A)	20.00	0.0229	mol/L	[99]
Irbesartan (form A)	5.15	0.001269	Mole F.	[29]
Irbesartan (form A)	9.70	0.001524	Mole F.	[29]
Irbesartan (form A)	14.95	0.001842	Mole F.	[29]
Irbesartan (form A)	19.80	0.002294	Mole F.	[29]
Irbesartan (form A)	24.55	0.002765	Mole F.	[29]
Irbesartan (form A)	30.00	0.003316	Mole F.	[29]
Irbesartan (form A)	34.90	0.003965	Mole F.	[29]
Irbesartan (form A)	39.55	0.004792	Mole F.	[29]
Irbesartan (form A)	45.20	0.005815	Mole F.	[29]
Irbesartan (form A)	49.60	0.006867	Mole F.	[29]
L-(+)-Ascorbic acid	19.85	0.00053	Mole F.	[102]
L-(+)-Ascorbic acid	24.85	0.00067	Mole F.	[102]
L-(+)-Ascorbic acid	29.85	0.00087	Mole F.	[102]
L-(+)-Ascorbic acid	34.85	0.00105	Mole F.	[102]
L-(+)-Ascorbic acid	39.85	0.00128	Mole F.	[102]
L-(+)-Ascorbic acid	44.85	0.00153	Mole F.	[102]
L-(+)-Ascorbic acid	49.85	0.00174	Mole F.	[102]
Lutein	Room	8000	mg/L	[6]
Methyl-3-(3,5-di-*tert*-butyl-4-hydroxyphenyl)-propionate	N/A	0.43	Mole F.	[4]
Naproxen	25.00	0.14180	Mole F.	[32]
Octadecyl-3-(3,5-di-*tert*-butyl-4-hydroxyphenyl)-propionate	N/A	0.25	Mole F.	[4]
Paracetamol	30.00	155.37	g/kg	[11]
Paracetamol	30.00	133	g/L	[8]
Paroxetine HCl hemihydrate	21.60	0.0000001	Mole F.	[57]
Paroxetine HCl hemihydrate	24.00	0.0000001	Mole F.	[57]

(continued)

Solubility Data of Drugs in Tetrahydrofurane at Various Temperatures   (continued)

Drug	T (°C)	Solubility	Solubility Unit	Reference
Paroxetine HCl hemihydrate	26.85	0.0000002	Mole F.	[57]
Paroxetine HCl hemihydrate	30.00	0.0000002	Mole F.	[57]
Paroxetine HCl hemihydrate	33.55	0.0000003	Mole F.	[57]
Paroxetine HCl hemihydrate	36.90	0.0000004	Mole F.	[57]
Paroxetine HCl hemihydrate	39.50	0.0000005	Mole F.	[57]
Paroxetine HCl hemihydrate	43.00	0.0000007	Mole F.	[57]
Paroxetine HCl hemihydrate	46.30	0.0000008	Mole F.	[57]
Paroxetine HCl hemihydrate	49.90	0.0000010	Mole F.	[57]
Paroxetine HCl hemihydrate	53.80	0.0000012	Mole F.	[57]
Paroxetine HCl hemihydrate	57.40	0.0000014	Mole F.	[57]
Paroxetine HCl hemihydrate	61.00	0.0000016	Mole F.	[57]
Pentaerythrittol tetrakis (3-(3,5-di-tert-butyl-4-hydroxyphenyl)-propionate)	N/A	0.053	Mole F.	[4]
Phenacetinum	9.88	0.018830	Mole F.	[58]
Phenacetinum	15.20	0.023073	Mole F.	[58]
Phenacetinum	20.73	0.027588	Mole F.	[58]
Phenacetinum	25.02	0.032940	Mole F.	[58]
Phenacetinum	30.00	0.039546	Mole F.	[58]
Phenacetinum	34.80	0.047304	Mole F.	[58]
Phenacetinum	40.00	0.057561	Mole F.	[58]
Phenacetinum	45.00	0.069547	Mole F.	[58]
Phenacetinum	50.30	0.084863	Mole F.	[58]
Phenacetinum	55.18	0.101650	Mole F.	[58]
Phenacetinum	60.40	0.122770	Mole F.	[58]
Salicylamide	25.00	0.1764	Mole F.	[37]
Temazepam	25.00	0.0785574	Mole F.	[15]
Trimethoprim	2.90	0.0006078	Mole F.	[65]
Trimethoprim	8.66	0.0007136	Mole F.	[65]
Trimethoprim	14.16	0.0008572	Mole F.	[65]
Trimethoprim	19.50	0.0009866	Mole F.	[65]
Trimethoprim	24.93	0.001171	Mole F.	[65]
Trimethoprim	29.97	0.001371	Mole F.	[65]
Trimethoprim	33.80	0.001546	Mole F.	[65]
Trimethoprim	38.54	0.001785	Mole F.	[65]
Trimethoprim	43.59	0.002084	Mole F.	[65]
Trimethoprim	48.19	0.002442	Mole F.	[65]
Trimethoprim	53.76	0.002958	Mole F.	[65]
Trimethoprim	60.72	0.003774	Mole F.	[65]

Solubility Data of Drugs in Toluene at Various Temperatures

Drug	T (°C)	Solubility	Solubility Unit	Reference
2,6-Diaminopyridine	0.10	0.00149	Mole F.	[93]
2,6-Diaminopyridine	4.00	0.00191	Mole F.	[93]
2,6-Diaminopyridine	8.10	0.00214	Mole F.	[93]
2,6-Diaminopyridine	12.40	0.00270	Mole F.	[93]
2,6-Diaminopyridine	16.30	0.00304	Mole F.	[93]
2,6-Diaminopyridine	20.50	0.00375	Mole F.	[93]
2,6-Diaminopyridine	24.40	0.00417	Mole F.	[93]
2,6-Diaminopyridine	28.30	0.00510	Mole F.	[93]
2,6-Diaminopyridine	32.20	0.00558	Mole F.	[93]
2,6-Diaminopyridine	36.40	0.00704	Mole F.	[93]
2,6-Diaminopyridine	40.30	0.00733	Mole F.	[93]
2,6-Diaminopyridine	44.20	0.00919	Mole F.	[93]
2,6-Diaminopyridine	48.10	0.00942	Mole F.	[93]
2,6-Diaminopyridine	52.40	0.01242	Mole F.	[93]
2,6-Diaminopyridine	56.30	0.01270	Mole F.	[93]
2,6-Diaminopyridine	60.20	0.01619	Mole F.	[93]
2,6-Diaminopyridine	64.10	0.01645	Mole F.	[93]
2,6-Diaminopyridine	68.00	0.02072	Mole F.	[93]
2,6-Diaminopyridine	72.50	0.02204	Mole F.	[93]
2,6-Diaminopyridine	76.40	0.02768	Mole F.	[93]
2,6-Diaminopyridine	80.30	0.02825	Mole F.	[93]
2,6-Diaminopyridine	84.20	0.03550	Mole F.	[93]
2,6-Diaminopyridine	88.10	0.03671	Mole F.	[93]
3,5-Di-*tert*-butyl-4-hydroxytoluene	N/A	0.48	Mole F.	[4]
Acetic acid	−26.80	0.180	Mole F.	[109]
Acetic acid	−21.10	0.260	Mole F.	[109]
Acetic acid	−15.20	0.300	Mole F.	[109]
Acetic acid	−9.00	0.400	Mole F.	[109]
Acetic acid	−5.50	0.462	Mole F.	[109]
Acetic acid	−4.00	0.485	Mole F.	[109]
Acetic acid	−2.00	0.562	Mole F.	[109]
Acetic acid	0.00	0.605	Mole F.	[109]
Acetic acid	2.40	0.678	Mole F.	[109]
Acetic acid	4.60	0.735	Mole F.	[109]
Acetic acid	7.20	0.805	Mole F.	[109]
Acetic acid	10.00	0.870	Mole F.	[109]
Acetylsalicylic acid	25.00	0.00129	Mole F.	[21]
Benzoic acid	25.00	0.0734	Mole F.	[24]
β-Carotene	Room	4000	mg/L	[6]
Diflunisal	25.00	0.000568	Mole F.	[26]
D-*p*-Hydroxyphenylglycine	5.13	0.0001566	Mole F.	[127]
D-*p*-Hydroxyphenylglycine	10.18	0.0001641	Mole F.	[127]
D-*p*-Hydroxyphenylglycine	15.23	0.0001652	Mole F.	[127]
D-*p*-Hydroxyphenylglycine	20.17	0.0001719	Mole F.	[127]

(*continued*)

Solubility Data of Drugs in Toluene at Various Temperatures   (continued)

Drug	T (°C)	Solubility	Solubility Unit	Reference
D-*p*-Hydroxyphenylglycine	25.08	0.0001749	Mole F.	[127]
D-*p*-Hydroxyphenylglycine	30.17	0.0001796	Mole F.	[127]
D-*p*-Hydroxyphenylglycine	35.24	0.0001885	Mole F.	[127]
D-*p*-Hydroxyphenylglycine	40.22	0.0001951	Mole F.	[127]
D-*p*-Hydroxyphenylglycine	45.02	0.0002050	Mole F.	[127]
D-*p*-Hydroxyphenylglycine	50.12	0.0002161	Mole F.	[127]
Ethyl *p*-aminobenzoate	25.00	0.0437178	Mole F.	[124]
Flubiprofen	25.00	0.0767000	Mole F.	[5]
Haloperidol	25.00	0.011768	Mole F.	[27]
Hydrocortisone	25.00	0.000268	mol/L	[80]
Ibuprofen	10.00	258.3	g/kg	[83]
Ibuprofen	15.00	343.1	g/kg	[83]
Ibuprofen	20.00	457.1	g/kg	[83]
Ibuprofen	30.00	749.7	g/kg	[83]
Ibuprofen	35.00	957	g/kg	[83]
Lutein	Room	500	mg/L	[6]
Methyl elaidate	−42.00	3.89	% Weight	[10]
Methyl elaidate	−11.40	44.86	% Weight	[10]
Methyl heptadecanoate	−24.30	3.18	% Weight	[10]
Methyl heptadecanoate	−9.00	14.48	% Weight	[10]
Methyl heptadecanoate	7.50	44.47	% Weight	[10]
Methyl linoleate	−56.40	45.94	% Weight	[10]
Methyl linoleate	−45.30	76.43	% Weight	[10]
Methyl oleate	−50.80	24.46	% Weight	[10]
Methyl oleate	−41.00	45.21	% Weight	[10]
Methyl palmitate	−26.20	2.94	% Weight	[10]
Methyl palmitate	−10.00	13.58	% Weight	[10]
Methyl palmitate	7.20	43.08	% Weight	[10]
Methyl palmitolate	−57.70	34.88	% Weight	[10]
Methyl palmitolate	−45.50	67.41	% Weight	[10]
Methyl *p*-aminobenzoate	25.00	0.0146986	Mole F.	[124]
Methyl petroselaidate	−35.70	2.59	% Weight	[10]
Methyl petroselaidate	−20.00	14.80	% Weight	[10]
Methyl petroselaidate	−3.00	45.38	% Weight	[10]
Methyl petroselinate	−52.30	2.59	% Weight	[10]
Methyl petroselinate	−37.80	24.58	% Weight	[10]
Methyl petroselinate	−21.50	45.30	Solubility	[10]
Methyl stearate	−32.40	0.12	% Weight	[10]
Methyl stearate	−13.00	1.19	% Weight	[10]
Methyl stearate	−1.70	4.04	% Weight	[10]
Methyl stearate	14.40	16.36	% Weight	[10]
Methyl stearate	23.50	32.29	% Weight	[10]
Methyl stearate	35.90	77.11	% Weight	[10]
Methyl-3-(3,5-di-*tert*-butyl-4-hydroxyphenyl)-propionate	N/A	0.33	Mole F.	[4]
n-Butyl *p*-aminobenzoate	25.00	0.3011942	Mole F.	[124]
n-Pentyl *p*-aminobenzoate	25.00	0.3362165	Mole F.	[124]
n-Propyl *p*-aminobenzoate	25.00	0.1261858	Mole F.	[124]

Solubility Data of Drugs in Toluene at Various Temperatures   (continued)

Drug	T (°C)	Solubility	Solubility Unit	Reference
Octadecyl-3-(3,5-di-*tert*-butyl-4-hydroxyphenyl)-propionate	N/A	0.21	Mole F.	[4]
p-Aminophenylacetic acid	16.00	1.2	g/kg	[83]
p-Aminophenylacetic acid	25.00	1.4	g/kg	[83]
Paracetamol	0.00	0.22	g/kg	[11]
Paracetamol	5.00	0.27	g/kg	[11]
Paracetamol	10.00	0.32	g/kg	[11]
Paracetamol	15.00	0.36	g/kg	[11]
Paracetamol	20.00	0.37	g/kg	[11]
Paracetamol	25.00	0.37	g/kg	[11]
Paracetamol	30.00	0.34	g/kg	[11]
Paracetamol	30.00	0.16	g/L	[8]
Pentaerythrittol tetrakis(3-(3,5-di-butyl-4-hydroxyphenyl)-propionate)	N/A	0.067	Mole F.	[4]
Phenylacetic acid	20.00	377.9	g/kg	[83]
p-Hydroxyphenylacetic acid	10.00	1.7	g/kg	[83]
p-Hydroxyphenylacetic acid	15.00	2.07	g/kg	[83]
p-Hydroxyphenylacetic acid	20.00	60.7	g/kg	[83]
p-Hydroxybenzoic acid	25.00	0.000029	Mole F.	[35]
p-Hydroxybenzoic acid	25.00	1.5	g/kg	[83]
Sulfadiazine	25.00	0.0000019	Mole F.	[39]
Sulfamethoxypyridazine	25.00	0.00004	Mole F.	[40]
Temazepam	25.00	0.0096185	Mole F.	[15]
Testosterone propionate	25.00	0.20	Mole F.	[16]
Vitamin K3	30.29	0.1194	Mole F.	[125]
Vitamin K3	35.69	0.1428	Mole F.	[125]
Vitamin K3	40.48	0.1655	Mole F.	[125]
Vitamin K3	44.50	0.1893	Mole F.	[125]
Vitamin K3	48.70	0.2157	Mole F.	[125]
Vitamin K3	52.84	0.2463	Mole F.	[125]
Vitamin K3	56.89	0.2768	Mole F.	[125]
Vitamin K3	60.71	0.3114	Mole F.	[125]
Vitamin K3	64.65	0.3507	Mole F.	[125]
Vitamin K3	69.44	0.4036	Mole F.	[125]
Xylitol	20.11	0.0000399	Mole F.	[66]
Xylitol	25.13	0.0000418	Mole F.	[66]
Xylitol	30.04	0.0000452	Mole F.	[66]
Xylitol	35.02	0.0000500	Mole F.	[66]
Xylitol	40.08	0.0000563	Mole F.	[66]
Xylitol	45.18	0.0000642	Mole F.	[66]
Xylitol	50.06	0.0000731	Mole F.	[66]
Xylitol	55.07	0.0000837	Mole F.	[66]
Xylitol	60.15	0.0000960	Mole F.	[66]
Xylitol	65.01	0.0001091	Mole F.	[66]
Xylitol	70.04	0.0001241	Mole F.	[66]

Solubility Data of Drugs in Trichloroethylene at Various Temperatures

Drug	T (°C)	Solubility	Solubility Unit	Reference
Stearic acid	19.60	0.023	Mole F.	[104]
Stearic acid	22.10	0.0299	Mole F.	[104]
Stearic acid	24.90	0.0437	Mole F.	[104]
Stearic acid	28.50	0.0603	Mole F.	[104]
Stearic acid	32.40	0.0908	Mole F.	[104]
Stearic acid	35.40	0.1178	Mole F.	[104]
Stearic acid	37.50	0.1367	Mole F.	[104]
Stearic acid	40.10	0.1644	Mole F.	[104]
Stearic acid	41.70	0.1854	Mole F.	[104]
Stearic acid	43.90	0.2122	Mole F.	[104]
Stearic acid	47.90	0.2643	Mole F.	[104]
Stearic acid	50.10	0.2985	Mole F.	[104]

Solubility Data of Drugs in Trichloromethane at Various Temperatures

Drug	T (°C)	Solubility	Solubility Unit	Reference
Imidacloprid	18.89	0.008485	Mole F.	[2]
Imidacloprid	21.34	0.009162	Mole F.	[2]
Imidacloprid	25.07	0.01016	Mole F.	[2]
Imidacloprid	30.69	0.01149	Mole F.	[2]
Imidacloprid	33.89	0.01230	Mole F.	[2]
Imidacloprid	36.90	0.01319	Mole F.	[2]
Imidacloprid	40.88	0.01427	Mole F.	[2]
Imidacloprid	43.60	0.01526	Mole F.	[2]
Imidacloprid	45.92	0.01674	Mole F.	[2]
Imidacloprid	48.59	0.01834	Mole F.	[2]
Imidacloprid	51.49	0.01984	Mole F.	[2]
Imidacloprid	53.56	0.02147	Mole F.	[2]
Imidacloprid	57.01	0.02403	Mole F.	[2]

Solubility Data of Drugs in Water-Saturated Chloroform at Various Temperatures

Drug	T (°C)	Solubility	Solubility Unit	Reference
Acetanilide	25.00	0.04136	mol/L	[73]
Acetanilide	30.00	0.0524	mol/L	[73]
Acetanilide	35.00	0.0624	mol/L	[73]
Acetanilide	40.00	0.0732	mol/L	[73]
Ibuprofen	25.00	2.82	mol/L	[81]
Ibuprofen	30.00	3.231	mol/L	[81]
Ibuprofen	35.00	3.628	mol/L	[81]
Ibuprofen	40.00	4.066	mol/L	[81]
Naproxen	20.00	0.2452	mol/L	[82]
Naproxen	25.00	0.289	mol/L	[82]
Naproxen	30.00	0.318	mol/L	[82]

Solubility Data of Drugs in Water-Saturated Chloroform at Various Temperatures   (continued)

Drug	T (°C)	Solubility	Solubility Unit	Reference
Naproxen	35.00	0.376	mol/L	[82]
Naproxen	40.00	0.440	mol/L	[82]
Paracetamol	25.00	0.1018	mol/L	[73]
Paracetamol	30.00	0.1180	mol/L	[73]
Paracetamol	35.00	0.1415	mol/L	[73]
Paracetamol	40.00	0.1692	mol/L	[73]
Phenacetin	25.00	0.00564	mol/L	[73]
Phenacetin	30.00	0.00615	mol/L	[73]
Phenacetin	35.00	0.007159	mol/L	[73]
Phenacetin	40.00	0.007904	mol/L	[73]

Solubility Data of Drugs in Water-Saturated Isopropyl Myristate at Various Temperatures

Drug	T (°C)	Solubility	Solubility Unit	Reference
Acetanilide	25.00	0.0450	mol/L	[73]
Acetanilide	30.00	0.0512	mol/L	[73]
Acetanilide	35.00	0.0620	mol/L	[73]
Acetanilide	40.00	0.0774	mol/L	[73]
Benzocaine	25.00	0.238	mol/L	[76]
Benzocaine	30.00	0.255	mol/L	[76]
Benzocaine	35.00	0.3188	mol/L	[76]
Benzocaine	40.00	0.478	mol/L	[76]
Ibuprofen	25.00	0.601	mol/L	[81]
Ibuprofen	30.00	0.746	mol/L	[81]
Ibuprofen	35.00	0.8792	mol/L	[81]
Ibuprofen	40.00	1.1012	mol/L	[81]
Naproxen	20.00	0.02066	mol/L	[82]
Naproxen	25.00	0.0244	mol/L	[82]
Naproxen	30.00	0.0274	mol/L	[82]
Naproxen	35.00	0.0308	mol/L	[82]
Naproxen	40.00	0.03533	mol/L	[82]
Paracetamol	25.00	0.1045	mol/L	[73]
Paracetamol	30.00	0.1187	mol/L	[73]
Paracetamol	35.00	0.1278	mol/L	[73]
Paracetamol	40.00	0.1483	mol/L	[73]
Phenacetin	25.00	0.00521	mol/L	[73]
Phenacetin	30.00	0.00596	mol/L	[73]
Phenacetin	35.00	0.00745	mol/L	[73]
Phenacetin	40.00	0.00813	mol/L	[73]

Solubility Data of Drugs in Water-Saturated Octanol at Various Temperatures

Drug	T (°C)	Solubility	Solubility Unit	Reference
Acetanilide	25.00	0.04335	mol/L	[73]
Acetanilide	30.00	0.0514	mol/L	[73]
Acetanilide	35.00	0.0641	mol/L	[73]
Acetanilide	40.00	0.0741	mol/L	[73]
Benzocaine	25.00	0.584	mol/L	[76]
Benzocaine	30.00	0.687	mol/L	[76]
Benzocaine	35.00	0.851	mol/L	[76]
Benzocaine	40.00	1.010	mol/L	[76]
Ibuprofen	25.00	1.779	mol/L	[81]
Ibuprofen	30.00	2.2002	mol/L	[81]
Ibuprofen	35.00	2.61	mol/L	[81]
Ibuprofen	40.00	3.157	mol/L	[81]
Naproxen	20.00	0.1163	mol/L	[82]
Naproxen	25.00	0.1450	mol/L	[82]
Naproxen	30.00	0.1598	mol/L	[82]
Naproxen	35.00	0.1943	mol/L	[82]
Naproxen	40.00	0.226	mol/L	[82]
Paracetamol	25.00	0.09588	mol/L	[73]
Paracetamol	30.00	0.1167	mol/L	[73]
Paracetamol	35.00	0.1346	mol/L	[73]
Paracetamol	40.00	0.1611	mol/L	[73]
Phenacetin	25.00	0.0056290	mol/L	[73]
Phenacetin	30.00	0.0068	mol/L	[73]
Phenacetin	35.00	0.00758	mol/L	[73]
Phenacetin	40.00	0.00876	mol/L	[73]
Sulfacetamide	25.00	0.0173	mol/L	[84]
Sulfacetamide	30.00	0.0202	mol/L	[84]
Sulfacetamide	35.00	0.0227	mol/L	[84]
Sulfacetamide	40.00	0.0258	mol/L	[84]
Sulfadiazine	25.00	0.0002032	mol/L	[84]
Sulfadiazine	30.00	0.0002460	mol/L	[84]
Sulfadiazine	35.00	0.0003070	mol/L	[84]
Sulfadiazine	40.00	0.0004040	mol/L	[84]
Sulfamerazine	25.00	0.000950	mol/L	[84]
Sulfamerazine	30.00	0.001140	mol/L	[84]
Sulfamerazine	35.00	0.001444	mol/L	[84]
Sulfamerazine	40.00	0.001640	mol/L	[84]
Sulfamethazine	25.00	0.00276	mol/L	[84]
Sulfamethazine	30.00	0.00343	mol/L	[84]
Sulfamethazine	35.00	0.00443	mol/L	[84]
Sulfamethazine	40.00	0.00520	mol/L	[84]
Sulfamethoxazole	25.00	0.00808	mol/L	[84]
Sulfamethoxazole	30.00	0.01000	mol/L	[84]
Sulfamethoxazole	35.00	0.01230	mol/L	[84]
Sulfamethoxazole	40.00	0.01397	mol/L	[84]
Sulfanilamide	25.00	0.00650	mol/L	[84]
Sulfanilamide	30.00	0.00730	mol/L	[84]
Sulfanilamide	35.00	0.00910	mol/L	[84]
Sulfanilamide	40.00	0.01070	mol/L	[84]

Solubility Data of Drugs in Water-Saturated Octanol at Various Temperatures (continued)

Drug	T (°C)	Solubility	Solubility Unit	Reference
Sulfapyridine	25.00	0.00091	mol/L	[84]
Sulfapyridine	30.00	0.00114	mol/L	[84]
Sulfapyridine	35.00	0.00142	mol/L	[84]
Sulfapyridine	40.00	0.00173	mol/L	[84]
Sulfathiazole	25.00	0.00153	mol/L	[84]
Sulfathiazole	30.00	0.00206	mol/L	[84]
Sulfathiazole	35.00	0.00249	mol/L	[84]
Sulfathiazole	40.00	0.00310	mol/L	[84]

Solubility Data of Drugs in Xyelene Octanol at Various Temperatures

Drug	T (°C)	Solubility	Solubility Unit	Reference
Salicylic acid	24.85	0.00616	Mole F.	[128]
Salicylic acid	29.85	0.00843	Mole F.	[128]
Salicylic acid	34.85	0.01101	Mole F.	[128]
Salicylic acid	39.85	0.01382	Mole F.	[128]
Salicylic acid	44.85	0.01751	Mole F.	[128]
Salicylic acid	49.85	0.02216	Mole F.	[128]
Salicylic acid	54.85	0.02704	Mole F.	[128]
Salicylic acid	59.85	0.03354	Mole F.	[128]
Salicylic acid	64.85	0.04013	Mole F.	[128]
Salicylic acid	69.85	0.04835	Mole F.	[128]
Salicylic acid	74.85	0.05782	Mole F.	[128]

Solubility Data of Drugs in Different Solvents at Various Temperatures

Solvent	Drug	T (°C)	Solubility	Solubility Unit	Reference
1,3-Butanediol	Sulfamethoxypyridazine	25.00	0.00301	Mole F.	[40]
1,3-Propanediol	Methyl *p*-hydroxybenzoate	25.00	0.0766	Mole F.	[35]
1,3-Propanediol	Sodium ibuprofen	25.00	0.1094705	Mole F.	[28]
1,4-Butanediol	Methyl *p*-hydroxybenzoate	25.00	0.1202	Mole F.	[35]
1,4-Butanediol	Sodium ibuprofen	25.00	0.2452935	Mole F.	[28]
1,4-Butanediol	Sulfadiazine	25.00	0.0003091	Mole F.	[39]
1,4-Butanediol	Sulfamethoxypyridazine	25.00	0.00422	Mole F.	[40]
1-Ethyl-2-pyrrolidone	Felodipine	25.00	0.099	Mole F.	[149]
1-Ethyl-2-pyrrolidone	Indomethacin	25.00	0.173	Mole F.	[149]
1-Ethyl-2-pyrrolidone	Ketoconazole	25.00	0.018	Mole F.	[149]
1-Ethyl-2-pyrrolidone	Nifedipine	25.00	0.083	Mole F.	[149]
1-Ethyl-2-pyrrolidone	Sucrose	25.00	0.026	Mole F.	[149]
2,2,4-Trimethyl pentane	2,4,8-Triiodophenol	25.00	0.00050	mol/L	[129]
2,2,4-Trimethyl pentane	Phenothiazine	25.00	0.000532	Mole F.	[34]
2,2,4-Trimethyl pentane	*p*-Iodophenol	25.00	0.0150	mol/L	[129]

(continued)

Solubility Data of Drugs in Different Solvents at Various Temperatures   (continued)

Solvent	Drug	T (°C)	Solubility	Solubility Unit	Reference
2,2,4-Trimethyl pentane	p-Nitrophenol	25.00	0.00033	mol/L	[129]
2,2,4-Trimethyl pentane	p-Phenylphenol	25.00	0.00124	mol/L	[129]
2,2,4-Trimethyl pentane	Xanthene	25.00	0.02451	Mole F.	[3]
2-Ethoxyethanol	Sufathiazole (form 2)	25.00	0.0224	Mole F.	[150]
2-Ethoxyethanol	Sulfamerazine	25.00	0.0109	Mole F.	[150]
2-Ethoxyethanol	Sulfameter	25.00	0.0119	Mole F.	[150]
2-Ethoxyethanol	Sulfamethazine	25.00	0.0184	Mole F.	[150]
2-Ethoxyethanol	Sulfamethoxazole	25.00	0.0911	Mole F.	[150]
2-Ethoxyethanol	Sulfisomidine	25.00	0.0111	Mole F.	[150]
2-Ethoxyethanol	Sulfisoxazole	25.00	0.0495	Mole F.	[150]
2-Ethyl-1-hexanol	Acetylsalicylic acid	25.00	0.03033	Mole F.	[22]
2-Ethyl-1-hexanol	Phenothiazine	25.00	0.01009	Mole F.	[34]
2-Ethyl-1-hexanol	Salicylamide	25.00	0.02409	Mole F.	[37]
2-Methyl-2-butanol	Glucose	40.00	1.08	g/L	[42]
2-Methyl-2-butanol	Glucose	60.00	2.40	g/L	[42]
2-Methyl-1-butanol	Acetylsalicylic acid	25.00	0.03444	Mole F.	[22]
2-Methyl-1-butanol	Phenothiazine	25.00	0.00726	Mole F.	[34]
2-Methyl-1-pentanol	Acetylsalicylic acid	25.00	0.03205	Mole F.	[22]
2-Methyl-1-pentanol	Phenothiazine	25.00	0.00866	Mole F.	[34]
2-Methyl-1-pentanol	Xanthene	25.00	0.01969	Mole F.	[3]
2-Phenylethanol	Clonazepam	25.00	24.5	g/L	[48]
2-Propylcyclohexane	Testosterone propionate	25.00	0.0155000	Mole F.	[131]
2-Pyrrolidone	Ricobendazole	25.00	17.65	g/L	[103]
3,7-Dimethyl-1-octanol	Acetylsalicylic acid	25.00	0.03039	Mole F.	[22]
3-Methyl-1-butanol	4-Aminobenzoic acid	25.00	0.01989	Mole F.	[20]
3-Methyl-1-butanol	Acetylsalicylic acid	25.00	0.03812	Mole F.	[22]
3-Methyl-1-butanol	Naproxen	25.00	0.01204	Mole F.	[32]
3-Methyl-1-butanol	Phenothiazine	25.00	0.00896	Mole F.	[34]
3-Methyl-1-butanol	Salicylamide	25.00	0.02561	Mole F.	[37]
3-Methyl-1-butanol	Xanthene	25.00	0.01633	Mole F.	[3]
4-Methyl-2-pentanol	Acetylsalicylic acid	25.00	0.03931	Mole F.	[22]
4-Methyl-2-pentanol	Phenothiazine	25.00	0.00728	Mole F.	[34]
4-Methyl-2-pentanol	Xanthene	25.00	0.01762	Mole F.	[3]
Acetic anhydride	Temazepam	25.00	0.0148120	Mole F.	[15]
Aniline	Sulfadiazine	25.00	0.0012229	Mole F.	[39]
Aniline	Sulfamethoxypyridazine	25.00	0.06301	Mole F.	[40]
Aniline	Temazepam	25.00	0.1801868	Mole F.	[15]
Anisole	Temazepam	25.00	0.0430890	Mole F.	[15]
Anisole	Testosterone propionate	25.00	0.25	Mole F.	[16]
Bicyclohexyl	Testosterone propionate	25.00	0.0165000	Mole F.	[131]
Butyl butyrate	p-Phenylphenol	25.00	0.62	mol/L	[129]
Butyl cyclohexane	Testosterone propionate	25.00	0.0141000	Mole F.	[131]
Butyl ether	2,4,8-Triiodophenol	25.00	0.10	mol/L	[129]
Butyl ether	p-Iodophenol	25.00	3.39	mol/L	[129]
Butyl ether	p-Nitrophenol	25.00	1.27	mol/L	[129]
Butyl ether	p-Phenylphenol	25.00	0.21	mol/L	[129]

Solubility Data of Drugs in Different Solvents at Various Temperatures (continued)

Solvent	Drug	T (°C)	Solubility	Solubility Unit	Reference
Butyronitrile	Phenothiazine	25.00	0.05741	Mole F.	[34]
Carbon disulfide	Testosterone propionate	25.00	0.18	Mole F.	[16]
cis-Decahydronaphthalene	Testosterone propionate	25.00	0.0170000	Mole F.	[131]
cis-Hexahydroindan	Testosterone propionate	25.00	0.0239000	Mole F.	[131]
Cyclooctane	Phenothiazine	25.00	0.001577	Mole F.	[34]
Cyclooctane	Testosterone propionate	25.00	0.0181	Mole F.	[131]
Cyclooctane	Xanthene	25.00	0.05414	Mole F.	[3]
Cyclopentanol	Phenothiazine	25.00	0.02119	Mole F.	[34]
Cyclopentanol	Xanthene	25.00	0.02886	Mole F.	[3]
Diacetin	Griseofulvin	100.00	0.0049	Mole F.	[142]
Dibutyl ether	2-Hydroxybenzoic acid	25.00	0.09185	Mole F.	[19]
Dibutyl ether	Acetylsalicylic acid	25.00	0.007095	Mole F.	[22]
Dibutyl ether	Methyl p-hydroxybenzoate	25.00	0.0268	Mole F.	[35]
Dibutyl ether	Naproxen	25.00	0.00493	Mole F.	[32]
Dibutyl ether	Phenothiazine	25.00	0.01144	Mole F.	[34]
Dibutyl ether	Salicylamide	25.00	0.003961	Mole F.	[37]
Dibutyl ether	Xanthene	25.00	0.08310	Mole F.	[3]
Diethyl acetamide	Methyl p-hydroxybenzoate	25.00	0.5299	Mole F.	[35]
Diethyl acetate	Ketoprofen	25.00	0.1112	Mole F.	[68]
Diethyl formamide	Methyl p-hydroxybenzoate	25.00	0.4907	Mole F.	[35]
Diethyl phthalate	Temazepam	25.00	0.0384696	Mole F.	[15]
Diethylamine	Paracetamol	30.00	1316.90	g/kg	[11]
Diethylene glycol	Sulfamethoxypyridazine	25.00	0.02891	Mole F.	[40]
Diethylene glycol-n-butyl ether	Ricobendazole	25.00	1.97	g/L	[103]
Diethyl ether	Temazepam	24.00	0.0044	Volume F.	[64]
Diisopropyl ether	Acetylsalicylic acid	25.00	0.01224	Mole F.	[22]
Diisopropyl ether	Naproxen	25.00	0.00585	Mole F.	[32]
Diisopropyl ether	Phenothiazine	25.00	0.01185	Mole F.	[34]
Dimethyl acetamide	Benzoic acid	25.00	0.5245	Mole F.	[24]
Dimethyl acetamide	Sulfadiazine	25.00	0.0275296	Mole F.	[39]
Dimethyl acetamide	Sulfadiazine	25.00	0.0239	Mole F.	[150]
Dimethyl acetamide	Methyl p-hydroxybenzoate	25.00	0.5418	Mole F.	[35]
Dimethyl acetamide	p-Hydroxybenzoic acid	25.00	0.2354	Mole F.	[35]
Diolein	Griseofulvin	130.00	0.021	Mole F.	[142]
Dipropyl ether	Methyl p-hydroxybenzoate	25.00	0.0314	Mole F.	[35]
Distearin	Griseofulvin	130.00	0.020	Mole F.	[142]
Dodecane	Ethyl p-aminobenzoate	25.00	0.0013202	Mole F.	[124]
Dodecane	Mestanolone	25.00	0.0003490	Mole F.	[69]
Dodecane	Methandienone	25.00	0.0004250	Mole F.	[69]
Dodecane	Methyl p-aminobenzoate	25.00	0.0006491	Mole F.	[124]
Dodecane	Methyltestosterone	25.00	0.0004690	Mole F.	[69]
Dodecane	Nandrolone	25.00	0.0004570	Mole F.	[69]
Dodecane	n-Butyl p-aminobenzoate	25.00	0.0037726	Mole F.	[124]
Dodecane	n-Pentyl p-aminobenzoate	25.00	0.0050418	Mole F.	[124]
Dodecane	n-Propyl p-aminobenzoate	25.00	0.0022654	Mole F.	[124]
Dodecane	Testosterone	25.00	0.0002880	Mole F.	[69]
Dodecane	Testosterone propionate	25.00	0.0112	Mole F.	[131]

(continued)

Solubility Data of Drugs in Different Solvents at Various Temperatures (continued)

Solvent	Drug	T (°C)	Solubility	Solubility Unit	Reference
Ethyl ether	β-Carotene	Room	1000	mg/L	[6]
Ethyl ether	Lutein	Room	2000	mg/L	[6]
Ethylcyclohexane	Testosterone propionate	25.00	0.0110	Mole F.	[131]
Ethylcyclopentane	Testosterone propionate	25.00	0.0131	Mole F.	[131]
Hexadecane	Mestanolone	25.00	0.0004800	Mole F.	[69]
Hexadecane	Methyltestosterone	25.00	0.0005680	Mole F.	[69]
Hexadecane	Phenothiazine	25.00	0.001661	Mole F.	[34]
Hexadecane	Testosterone propionate	25.00	0.0100	Mole F.	[131]
Hexadecane	Xanthene	25.00	0.06835	Mole F.	[3]
Hexyl acetate	Methyl p-hydroxybenzoate	25.00	0.1164	Mole F.	[35]
Methyl butyrate	Acetylsalicylic acid	25.00	0.02970	Mole F.	[22]
Methyl cyclohexane	Phenothiazine	25.00	0.001027	Mole F.	[34]
Methyl cyclohexane	Testosterone propionate	25.00	0.0145	Mole F.	[131]
Methyl cyclohexane	Xanthene	25.00	0.04275	Mole F.	[3]
Methyl cyclopentane	Testosterone propionate	25.00	0.0107	Mole F.	[131]
Methyl formate	Benzoic acid	25.00	0.3428	Mole F.	[24]
Methyl formate	Betulin	5.05	0.000187	Mole F.	[7]
Methyl formate	Betulin	10.05	0.000212	Mole F.	[7]
Methyl formate	Betulin	15.05	0.000257	Mole F.	[7]
Methyl formate	Betulin	25.05	0.000316	Mole F.	[7]
Methyl formate	Methyl p-hydroxybenzoate	25.00	0.2981	Mole F.	[35]
Methyl formate	p-Hydroxybenzoic acid	25.00	0.1025	Mole F.	[35]
Methyl formate	Riboflavin	30.00	2.47	g/L	[118]
Methyl tert-butyl ether	Acetylsalicylic acid	25.00	0.04013	Mole F.	[22]
Methyl tert-butyl ether	β-Carotene	Room	1000	mg/L	[6]
Methyl tert-butyl ether	Lutein	Room	2000	mg/L	[6]
Methyl tert-butyl ether	Phenothiazine	25.00	0.02339	Mole F.	[34]
Methyl tert-butyl ether	Xanthene	25.00	0.07846	Mole F.	[3]
MIBK (4-methyl-2-pentanone)	Phenylacetic acid	20.00	803.5	g/kg	[83]
MIBK (4-methyl-2-pentanone)	p-Hydroxybenzoic acid	25.00	153.4	g/kg	[83]
Mineral oil	Benztropine	37.00	563.70	g/L	[145]
Monolaurin	Griseofulvin	100.00	0.0027	Mole F.	[142]
Monoolein	Griseofulvin	130.00	0.0062	Mole F.	[142]
Monoricinolein	Griseofulvin	130.00	0.0060	Mole F.	[142]
Monostearin	Griseofulvin	130.00	0.0062	Mole F.	[142]
N,N-Dimethylacetamide	Sulfamethoxypyridazine	25.00	0.09389	Mole F.	[40]
N-Methyl formamide	Sodium ibuprofen	25.00	0.0019006	Mole F.	[28]
N-Methyl formamide	Sulfadiazine	25.00	0.0045867	Mole F.	[39]
N-Methyl formamide	Sulfamethoxypyridazine	25.00	0.08721	Mole F.	[40]
Nonane	Benzoic acid	25.00	0.0141	Mole F.	[24]
Nonane	Mestanolone	25.00	0.0002690	Mole F.	[69]
Nonane	Methandienone	25.00	0.0003330	Mole F.	[69]
Nonane	Methyl p-hydroxybenzoate	25.00	0.000106	Mole F.	[35]
Nonane	Methyltestosterone	25.00	0.0003710	Mole F.	[69]
Nonane	Nandrolone	25.00	0.0003530	Mole F.	[69]
Nonane	Testosterone	25.00	0.0002090	Mole F.	[69]
Nonane	Testosterone propionate	25.00	0.0074	Mole F.	[131]

Solubility Data of Drugs in Different Solvents at Various Temperatures   (continued)

Solvent	Drug	T (°C)	Solubility	Solubility Unit	Reference
Nonane	Xanthene	25.00	0.04306	Mole F.	[3]
PEG 300	Ricobendazole	25.00	1.45	g/L	[103]
PEG 300	Salicylic acid	25.00	0.4931	Mole F.	[38]
Pentane	Benzoic acid	25.00	0.0059	Mole F.	[24]
Pentane	Flubiprofen	25.00	0.000350	Mole F.	[26]
Pentane	Methyl p-hydroxybenzoate	25.00	0.000063	Mole F.	[35]
Pentane	Testosterone propionate	25.00	0.003	Mole F.	[16]
Pentyl acetate	Acetylsalicylic acid	25.00	0.02677	Mole F.	[22]
Pentyl ether	p-Phenylphenol	25.00	0.12	mol/L	[129]
Propionitrile	Phenothiazine	25.00	0.03872	Mole F.	[34]
Propylcyclohexane	Testosterone propionate	25.00	0.0100	Mole F.	[131]
Propylene carbonate	Acetylsalicylic acid	25.00	0.03133	Mole F.	[22]
Pyridine	Benzoic acid	25.00	0.5348	Mole F.	[24]
Pyridine	Methyl p-hydroxybenzoate	25.00	0.3243	Mole F.	[35]
Pyridine	p-Hydroxybenzoic acid	25.00	0.1044	Mole F.	[35]
Pyridine	Sulfamethoxypyridazine	25.00	0.02329	Mole F.	[40]
tert-Butylcyclohexane	Testosterone propionate	25.00	0.0153	Mole F.	[131]
tert-Pentyl alcohol	Glucose	40.00	1.08	g/L	[42]
tert-Pentyl alcohol	Glucose	60.00	2.40	g/L	[42]
Tetrahydronaphthalene	Testosterone propionate	25.00	0.20	Mole F.	[16]
trans-1,2-Dichloroethylene	Testosterone propionate	25.00	0.32	Mole F.	[16]
Triacetin	Griseofulvin	100.00	0.0031	Mole F.	[142]
Tributyrin	Griseofulvin	100.00	0.0064	Mole F.	[142]
Tricaproin	Griseofulvin	100.00	0.0035	Mole F.	[142]
Tricaprylin	Griseofulvin	100.00	0.0060	Mole F.	[142]
Trilaurin	Griseofulvin	130.00	0.016	Mole F.	[142]
Trimyristin	Griseofulvin	130.00	0.0098	Mole F.	[142]
Triolein	Griseofulvin	130.00	0.0031	Mole F.	[142]
Tripalmitin	Griseofulvin	130.00	0.0093	Mole F.	[142]
Triricinolein	Griseofulvin	130.00	0.011	Mole F.	[142]
Tristearin	Griseofulvin	130.00	0.0082	Mole F.	[142]
Undecane	Mestanolone	25.00	0.0003570	Mole F.	[69]
Undecane	Methandienone	25.00	0.0003870	Mole F.	[69]
Undecane	Methyltestosterone	25.00	0.0004640	Mole F.	[69]
Undecane	Nandrolone	25.00	0.0004200	Mole F.	[69]
Undecane	Testosterone	25.00	0.0002780	Mole F.	[69]
Undecane	Testosterone propionate	25.00	0.0085	Mole F.	[131]

REFERENCES

1. Baluja, S., Bhalodia, R., Bhatt, M., Vekariya, N., and Gajera, R., Solubility of enrofloxacin sodium in various solvents at various temperatures. *Journal of Chemical and Engineering Data*, 2008. 53: 2897–2899.
2. Kong, M.Z., Shi, X.H., Cao, Y.C., and Zhou, C.R., Solubility of imidacloprid in different solvents. *Journal of Chemical and Engineering Data*, 2008. 53: 615–618.
3. Monarrez, C.I., Stovall, D.M., Woo, J.H., Taylor, P., and Acree, W.E. Jr., Solubility of xanthene in organic nonelectrolyte solvents: Comparison of observed versus predicted values based upon mobile order theory. *Physics and Chemistry of Liquids*, 2002. 40: 703–714.
4. Boersma, A., Mobility and solubility of antioxidants and oxygen in glassy polymers. I. Concentration and temperature dependence of antioxidant sorption. *Journal of Applied Polymer Science*, 2003. 89: 2163–2178.

5. Wang, L., Yin, Q., Zhang, M., and Wang, J., Solubility of acephate in different solvents from (292.90 to 327.60) K. *Journal of Chemical and Engineering Data*, 2007. 52: 426–428.

6. Craft, N.E. and Scares, J.H. Jr., Relative solubility, stability, and absorptivity of lutein and β-carotene in organic solvents. *Journal of Agricultural and Food Chemistry*, 1992. 40: 431–434.

7. Cao, D., Zhao, G., and Yan, W., Solubilities of betulin in fourteen organic solvents at different temperatures. *Journal of Chemical and Engineering Data*, 2007. 52: 1366–1368.

8. Greene, L.R., Blackburn, A.C., and Miller, J.M., Rapid, small-scale determination of organic solvent solubility using a thermogravimetric analyzer. *Journal of Pharmaceutical and Biomedical Analysis*, 2005. 39: 344–347.

9. Wu, J., Wang, J., and Zhang, M., Solubility of cefazolin sodium pentahydrate in different solvents between 275 K and 310 K. *Journal of Chemical and Engineering Data*, 2005. 50: 2026–2027.

10. Bailey, A.V., Harris, J.A., and Skau, E.L., Solubilities of some normal saturated and unsaturated long-chain fatty acid methyl esters in acetone, *n*-hexane, toluene, and 1,2-dichloroethane. *Journal of Chemical and Engineering Data*, 1970. 15: 583–585.

11. Granberg, R.A. and Rasmuson, A.C., Solubility of paracetamol in pure solvents. *Journal of Chemical and Engineering Data*, 1999. 44: 1391–1395.

12. Wang, W., Lu, X.H., Qin, X.J., Zhang, X.H., and Xu, Y.Q., Solubility of pyoluteorin in water, dichloromethane, chloroform, and carbon tetrachloride from (278.2 to 333.2) K. *Journal of Chemical and Engineering Data*, 2008. 53: 2241–2243.

13. Zhou, K., Li, J., and Ren, Y.S., Solubility of rifapentine in different organic solvents. *Journal of Chemical and Engineering Data*, 2008. 53: 998–999.

14. Brandreth, D.A. and Johnson, R.E., Solubility of stearic acid in some halofluorocarbons, chlorocarbons, ethanol, and their azeotropes. *Journal of Chemical and Engineering Data*, 1971. 16: 325–327.

15. Richardson, P.J., McCafferty, D.F., and Woolfson, A.D., Determination of three-component partial solubility parameters for temazepam and the effects of change in partial molal volume on the thermodynamics of drug solubility. *International Journal of Pharmaceutics*, 1992. 78: 189–198.

16. Bowen, D.B., PhD dissertation, University of Wales, Cardiff, Great Britain,1969, (cited in: James, K.C., Ng, C.T., and Noyce, P.R., Solubilities of testosterone propionate and related esters in organic solvents, *Journal of Pharmaceutical Sciences* 65(5), 656–659, 1976).

17. Nie, Q. and Wang, J.K., Solubility of 16α,17α-epoxyprogesterone in six different solvents. *Journal of Chemical and Engineering Data*, 2005. 50: 1750–1752.

18. Li, Q.S., Su, M.G., and Wang, S., Solubility of 2-(4-ethylbenzoyl)benzoic acid in eleven organic solvents between 279.55 K and 343.15 K. *Journal of Chemical and Engineering Data*, 2007. 52: 2477–2479.

19. De Fina, K.M., Sharp, T.L., Roy, L.E., and Acree, W.E. Jr., Solubility of 2-hydroxybenzoic acid in select organic solvents at 298.15 K. *Journal of Chemical and Engineering Data*, 1999. 44: 1262–1264.

20. Daniels, C.R., Charlton, A.K., Wold, R.M., Moreno, R.J., Acree, W.E. Jr., and Abraham, M.H., Mathematical correlation of 4-aminobenzoic acid solubilities in organic solvents with the Abraham solvation parameter model. *Physics and Chemistry of Liquids*, 2004. 42: 633–641.

21. Perlovich, G.L. and Bauer-Brandl, A., Thermodynamics of solutions I: Benzoic acid and acetylsalicylic acid as models for drug substances and the prediction of solubility. *Pharmaceutical Research*, 2003. 20: 471–478.

22. Charlton, A.K., Daniels, C.R., Acree, W.E. Jr., and Abraham, M.H., Solubility of crystalline nonelectrolyte solutes in organic solvents: Mathematical correlation of acetylsalicylic acid solubilities with the Abraham general solvation model. *Journal of Solution Chemistry*, 2003. 32: 1087–1102.

23. Barra, J., Lescure, F., Doelker, E., and Bustamante, P., The expanded Hansen approach to solubility parameters. Paracetamol and citric acid in individual solvents. *Journal of Pharmacy and Pharmacology*, 1997. 49: 644–651.

24. Beerbower, A., Wu, P.L., and Martin, A., Expanded solubility parameter approach. I. Naphthalene and benzoic acid in individual solvents. *Journal of Pharmaceutical Sciences*, 1984. 73: 179–188.

25. Barra, J., Pena, M.A., and Bustamante, P., Proposition of group molar constants for sodium to calculate the partial solubility parameters of sodium salts using the van Krevelen group contribution method. *European Journal of Pharmaceutical Sciences*, 2000. 10: 153–161.

26. Perlovich, G.L., Kurkov, S.V., and Bauer-Brandl, A., Thermodynamics of solutions. II. Flurbiprofen and diflunisal as models for studying solvation of drug substances. European *Journal of Pharmaceutical Sciences*, 2003. 19: 423–432.

27. Subrahmanyam, C.V.S. and Suresh, S., Solubility behaviour of haloperidol in individual solvents determination of partial solubility parameters. *European Journal of Pharmaceutics and Biopharmaceutics*, 1999. 47: 289–294.

28. Bustamante, P., Pena, M.A., and Barra, J., The modified extended Hansen method to determine partial solubility parameters of drugs containing a single hydrogen bonding group and their sodium derivatives: Benzoic acid/Na and ibuprofen/Na. *International Journal of Pharmaceutics*, 2000. 194: 117–124.

29. Wang, L., Wang, J., Bao, Y., and Li, T., Solubility of irbesartan (form A) in different solvents between 278 K and 323 K. *Journal of Chemical and Engineering Data*, 2007. 52: 2016–2017.

30. Pena, M.A., Daali, Y., Barra, J., and Bustamante, P., Partial solubility parameters of lactose, mannitol and saccharose using the modified extended Hansen method and evaporation light scattering detection. *Chemical and Pharmaceutical Bulletin*, 2000. 48: 179–183.

31. Bustamante, P., Pena, M.A., and Barra, J., Partial-solubility parameters of naproxen and sodium diclofenac. *Journal of Pharmacy and Pharmacology*, 1998. 50: 975–982.

32. Daniels, C.R., Charlton, A.K., Wold, R.M., Pustejovsky, E., Furman, A.N., Bilbrey, A.C., Love, J.N., Garza, J.A., Acree, W.E. Jr., and Abraham, M.H., Mathematical correlation of naproxen solubilities in organic solvents with the Abraham solvation parameter model. *Physics and Chemistry of Liquids*, 2004. 42: 481–491.

33. Bustamante, P., Pena, M.A., and Barra, J., Partial solubility parameters of piroxicam and niflumic acid. *International Journal of Pharmaceutics*, 1998. 174: 141–150.

34. Hoover, K.R., Acree, W.E. Jr., and Abraham, M.H., Mathematical correlation of phenothiazine solubilities in organic solvents with the Abraham solvation parameter model. *Physics and Chemistry of Liquids*, 2006. 44: 367–376.

35. Martin, A., Wu, P.L., and Beerbower, A., Expanded solubility parameter approach II: *p*-Hydroxybenzoic acid and methyl *p*-hydroxybenzoate in individual solvents. *Journal of Pharmaceutical Sciences*, 1984. 73: 188–194.

36. Thimmasetty, J., Subrahmanyam, C.V.S., Sathesh Babu, P.R., Maulik, M.A., and Viswanath, B.A., Solubility behavior of pimozide in polar and nonpolar solvents: Partial solubility parameters approach. *Journal of Solution Chemistry*, 2008. 37: 1365–1378.

37. Blake-Taylor, B.H., Deleon, V.H., Acree, W.E., and Abraham, M.H., Mathematical correlation of salicylamide solubilities in organic solvents with the Abraham solvation parameter model. *Physics and Chemistry of Liquids*, 2007. 45: 389–398.

38. Matsuda, H., Kaburagi, K., Matsumoto, S., Kurihara, K., Tochigi, K., and Tomono, K., Solubilities of salicylic acid in pure solvents and binary mixtures containing cosolvent. *Journal of Chemical and Engineering Data*, 2009. 54: 480–484.

39. Bustamante, P., Martin, A., and Gonzalez-Guisandez, M.A., Partial solubility parameters and solvatochromic parameters for predicting the solubility of single and multiple drugs in individual solvents. *Journal of Pharmaceutical Sciences*, 1993. 82: 635–640.

40. Bustamante, P., Escalera, B., Martin, A., and Selles, E., Predicting the solubility of sulfamethoxypyridazine in individual solvents I: Calculating partial solubility parameters. *Journal of Pharmaceutical Sciences*, 1989. 78: 567–573.

41. Laprade, B., Mauger, J.W., and Petersen, H. Jr., Solubility of straight chain and branched alkyl barbiturates in straight chain alcohols. *Journal of Pharmaceutical Sciences*, 1976. 65: 277–280.

42. Leontarakis, G., Tsavas, P., Voutsas, E., Magoulas, K., and Tassios, D., Experimental and predicted results of anomeric equilibrium of glucose in alcohols. *Journal of Chemical and Engineering Data*, 2005. 50: 1924–1927.

43. Restaino, F.A. and Martin, A.N., Solubility of Benzoic acid and related compound in a series of *n*-alkanols. *Journal of Pharmaceutical Science*, 1964. 53: 636–639.

44. Lu, Y.C., Lin, Q., Luo, G.S., and Dai, Y.Y., Solubility of berberine chloride in various solvents. *Journal of Chemical and Engineering Data*, 2006. 51: 642–644.

45. Alexander, K.S., Mauger, J.W., Petersen, H. Jr., and Paruta, A.N., Solubility profiles and thermodynamics of parabens in aliphatic alcohols. *Journal of Pharmaceutical Sciences*, 1977. 66: 42–48.

46. Liu, W., Dang, L., Black, S., and Wei, H., Solubility of carbamazepine (form III) in different solvents from (275 to 343) K. *Journal of Chemical and Engineering Data*, 2008. 53: 2204–2206.

47. Li, Q.S., Tian, Y.M., and Wang, S., Solubility of chlorpheniramine maleate in ethanol, 1-propanol, 1-butanol, benzene, ethyl acetate, ethyl formate, and butyl acetate between 283 K and 333 K. *Journal of Chemical and Engineering Data*, 2007. 52: 2163–2165.

48. Hammad, M.A. and Muller, B.W., Solubility and stability of clonazepam in mixed micelles. *International Journal of Pharmaceutics*, 1998. 169: 55–64.

49. Zhou, Z., Wang, J., Qu, Y., and Wei, H., Solubility of D(−)-*p*-hydroxyphenylglycine dane salt in eight alcohols between (293 and 343) K. *Journal of Chemical and Engineering Data*, 2008. 53: 2230–2232.

50. Liu, L. and Chen, J., Solubility of hesperetin in various solvents from (288.2 to 323.2) K. *Journal of Chemical and Engineering Data*, 2008. 53: 1649–1650.

51. Perlovich, G.L., Kurkov, S.V., Kinchin, A.N., and Bauer-Brandl, A., Thermodynamics of solutions IV: Solvation of ketoprofen in comparison with other NSAIDs. *Journal of Pharmaceutical Sciences*, 2003. 92: 2502–2511.

52. Jozwiakowski, M.J., Nguyen, N.A.T., Sisco, J.M., and Spancake, C.W., Solubility behavior of lamivudine crystal forms in recrystallization solvents. *Journal of Pharmaceutical Sciences*, 1996. 85: 193–199.

53. Nti-Gyabaah, J., Chmielowski, R., Chan, V., and Chiew, Y.C., Solubility of lovastatin in a family of six alcohols: Ethanol, 1-propanol, 1-butanol, 1-pentanol, 1-hexanol, and 1-octanol. *International Journal of Pharmaceutics*, 2008. 359: 111–117.

54. Peng, B., Zi, J., and Yan, W., Measurement and correlation of solubilities of luteolin in organic solvents at different temperatures. *Journal of Chemical and Engineering Data*, 2006. 51: 2038–2040.

55. Seedher, N. and Bhatia, S., Solubility enhancement of cox-2 inhibitors using various solvent systems. *AAPS PharmSciTech*, 2003. 4: Article 33.

56. Liu, L. and Wang, X., Solubility of oleanolic acid in various solvents from (288.3 to 328.3) K. *Journal of Chemical and Engineering Data*, 2007. 52: 2527–2528.

57. Ren, G.B., Wang, J.K., Yin, Q.X., and Zhang, M.J., Solubilities of proxetine hydrochloride hemihydrate between 286 K and 363 K. *Journal of Chemical and Engineering Data*, 2004. 49: 1671–1674.

58. Chang, Q.L., Li, Q.S., Wang, S., and Tian, Y.M., Solubility of phenacetinum in methanol, ethanol, 1-propanol, 1-butanol, 1-pentanol, tetrahydrofuran, ethyl acetate, and benzene between 282.65 K and 333.70 K. *Journal of Chemical and Engineering Data*, 2007. 52: 1894–1896.

59. Liu, B.S., Gong, J.B., Wang, J.K., and Jia, C.Y., Solubility of potassium clavulanate in ethanol, 1-propanol, 1-butanol, 2-propanol, and 2-methyl-1-propanol between 273 K and 305 K. *Journal of Chemical and Engineering Data*, 2005. 50: 1684–1686.

60. Jia, C., Yin, Q., Song, J., Hou, G., and Zhang, M., Solubility of pravastatin sodium in water, methanol, ethanol, 2-propanol, 1-propanol, and 1-butanol from (278 to 333) K. *Journal of Chemical and Engineering Data*, 2008. 53: 2466–2468.

61. Zi, J., Peng, B., and Yan, W., Solubilities of rutin in eight solvents at $T = 283.15, 298.15, 313.15, 323.15,$ and 333.15 K. *Fluid Phase Equilibria*, 2007. 261: 111–114.

62. Regosz, A., Pelplinska, T., Kowalski, P., and Thiel, Z., Prediction of solubility of sulfonamides in water and organic solvents based on the extended regular solution theory. *International Journal of Pharmaceutics*, 1992. 88: 437–442.

63. Mauger, J.W., Paruta, A.N., and Gerraughty, R.J., Solubilities of sulfadiazine, sulfisomidine, and sulfadimethoxine in several normal alcohols. *Journal of Pharmaceutical Sciences*, 1972. 61: 94–97.

64. Van den Mooter, G., Augustijns, P., and Kinget, R., Application of the thermodynamics of mobile order and disorder to explain the solubility of temazepam in aqueous solutions of polyethylene glycol 6000. *International Journal of Pharmaceutics*, 1998. 164: 81–89.

65. Li, Q.S., Li, Z., and Wang, S., Solubility of trimethoprim (TMP) in different organic solvents from (278 to 333) K. *Journal of Chemical and Engineering Data*, 2008. 53: 286–287.

66. Wang, S., Li, Q.S., Li, Z., and Su, M.G., Solubility of xylitol in ethanol, acetone, *N*,*N*-dimethylformamide, 1-butanol, 1-pentanol, toluene, 2-propanol, and water. *Journal of Chemical and Engineering Data*, 2007. 52: 186–188.

67. Stovall, D.M., Givens, C., Keown, S., Hoover, K.R., Rodriguez, E., Acree, W.E. Jr., and Abraham, M.H., Solubility of crystalline nonelectrolyte solutes in organic solvents: Mathematical correlation of ibuprofen solubilities with the Abraham solvation parameter model. *Physics and Chemistry of Liquids*, 2005. 43: 261–268.

68. Daniels, C.R., Charlton, A.K., Acree, W.E. Jr., and Abraham, M.H., Thermochemical behavior of dissolved carboxylic acid solutes: Part 2—Mathematical correlation of ketoprofen solubilities with the Abraham general solvation model. *Physics and Chemistry of Liquids*, 2004. 42: 305–312.

69. Gharavi, M., James, K.C., and Sanders, L.M., Solubilities of mestanolone, methandienone, methyltestosterone, nandrolone and testosterone in homologous series of alkanes and alkanols. *International Journal of Pharmaceutics*, 1983. 14: 333–341.

70. Yalkowsky, S.H., Valvani, S.C., and Roseman, T.J., Solubility and partitioning. VI: Octanol solubility and octanol–water partition coefficients. *Journal of Pharmaceutical Sciences*, 1983. 72: 866–870.

71. Perlovich, G.L., Volkova, T.V., and Bauer-Brandl, A., Towards an understanding of the molecular mechanism of solvation of drug molecules: A thermodynamic approach by crystal lattice energy, sublimation, and solubility exemplified by hydroxybenzoic acids. *Journal of Pharmaceutical Sciences*, 2006. 95: 1448–1458.

72. Perlovich, G.L., Volkova, T.V., and Bauer-Brandl, A., Towards an understanding of the molecular mechanism of solvation of drug molecules: A thermodynamic approach by crystal lattice energy, sublimation, and solubility exemplified by paracetamol, acetanilide, and phenacetin. *Journal of Pharmaceutical Sciences*, 2006. 95: 2158–2169.

73. Baena, Y., Pinzon, J.A., Barbosa, H.J., and Martinez, F., Temperature-dependence of the solubility of some acetanilide derivatives in several organic and aqueous solvents. *Physics and Chemistry of Liquids*, 2004. 42: 603–613.

74. Fini, A., Laus, M., Orienti, I., and Zecchi, V., Dissolution and partition thermodynamic functions of some nonsteroidal anti-inflammatory drugs. *Journal of Pharmaceutical Sciences*, 1986. 75: 23–25.

75. Perlovich, G.L., Volkova, T.V., and Bauer-Brandl, A., Thermodynamic study of sublimation, solubility, solvation, and distribution processes of atenolol and pindolol. *Molecular Pharmaceutics*, 2007. 4: 929–935.

76. Avila, C.M. and Martinez, F., Thermodynamic study of the solubility of benzocaine in some organic and aqueous solvents. *Journal of Solution Chemistry*, 2002. 31: 975–985.

77. Kurkov, S.V. and Perlovich, G.L., Thermodynamic studies of fenbufen, diflunisal, and flurbiprofen: Sublimation, solution and solvation of biphenyl substituted drugs. *International Journal of Pharmaceutics*, 2008. 357: 100–107.

78. Perlovich, G.L., Surov, A.O., and Bauer-Brandl, A., Thermodynamic properties of flufenamic and niflumic acids-specific and non-specific interactions in solution and in crystal lattices, mechanism of solvation, partitioning and distribution. *Journal of Pharmaceutical and Biomedical Analysis*, 2007. 45: 679–687.

79. Pinsuwan, S., Li, A., and Yalkowsky, S.H., Correlation of octanol/water solubility ratios and partition coefficients. *Journal of Chemical and Engineering Data*, 1995. 40: 623–626.

80. Hagen, T.A. and Flynn, G.L., Solubility of hydrocortisone in organic and aqueous media: Evidence for regular solution behavior in apolar solvents. *Journal of Pharmaceutical Sciences*, 1983. 72: 409–414.

81. Garzon, L.C. and Martinez, F., Temperature dependence of solubility for ibuprofen in some organic and aqueous solvents. *Journal of Solution Chemistry*, 2004. 33: 1379–1395.

82. Mora, C.P. and Martinez, F., Solubility of naproxen in several organic solvents at different temperatures. *Fluid Phase Equilibria*, 2007. 255: 70–77.

83. Gracin, S. and Rasmuson, A.C., Solubility of phenylacetic acid, *p*-hydroxyphenylacetic acid, *p*-aminophenylacetic acid, *p*-hydroxybenzoic acid, and ibuprofen in pure solvents. *Journal of Chemical and Engineering Data*, 2002. 47: 1379–1383.

84. Martinez, F. and Gomez, A., Thermodynamic study of the solubility of some sulfonamides in octanol, water, and the mutually saturated solvents. *Journal of Solution Chemistry*, 2001. 30: 909–923.

85. Aragón, D.M., Ruidiaz, M.A., Vargas, E.F., Bregni, C., Chiappetta, D.A., Sosnik, A., and Martinez, F., Solubility of the antimicrobial agent triclosan in organic solvents of different hydrogen bonding capabilities at several temperatures. *Journal of Chemical and Engineering Data*, 2008. 53: 2576–2580.

86. Guo, K., Yin, Q., Yang, Y., Zhang, M., and Wang, J., Solubility of losartan potassium in different pure solvents from (293.15 to 343.15) K. *Journal of Chemical and Engineering Data*, 2008. 53: 1467–1469.

87. Mirmehrabi, M., Rohani, S., Murthy, K.S.K., and Radatus, B., Solubility, dissolution rate and phase transition studies of Ranitidine HCl tautomeric forms. *International Journal of Pharmaceutics*, 2004. 282: 73–85.

88. Crocker, L.S. and McCauley, J.A., Solubilities of losartan polymorphs. *Die Pharmazie*, 1997. 52: 72.

89. Nti-Gyabaah, J. and Chiew, Y.C., Solubility of lovastatin in ethyl acetate, propyl acetate, isopropyl acetate, butyl acetate, *sec*-Butyl acetate, isobutyl acetate, *tert*-Butyl acetate, and 2-butanone, between (285 and 313) K. *Journal of Chemical and Engineering Data*, 2008. 53: 2060–2065.

90. Beckmann, W., Boistelle, R., and Sato, K., Solubility of the A, B, and C polymorphs of stearic acid in decane, methanol, and butanone. *Journal of Chemical and Engineering Data*, 1984. 29: 211–214.

91. Nandi, I., Bari, M., and Joshi, H., Study of isopropyl myristate microemulsion systems containing cyclodextrins to improve the solubility of 2 model hydrophobic drugs. *AAPS PharmSciTech*, 2003. 4: Article 10.

92. Wang, S., Li, Q.S., and Su, M.G., Solubility of 1H-1,2,4-triazole in ethanol, 1-propanol, 2-propanol, 1,2-propanediol, ethyl formate, methyl acetate, ethyl acetate, and butyl acetate at (283 to 363) K. *Journal of Chemical and Engineering Data*, 2007. 52: 856–858.

93. Chen, X.H., Zeng, Z.X., Xue, W.L., and Pu, T., Solubility of 2,6-diaminopyridine in toluene, *o*-xylene, ethylbenzene, methanol, ethanol, 2-propanol, and sodium hydroxide solutions. *Journal of Chemical and Engineering Data*, 2007. 52: 1911–1915.

94. Maia, G.D. and Giulietti, M., Solubility of acetylsalicylic acid in ethanol, acetone, propylene glycol, and 2-propanol. *Journal of Chemical and Engineering Data*, 2008. 53: 256–258.

95. Sheikhzadeh, M., Rohani, S., Taffish, M., and Murad, S., Solubility analysis of buspirone hydrochloride polymorphs: Measurements and prediction. *International Journal of Pharmaceutics*, 2007. 338: 55–63.

96. Park, K., Evans, J.M.B., and Myerson, A.S., Determination of solubility of polymorphs using differential scanning calorimetry. *Crystal Growth and Design*, 2003. 3: 991–995.

97. Bar, L.K., Garti, N., Sarig, S., and Bar, R., Solubilities of cholesterol, sitosterol, and cholestryl acetate in polar organic solvents. *Journal of Chemical and Engineering Data*, 1984. 29: 440–443.

98. Caço, A.I., Varanda, F., De Melo, M.J.P., Dias, A.M.A., Dohrn, R., and Marrucho, I.M., Solubility of antibiotics in different solvents. Part II. Non-hydrochloride forms of tetracycline and ciprofloxacin. *Industrial and Engineering Chemistry Research*, 2008. 47: 8083–8089.

99. Teychene, S., Autret, J.M., and Biscans, B., Determination of solubility profiles of eflucimibe polymorphs: Experimental and modeling. *Journal of Pharmaceutical Sciences*, 2006. 95: 871–882.

100. Wang, Z., Wang, J., Zhang, M., and Dang, L., Solubility of erythromycin A dihydrate in different pure solvents and acetone + water binary mixtures between 293 K and 323 K. *Journal of Chemical and Engineering Data*, 2006. 51: 1062–1065.

101. Li, X., Yin, Q., Chen, W., and Wang, J., Solubility of hydroquinone in different solvents from 276.65 K to 345.10 K. *Journal of Chemical and Engineering Data*, 2006. 51: 127–129.

102. Shalmashi, A. and Eliassi, A., Solubility of L-(+)-ascorbic acid in water, ethanol, methanol, propan-2-ol, acetone, acetonitrile, ethyl acetate, and tetrahydrofuran from (293 to 323) K. *Journal of Chemical and Engineering Data*, 2008. 53: 1332–1334.

103. Wu, Z., Razzak, M., Tucker, I.G., and Medlicott, N.J., Physicochemical characterization of ricobendazole: I. Solubility, lipophilicity, and ionization characteristics. *Journal of Pharmaceutical Sciences*, 2005. 94: 983–993.

104. Calvo, B. and Cepeda, E.A., Solubilities of stearic acid in organic solvents and in azeotropic solvent mixtures. *Journal of Chemical and Engineering Data*, 2008. 53: 628–633.

105. Nie, Q., Wang, J.K., Wang, Y.L., and Wang, S., Solubility of 11α-hydroxy-16α, 17α-epoxyprogesterone in different solvents between 283 K and 323 K. *Journal of Chemical and Engineering Data*, 2005. 50: 989–992.

106. Li, Q.S., Yi, Z.M., Su, M.G., Wang, S., and Wu, X.H., Solubility of dioxopromethazine HCl in different solvents. *Journal of Chemical and Engineering Data*, 2008. 53: 301–302.

107. Nordstrom, F.L. and Rasmuson, A.C., Solubility and melting properties of salicylamide. *Journal of Chemical and Engineering Data*, 2006. 51: 1775–1777.

108. Nordstrom, F.L. and Rasmuson, A.C., Solubility and melting properties of salicylic acid. *Journal of Chemical and Engineering Data*, 2006. 51: 1668–1671.

109. Carta, R. and Dernini, S., Solubility of solid acetic acid in liquid organic solvents. *Journal of Chemical and Engineering Data*, 1983. 28: 328–330.

110. Wang, Y.Q., Ji, M.J., Cao, Y., and Xu, W.L., Solubility of β-sitosteryl maleate and Stigmasteryl maleate in acetone and ethyl acetate. *Journal of Chemical and Engineering Data*, 2007. 52: 2110–2111.

111. Pardillo-Fontdevila, E., Acosta-Esquijarosa, J., Nuevas-Paz, L., Gago-Alvarez, A., and Jauregui-Haza, U., Solubility of cefotaxime sodium salt in seven solvents used in the pharmaceutical industry. *Journal of Chemical and Engineering Data*, 1998. 43: 49–50.

112. Zhang, C., Wang, J., and Wang, Y., Solubility of ceftriaxone disodium in acetone, methanol, ethanol, N,N-dimethylformamide, and formamide between 278 and 318 K. *Journal of Chemical and Engineering Data*, 2005. 50: 1757–1760.

113. Hao, H.X., Hou, B.H., Wang, J.K., and Zhang, M.J., Solubility of erythritol in different solvents. *Journal of Chemical and Engineering Data*, 2005. 50: 1454–1456.

114. Heryanto, R., Hasan, M., and Abdullah, E.C., Solubility of isoniazid in various organic solvents from (301 to 313) K. *Journal of Chemical and Engineering Data*, 2008. 53: 1962–1964.

115. Chebil, L., Humeau, C., Anthony, J., Dehez, F., Engasser, J.M., and Ghoul, M., Solubility of flavonoids in organic solvents. *Journal of Chemical and Engineering Data*, 2007. 52: 1552–1556.

116. Sun, H., Gong, J.B., and Wang, J.K., Solubility of Lovastatin in acetone, methanol, ethanol, ethyl acetate, and butyl acetate between 283 K and 323 K. *Journal of Chemical and Engineering Data*, 2005. 50: 1389–1391.

117. Wang, L.H. and Cheng, Y.Y., Solubility of puerarin in water, ethanol, and acetone from (288.2 to 328.2) K. *Journal of Chemical and Engineering Data*, 2005. 50: 1375–1376.

118. Coffman, R.E. and Kildsig, D.O., Effect of nicotinamide and urea on the solubility of riboflavin in various solvents. *Journal of Pharmaceutical Sciences*, 1996. 85: 951–954.

119. Zilnik, L.F., Jazbinsek, A., Hvala, A., Vrecer, F., and Klamt, A., Solubility of sodium diclofenac in different solvents. *Fluid Phase Equilibria*, 2007. 261: 140–145.

120. Gu, C.H. and Grant, D.J.W., Estimating the relative stability of polymorphs and hydrates from heats of solution and solubility data. *Journal of Pharmaceutical Sciences*, 2001. 90: 1277–1287.

121. Zhao, X., Gan, L., and Zhou, C., Solubility of 1,6,6-trimethyl-6,7,8,9-tetrahydrophenanthro[1,2-b]furan-10, 11-dione in four organic solvents from (283.2 to 323.3) K. *Journal of Chemical and Engineering Data*, 2008. 53: 1975–1977.

122. Li, J., Masso, J.J., and Guertin, J.A., Prediction of drug solubility in an acrylate adhesive based on the drug-polymer interaction parameter and drug solubility in acetonitrile. *Journal of Controlled Release*, 2002. 83: 211–221.

123. Rao, Y.J. and Viswanath, D.S., Solubility of chloramine in organic liquids. *Journal of Chemical and Engineering Data*, 1975. 20: 29–30.

124. Neau, S.H., Flynn, G.L., and Yalkowsky, S.H., The influence of heat capacity assumptions on the estimation of solubility parameters from solubility data. *International Journal of Pharmaceutics*, 1989. 49: 223–229.

125. Song, C.Y., Shen, H.Z., Wang, L.C., Zhao, J.H., and Wang, F.A., Solubilities of vitamin K3 in benzene, toluene, ethylbenzene, *o*-xylene, *m*-xylene, and *p*-xylene between (299.44 and 344.24) K. *Journal of Chemical and Engineering Data*, 2008. 53: 283–285.

126. Hammad, M.A. and Muller, B.W., Increasing drug solubility by means of bile salt–phosphatidylcholine-based mixed micelles. *European Journal of Pharmaceutics and Biopharmaceutics*, 1998. 46: 361–367.

127. Wang, S., Li, Q.S., and Li, Y.L., Solubility of D-*p*-hydroxyphenylglycine in water, methanol, ethanol, carbon tetrachloride, toluene, and *N,N*-dimethylformamide between 278 K and 323 K. *Journal of Chemical and Engineering Data*, 2006. 51: 2201–2202.

128. Shalmashi, A. and Eliassi, A., Solubility of salicylic acid in water, ethanol, carbon tetrachloride, ethyl acetate, and xylene. *Journal of Chemical and Engineering Data*, 2008. 53: 199–200.

129. Anderson, B.D., Rytting, J.H., and Higuchi, T., Solubility of polar organic solutes in nonaqueous systems: Role of specific interactions. *Journal of Pharmaceutical Sciences*, 1980. 69: 676–680.

130. Dearden, J.C. and O'Sullivan, J.G., Solubility of pharmaceutical in cyclohexane. *Journal of Pharmacy and Pharmacology*, 1988. 40: 77P.

131. James, K.C., Ng, C.T., and Noyce, P.R., Solubilities of testosterone propionate and related esters in organic solvents. *Journal of Pharmaceutical Sciences*, 1976. 65: 656–659.

132. Zhao, C.X. and He, C.H., Solubility of atractylenolide III in hexane, ethyl acetate, diethyl ether, and ethanol from (283.2 to 323.2) K. *Journal of Chemical and Engineering Data*, 2007. 52: 1223–1225.

133. Zhang, H., Wang, J., Chen, Y., and Zhang, M., Solubility of sodium cefotaxime in different solvents. *Journal of Chemical and Engineering Data*, 2007. 52: 982–985.

134. Zhu, L., Chen, Y., and Wang, J.K., Solubility of Clindamycin phosphate in methanol, ethanol, and dimethyl sulfoxide. *Journal of Chemical and Engineering Data*, 2008. 53: 1984–1985.

135. Chellquist, E.M. and Gorman, W.G., Benzoyl peroxide solubility and stability in hydric solvents. *Pharmaceutical Research*, 1992. 9: 1341–1346.

136. Wyttenbach, N., Alsenz, J., and Grassmann, O., Miniaturized assay for solubility and residual solid screening (SORESOS) in early drug development. *Pharmaceutical Research*, 2007. 24: 888–898.

137. Alsenz, J., Meister, E., and Haenel, E., Development of a partially automated solubility screening (PASS) assay for early drug development. *Journal of Pharmaceutical Sciences*, 2007. 96: 1748–1762.

138. Hao, H.X., Wang, J.K., and Wang, Y.L., Solubility of dexamethasone sodium phosphate in different solvents. *Journal of Chemical and Engineering Data*, 2004. 49: 1697–1698.

139. Daneshfar, A., Ghaziaskar, H.S., and Homayoun, N., Solubility of gallic acid in methanol, ethanol, water, and ethyl acetate. *Journal of Chemical and Engineering Data*, 2008. 53: 776–778.

140. Dong, Y., Ng, W.K., Surana, U., and Tan, R.B.H., Solubilization and preformulation of poorly water soluble and hydrolysis susceptible *N*-epoxymethyl-1,8-naphthalimide (ENA) compound. *International Journal of Pharmaceutics*, 2008. 356: 130–136.

141. Yang, Z.J., Hu, H.B., Zhang, X.H., and Xu, Y.Q., Solubility of phenazine-1-carboxylic acid in water, methanol, and ethanol from (278.2 to 328.2) K. *Journal of Chemical and Engineering Data*, 2007. 52: 184–185.

142. Grant, D.J.W. and Abougela, I.K.A., Prediction of the solubility of griseofulvin in glycerides and the solvents of relatively low polarity from simple regular solution theory. *International Journal of Pharmaceutics*, 1983. 17: 77–89.

143. Goosen, C., Laing, T.J., Plessis, J.D., Goosen, T.C., and Flynn, G.L., Physicochemical characterization and solubility analysis of thalidomide and its *N*-alkyl analogs. *Pharmaceutical Research*, 2002. 19: 13–19.

144. Roberts, W.J. and Sloan, K.B., Correlation of aqueous and lipid solubilities with flux for prodrugs of 5-fluorouracil, theophylline, and 6-mercaptopurine: A Potts–Guy approach. *Journal of Pharmaceutical Sciences*, 1999. 88: 515–522.
145. Gorukanti, S.R., Li, L., and Kim, K.H., Transdermal delivery of antiparkinsonian agent, benztropine. I. Effect of vehicles on skin permeation. *International Journal of Pharmaceutics*, 1999. 192: 159–172.
146. Bala, I., Bhardwaj, V., Hariharan, S., and Kumar, M.N.V.R., Analytical methods for assay of ellagic acid and its solubility studies. *Journal of Pharmaceutical and Biomedical Analysis*, 2006. 40: 206–210.
147. Wilkes, J.B., Solubility of benzamide in *m*-xylene. *Journal of Chemical and Engineering Data*, 1963. 8: 234.
148. Li, D.Q., Liu, D.Z., and Wang, F.A., Solubilities of terephthalaldehydic, *p*-toluic, benzoic, terephthalic, and isophthalic acids in *N*-methyl-2-pyrrolidone from 295.65 K to 371.35 K. *Journal of Chemical and Engineering Data*, 2001. 46: 172–173.
149. Marsac, P.J., Li, T., and Taylor, L.S., Estimation of drug-polymer miscibility and solubility in amorphous solid dispersions using experimentally determined interaction parameters. *Pharmaceutical Research*, 2009. 26: 139–151.
150. Sunwoo, C. and Eisen, H., Solubility parameter of selected sulfonamides. *Journal of Pharmaceutical Sciences*, 1971. 60: 238–244.

3 Solubility Data in Binary Solvent Mixtures

This chapter reports the solubility data of pharmaceutically interested solutes in aqueous and nonaqueous binary solvent mixtures collected from the literature. The solvent compositions and the solubilities in mixed solvents are reported using different units which included in the tables concerning the originally reported data. The model constants of the Jouyban–Acree model for predicting the solubility at all solvent compositions are reported when the solubility data of the drug in monosolvents and enough number of data points in mixed solvents were available to calculate the statistically significant model constants. The model for binary solvent mixtures at various temperatures could be written as the following equation.

$$\log X_{m,T} = f_1 \log X_{1,T} + f_2 \log X_{2,T} + \frac{J_0 f_1 f_2}{T} + \frac{J_1 f_1 f_2 (f_1 - f_2)}{T} + \frac{J_2 f_1 f_2 (f_1 - f_2)^2}{T}$$

where
    X is the solute solubility
    f denotes the fraction of the solvent in the mixtures in the absence of the solute
    T is the absolute temperature
    J terms are the model constants
    Subscripts m, 1, and 2 are the solvent mixture and neat solvents 1–2

Solute	Solvent 1	Solvent 2	T (°C)
Acetaminophen	Dioxane	Water	20
Solubility (mole F.)	Vol. F. 1		
0.00170	0.000		
0.00410	0.100		
0.01090	0.200		
0.02250	0.300		
0.03750	0.400		
0.04650	0.500		
0.07360	0.600		
0.09690	0.700		
0.11350	0.800		
0.11380	0.850		
0.08840	0.900		
0.02470	1.000		

Constants of Jouyban–Acree model

J_0	1151.760
J_1	480.652
J_2	773.079

Reference of data: [1].

Solute	Solvent 1	Solvent 2	T (°C)
Acetaminophen	Dioxane	Water	25
Solubility (mole F.)	Vol. F. 1		
0.00190	0.000		
0.00500	0.100		
0.01290	0.200		
0.02440	0.300		
0.04240	0.400		
0.06400	0.500		
0.08030	0.600		
0.10780	0.700		
0.12070	0.800		
0.13530	0.850		
0.09910	0.900		
0.02670	1.000		

Constants of Jouyban–Acree model

J_0	1151.760
J_1	480.652
J_2	773.079

Reference of data: [1].

Solute	Solvent 1	Solvent 2	T (°C)
Acetaminophen	Dioxane	Water	30
Solubility (mole F.)	Vol. F. 1		
0.00230	0.000		
0.00610	0.100		
0.01420	0.200		
0.02920	0.300		
0.05540	0.400		
0.07580	0.500		
0.09850	0.600		
0.11830	0.700		
0.13070	0.800		
0.14410	0.850		
0.10650	0.900		
0.02940	1.000		

Constants of Jouyban–Acree model

J_0	1151.760
J_1	480.652
J_2	773.079

Reference of data: [1].

Solute	Solvent 1	Solvent 2	T (°C)
Acetaminophen	Dioxane	Water	35
Solubility (mole F.)	Vol. F. 1		
0.00260	0.000		
0.00660	0.100		
0.01740	0.200		
0.03660	0.300		
0.06400	0.400		
0.09320	0.500		
0.11770	0.600		
0.13400	0.700		
0.14020	0.800		
0.15230	0.850		
0.11740	0.900		
0.03170	1.000		

Constants of Jouyban–Acree model

J_0	1151.760
J_1	480.652
J_2	773.079

Reference of data: [1].

Solute	Solvent 1	Solvent 2	T (°C)
Acetaminophen	Dioxane	Water	40
Solubility (mole F.)	Vol. F. 1		
0.00300	0.000		
0.00740	0.100		
0.02060	0.200		
0.04130	0.300		
0.07420	0.400		
0.10380	0.500		
0.13790	0.600		
0.15090	0.700		
0.15470	0.800		
0.16590	0.850		
0.13320	0.900		
0.03560	1.000		

Constants of Jouyban–Acree model

J_0	1151.760
J_1	480.652
J_2	773.079

Reference of data: [1].

Solute	Solvent 1	Solvent 2	T (°C)
Acetaminophen	Ethanol	Propylene glycol	20
Solubility (mole F.)	Mass F. 1		
0.05045	1.000		
0.05200	0.900		
0.05450	0.800		
0.05640	0.700		
0.05690	0.600		
0.05750	0.500		
0.05680	0.400		
0.05590	0.300		
0.05440	0.200		
0.05220	0.100		
0.04830	0.000		

Constants of Jouyban–Acree model

J_0	66.497
J_1	0
J_2	0

Reference of data: [2].

Solute	Solvent 1	Solvent 2	T (°C)
Acetaminophen	Ethanol	Propylene glycol	25
Solubility (mole F.)	Mass F. 1		
0.05460	1.000		
0.05910	0.900		
0.06040	0.800		
0.06180	0.700		
0.06260	0.600		
0.06260	0.500		
0.06240	0.400		
0.06200	0.300		
0.05900	0.200		
0.05680	0.100		
0.05160	0.000		

Constants of Jouyban–Acree model

J_0	66.497
J_1	0
J_2	0

Reference of data: [2].

Solute	Solvent 1	Solvent 2	T (°C)
Acetaminophen	Ethanol	Propylene glycol	30
Solubility (mole F.)	Mass F. 1		
0.06200	1.000		
0.06360	0.900		
0.06590	0.800		
0.06710	0.700		
0.06800	0.600		
0.06760	0.500		
0.06690	0.400		
0.06730	0.300		
0.06640	0.200		
0.06460	0.100		
0.06030	0.000		

Constants of Jouyban–Acree model

J_0	66.497
J_1	0
J_2	0

Reference of data: [2].

Solute	Solvent 1	Solvent 2	T (°C)
Acetaminophen	Ethanol	Propylene glycol	35
Solubility (mole F.)	Mass F. 1		
0.06700	1.000		
0.06800	0.900		
0.06960	0.800		
0.07230	0.700		
0.07300	0.600		
0.07370	0.500		
0.07380	0.400		
0.07380	0.300		
0.07250	0.200		
0.07170	0.100		
0.06640	0.000		

Constants of Jouyban–Acree model

J_0	66.497
J_1	0
J_2	0

Reference of data: [2].

Solute	Solvent 1	Solvent 2	T (°C)
Acetaminophen	Ethanol	Propylene glycol	40
Solubility (mole F.)	Mass F. 1		
0.07050	1.000		
0.07170	0.900		
0.07310	0.800		
0.07690	0.700		
0.07810	0.600		
0.07900	0.500		
0.07990	0.400		
0.08060	0.300		
0.08120	0.200		
0.07910	0.100		
0.07520	0.000		

Constants of Jouyban–Acree model

J_0	66.497
J_1	0
J_2	0

Reference of data: [2].

Solute	Solvent 1	Solvent 2	T (°C)
Acetaminophen	Ethanol	Propylene glycol	25
Solubility (mole F.)	Vol. F. 1		
0.05451	1.000		
0.05913	0.900		
0.06044	0.800		
0.06182	0.700		
0.06258	0.600		
0.06259	0.500		
0.06241	0.400		
0.06202	0.300		
0.05905	0.200		
0.05681	0.100		
0.05163	0.000		

Constants of Jouyban–Acree model

J_0	85.859
J_1	−9.873
J_2	48.628

Reference of data: [3].

Solute	Solvent 1	Solvent 2	T (°C)
Acetaminophen	Ethanol	Water	30
Solubility (g/L)	Vol. F. 1		
21.0	0.000		
31.0	0.100		
49.0	0.200		
77.0	0.300		
119.0	0.400		
160.0	0.500		
196.0	0.600		
224.0	0.700		
243.0	0.800		
231.0	0.900		
202.0	1.000		

Constants of Jouyban–Acree model

J_0	470.525
J_1	147.945
J_2	−140.178

Reference of data: [4].

Solute	Solvent 1	Solvent 2	T (°C)
Acetaminophen	Ethanol	Water	20
Solubility (mole F.)	Vol. F. 1		
0.00174151	0.000		
0.00240309	0.100		
0.00495182	0.200		
0.03916389	0.600		
0.07236734	0.850		
0.06253680	0.950		
0.05213076	1.000		

Constants of Jouyban–Acree model

J_0	701.144
J_1	307.343
J_2	0

Reference of data: [5].

Solute	Solvent 1	Solvent 2	T (°C)
Acetaminophen	Ethanol	Water	25
Solubility (mole F.)	Vol. F. 1		
0.00191507	0.000		
0.00273944	0.100		
0.00637736	0.200		
0.04958831	0.600		
0.07989834	0.850		
0.06987830	0.950		
0.05420402	1.000		

Constants of Jouyban–Acree model

J_0	701.144
J_1	307.343
J_2	0

Reference of data: [5].

Solute	Solvent 1	Solvent 2	T (°C)
Acetaminophen	Ethanol	Water	30
Solubility (mole F.)	Vol. F. 1		
0.00230425	0.000		
0.00321798	0.100		
0.00737986	0.200		
0.05854262	0.600		
0.08716084	0.850		
0.08029884	0.950		
0.06166738	1.000		

Constants of Jouyban–Acree model

J_0	701.144
J_1	307.343
J_2	0

Reference of data: [5].

Solute	Solvent 1	Solvent 2	T (°C)
Acetaminophen	Ethanol	Water	35
Solubility (mole F.)	Vol. F. 1		
0.00262414	0.000		
0.00395411	0.100		
0.00978406	0.200		
0.06932150	0.600		
0.09827357	0.850		
0.09451468	0.950		
0.06587475	1.000		

Constants of Jouyban–Acree model

J_0	701.144
J_1	307.343
J_2	0

Reference of data: [5].

Solute	Solvent 1	Solvent 2	T (°C)
Acetaminophen	Ethanol	Water	40
Solubility (mole F.)	Vol. F. 1		
0.00283135	0.000		
0.00492220	0.100		
0.01073764	0.200		
0.07886639	0.600		
0.10603351	0.850		
0.09837189	0.950		
0.07051005	1.000		

Constants of Jouyban–Acree model

J_0	701.144
J_1	307.343
J_2	0

Reference of data: [5].

Solute	Solvent 1	Solvent 2	T (°C)
Acetaminophen	Polyethylene glycol 200	Water	30
Solubility (g/L)	Vol. F. 1		
22.0	0.000		
37.0	0.100		
57.0	0.200		
85.0	0.300		
120.0	0.400		
159.0	0.500		
227.0	0.600		
237.0	0.700		
263.0	0.800		
235.0	0.900		
197.0	1.000		

Constants of Jouyban–Acree model

J_0	485.087
J_1	136.843
J_2	48.332

Reference of data: [4].

Solute	Solvent 1	Solvent 2	T (°C)
Acetaminophen	Polyethylene glycol 400	Water	30
Solubility (g/L)	Vol. F. 1		
22.0	0.000		
42.0	0.100		
56.0	0.200		
85.0	0.300		
122.0	0.400		
166.0	0.500		
210.0	0.600		

Constants of Jouyban–Acree model

J_0
J_1
J_2

Reference of data: [4].

Solute	Solvent 1	Solvent 2	T (°C)
Acetaminophen	Propanol (2)	Water	5.0
Solubility (g/kg)	Mass F. 1		
8.090	0.000		
11.44	0.100		
23.58	0.200		
48.94	0.300		
82.55	0.400		
114.16	0.500		
145.57	0.600		
163.94	0.700		
168.5	0.800		
126.77	0.900		
79.04	1.000		

Constants of Jouyban–Acree model

J_0	801.180
J_1	142.955
J_2	0

Reference of data: [6].

Solute	Solvent 1	Solvent 2	T (°C)
Acetaminophen	Propanol (2)	Water	10.0
Solubility (g/kg)	Mass F. 1		
9.12	0.000		
14.20	0.100		
29.24	0.200		
58.82	0.300		
95.96	0.400		
131.67	0.500		
166.44	0.600		
181.67	0.700		
183.45	0.800		
139.47	0.900		
88.32	1.000		

Constants of Jouyban–Acree model

J_0	801.180
J_1	142.955
J_2	0

Reference of data: [6].

Solute	Solvent 1	Solvent 2	T (°C)
Acetaminophen	Propanol (2)	Water	15.0
Solubility (g/kg)	Mass F. 1		
10.71	0.000		
18.32	0.100		
36.95	0.200		
72.28	0.300		
114.03	0.400		
150.31	0.500		
187.28	0.600		
205.10	0.700		
203.52	0.800		
158.07	0.900		
97.80	1.000		

Constants of Jouyban–Acree model

J_0	801.180
J_1	142.955
J_2	0

Reference of data: [6].

Solute	Solvent 1	Solvent 2	T (°C)
Acetaminophen	Propanol (2)	Water	20.0
Solubility (g/kg)	Mass F. 1		
12.22	0.000		
23.25	0.100		
45.69	0.200		
87.01	0.300		
133.15	0.400		
172.94	0.500		
212.27	0.600		
226.68	0.700		
224.95	0.800		
178.55	0.900		
110.70	1.000		

Constants of Jouyban–Acree model

J_0	801.180
J_1	142.955
J_2	0

Reference of data: [6].

Solute	Solvent 1	Solvent 2	T (°C)
Acetaminophen	Propanol (2)	Water	25.0
Solubility (g/kg)	Mass F. 1		
14.98	0.000		
30.01	0.100		
56.53	0.200		
105.53	0.300		
154.74	0.400		
199.05	0.500		
239.80	0.600		
254.31	0.700		
247.50	0.800		
198.95	0.900		
122.34	1.000		

Constants of Jouyban–Acree model

J_0	801.180
J_1	142.955
J_2	0

Reference of data: [6].

Solute	Solvent 1	Solvent 2	T (°C)
Acetaminophen	Propanol (2)	Water	30.0
Solubility (g/kg)	Mass F. 1		
17.36	0.000		
37.66	0.100		
70.45	0.200		
125.34	0.300		
182.58	0.400		
230.02	0.500		
268.81	0.600		
282.61	0.700		
274.01	0.800		
223.11	0.900		
135.74	1.000		

Constants of Jouyban–Acree model

J_0	801.180
J_1	142.955
J_2	0

Reference of data: [6].

Solute	Solvent 1	Solvent 2	T (°C)
Acetaminophen	Propanol (2)	Water	35.0
Solubility (g/kg)	Mass F. 1		
20.80	0.000		
47.43	0.100		
86.50	0.200		
152.03	0.300		
213.08	0.400		
263.19	0.500		
303.33	0.600		
312.54	0.700		
300.63	0.800		
250.71	0.900		
151.57	1.000		

Constants of Jouyban–Acree model

J_0	801.180
J_1	142.955
J_2	0

Reference of data: [6].

Solute	Solvent 1	Solvent 2	T (°C)
Acetaminophen	Propanol (2)	Water	40.0
Solubility (g/kg)	Mass F. 1		
24.75	0.000		
61.14	0.100		
107.67	0.200		
183.25	0.300		
247.44	0.400		
302.26	0.500		
341.83	0.600		
348.46	0.700		
331.72	0.800		
280.03	0.900		
169.69	1.000		

Constants of Jouyban–Acree model

J_0	801.180
J_1	142.955
J_2	0

Reference of data: [6].

Solute	Solvent 1	Solvent 2	T (°C)
Acetaminophen	Propylene glycol	Water	20
Solubility (mole F.)	Mass F. 1		
0.00152	0.000		
0.00188	0.100		
0.00287	0.200		
0.00403	0.300		
0.00676	0.400		
0.01086	0.500		
0.01640	0.600		
0.02430	0.700		
0.03360	0.800		
0.04260	0.900		
0.04860	1.000		

Constants of Jouyban–Acree model

J_0	188.925
J_1	273.486
J_2	−221.284

Reference of data: [7].

Solute	Solvent 1	Solvent 2	T (°C)
Acetaminophen	Propylene glycol	Water	25
Solubility (mole F.)	Mass F. 1		
0.00185	0.000		
0.00237	0.100		
0.00347	0.200		
0.00522	0.300		
0.00803	0.400		
0.01294	0.500		
0.08600	0.600		
0.02800	0.700		
0.03740	0.800		
0.04640	0.900		
0.05160	1.000		

Constants of Jouyban–Acree model

J_0	188.925
J_1	273.486
J_2	−221.284

Reference of data: [7].

Solute	Solvent 1	Solvent 2	T (°C)
Acetaminophen	Propylene glycol	Water	30
Solubility (mole F.)	Mass F. 1		
0.00209	0.000		
0.00276	0.100		
0.00438	0.200		
0.00581	0.300		
0.00924	0.400		
0.01385	0.500		
0.02080	0.600		
0.03000	0.700		
0.04080	0.800		
0.05160	0.900		
0.06030	1.000		

Constants of Jouyban–Acree model

J_0	188.925
J_1	273.486
J_2	−221.284

Reference of data: [7].

Solute	Solvent 1	Solvent 2	T (°C)
Acetaminophen	Propylene glycol	Water	35
Solubility (mole F.)	Mass F. 1		
0.00256	0.000		
0.00319	0.100		
0.00506	0.200		
0.00677	0.300		
0.01112	0.400		
0.01660	0.500		
0.02380	0.600		
0.03440	0.700		
0.04560	0.800		
0.05670	0.900		
0.06640	1.000		

Constants of Jouyban–Acree model

J_0	188.925
J_1	273.486
J_2	−221.284

Reference of data: [7].

Solute	Solvent 1	Solvent 2	T (°C)
Acetaminophen	Propylene glycol	Water	40
Solubility (mole F.)	Mass F. 1		
0.00315	0.000		
0.00385	0.100		
0.00600	0.200		
0.00780	0.300		
0.01249	0.400		
0.01890	0.500		
0.02680	0.600		
0.03770	0.700		
0.05100	0.800		
0.06420	0.900		
0.07520	1.000		

Constants of Jouyban–Acree model

J_0	188.925
J_1	273.486
J_2	−221.284

Reference of data: [7].

Solute	Solvent 1	Solvent 2	T (°C)
Acetaminophen	Propylene glycol	Water	25
Solubility (mole F.)	Vol. F. 1		
0.00175160	0.000		
0.00284414	0.100		
0.00469738	0.200		
0.00741721	0.300		
0.01249243	0.400		
0.01933690	0.500		
0.02995951	0.600		
0.03678411	0.700		
0.04811030	0.800		
0.04799956	0.900		
0.04836985	1.000		

Constants of Jouyban–Acree model

J_0	383.680
J_1	211.658
J_2	0

Reference of data: [8].

Solute	Solvent 1	Solvent 2	T (°C)
Acetaminophen	Glucose	Water	37.0
Solubility (mole F.)	W/V% 1		
0.0108	10		
0.0153	20		
0.0168	30		
0.0176	40		
0.0180	50		
0.0159	60		

Constants of Jouyban–Acree model

J_0

J_1

J_2

Reference of data: [9].

Solute	Solvent 1	Solvent 2	T (°C)
Acetaminophen	Sorbitol	Water	37.0
Solubility (mole F.)	W/V% 1		
0.0164	10		
0.0155	20		
0.0170	30		
0.0164	40		
0.0193	50		
0.0188	60		
0.182	70		

Constants of Jouyban–Acree model

J_0

J_1

J_2

Reference of data: [9].

Solute	Solvent 1	Solvent 2	T (°C)
Acetaminophen	Sucrose	Water	37.0
Solubility (mole F.)	W/V% 1		
0.0051	20		
0.0069	30		
0.0171	40		
0.0171	50		
0.0205	60		

Constants of Jouyban–Acree model

J_0

J_1

J_2

Reference of data: [9].

Solute	Solvent 1	Solvent 2	T (°C)
Acetaminophen	Water	Glucose	20.0
Solubility (mole F.)		W/V% 2	
0.0198		10	
0.0126		20	
0.0114		30	
0.0108		40	
0.0112		50	
0.0099		60	

Constants of Jouyban–Acree model

J_0
J_1
J_2

Reference of data: [9].

Solute	Solvent 1	Solvent 2	T (°C)
Acetaminophen	Water	Sorbitol	20.0
Solubility (mole F.)		W/V% 2	
0.0146		10	
0.0137		20	
0.0152		30	
0.0132		40	
0.0139		50	
0.0137		60	
0.0135		70	

Constants of Jouyban–Acree model

J_0
J_1
J_2

Reference of data: [9].

Solute	Solvent 1	Solvent 2	T (°C)
Acetaminophen	Water	Sucrose	20.0
Solubility (mole F.)		W/V% 2	
0.0274		10	
0.0171		20	
0.0137		30	
0.0146		40	
0.0130		50	
0.0137		60	
0.0127		70	

Constants of Jouyban–Acree model

J_0
J_1
J_2

Reference of data: [9].

Solute	Solvent 1	Solvent 2	T (°C)
Acetanilide	Dioxane	Water	20
Solubility (mole F.)	Vol. F. 1		
0.00070	0.000		
0.00330	0.100		
0.01060	0.200		
0.00880	0.300		
0.02010	0.400		
0.05060	0.500		
0.07030	0.600		
0.13240	0.700		
0.13900	0.800		
0.15880	0.900		
0.09390	1.000		

Constants of Jouyban–Acree model

J_0	812.964
J_1	0
J_2	147.541

Reference of data: [1].

Solute	Solvent 1	Solvent 2	T (°C)
Acetanilide	Dioxane	Water	25
Solubility (mole F.)	Vol. F. 1		
0.00080	0.000		
0.00400	0.100		
0.01220	0.200		
0.01030	0.300		
0.02280	0.400		
0.05890	0.500		
0.07820	0.600		
0.14920	0.700		
0.16410	0.800		
0.18070	0.900		
0.11990	1.000		

Constants of Jouyban–Acree model

J_0	812.964
J_1	0
J_2	147.541

Reference of data: [1].

Solute	Solvent 1	Solvent 2	T (°C)
Acetanilide	Dioxane	Water	30
Solubility (mole F.)	Vol. F. 1		
0.00090	0.000		
0.00450	0.100		
0.01440	0.200		
0.01220	0.300		
0.02620	0.400		
0.07520	0.500		
0.08750	0.600		
0.17150	0.700		
0.19820	0.800		
0.19760	0.900		
0.13360	1.000		

Constants of Jouyban–Acree model

J_0	812.964
J_1	0
J_2	147.541

Reference of data: [1].

Solute	Solvent 1	Solvent 2	T (°C)
Acetanilide	Dioxane	Water	35
Solubility (mole F.)	Vol. F. 1		
0.00090	0.000		
0.00480	0.100		
0.01760	0.200		
0.01420	0.300		
0.03170	0.400		
0.09830	0.500		
0.10220	0.600		
0.18240	0.700		
0.24180	0.800		
0.21450	0.900		
0.16330	1.000		

Constants of Jouyban–Acree model

J_0	812.964
J_1	0
J_2	147.541

Reference of data: [1].

Solute	Solvent 1	Solvent 2	T (°C)
Acetanilide	Dioxane	Water	40
Solubility (mole F.)	Vol. F. 1		
0.00110	0.000		
0.00560	0.100		
0.02050	0.200		
0.01630	0.300		
0.03920	0.400		
0.00530[a]	0.500		
0.11650	0.600		
0.20380	0.700		
0.28670	0.800		
0.24040	0.900		
0.21290	1.000		

Constants of Jouyban–Acree model

J_0	812.964
J_1	0
J_2	147.541

Reference of data: [1].

[a] This data looks like a mistyped datum in the original references. Considering the solubility data at the same solvent composition at temperatures 20°C–35°C and solubility data at the same temperature and different solvent compositions, it should be 0.10530.

Solute	Solvent 1	Solvent 2	T (°C)
Acetanilide	Ethanol	Water	20
Solubility (mass F.)	Mass F. 1		
0.00520	0.000		
0.02100	0.263		
0.12050	0.525		
0.22480	0.748		
0.24870	0.842		
0.24410	0.904		
0.23840	0.952		
0.22360	1.000		

Constants of Jouyban–Acree model

J_0	545.946
J_1	365.352
J_2	−285.966

Reference of data: [10].

Solute	Solvent 1	Solvent 2	T (°C)
Acetanilide	Ethanol	Water	25
Solubility (mass F.)	Mass F. 1		
0.00540	0.000		
0.00930	0.100		
0.01280	0.200		
0.02300	0.300		
0.04850	0.400		
0.08870	0.500		
0.14170	0.600		
0.19840	0.700		
0.25170	0.800		
0.26930	0.850		
0.27650	0.900		
0.26820	0.950		
0.24770	1.000		

Constants of Jouyban–Acree model

J_0	545.946
J_1	365.352
J_2	−285.966

Reference of data: [10].

Solute	Solvent 1	Solvent 2	T (°C)
Acetanilide	Ethanol	Water	30
Solubility (mass F.)	Mass F. 1		
0.0069	0.000		
0.0100	0.100		
0.0220	0.200		
0.0480	0.300		
0.0940	0.400		
0.1540	0.500		
0.2200	0.600		
0.2760	0.700		
0.3120	0.800		
0.3170	0.850		
0.3160	0.900		
0.3080	0.950		
0.2900	1.000		

Constants of Jouyban–Acree model

J_0	545.946
J_1	365.352
J_2	−285.966

Reference of data: [10].

Solute	Solvent 1	Solvent 2	T (°C)
Acetanilide	Ethanol	Water	25
Solubility (mole F.)	Vol. F. 1		
0.00080	0.000		
0.00120	0.100		
0.00270	0.200		
0.00580	0.300		
0.01290	0.400		
0.02020	0.500		
0.04030	0.600		
0.05780	0.700		
0.06490	0.800		
0.07140	0.900		
0.08190	1.000		

Constants of Jouyban–Acree model

J_0	550.462
J_1	310.795
J_2	−454.579

Reference of data: [11].

Solute	Solvent 1	Solvent 2	T (°C)
Acetanilide	Ethyl acetate	Ethanol	25
Solubility (mole F.)	Vol. F. 1		
0.08190	0.000		
0.11380	0.400		
0.13450	0.500		
0.15760	0.700		
0.15020	0.800		
0.11250	1.000		

Constants of Jouyban–Acree model

J_0	175.891
J_1	298.347
J_2	−189.429

Reference of data: [11].

Solute	Solvent 1	Solvent 2	T (°C)
Acetanilide	Methanol	Water	20
Solubility (mass F.)	Mass F. 1		
0.00520	0.000		
0.01720	0.254		
0.06760	0.498		
0.18800	0.717		
0.27270	0.866		
0.29630	0.935		
0.30360	0.962		
0.31510	1.000		

Constants of Jouyban–Acree model

J_0	276.041
J_1	308.741
J_2	−106.830

Reference of data: [10].

Solute	Solvent 1	Solvent 2	T (°C)
Alanine (β)	Water	Ethanol	25
Solubility (mole F.)	Vol. F. 1		
0.15276	1.000		
0.13305	0.800		
0.09750	0.600		
0.05212	0.400		
0.00927	0.200		
0.00174	0.100		
0.00011	0.000		

Constants of Jouyban–Acree model

J_0	1507.274
J_1	1141.319
J_2	684.668

Reference of data: [12].

Solute	Solvent 1	Solvent 2	T (°C)
Alanine (DL)	Water	Ethanol	0
Solubility (mole F.)	Vol. F. 1		
0.00878	0.751		
0.00315	0.499		
0.00104	0.255		
0.00008	0.049		

Constants of Jouyban–Acree model

J_0	
J_1	
J_2	

Reference of data: [13].

Solute	Solvent 1	Solvent 2	T (°C)
Alanine (DL)	Water	Ethanol	25
Solubility (mole F.)	Vol. F. 1		
0.01610	0.751		
0.00681	0.499		
0.00195	0.258		
0.00015	0.049		

Constants of Jouyban–Acree model

J_0
J_1
J_2

Reference of data: [13].

Solute	Solvent 1	Solvent 2	T (°C)
Alanine (DL)	Water	Ethanol	45
Solubility (mole F.)	Vol. F. 1		
0.0239	0.747		
0.0114	0.499		
0.00321	0.258		
0.00252	0.049		

Constants of Jouyban–Acree model

J_0
J_1
J_2

Reference of data: [13].

Solute	Solvent 1	Solvent 2	T (°C)
Alanine (DL)	Water	Ethanol	65
Solubility (mole F.)	Vol. F. 1		
0.035300	0.751		
0.017900	0.499		
0.005020	0.258		
0.000393	0.049		

Constants of Jouyban–Acree model

J_0
J_1
J_2

Reference of data: [13].

Solute	Solvent 1	Solvent 2	T (°C)
Alanine (L)	Water	1-Propanol	25
Solubility (mole F.)	Mole F. 1		
0.03240	1.000		
0.02540	0.986		
0.02030	0.950		
0.01460	0.889		
0.00984	0.805		
0.00529	0.690		
0.00174	0.529		
0.00003	0.000		

Constants of Jouyban–Acree model

J_0	149.451
J_1	696.293
J_2	−1245.047

Reference of data: [14].

Solute	Solvent 1	Solvent 2	T (°C)
Alanine (L)	Water	2-Propanol	25
Solubility (mole F.)	Mole F. 1		
0.03240	1.000		
0.02870	0.986		
0.02130	0.951		
0.01190	0.886		
0.00717	0.803		
0.00350	0.690		
0.00110	0.527		
0.00002	0.000		

Constants of Jouyban–Acree model

J_0	124.463
J_1	131.734
J_2	−570.647

Reference of data: [14].

Solute	Solvent 1	Solvent 2	T (°C)
Alanine (α)	Water	Methanol	15
Solubility (mole/L)	Mass F. 1		
1.640	1.000		
1.500	0.920		
1.310	0.840		
0.870	0.748		
0.375	0.656		
0.345	0.553		
0.155	0.458		
0.143	0.359		
0.020	0.241		
0.010	0.121		
0.005	0.000		

Constants of Jouyban–Acree model

J_0	417.490
J_1	309.166
J_2	−383.077

Reference of data: [15].

Solute	Solvent 1	Solvent 2	T (°C)
Alanine (α)	Water	Methanol	20
Solubility (mole/L)	Mass F. 1		
1.7800	1.000		
1.6200	0.920		
1.4200	0.840		
0.8900	0.748		
0.4050	0.656		
0.3750	0.553		
0.1695	0.458		
0.1585	0.359		
0.0215	0.241		
0.0130	0.121		
0.0070	0.000		

Constants of Jouyban–Acree model

J_0	417.490
J_1	309.166
J_2	−383.077

Reference of data: [15].

Solute	Solvent 1	Solvent 2	T (°C)
Alanine (α)	Water	Methanol	25
Solubility (mole/L)	Mass F. 1		
1.8700	1.000		
1.7000	0.920		
1.4900	0.840		
0.9100	0.748		
0.4250	0.656		
0.4000	0.553		
0.1825	0.458		
0.1675	0.359		
0.0240	0.241		
0.0160	0.121		
0.0090	0.000		

Constants of Jouyban–Acree model

J_0	417.490
J_1	309.166
J_2	−383.077

Reference of data: [15].

Solute	Solvent 1	Solvent 2	T (°C)
Chlorobenzoic acid (ortho)	Dimethyl formamide	Water	30.0
Solubility (mole F.)	Mole F. 1		
0.000240	0.000		
0.000320	0.012		
0.000400	0.025		
0.000830	0.040		
0.001120	0.053		
0.001810	0.072		
0.005510	0.104		
0.006920	0.123		
0.015420	0.160		
0.024050	0.189		
0.036310	0.222		
0.044720	0.250		
0.062440	0.337		
0.079990	0.402		
0.093370	0.497		
0.103380	0.569		
0.117790	0.678		
0.135090	0.816		
0.160840	1.000		

Constants of Jouyban–Acree model

J_0	1457.816
J_1	−1527.777
J_2	990.990

Reference of data: [16].

Solute	Solvent 1	Solvent 2	T (°C)
Chlorobenzoic acid (para)	Dimethyl formamide	Water	30.0
Solubility (mole F.)	Mole F. 1		
0.000009	0.000		
0.000120	0.012		
0.000230	0.025		
0.000480	0.053		
0.002140	0.072		
0.003930	0.104		
0.013970	0.135		
0.028330	0.189		
0.057010	0.250		
0.081090	0.302		
0.128340	0.402		
0.162790	0.497		
0.220350	0.569		
0.254460	0.631		
0.286120	0.678		
0.350230	0.816		
0.470670	1.000		

Constants of Jouyban–Acree model

J_0	2169.957
J_1	−2773.528
J_2	4306.136

Reference of data: [16].

Solute	Solvent 1	Solvent 2	T (°C)
Cyclodextrin (α)	Water	2-Propanol	25
Solubility (mole F.)	Mole F. 1		
0.12540[a]	1.000		
0.1800	0.998		
0.2580	0.998		
0.2780	0.997		
0.2020	0.996		
0.2310	0.995		
0.2180	0.994		
0.2540	0.992		
0.2790	0.990		
0.2940	0.988		
0.2740	0.980		
0.2250	0.967		
0.1120	0.876		
0.0680	0.829		
0.03230	0.770		
0.00715	0.650		
0.00320	0.586		
0.00250	0.540		
0.00016	0.358		

Constants of Jouyban–Acree model

J_0

J_1

Reference of data: [17].

[a] This value is average of five determinations reported in the original paper.

Solute	Solvent 1	Solvent 2	T (°C)
Cyclodextrin (α)	Water	Acetone	25
Solubility (mole F.)	Mole F. 1		
0.12540[a]	1.000		
0.1560	0.998		
0.1530	0.997		
0.2230	0.993		
0.2310	0.991		
0.2260	0.988		
0.1850	0.982		
0.1740	0.977		

(*continued*)

(continued)

Solute	Solvent 1	Solvent 2	T (°C)
0.1490	0.961		
0.1040	0.934		
0.0730	0.895		
0.0380	0.860		
0.0240	0.797		
0.00810	0.751		
0.00640	0.697		
0.00150	0.607		

Constants of Jouyban–Acree model

J_0

J_1

J_2

Reference of data: [17].

[a] This value is average of five determinations reported in the original paper.

Solute	Solvent 1	Solvent 2	T (°C)
Cyclodextrin (α)	Water	Ethylene glycol	25
Solubility (mole F.)	Mole F. 1		
0.12540[a]	1.000		
0.1460	0.994		
0.1760	0.991		
0.1760	0.988		
0.1930	0.985		
0.2290	0.982		
0.1990	0.969		
0.1470	0.947		
0.1530	0.916		
0.1110	0.880		
0.1040	0.838		
0.1140	0.715		
0.1320	0.661		
0.1330	0.658		
0.0860	0.597		
0.0490	0.465		
0.0070	0.000		

Constants of Jouyban–Acree model

J_0	378.367
J_1	322.249
J_2	42.021

Reference of data: [17].

[a] This value is average of five determinations reported in the original paper.

Solute	Solvent 1	Solvent 2	T (°C)
Cyclodextrin (α)	Water	Methanol	25
Solubility (mole F.)	Mass F. 1		
0.12540[a]	1.000		
0.1340	0.998		
0.1310	0.996		
0.1280	0.995		
0.1460	0.994		
0.1520	0.993		
0.1270	0.989		
0.1120	0.986		
0.0870	0.983		
0.0950	0.977		
0.0730	0.973		
0.07140	0.965		
0.05630	0.942		
0.04730	0.940		
0.02580	0.840		
0.02200	0.767		
0.01270	0.641		
0.00840	0.551		
0.00240	0.316		
0.00049	0.000		

Constants of Jouyban–Acree model

J_0	−143.239
J_1	−539.813
J_2	0

Reference of data: [17].

[a] This value is average of five determinations reported in the original paper.

Solute	Solvent 1	Solvent 2	T (°C)
Aminoacetophenone (para)	Propylene glycol	Water	37
Solubility (g/L)	Vol. F. 1		
10.0	0.00		
15.6	0.10		
23.5	0.20		
35.6	0.30		
56.1	0.40		
85.6	0.50		
94.0	0.60		
105.0	0.70		
125.0	0.80		
147.0	0.90		
181.0	1.00		

Constants of Jouyban–Acree model

J_0	321.266
J_1	−59.886
J_2	−318.764

Reference of data: [18].

Solute	Solvent 1	Solvent 2	T (°C)
Aminocaproic acid (ε)	Water	Ethanol	25
Solubility (mole F.)	Vol. F. 1		
0.10593	1.000		
0.10046	0.800		
0.08913	0.600		
0.06531	0.400		
0.01910	0.200		
0.00326	0.100		
0.00011	0.000		

Constants of Jouyban–Acree model

J_0	1610.618
J_1	1631.964
J_2	1312.079

Reference of data: [12].

Solute	Solvent 1	Solvent 2	T (°C)
Aminopyrine	Ethanol	Water	25
Solubility (g/L)	Mass F. 1		
53.0	0.000		
92.0	0.100		
157.0	0.200		
253.0	0.300		
335.0	0.400		
389.0	0.500		
425.0	0.600		
435.0	0.700		
435.0	0.800		
409.0	0.900		
357.0	1.000		

Constants of Jouyban–Acree model

J_0	546.653
J_1	−108.832
J_2	−70.467

Reference of data: [19].

Solute	Solvent 1	Solvent 2	T (°C)
Amobarbital	Ethanol	Water	25
Solubility (g/L)	Mass F. 1		
219.6	1.000		
217.1	0.975		
213.4	0.950		
210.3	0.925		
205.6	0.900		
196.6	0.875		
191.9	0.850		
182.2	0.825		
172.1	0.800		
160.1	0.775		
145.8	0.750		
137.5	0.725		
123.6	0.700		
110.9	0.675		
104.2	0.650		
88.2	0.625		
81.0	0.600		
67.0	0.575		
62.3	0.550		
51.8	0.525		
40.0	0.500		
33.4	0.475		
25.5	0.450		
19.5	0.425		
16.4	0.400		
11.2	0.375		
9.6	0.350		
7.8	0.325		
7.4	0.300		
5.3	0.275		
2.6	0.250		
2.2	0.225		
1.7	0.200		
1.2	0.175		
1.1	0.150		
0.96	0.125		
0.84	0.100		
0.70	0.075		
0.68	0.050		
0.64	0.025		
0.56	0.000		

Constants of Jouyban–Acree model

J_0	665.731
J_1	619.060
J_2	−773.621

Reference of data: [20].

Solute	Solvent 1	Solvent 2	T (°C)
Amoxycillin trihydrate	Propylene glycol	Ethyl acetate	25
Solubility (mole F.)	Mass F. 1		
0.0004042	1.000		
0.001356	0.750		
0.001228	0.500		
0.0004267	0.250		
0.0000220	0.000		

Constants of Jouyban–Acree model

J_0
J_1
J_2

Reference of data: [21].

Solute	Solvent 1	Solvent 2	T (°C)
Amoxycillin trihydrate	Propylene glycol	Water	25
Solubility (mole F.)	Vol. F. 1		
0.0001263	0.000		
0.0001786	0.250		
0.0002640	0.500		
0.0003619	0.750		
0.0004042	1.000		

Constants of Jouyban–Acree model

J_0	81.073
J_1	86.043
J_2	0

Reference of data: [21].

Solute	Solvent 1	Solvent 2	T (°C)
Antipyrine	Water	Ethanol	25
Solubility (g/L)	Mass F. 1		
620.0	1.000		
650.0	0.900		
665.0	0.800		
670.0	0.700		
671.0	0.600		
660.0	0.500		
638.0	0.400		
620.0	0.300		
573.0	0.200		
510.0	0.100		
425.0	0.000		

Constants of Jouyban–Acree model

J_0	129.757
J_1	−52.105
J_2	56.063

Reference of data: [19].

Solute	Solvent 1	Solvent 2	T (°C)
Artemisinin	Ethanol	Water	5.05
Solubility (mole F.)	Mole F. 1		
0.0000408	0.138		
0.0000987	0.273		
0.0004170	0.467		
0.0007000	0.771		
0.0009980	1.000		

Constants of Jouyban–Acree model

J_0

J_1

J_2

Reference of data: [22].

Solute	Solvent 1	Solvent 2	T (°C)
Artemisinin	Ethanol	Water	13.05
Solubility (mole F.)	Mole F. 1		
0.0000719	0.138		
0.0001728	0.273		
0.0007020	0.467		
0.0014590	0.771		
0.0018110	1.000		

Constants of Jouyban–Acree model

J_0

J_1

J_2

Reference of data: [22].

Solute	Solvent 1	Solvent 2	T (°C)
Artemisinin	Ethanol	Water	30.05
Solubility (mole F.)	Mole F. 1		
0.0001483	0.138		
0.0003550	0.273		
0.0014220	0.467		
0.0021400	0.771		
0.0029730	1.000		

Constants of Jouyban–Acree model

J_0

J_1

J_2

Reference of data: [22].

Solute	Solvent 1	Solvent 2	T (°C)
Artemisinin	Ethanol	Water	40.05
Solubility (mole F.)	Mole F. 1		
0.0002320	0.138		
0.0005330	0.273		
0.0024700	0.467		
0.0035880	0.771		
0.0045670	1.000		

Constants of Jouyban–Acree model

J_0

J_1

J_2

Reference of data: [22].

Solute	Solvent 1	Solvent 2	T (°C)
Artemisinin	Ethanol	Water	50.05
Solubility (mole F.)	Mole F. 1		
0.000316	0.138		
0.000884	0.273		
0.004209	0.467		
0.006063	0.771		
0.008006	1.000		

Constants of Jouyban–Acree model

J_0

J_1

J_2

Reference of data: [22].

Solute	Solvent 1	Solvent 2	T (°C)
Artemisinin	Ethanol	Water	70.05
Solubility (mole F.)	Mole F. 1		
0.000538	0.138		
0.001644	0.273		
0.008210	0.467		
0.012780	0.771		
0.015880	1.000		

Constants of Jouyban–Acree model

J_0

J_1

J_2

Reference of data: [22].

Solute	Solvent 1	Solvent 2	T (°C)
Asparagine	Water	Dioxane	25
Solubility (g/100 g)	Mass F. 1		
2.510	1.000		
1.190	0.800		
0.410	0.600		
0.095	0.400		

Constants of Jouyban–Acree model

J_0

J_1

J_2

Reference of data: [23].

Solute	Solvent 1	Solvent 2	T (°C)
Asparagine	Water	Ethanol	25
Solubility (g/100 g)	Mass F. 1		
2.510	1.000		
1.070	0.800		
0.428	0.600		
0.151	0.400		

Constants of Jouyban–Acree model

J_0

J_1

J_2

Reference of data: [23].

Solute	Solvent 1	Solvent 2	T (°C)
Asparagine (L)	Water	Ethanol	25
Solubility (mole F.)	Vol. F. 1		
0.00340408200	1.000		
0.00154881700	0.800		
0.00073536000	0.600		
0.00030619600	0.400		
0.00000134896	0.000		

Constants of Jouyban–Acree model

J_0	1032.624
J_1	931.407
J_2	435.578

Reference of data: [12].

Solute	Solvent 1	Solvent 2	T (°C)
Asparagine (L)	Water	1-Propanol	25
Solubility (mole F.)	Mole F. 1		
0.0034300	1.000		
0.0018500	0.950		
0.0011800	0.886		
0.0008430	0.833		
0.0005610	0.769		
0.0003560	0.690		
0.0001760	0.588		
0.0000699	0.455		
0.0000140	0.270		

Constants of Jouyban–Acree model

J_0

J_1

J_2

Reference of data: [14].

Solute	Solvent 1	Solvent 2	T (°C)
Asparagine (L)	Water	2-Propanol	25
Solubility (mole F.)	Mole F. 1		
0.0034300	1.000		
0.0016500	0.950		
0.0009060	0.886		
0.0006300	0.833		
0.0004190	0.769		
0.0002460	0.690		
0.0001360	0.588		
0.0000084	0.270		

Constants of Jouyban–Acree model

J_0

J_1

J_2

Reference of data: [14].

Solute	Solvent 1	Solvent 2	T (°C)
Asparagine (L)	Water	Methanol	15
Solubility (mole/L)	Mass F. 1		
0.21800	1.000		
0.17200	0.920		
0.07650	0.840		
0.03420	0.748		
0.01300	0.656		
0.00614	0.553		
0.00295	0.458		
0.00170	0.359		
0.00118	0.241		
0.00046	0.121		
0.00027	0.000		

Constants of Jouyban–Acree model

J_0	−374.180
J_1	140.973
J_2	644.010

Reference of data: [15].

Solute	Solvent 1	Solvent 2	T (°C)
Asparagine (L)	Water	Methanol	20
Solubility (mole/L)	Mass F. 1		
0.22100	1.000		
0.17500	0.920		
0.07820	0.840		
0.03550	0.748		
0.01350	0.656		
0.00617	0.553		
0.00315	0.458		
0.00191	0.359		
0.00132	0.241		
0.00054	0.121		
0.00035	0.000		

Constants of Jouyban–Acree model

J_0	−374.180
J_1	140.973
J_2	644.010

Reference of data: [15].

Solute	Solvent 1	Solvent 2	T (°C)
Asparagine (L)	Water	Methanol	25
Solubility (mole/L)	Mass F. 1		
0.22600	1.000		
0.17700	0.920		
0.08000	0.840		
0.03630	0.748		
0.01410	0.656		
0.00620	0.553		
0.00335	0.458		
0.00215	0.359		
0.00155	0.241		
0.00065	0.121		
0.00045	0.000		

Constants of Jouyban–Acree model

J_0	−374.180
J_1	140.973
J_2	644.010

Reference of data: [15].

Solute	Solvent 1	Solvent 2	T (°C)
Aspartic acid (L)	Water	Ethanol	25
Solubility (mole F.)	Mass F. 1		
0.0006792040	1.000		
0.0003069020	0.800		
0.0001610650	0.600		
0.0000769130	0.400		
0.0000263027	0.200		
0.0000095719	0.100		
0.0000006808	0.000		

Constants of Jouyban–Acree model

J_0	821.391
J_1	1167.255
J_2	1179.541

Reference of data: [12].

Solute	Solvent 1	Solvent 2	T (°C)
Barbital	Ethanol	Water	25
Solubility (g/L)	Mass F.1		
92.3	1.000		
98.6	0.975		
103.1	0.950		

(continued)

Solute	Solvent 1	Solvent 2	T (°C)
110.0	0.925		
113.3	0.900		
118.3	0.875		
120.7	0.850		
117.2	0.825		
112.5	0.800		
107.7	0.775		
100.1	0.750		
94.3	0.725		
90.2	0.700		
85.1	0.675		
80.8	0.650		
75.6	0.625		
70.0	0.600		
66.3	0.575		
60.2	0.550		
56.5	0.525		
51.6	0.500		
47.7	0.475		
43.1	0.450		
39.2	0.425		
34.1	0.400		
30.6	0.375		
28.3	0.350		
24.1	0.325		
20.9	0.300		
17.1	0.275		
15.6	0.250		
14.2	0.225		
13.3	0.200		
12.5	0.175		
11.1	0.150		
10.1	0.125		
9.0	0.100		
8.0	0.075		
7.5	0.050		
7.4	0.025		
7.3	0.000		

Constants of Jouyban–Acree model

J_0	347.680
J_1	396.223
J_2	−94.682

Reference of data: [20].

Solute	Solvent 1	Solvent 2	T (°C)
Benzocaine	Dioxane	Water	5
Solubility (mole/L)	Vol. F. 1		
0.002	0.000		
0.005	0.100		
0.02	0.200		
0.05	0.300		
0.17	0.400		
0.38	0.500		
0.50	0.600		
1.02	0.700		
1.53	0.800		
2.10	0.900		
0.62	1.000		

Constants of Jouyban–Acree model

J_0	946.982
J_1	441.345
J_2	0

Reference of data: [24].

Solute	Solvent 1	Solvent 2	T (°C)
Benzocaine	Dioxane	Water	10
Solubility (mole/L)	Vol. F. 1		
0.0030	0.000		
0.0057	0.100		
0.031	0.200		
0.07	0.300		
0.20	0.400		
0.44	0.500		
0.56	0.600		
1.31	0.700		
1.79	0.800		
2.47	0.900		
1.15	1.000		

Constants of Jouyban–Acree model

J_0	946.982
J_1	441.345
J_2	0

Reference of data: [24].

Solute	Solvent 1	Solvent 2	T (°C)
Benzocaine	Dioxane	Water	15
Solubility (mole/L)	Vol. F. 1		
0.004	0.000		
0.009	0.100		
0.032	0.200		
0.077	0.300		
0.25	0.400		
0.50	0.500		
0.63	0.600		
1.54	0.700		
2.05	0.800		
2.83	0.900		
1.58	1.000		

Constants of Jouyban–Acree model

J_0	946.982
J_1	441.345
J_2	0

Reference of data: [24].

Solute	Solvent 1	Solvent 2	T (°C)
Benzocaine	Dioxane	Water	20
Solubility (mole/L)	Vol. F. 1		
0.0041	0.000		
0.013	0.100		
0.038	0.200		
0.09	0.300		
0.28	0.400		
0.56	0.500		
0.69	0.600		
1.95	0.700		
2.40	0.800		
3.16	0.900		
1.89	1.000		

Constants of Jouyban–Acree model

J_0	946.982
J_1	441.345
J_2	0

Reference of data: [24].

Solute	Solvent 1	Solvent 2	T (°C)
Benzocaine	Dioxane	Water	25
Solubility (mole/L)	Vol. F. 1		
0.0056	0.000		
0.015	0.100		
0.043	0.200		
0.13	0.300		
0.33	0.400		
2.64	0.700		
2.95	0.800		
3.91	0.900		
2.91	1.000		

Constants of Jouyban–Acree model

J_0	946.982
J_1	441.345
J_2	0

Reference of data: [24].

Solute	Solvent 1	Solvent 2	T (°C)
Benzocaine	Dioxane	Water	30
Solubility (mole/L)	Vol. F. 1		
0.007	0.000		
0.019	0.100		
0.051	0.200		
0.15	0.300		
0.35	0.400		
3.3	0.800		
4.7	0.900		
3.33	1.000		

Constants of Jouyban–Acree model

J_0	946.982
J_1	441.345
J_2	0

Reference of data: [24].

Solute	Solvent 1	Solvent 2	T (°C)
Benzocaine	Dioxane	Water	35
Solubility (mole/L)	Vol. F. 1		
0.0097	0.000		
0.02	0.100		
0.06	0.200		
0.20	0.300		
0.39	0.400		
3.52	0.800		
5.22	0.900		
3.82	1.000		

Constants of Jouyban–Acree model

J_0	946.982
J_1	441.345
J_2	0

Reference of data: [24].

Solute	Solvent 1	Solvent 2	T (°C)
Benzocaine	Dioxane	Water	40
Solubility (mole/L)	Vol. F. 1		
0.01	0.000		
0.03	0.100		
0.07	0.200		
0.23	0.300		
3.89	0.800		
5.93	0.900		
4.55	1.000		

Constants of Jouyban–Acree model

J_0	946.982
J_1	441.345
J_2	0

Reference of data: [24].

Solute	Solvent 1	Solvent 2	T (°C)
Benzocaine	Ethanol	Water	25
Solubility (mole F.)	Vol. F. 1		
0.0006	0.000		
0.0009	0.100		
0.0029	0.200		
0.0043	0.300		
0.0047	0.400		
0.0095	0.500		
0.0209	0.600		
0.0655	0.700		
0.0847	0.800		
0.1230	0.900		
0.1530	1.000		

Constants of Jouyban–Acree model

J_0	164.668
J_1	274.293
J_2	0

Reference of data: [11].

Solute	Solvent 1	Solvent 2	T (°C)
Benzocaine	Ethyl acetate	Ethanol	25
Solubility (mole F.)	Vol. F. 1		
0.1530	0.000		
0.2497	0.300		
0.3123	0.500		
0.2332	0.800		
0.2021	0.900		
0.1650	1.000		

Constants of Jouyban–Acree model

J_0	345.406
J_1	58.625
J_2	−195.572

Reference of data: [11].

Solute	Solvent 1	Solvent 2	T (°C)
Benzoic acid	Acetic acid	Water	25.2
Solubility (g/100 g)	Mass F. 1		
22.02	0.800		
23.15	0.850		
25.09	0.950		
25.06	1.000		

Constants of Jouyban–Acree model

J_0
J_1
J_2

Reference of data: [25].

Solute	Solvent 1	Solvent 2	T (°C)
Benzoic acid	Acetic acid	Water	35.3
Solubility (g/100 g)	Mass F. 1		
25.29	0.800		
27.88	0.850		
30.35	0.950		
30.68	1.000		

Constants of Jouyban–Acree model

J_0
J_1
J_2

Reference of data: [25].

Solute	Solvent 1	Solvent 2	T (°C)
Benzoic acid	Acetic acid	Water	45.2
Solubility (g/100 g)	Mass F. 1		
29.94	0.800		
31.73	0.850		
36.22	0.950		
38.05	1.000		

Constants of Jouyban–Acree model

J_0

J_1

J_2

Reference of data: [25].

Solute	Solvent 1	Solvent 2	T (°C)
Benzoic acid	Acetic acid	Water	55.1
Solubility (g/100 g)	Mass F. 1		
33.54	0.800		
36.16	0.850		
41.54	0.950		
44.68	1.000		

Constants of Jouyban–Acree model

J_0

J_1

J_2

Reference of data: [25].

Solute	Solvent 1	Solvent 2	T (°C)
Benzoic acid	Acetic acid	Water	64.9
Solubility (g/100 g)	Mass F. 1		
36.55	0.800		
40.75	0.850		
49.06	0.950		
53.15	1.000		

Constants of Jouyban–Acree model

J_0

J_1

J_2

Reference of data: [25].

Solute	Solvent 1	Solvent 2	T (°C)
Benzoic acid	Acetic acid	Water	75.2
Solubility (g/100 g)	Mass F. 1		
40.43	0.800		
45.03	0.850		
53.64	0.950		
59.88	1.000		

Constants of Jouyban–Acree model

J_0

J_1

J_2

Reference of data: [25].

Solute	Solvent 1	Solvent 2	T (°C)
Benzoic acid	Acetic acid	Water	85.5
Solubility (g/100 g)	Mass F. 1		
45.17	0.800		
50.48	0.850		
62.16	0.950		
69.76	1.000		

Constants of Jouyban–Acree model

J_0

J_1

J_2

Reference of data: [25].

Solute	Solvent 1	Solvent 2	T (°C)
Benzoic acid	Ethanol	Water	15
Solubility (mole/L)	Mass F. 1		
0.0201	0.000		
0.0214	0.080		
0.0247	0.164		
0.0531	0.253		
0.0982	0.344		
0.1895	0.440		
0.8004	0.541		
1.3826	0.647		
1.7990	0.760		
2.0391	0.876		
2.2439	1.000		

Constants of Jouyban–Acree model

J_0	524.038
J_1	633.729
J_2	−727.396

Reference of data: [26].

Solute	Solvent 1	Solvent 2	T (°C)
Benzoic acid	Ethanol	Water	20
Solubility (mole/L)	Mass F. 1		
0.0240	0.000		
0.0291	0.080		
0.0367	0.164		
0.0856	0.253		
0.1920	0.344		
0.4052	0.440		
1.0256	0.541		
1.5467	0.647		
1.9921	0.760		
2.3403	0.876		
2.5042	1.000		

Constants of Jouyban–Acree model

J_0	524.038
J_1	633.729
J_2	−727.396

Reference of data: [26].

Solute	Solvent 1	Solvent 2	T (°C)
Benzoic acid	Ethanol	Water	25
Solubility (mole/L)	Mass F. 1		
0.0284	0.000		
0.0389	0.080		
0.0577	0.164		
0.1201	0.253		
0.3593	0.344		
0.7785	0.440		
1.3061	0.541		
1.7316	0.647		
2.2034	0.760		
2.6656	0.876		
2.7481	1.000		

Constants of Jouyban–Acree model

J_0	524.038
J_1	633.729
J_2	−727.396

Reference of data: [26].

Solute	Solvent 1	Solvent 2	T (°C)
Benzoic acid	Ethylene glycol	Water	25
Solubility (mole/L)	Vol. F. 1		
0.0279	0.00		
0.0406	0.20		
0.0897	0.40		
0.2070	0.60		
0.5190	0.80		
1.5010	1.00		

Constants of Jouyban–Acree model

J_0	−212.087
J_1	95.267
J_2	−182.879

Reference of data: [27].

Solute	Solvent 1	Solvent 2	T (°C)
Benzoyl peroxide	Polyethylene glycol 400	Water	25.0
Solubility (g/kg)	Mass F. 1		
39.8	1.000		
14.2	0.900		
2.49	0.800		
1.5	0.700		
0.338	0.600		
0.128	0.500		
0.0377	0.400		
0.0124	0.300		
0.00327	0.200		
0.000578	0.100		
0.000155	0.000		

Constants of Jouyban–Acree model

J_0	232.342
J_1	−272.680
J_2	0

Reference of data: [28].

Solute	Solvent 1	Solvent 2	T (°C)
Berberine chloride	Ethanol	1-Butanol	24.85
Solubility (mole F.)	Mole F. 1		
0.000054	0.000		
0.000317	0.482		
0.000340	0.689		
0.000352	0.858		
0.000444	1.000		

Constants of Jouyban–Acree model

J_0	202.172
J_1	0
J_2	0

Reference of data: [29].

Solute	Solvent 1	Solvent 2	T (°C)
Berberine chloride	Ethanol	1-Butanol	29.85
Solubility (mole F.)	Mole F. 1		
0.000145	0.000		
0.000380	0.482		
0.000459	0.689		
0.000551	0.858		
0.000528	1.000		

Constants of Jouyban–Acree model

J_0	202.172
J_1	0
J_2	0

Reference of data: [29].

Solute	Solvent 1	Solvent 2	T (°C)
Berberine chloride	Ethanol	1-Butanol	34.85
Solubility (mole F.)	Mole F. 1		
0.000179	0.000		
0.000450	0.482		
0.000569	0.689		
0.000653	0.858		
0.000637	1.000		

Constants of Jouyban–Acree model

J_0	202.172
J_1	0
J_2	0

Reference of data: [29].

Solute	Solvent 1	Solvent 2	T (°C)
Berberine chloride	Ethanol	1-Butanol	39.85
Solubility (mole F.)	Mole F. 1		
0.000230	0.000		
0.000567	0.482		
0.000648	0.689		
0.000742	0.858		
0.000778	1.000		

Constants of Jouyban–Acree model

J_0	202.172
J_1	0
J_2	0

Reference of data: [29].

Solute	Solvent 1	Solvent 2	T (°C)
Berberine chloride	Ethanol	1-Octanol	24.85
Solubility (mole F.)	Mole F. 1		
0.000297	0.000		
0.000254	0.613		
0.000287	0.855		
0.000326	0.924		
0.000444	1.000		

Constants of Jouyban–Acree model

J_0	−194.201
J_1	−423.004
J_2	0

Reference of data: [29].

Solute	Solvent 1	Solvent 2	T (°C)
Berberine chloride	Ethanol	1-Octanol	29.85
Solubility (mole F.)	Mole F. 1		
0.000313	0.000		
0.000272	0.613		
0.000316	0.855		
0.000367	0.924		
0.000528	1.000		

Constants of Jouyban–Acree model

J_0	−194.201
J_1	−423.004
J_2	0

Reference of data: [29].

Solute	Solvent 1	Solvent 2	T (°C)
Berberine chloride	Ethanol	1-Octanol	34.85
Solubility (mole F.)	Mole F. 1		
0.000357	0.000		
0.000289	0.613		
0.000355	0.855		
0.000418	0.924		
0.000637	1.000		

Constants of Jouyban–Acree model

J_0	−194.201
J_1	−423.004
J_2	0

Reference of data: [29].

Solute	Solvent 1	Solvent 2	T (°C)
Berberine chloride	Ethanol	1-Octanol	39.85
Solubility (mole F.)	Mole F. 1		
0.000435	0.000		
0.000323	0.613		
0.000432	0.855		
0.000586	0.924		
0.000778	1.000		

Constants of Jouyban–Acree model

J_0	−194.201
J_1	−423.004
J_2	0

Reference of data: [29].

Solute	Solvent 1	Solvent 2	T (°C)
Berberine chloride	Ethanol	2-Propanol	24.85
Solubility (mole F.)	Mole F. 1		
0.000171	0.000		
0.000206	0.346		
0.000313	0.517		
0.000369	0.694		
0.000413	0.807		
0.000444	1.000		

Constants of Jouyban–Acree model

J_0	28.015
J_1	0
J_2	0

Reference of data: [29].

Solute	Solvent 1	Solvent 2	T (°C)
Berberine chloride	Ethanol	2-Propanol	29.85
Solubility (mole F.)	Mole F. 1		
0.000211	0.000		
0.000257	0.346		
0.000384	0.517		
0.000440	0.694		
0.000479	0.807		
0.000528	1.000		

Constants of Jouyban–Acree model

J_0	28.015
J_1	0
J_2	0

Reference of data: [29].

Solute	Solvent 1	Solvent 2	T (°C)
Berberine chloride	Ethanol	2-Propanol	34.85
Solubility (mole F.)	Mole F. 1		
0.000252	0.000		
0.000331	0.346		
0.000465	0.517		
0.000520	0.694		
0.000570	0.807		
0.000637	1.000		

Constants of Jouyban–Acree model

J_0	28.015
J_1	0
J_2	0

Reference of data: [29].

Solute	Solvent 1	Solvent 2	T (°C)
Berberine chloride	Ethanol	2-Propanol	39.85
Solubility (mole F.)	Mole F. 1		
0.000333	0.000		
0.000387	0.346		
0.000567	0.517		
0.000671	0.694		
0.000740	0.807		
0.000778	1.000		

Constants of Jouyban–Acree model

J_0	28.015
J_1	0
J_2	0

Reference of data: [29].

Solute	Solvent 1	Solvent 2	T (°C)
Biphenyl	Carbon tetrachloride	Cyclohexane	25
Solubility (mole F.)	Mole F. 1		
0.3421	1.000		
0.3213	0.835		
0.3015	0.661		
0.2740	0.479		
0.2474	0.317		
0.2184	0.149		
0.1921	0.000		

Constants of Jouyban–Acree model

J_0	40.048
J_1	−6.767
J_2	0

Reference of data: [30].

Solute	Solvent 1	Solvent 2	T (°C)
Biphenyl	Carbon tetrachloride	Heptane (*n*)	25
Solubility (mole F.)	Mole F. 1		
0.1381	0.000		
0.1594	0.137		
0.1938	0.319		
0.2340	0.505		
0.2815	0.710		
0.3051	0.813		
0.3421	1.000		

Constants of Jouyban–Acree model

J_0	35.715
J_1	17.916
J_2	0

Reference of data: [30].

Solute	Solvent 1	Solvent 2	T (°C)
Biphenyl	Carbon tetrachloride	Hexane (*n*)	25
Solubility (mole F.)	Mole F. 1		
0.1233	0.000		
0.1538	0.173		
0.1843	0.298		
0.2268	0.473		
0.2691	0.639		
0.3103	0.832		
0.3421	1.000		

Constants of Jouyban–Acree model

J_0	68.920
J_1	18.741
J_2	−30.223

Reference of data: [30].

Solute	Solvent 1	Solvent 2	T (°C)
Biphenyl	Cyclohexane	Hexane (n)	25
Solubility (mole F.)	Mole F. 1		
0.1233	0.000		
0.1378	0.213		
0.1476	0.362		
0.1609	0.531		
0.1727	0.684		
0.1817	0.818		
0.1921	1.000		

Constants of Jouyban–Acree model

J_0	15.586
J_1	9.710
J_2	0

Reference of data: [30].

Solute	Solvent 1	Solvent 2	T (°C)
Biphenyl	Cyclohexane	Octane (n)	25
Solubility (mole F.)	Mole F. 1		
0.1480	0.000		
0.1539	0.146		
0.1601	0.331		
0.1675	0.502		
0.1789	0.732		
0.1825	0.829		
0.1921	1.000		

Constants of Jouyban–Acree model

J_0	−3.194
J_1	0
J_2	0

Reference of data: [30].

Solute	Solvent 1	Solvent 2	T (°C)
Biphenyl	Cyclohexane	Heptane (n)	25
Solubility (mole F.)	Mole F. 1		
0.1381	0.000		
0.1477	0.183		
0.1571	0.365		
0.1663	0.546		
0.1765	0.739		
0.1845	0.851		
0.1921	1.000		

Constants of Jouyban–Acree model

J_0	3.926
J_1	0
J_2	0

Reference of data: [30].

Solute	Solvent 1	Solvent 2	T (°C)
Biphenyl	Methanol	Water	25
Solubility (mole/L)	Mass F. 1		
0.00004	0.000		
0.00005	0.018		
0.00008	0.057		
0.00010	0.077		
0.00013	0.099		
0.00018	0.121		
0.00025	0.144		
0.00037	0.169		
0.00036	0.185		
0.00056	0.195		
0.00081	0.221		
0.00127	0.250		
0.00248	0.297		
0.00416	0.337		
0.00708	0.383		
0.01204	0.432		
0.01997	0.487		
0.03297	0.543		
0.05267	0.611		
0.08776	0.688		
0.09112	0.693		
0.11860	0.744		
0.14390	0.771		
0.17340	0.809		
0.24500	0.873		
0.31670	0.929		
0.36210	0.956		
0.42735	1.000		

Constants of Jouyban–Acree model

J_0	799.292
J_1	88.176
J_2	0

Reference of data: [31].

Solute	Solvent 1	Solvent 2	T (°C)
Butabarbital	Ethanol	Water	25
Solubility (g/L)	Mass F. 1		
84.0	1.000		
85.9	0.975		
87.9	0.950		

(continued)

(continued)

Solute	Solvent 1	Solvent 2	T (°C)
89.3	0.925		
90.1	0.900		
90.6	0.875		
89.6	0.850		
88.5	0.825		
85.9	0.800		
82.6	0.775		
79.2	0.750		
73.4	0.725		
68.6	0.700		
63.6	0.675		
58.6	0.650		
53.7	0.625		
48.2	0.600		
43.0	0.575		
38.4	0.550		
33.4	0.525		
29.4	0.500		
24.8	0.475		
21.1	0.450		
17.7	0.425		
14.5	0.400		
11.9	0.375		
9.6	0.350		
7.5	0.325		
6.4	0.300		
4.8	0.275		
3.7	0.250		
2.9	0.225		
2.4	0.200		
2.0	0.175		
1.7	0.150		
1.5	0.125		
1.4	0.100		
1.2	0.075		
1.1	0.050		
1.0	0.025		
0.9	0.000		

Constants of Jouyban–Acree model

J_0	616.685
J_1	513.981
J_2	−508.591

Reference of data: [20].

Solute	Solvent 1	Solvent 2	T (°C)
Butyl p-aminobenzoate	Propylene glycol	Water	27.0
Solubility (mole F.)	Vol. F. 1		
0.0000164	0.000		
0.0000289	0.100		
0.0000507	0.200		

(continued)

Solute	Solvent 1	Solvent 2	T (°C)
0.0001052	0.300		
0.0002859	0.400		
0.0008169	0.500		
0.0022429	0.600		
0.0058577	0.700		
0.0172490	0.800		
0.0488012	0.900		
0.0898153	1.000		

Constants of Jouyban–Acree model

J_0	−246.012
J_1	460.027
J_2	164.985

Reference of data: [32].

Solute	Solvent 1	Solvent 2	T (°C)
Butyl *p*-aminobenzoate	Propylene glycol	Water	37.0
Solubility (mole F.)	Vol. F. 1		
0.00003162	0.000		
0.00012331	0.200		
0.00077983	0.400		
0.00562341	0.600		
0.04518559	0.800		
0.19498446	1.000		

Constants of Jouyban–Acree model

J_0	−102.274
J_1	439.348
J_2	165.540

Reference of data: [33].

Solute	Solvent 1	Solvent 2	T (°C)
Butyl *p*-hydroxybenzoate	Propylene glycol	Water	27.0
Solubility (mole F.)	Vol. F. 1		
0.0000289	0.000		
0.0000477	0.100		
0.0000803	0.200		
0.0001666	0.300		
0.0004620	0.400		
0.0016125	0.500		
0.0059166	0.600		
0.0275983	0.700		
0.0820850	0.800		
0.1495686	0.900		
0.1720448	1.000		

Constants of Jouyban–Acree model

J_0	−171.980
J_1	1151.475
J_2	757.902

Reference of data: [32].

Solute	Solvent 1	Solvent 2	T (°C)
Caffeine	Dimethyl formamide	Water	25
Solubility (mole F.)	Vol. F. 1		
0.00226	0.000		
0.00261	0.100		
0.00294	0.200		
0.00330	0.300		
0.00382	0.400		
0.00495	0.500		
0.00614	0.600		
0.01132	0.700		
0.01395	0.800		
0.01386	0.900		
0.01194	1.000		

Constants of Jouyban–Acree model

J_0	0
J_1	363.539
J_2	492.986

Reference of data: [34].

Solute	Solvent 1	Solvent 2	T (°C)
Caffeine	Dioxane	Water	25
Solubility (mole/L)	Mass F. 1		
0.00229	0.000		
0.00453	0.100		
0.00780	0.200		
0.01195	0.300		
0.01465	0.350		
0.01617	0.400		
0.01997	0.450		
0.02137	0.500		
0.02429	0.550		
0.02628	0.600		
0.02647	0.650		
0.02823	0.700		
0.02605	0.750		
0.02698	0.800		
0.02255	0.900		
0.00849	1.000		

Constants of Jouyban–Acree model

J_0	815.475
J_1	286.227
J_2	446.336

Reference of data: [35].

Solute	Solvent 1	Solvent 2	T (°C)
Caffeine	Ethyl acetate	Ethanol	5
Solubility (mole F.)	Mole F. 1		
0.00104	0.000		
0.00129	0.100		
0.00150	0.200		
0.00186	0.300		
0.00226	0.400		
0.00264	0.500		
0.00310	0.600		
0.00374	0.700		
0.00380	0.800		
0.00310	0.900		
0.00283	1.000		

Constants of Jouyban–Acree model

J_0	304.380
J_1	128.797
J_2	98.633

Reference of data: [36].

Solute	Solvent 1	Solvent 2	T (°C)
Caffeine	Ethyl acetate	Ethanol	15
Solubility (mole F.)	Mole F. 1		
0.00132	0.000		
0.00167	0.100		
0.00207	0.200		
0.00252	0.300		
0.00310	0.400		
0.00345	0.500		
0.00440	0.600		
0.00484	0.700		
0.00467	0.800		
0.00400	0.900		
0.00343	1.000		

Constants of Jouyban–Acree model

J_0	304.380
J_1	128.797
J_2	98.633

Reference of data: [36].

Solute	Solvent 1	Solvent 2	T (°C)
Caffeine	Ethyl acetate	Ethanol	25
Solubility (mole F.)	Mole F. 1		
0.00170	0.000		
0.00227	0.100		
0.00291	0.200		
0.00376	0.300		
0.00411	0.400		
0.00473	0.500		
0.00557	0.600		
0.00578	0.700		
0.00578	0.800		
0.00513	0.900		
0.00409	1.000		

Constants of Jouyban–Acree model

J_0	304.380
J_1	128.797
J_2	98.633

Reference of data: [36].

Solute	Solvent 1	Solvent 2	T (°C)
Caffeine	Ethyl acetate	Ethanol	35
Solubility (mole F.)	Mole F. 1		
0.00216	0.000		
0.00303	0.100		
0.00407	0.200		
0.00478	0.300		
0.00560	0.400		
0.00621	0.500		
0.00730	0.600		
0.00765	0.700		
0.00746	0.800		
0.00654	0.900		
0.00489	1.000		

Constants of Jouyban–Acree model

J_0	304.380
J_1	128.797
J_2	98.633

Reference of data: [36].

Solute	Solvent 1	Solvent 2	T (°C)
Caffeine	Ethyl acetate	Ethanol	40
Solubility (mole F.)	Mole F. 1		
0.00260	0.000		
0.00358	0.100		
0.00461	0.200		
0.00537	0.300		
0.00666	0.400		
0.00764	0.500		
0.00801	0.600		
0.00826	0.700		
0.00808	0.800		
0.00741	0.900		
0.00547	1.000		

Constants of Jouyban–Acree model

J_0	304.380
J_1	128.797
J_2	98.633

Reference of data: [36].

Solute	Solvent 1	Solvent 2	T (°C)
Caffeine	Water	Ethanol	5
Solubility (mole F.)	Vol. F. 1		
0.00122	1.000		
0.00162	0.900		
0.00201	0.800		
0.00301	0.700		
0.00390	0.600		
0.00444	0.500		
0.00520	0.400		
0.00580	0.300		
0.00506	0.200		
0.00350	0.100		
0.00104	0.000		

Constants of Jouyban–Acree model

J_0	941.111
J_1	−591.566
J_2	379.628

Reference of data: [36].

Solute	Solvent 1	Solvent 2	T (°C)
Caffeine	Water	Ethanol	15
Solubility (mole F.)	Vol. F. 1		
0.00155	1.000		
0.00217	0.900		
0.00286	0.800		
0.00396	0.700		
0.00623	0.600		
0.00726	0.500		
0.00828	0.400		
0.00869	0.300		
0.00790	0.200		
0.00456	0.100		
0.00132	0.000		

Constants of Jouyban–Acree model

J_0	941.111
J_1	−591.566
J_2	379.628

Reference of data: [36].

Solute	Solvent 1	Solvent 2	T (°C)
Caffeine	Water	Ethanol	25
Solubility (mole F.)	Vol. F. 1		
0.00209	1.000		
0.00332	0.900		
0.00507	0.800		
0.00707	0.700		
0.00982	0.600		
0.01138	0.500		
0.01470	0.400		
0.01439	0.300		
0.01151	0.200		
0.00603	0.100		
0.00170	0.000		

Constants of Jouyban–Acree model

J_0	941.111
J_1	−591.566
J_2	379.628

Reference of data: [36].

Solute	Solvent 1	Solvent 2	T (°C)
Caffeine	Water	Ethanol	35
Solubility (mole F.)	Vol. F. 1		
0.00307	1.000		
0.00575	0.900		
0.00942	0.800		
0.01358	0.700		
0.02004	0.600		
0.02285	0.500		
0.02489	0.400		
0.02321	0.300		
0.01532	0.200		
0.00772	0.100		
0.00216	0.000		

Constants of Jouyban–Acree model

J_0	941.111
J_1	−591.566
J_2	379.628

Reference of data: [36].

Solute	Solvent 1	Solvent 2	T (°C)
Caffeine	Water	Ethanol	40
Solubility (mole F.)	Vol. F. 1		
0.00431	1.000		
0.00735	0.900		
0.01191	0.800		
0.01789	0.700		
0.02384	0.600		
0.02766	0.500		
0.03171	0.400		
0.02575	0.300		
0.01802	0.200		
0.00924	0.100		
0.00260	0.000		

Constants of Jouyban–Acree model

J_0	941.111
J_1	−591.566
J_2	379.628

Reference of data: [36].

Solute	Solvent 1	Solvent 2	T (°C)
Calcium oxalate	Water	Ethanol	20.0
Solubility (mole/L)	Vol. F. 1		
0.00000035	0.000		
0.00000042	0.100		
0.00000106	0.200		
0.00000169	0.300		
0.00000303	0.400		
0.00000458	0.500		
0.00000613	0.600		
0.00000986	0.700		
0.00001710	0.800		
0.00002750	0.900		
0.00003750	1.000		

Constants of Jouyban–Acree model

J_0	88.703
J_1	0
J_2	0

Reference of data: [37].

Solute	Solvent 1	Solvent 2	T (°C)
Carbamazepine	Dioxane	Water	25
Solubility (mole F.)	Vol. F. 1		
0.02253	1.000		
0.04395	0.950		
0.02454	0.900		
0.03668	0.850		
0.01824	0.800		
0.01060	0.750		
0.00905	0.700		
0.00926	0.650		
0.00323	0.600		
0.00268	0.500		
0.00118	0.400		
0.00042	0.300		
0.00015	0.200		
0.00002	0.000		

Constants of Jouyban–Acree model

J_0	599.157
J_1	298.291
J_2	759.657

Reference of data: [38].

Solute	Solvent 1	Solvent 2	T (°C)
Carotene (β)	2-Butanone	n-Hexane	20.0
Solubility (mole/L)	Mole F. 1		
0.00102	1.000		
0.00282	0.747		
0.00336	0.502		
0.00288	0.246		
0.000663	0.000		

Constants of Jouyban–Acree model

J_0	716.600
J_1	−167.637
J_2	511.873

Reference of data: [39].

Solute	Solvent 1	Solvent 2	T (°C)
Carotene (β)	Cyclohexane	1,2-Dimethoxy ethane	20.0
Solubility (mole/L)	Mole F. 1		
0.00061	0.000		
0.00124	0.095		
0.00263	0.197		
0.00379	0.302		
0.00387	0.304		
0.00537	0.396		
0.00570	0.499		
0.00612	0.615		
0.00603	0.701		
0.00519	0.802		
0.00358	0.937		
0.00326	1.000		

Constants of Jouyban–Acree model

J_0	737.002
J_1	−196.046
J_2	0

Reference of data: [40].

Solute	Solvent 1	Solvent 2	T (°C)
Carotene (β)	Cyclohexane	1-Octanol	20.0
Solubility (mole/L)	Mole F. 1		
0.000177	0.000		
0.000166	0.199		
0.000171	0.301		
0.000566	0.395		
0.000516	0.401		

(continued)

(continued)

Solute	Solvent 1	Solvent 2	T (°C)
0.000493	0.502		
0.000809	0.609		
0.001015	0.697		
0.001228	0.797		
0.001438	0.800		
0.002298	0.899		
0.003262	1.000		

Constants of Jouyban–Acree model

J_0	−202.852
J_1	0
J_2	0

Reference of data: [41].

Solute	Solvent 1	Solvent 2	T (°C)
Carotene (β)	Cyclohexane	2-Butanone	20.0
Solubility (mole/L)	Mole F. 1		
0.00102	0.000		
0.00431	0.244		
0.00940	0.483		
0.00947	0.501		
0.00889	0.754		
0.00902	0.759		
0.00266	1.000		

Constants of Jouyban–Acree model

J_0	895.485
J_1	162.861
J_2	81.299

Reference of data: [39].

Solute	Solvent 1	Solvent 2	T (°C)
Carotene (β)	Cyclohexane	2-Propanone	20.0
Solubility (mole/L)	Mole F. 1		
0.000222	0.000		
0.001510	0.252		
0.004260	0.503		
0.005530	0.760		
0.002660	1.000		

Constants of Jouyban–Acree model

J_0	884.632
J_1	0
J_2	0

Reference of data: [39].

Solute	Solvent 1	Solvent 2	T (°C)
Carotene (β)	Cyclohexane	Cyclohexanone	20.0
Solubility (mole/L)	Mole F. 1		
0.00308	0.000		
0.00287	0.099		
0.00684	0.300		
0.00842	0.403		
0.00855	0.500		
0.00803	0.596		
0.00813	0.708		
0.00715	0.787		
0.00623	0.801		
0.00453	0.904		
0.00326	1.000		

Constants of Jouyban–Acree model

J_0	536.095
J_1	216.617
J_2	−366.180

Reference of data: [41].

Solute	Solvent 1	Solvent 2	T (°C)
Carotene (β)	Cyclohexane	Dibutyl ether	20.0
Solubility (mole/L)	Mole F. 1		
0.00093	0.000		
0.00134	0.103		
0.00165	0.202		
0.00179	0.302		
0.00229	0.404		
0.00235	0.498		
0.00252	0.600		
0.00287	0.700		
0.00316	0.799		
0.00303	0.825		
0.00324	0.901		
0.00326	1.000		

Constants of Jouyban–Acree model

J_0	155.330
J_1	−79.998
J_2	123.208

Reference of data: [40].

Solute	Solvent 1	Solvent 2	T (°C)
Carotene (β)	Cyclohexane	*tert*-Amyl methyl ether	20.0
Solubility (mole/L)	Mole F. 1		
0.00273	1.000		
0.00278	0.899		
0.00263	0.796		
0.00293	0.687		
0.00240	0.607		
0.00229	0.490		
0.00251	0.401		
0.00198	0.292		
0.00174	0.196		
0.00153	0.101		
0.00136	0.000		

Constants of Jouyban–Acree model

J_0	115.900
J_1	0
J_2	0

Reference of data: [42].

Solute	Solvent 1	Solvent 2	T (°C)
Carotene (β)	Cyclohexane	*tert*-Butyl methyl ether	20.0
Solubility (mole/L)	Mole F. 1		
0.00273	1.000		
0.00279	0.898		
0.00284	0.800		
0.00265	0.696		
0.00245	0.605		
0.00245	0.501		
0.00232	0.398		
0.00218	0.305		
0.00172	0.203		
0.00142	0.100		
0.00099	0.000		

Constants of Jouyban–Acree model

J_0	200.115
J_1	−117.617
J_2	109.510

Reference of data: [42].

Solute	Solvent 1	Solvent 2	T (°C)
Carotene (β)	Cyclohexanone	n-Hexane	20.0
Solubility (mole/L)	Mole F. 1		
0.00308	1.000		
0.00485	0.050		
0.00456	0.798		
0.00502	0.798		
0.00465	0.590		
0.00336	0.500		
0.00377	0.498		
0.00275	0.397		
0.00240	0.306		
0.00199	0.199		
0.00089	0.105		
0.00063	0.000		

Constants of Jouyban–Acree model

J_0	507.933
J_1	0
J_2	0

Reference of data: [41].

Solute	Solvent 1	Solvent 2	T (°C)
Carotene (β)	Dibutyl ether	n-Hexane	20.0
Solubility (mole/L)	Mole F. 1		
0.00093	1.000		
0.00089	0.893		
0.00083	0.852		
0.00079	0.798		
0.00081	0.695		
0.00078	0.695		
0.00073	0.585		
0.00075	0.509		
0.00071	0.480		
0.00069	0.397		
0.00066	0.286		
0.00069	0.203		
0.00064	0.200		
0.00062	0.167		
0.00065	0.100		
0.00065	0.000		
0.00093	1.000		

Constants of Jouyban–Acree model

J_0	−42.580
J_1	0
J_2	0

Reference of data: [40].

Solute	Solvent 1	Solvent 2	T (°C)
Carotene (β)	n-Hexane	1,2-Dimethoxy ethane	20.0
Solubility (mole/L)	Mole F. 1		
0.00061	0.000		
0.00082	0.101		
0.00126	0.199		
0.00169	0.299		
0.00170	0.400		
0.00193	0.504		
0.00173	0.603		
0.00171	0.695		
0.00152	0.798		
0.00143	0.798		
0.00100	0.900		
0.00065	1.000		

Constants of Jouyban–Acree model

J_0	572.074
J_1	067.282
J_2	0

Reference of data: [40].

Solute	Solvent 1	Solvent 2	T (°C)
Carotene (β)	n-Hexane	1-Octanol	20.0
Solubility (mole/L)	Mole F. 1		
0.000177	0.000		
0.000090	0.200		
0.000273	0.293		
0.000169	0.398		
0.000359	0.495		
0.000258	0.598		
0.000318	0.605		
0.000486	0.697		
0.000600	0.798		
0.000768	0.900		
0.000646	1.000		

Constants of Jouyban–Acree model

J_0	−125.462
J_1	423.363
J_2	0

Reference of data: [41].

Solute	Solvent 1	Solvent 2	T (°C)
Carotene (β)	n-Hexane	2-Propanone	20.0
Solubility (mole/L)	Mole F. 1		
0.000222	0.000		
0.000664	0.246		
0.001340	0.509		
0.001720	0.753		
0.000663	1.000		

Constants of Jouyban–Acree model

J_0	667.010
J_1	262.382
J_2	0

Reference of data: [39].

Solute	Solvent 1	Solvent 2	T (°C)
Carotene (β)	$tert$-Amyl methyl ether	n-Hexane	20.0
Solubility (mole/L)	Mole F. 1		
0.00069	0.000		
0.00075	0.102		
0.00080	0.200		
0.00080	0.201		
0.00078	0.302		
0.00106	0.400		
0.00114	0.498		
0.00120	0.601		
0.00146	0.694		
0.00141	0.799		
0.00149	0.895		
0.00136	1.000		

Constants of Jouyban–Acree model

J_0	77.908
J_1	142.531
J_2	0

Reference of data: [42].

Solute	Solvent 1	Solvent 2	T (°C)
Carotene (β)	*tert*-Amyl methyl ether	*n*-Hexane	20.0
Solubility (mole/L)	Mole F. 1		
0.00069	0.000		
0.00078	0.193		
0.00086	0.202		
0.00093	0.399		
0.00114	0.498		
0.00112	0.596		
0.00114	0.796		
0.00099	1.000		

Constants of Jouyban–Acree model

J_0	130.002
J_1	86.888
J_2	0

Reference of data: [42].

Solute	Solvent 1	Solvent 2	T (°C)
Carotene (β)	Toluene	1,2-Dimethoxy ethane	20.0
Solubility (mole/L)	Mole F. 1		
0.00061	0.000		
0.00107	0.115		
0.00150	0.200		
0.00192	0.309		
0.00252	0.406		
0.00305	0.497		
0.00290	0.608		
0.00434	0.704		
0.00699	0.803		
0.00773	0.908		
0.01015	1.000		

Constants of Jouyban–Acree model

J_0	92.900
J_1	−208.410
J_2	0

Reference of data: [40].

Solute	Solvent 1	Solvent 2	T (°C)
Carotene (β)	Toluene	1-Octanol	20.0
Solubility (mole/L)	Mole F. 1		
0.000177	0.000		
0.000309	0.203		
0.000593	0.310		
0.000615	0.401		
0.001185	0.492		
0.001877	0.594		
0.002706	0.703		
0.004927	0.755		
0.008230	0.900		
0.010596	1.000		

Constants of Jouyban–Acree model

J_0	−68.957
J_1	223.009
J_2	0

Reference of data: [41].

Solute	Solvent 1	Solvent 2	T (°C)
Carotene (β)	Toluene	2-Butanone	20.0
Solubility (mole/L)	Mole F. 1		
0.001020	0.000		
0.003390	0.253		
0.006360	0.507		
0.015300	0.770		
0.010100	1.000		

Constants of Jouyban–Acree model

J_0	443.349
J_1	0
J_2	0

Reference of data: [39].

Solute	Solvent 1	Solvent 2	T (°C)
Carotene (β)	Toluene	Cyclohexanone	20.0
Solubility (mole/L)	Mole F. 1		
0.00308	0.000		
0.00352	0.097		
0.00352	0.019		
0.00440	0.299		
0.00571	0.395		
0.00674	0.501		
0.00743	0.599		
0.00723	0.694		
0.00868	0.791		
0.00917	0.897		
0.01059	1.000		

Constants of Jouyban–Acree model

J_0	46.671
J_1	0
J_2	0

Reference of data: [41].

Solute	Solvent 1	Solvent 2	T (°C)
Carotene (β)	Toluene	Dibutyl ether	20.0
Solubility (mole/L)	Mole F. 1		
0.00093	0.000		
0.00099	0.105		
0.00114	0.202		
0.00147	0.299		
0.00145	0.407		
0.00229	0.510		
0.00287	0.603		
0.00348	0.680		
0.00391	0.704		
0.00470	0.798		
0.00658	0.902		
0.01015	1.000		

Constants of Jouyban–Acree model

J_0	193.858
J_1	0
J_2	0

Reference of data: [40].

Solute	Solvent 1	Solvent 2	T (°C)
Carotene (β)	Toluene	*tert*-Amyl methyl ether	20.0
Solubility (mole/L)	Mole F. 1		
0.00136	0.000		
0.00173	0.109		
0.00220	0.201		
0.00249	0.295		
0.00282	0.416		
0.00352	0.497		
0.00355	0.497		
0.00433	0.589		
0.00457	0.691		
0.00563	0.817		
0.00707	0.900		
0.01009	1.000		

Constants of Jouyban–Acree model

J_0	−38.470
J_1	−146.283
J_2	0

Reference of data: [42].

Solute	Solvent 1	Solvent 2	T (°C)
Carotene (β)	Toluene	*tert*-Butyl methyl ether	20.0
Solubility (mole/L)	Mole F. 1		
0.01009	1.000		
0.00707	0.899		
0.00593	0.810		
0.00503	0.698		
0.00430	0.601		
0.00373	0.495		
0.00267	0.400		
0.00204	0.301		
0.00181	0.205		
0.00155	0.100		
0.00099	0.000		

Constants of Jouyban–Acree model

J_0	37.323
J_1	−125.173
J_2	0

Reference of data: [42].

Solute	Solvent 1	Solvent 2	T (°C)
Ceftriaxone disodium hemiheptahydrate	Water	Acetone	25.0
Solubility (g/kg)	Mass F. 1		
0.215	0.100		
1.91	0.200		
10.72	0.300		
35.29	0.400		
97.17	0.500		
173.05	0.600		
252.88	0.700		
355.72	0.800		

Constants of Jouyban–Acree model

J_0
J_1
J_2

Reference of data: [43].

Solute	Solvent 1	Solvent 2	T (°C)
Ceftriaxone disodium hemiheptahydrate	Water	Acetone	35.0
Solubility (g/kg)	Mass F. 1		
0.188	0.100		
2.32	0.200		
13.15	0.300		
41.83	0.400		
110.61	0.500		
213.2	0.600		
295.73	0.700		
374.28	0.800		

Constants of Jouyban–Acree model

J_0
J_1
J_2

Reference of data: [43].

Solute	Solvent 1	Solvent 2	T (°C)
Ceftriaxone disodium hemiheptahydrate	Water	Acetone	45.0
Solubility (g/kg)	Mass F. 1		
1.492	0.100		
3.640	0.200		
15.68	0.300		
51.30	0.400		
142.45	0.500		
236.6	0.600		
331.39	0.700		
395.72	0.800		

Constants of Jouyban–Acree model

J_0
J_1
J_2

Reference of data: [43].

Solute	Solvent 1	Solvent 2	T (°C)
Cefazolin sodium pentahydrate	Water	Ethanol	4.2
Solubility (mole F.)	Mole F. 1		
0.0024690	0.900		
0.0021990	0.806		
0.0018060	0.699		
0.0013680	0.600		
0.0007914	0.498		
0.0005051	0.408		
0.0002759	0.298		
0.0001463	0.207		

Constants of Jouyban–Acree model

J_0
J_1
J_2

Reference of data: [44,45].

Solute	Solvent 1	Solvent 2	T (°C)
Cefazolin sodium pentahydrate	Water	Ethanol	8.5
Solubility (mole F.)	Mole F. 1		
0.0034870	0.900		
0.0030240	0.806		
0.0022930	0.699		
0.0016040	0.600		
0.0009937	0.498		
0.0006314	0.408		
0.0003128	0.298		
0.0002050	0.207		

Constants of Jouyban–Acree model

J_0
J_1
J_2

Reference of data: [44,45].

Solute	Solvent 1	Solvent 2	T (°C)
Cefazolin sodium pentahydrate	Water	Ethanol	12.0
Solubility (mole F.)	Mole F. 1		
0.0044600	0.900		
0.0039750	0.806		
0.0028110	0.699		
0.0018890	0.600		
0.0011950	0.498		
0.0007542	0.408		
0.0003581	0.298		
0.0002718	0.207		

Constants of Jouyban–Acree model

J_0
J_1
J_2

Reference of data: [44,45].

Solute	Solvent 1	Solvent 2	T (°C)
Cefazolin sodium pentahydrate	Water	Ethanol	16.1
Solubility (mole F.)	Mole F. 1		
0.0061800	0.900		
0.0052590	0.806		
0.0037580	0.699		
0.0025880	0.600		
0.0015010	0.498		
0.0008666	0.408		
0.0004986	0.298		
0.0003260	0.207		

Constants of Jouyban–Acree model

J_0

J_1

J_2

Reference of data: [44,45].

Solute	Solvent 1	Solvent 2	T (°C)
Cefazolin sodium pentahydrate	Water	Ethanol	19.6
Solubility (mole F.)	Mole F. 1		
0.0081190	0.900		
0.0067350	0.806		
0.0048980	0.699		
0.0033230	0.600		
0.0018070	0.498		
0.0009962	0.408		
0.0006029	0.298		
0.0003808	0.207		

Constants of Jouyban–Acree model

J_0

J_1

J_2

Reference of data: [44,45].

Solute	Solvent 1	Solvent 2	T (°C)
Cefazolin sodium pentahydrate	Water	Ethanol	24.4
Solubility (mole F.)	Mole F. 1		
0.0104700	0.900		
0.0085830	0.806		
0.0064910	0.699		
0.0042050	0.600		
0.0022060	0.498		
0.0014190	0.408		
0.0007797	0.298		
0.0005126	0.207		

Constants of Jouyban–Acree model

J_0

J_1

J_2

Reference of data: [44,45].

Solute	Solvent 1	Solvent 2	T (°C)
Cefazolin sodium pentahydrate	Water	Ethanol	27.8
Solubility (mole F.)	Mole F. 1		
0.0131300	0.900		
0.0109400	0.806		
0.0081310	0.699		
0.0054160	0.600		
0.0026990	0.498		
0.0018190	0.408		
0.0009345	0.298		
0.0006532	0.207		

Constants of Jouyban–Acree model

J_0

J_1

J_2

Reference of data: [44,45].

Solute	Solvent 1	Solvent 2	T (°C)
Cefazolin sodium pentahydrate	Water	Ethanol	30.8
Solubility (mole F.)	Mole F. 1		
0.016090	0.900		
0.013800	0.806		
0.009631	0.699		
0.006698	0.600		
0.003592	0.498		
0.002171	0.408		
0.001127	0.298		
0.000813	0.207		

Constants of Jouyban–Acree model

J_0

J_1

J_2

Reference of data: [44,45].

Solute	Solvent 1	Solvent 2	T (°C)
Cefazolin sodium pentahydrate	Water	Ethanol	33.5
Solubility (mole F.)	Mole F. 1		
0.021090	0.900		
0.018220	0.806		
0.012270	0.699		
0.007855	0.600		
0.004267	0.498		
0.002593	0.408		
0.001385	0.298		
0.001042	0.207		

Constants of Jouyban–Acree model

J_0

J_1

J_2

Reference of data: [44,45].

Solute	Solvent 1	Solvent 2	T (°C)
Cefazolin sodium pentahydrate	Water	Ethanol	36.9
Solubility (mole F.)	Mole F. 1		
0.026190	0.900		
0.022130	0.806		
0.015400	0.699		
0.009785	0.600		
0.005370	0.498		
0.003156	0.408		
0.001965	0.298		
0.001626	0.207		

Constants of Jouyban–Acree model

J_0

J_1

J_2

Reference of data: [44,45].

Solute	Solvent 1	Solvent 2	T (°C)
Cefazolin sodium pentahydrate	Water	2-Propanol	0
Solubility (mole F.)	Mole F. 1		
0.00001	0.000		
0.00004	0.104		
0.00006	0.205		
0.00009	0.304		
0.00016	0.439		
0.00021	0.500		
0.00038	0.602		
0.00076	0.694		
0.00175	0.801		
0.00360	0.905		
0.00553	1.000		

Constants of Jouyban–Acree model

J_0	−52.873
J_1	−102.633
J_2	575.656

Reference of data: [46].

Solute	Solvent 1	Solvent 2	T (°C)
Cefazolin sodium pentahydrate	Water	2-Propanol	5
Solubility (mole F.)	Mole F. 1		
0.00002	0.000		
0.00005	0.104		
0.00009	0.205		
0.00013	0.304		
0.00021	0.439		
0.00031	0.500		
0.00053	0.602		
0.00100	0.694		
0.00206	0.801		
0.00405	0.905		
0.00658	1.000		

Constants of Jouyban–Acree model

J_0	−52.873
J_1	−102.633
J_2	575.656

Reference of data: [46].

Solute	Solvent 1	Solvent 2	T (°C)
Cefazolin sodium pentahydrate	Water	2-Propanol	10
Solubility (mole F.)	Mole F. 1		
0.00003	0.000		
0.00007	0.104		
0.00013	0.205		
0.00018	0.304		
0.00031	0.439		
0.00043	0.500		
0.00073	0.602		
0.00132	0.694		
0.00262	0.801		
0.00499	0.905		
0.00771	1.000		

Constants of Jouyban–Acree model

J_0	−52.873
J_1	−102.633
J_2	575.656

Reference of data: [46].

Solute	Solvent 1	Solvent 2	T (°C)
Cefazolin sodium pentahydrate	Water	2-Propanol	15
Solubility (mole F.)	Mole F. 1		
0.00003	0.000		
0.00009	0.104		
0.00015	0.205		
0.00024	0.304		
0.00042	0.439		
0.00057	0.500		
0.00100	0.602		
0.00181	0.694		
0.00370	0.801		
0.00692	0.905		
0.01017	1.000		

Constants of Jouyban–Acree model

J_0	−52.873
J_1	−102.633
J_2	575.656

Reference of data: [46].

Solute	Solvent 1	Solvent 2	T (°C)
Cefazolin sodium pentahydrate	Water	2-Propanol	20
Solubility (mole F.)	Mole F. 1		
0.00005	0.000		
0.00011	0.104		
0.00021	0.205		
0.00031	0.304		
0.00053	0.439		
0.00072	0.500		
0.00130	0.602		
0.00245	0.694		
0.00511	0.801		
0.00957	0.905		
0.01355	1.000		

Constants of Jouyban–Acree model

J_0	−52.873
J_1	−102.633
J_2	575.656

Reference of data: [46].

Solute	Solvent 1	Solvent 2	T (°C)
Cefazolin sodium pentahydrate	Water	2-Propanol	25
Solubility (mole F.)	Mole F. 1		
0.0007576	0.000		
0.0001578	0.104		
0.0002566	0.205		
0.0003750	0.304		
0.0007053	0.439		
0.0009204	0.500		
0.001730	0.602		
0.003441	0.694		
0.006884	0.801		
0.01213	0.905		
0.01692	1.000		

Constants of Jouyban–Acree model

J_0	−52.873
J_1	−102.633
J_2	575.656

Reference of data: [46].

Solute	Solvent 1	Solvent 2	T (°C)
Cefazolin sodium pentahydrate	Water	2-Propanol	30

Solubility (mole F.)	Mole F. 1
0.00011	0.000
0.00023	0.104
0.00036	0.205
0.00054	0.304
0.00098	0.439
0.00135	0.500
0.00249	0.602
0.00457	0.694
0.00918	0.801
0.01583	0.905
0.02211	1.000

Constants of Jouyban–Acree model

J_0	−52.873
J_1	−102.633
J_2	575.656

Reference of data: [46].

Solute	Solvent 1	Solvent 2	T (°C)
Cefazolin sodium pentahydrate	Water	2-Propanol	35

Solubility (mole F.)	Mole F. 1
0.00015	0.000
0.00036	0.104
0.00053	0.205
0.00083	0.304
0.00149	0.439
0.00218	0.500
0.00400	0.602
0.00666	0.694
0.01117	0.801
0.01806	0.905
0.02644	1.000

Constants of Jouyban–Acree model

J_0	−52.873
J_1	−102.633
J_2	575.656

Reference of data: [46].

Solute	Solvent 1	Solvent 2	T (°C)
Ceftriaxone disodium hemiheptahydrate	Water	Ethanol	10.0
Solubility (mass F.)	Mass F. 1		
0.124193	0.713		
0.068376	0.602		
0.035141	0.497		
0.019814	0.401		
0.010381	0.316		
0.002510	0.198		
0.001808	0.155		
0.000962	0.111		

Constants of Jouyban–Acree model

J_0
J_1
J_2

Reference of data: [47].

Solute	Solvent 1	Solvent 2	T (°C)
Ceftriaxone disodium hemiheptahydrate	Water	Ethanol	20.0
Solubility (mass F.)	Mass F. 1		
0.159466	0.713		
0.091917	0.602		
0.051892	0.497		
0.028402	0.401		
0.014835	0.316		
0.004418	0.198		
0.002891	0.155		
0.001199	0.111		

Constants of Jouyban–Acree model

J_0
J_1
J_2

Reference of data: [47].

Solute	Solvent 1	Solvent 2	T (°C)
Ceftriaxone disodium hemiheptahydrate	Water	Ethanol	30.0
Solubility (mass F.)	Mass F. 1		
0.201844	0.713		
0.121828	0.602		
0.071833	0.497		
0.038296	0.401		
0.021184	0.316		
0.005785	0.198		
0.003252	0.155		
0.001517	0.111		

Constants of Jouyban–Acree model

J_0

J_1

J_2

Reference of data: [47].

Solute	Solvent 1	Solvent 2	T (°C)
Ceftriaxone disodium hemiheptahydrate	Water	Methanol	25.0
Solubility (mole F.)	Mass F. 1		
0.0589	0.100		
0.0757	0.200		
0.1220	0.300		
0.1937	0.400		
0.2543	0.500		
0.3274	0.600		
0.4322	0.700		
0.5223	0.800		

Constants of Jouyban–Acree model

J_0

J_1

J_2

Reference of data: [48].

Solute	Solvent 1	Solvent 2	T (°C)
Ceftriaxone disodium hemiheptahydrate	Water	Methanol	35.0
Solubility (mole F.)	Mass F. 1		
0.0626	0.100		
0.0828	0.200		
0.1295	0.300		
0.2167	0.400		
0.3139	0.500		
0.3963	0.600		
0.4807	0.700		
0.5695	0.800		

Constants of Jouyban–Acree model

J_0

J_1

J_2

Reference of data: [48].

Solute	Solvent 1	Solvent 2	T (°C)
Ceftriaxone disodium hemiheptahydrate	Water	Methanol	45.0
Solubility (mole F.)	Mass F. 1		
0.0714	0.100		
0.0991	0.200		
0.1691	0.300		
0.2819	0.400		
0.3960	0.500		
0.4924	0.600		
0.5642	0.700		
0.6278	0.800		

Constants of Jouyban–Acree model

J_0
J_1
J_2

Reference of data: [48].

Solute	Solvent 1	Solvent 2	T (°C)
Celecoxib	Ethanol	Glycerol	25.0
Solubility (g/L)	Vol. F. 1		
28.85	0.400		
37.69	0.600		
58.85	0.800		
61.54	0.900		
63.346	1.000		

Constants of Jouyban–Acree model

J_0
J_1
J_2

Reference of data: [49].

Solute	Solvent 1	Solvent 2	T (°C)
Celecoxib	Ethanol	Water	25
Solubility (g/L)	Vol. F. 1		
0.007	0.000		
0.010	0.100		
0.015	0.200		
0.294	0.400		
5.062	0.600		
31.904	0.800		
48.654	0.900		
63.346	1.000		

Constants of Jouyban–Acree model

J_0	335.383
J_1	1342.937
J_2	−627.146

Reference of data: [49].

Solute	Solvent 1	Solvent 2	T (°C)
Celecoxib	Polyethylene glycol 400	Ethanol	25
Solubility (g/L)	Vol. F. 1		
63.346	0.000		
171.778	0.200		
251.810	0.400		
327.939	0.600		
366.980	0.800		
391.585	0.900		
414.804	1.000		

Constants of Jouyban–Acree model

J_0	304.583
J_1	−232.426
J_2	127.193

Reference of data: [49].

Solute	Solvent 1	Solvent 2	T (°C)
Chlordiazepoxide	Ethanol	Water	30
Solubility (mole F.)	Vol. F. 1		
0.00000615	0.000		
0.00001180	0.100		
0.00002080	0.200		
0.00006630	0.300		
0.00027000	0.400		
0.00076900	0.500		
0.00145000	0.600		
0.00294000	0.700		
0.00443000	0.800		
0.00499000	0.900		
0.00335000	1.000		

Constants of Jouyban–Acree model

J_0	815.761
J_1	973.330
J_2	−367.813

Reference of data: [50].

Solute	Solvent 1	Solvent 2	T (°C)
Chlordiazepoxide	Polyethylene glycol 200	Water	30
Solubility (mole F.)	Vol. F. 1		
0.00000595	0.000		
0.00003390	0.100		
0.00009140	0.200		
0.00017200	0.300		
0.00033100	0.400		
0.00070400	0.500		
0.00160000	0.600		
0.00295000	0.700		
0.00640000	0.800		
0.01380000	0.900		
0.02318000	1.000		

Constants of Jouyban–Acree model

J_0	320.533
J_1	−428.598
J_2	792.355

Reference of data: [50].

Solute	Solvent 1	Solvent 2	T (°C)
Chlordiazepoxide	Propylene glycol	Water	30
Solubility (mole F.)	Vol. F. 1		
0.00000637	0.000		
0.00001400	0.100		
0.00002350	0.200		
0.00004400	0.300		
0.00010400	0.400		
0.00019500	0.500		
0.00047000	0.600		
0.00075900	0.700		
0.00141000	0.800		
0.00269000	0.900		
0.00488000	1.000		

Constants of Jouyban–Acree model

J_0	69.662
J_1	86.676
J_2	0

Reference of data: [50].

Solute	Solvent 1	Solvent 2	T (°C)
Chlorpyrifos	Methanol	Water	25.0
Solubility (mg/L)	Mass F. 1		
6.36	0.000		
8.25	0.100		
11.60	0.200		
22.20	0.300		
43.30	0.400		
290.0	0.500		
456	0.600		
4608	0.800		
116498	1.000		

Constants of Jouyban–Acree model

J_0	−938.257
J_1	−248.612
J_2	−608.990

Reference of data: [51].

Solute	Solvent 1	Solvent 2	T (°C)
Chlorpyrifos	Methanol	Water	30.0
Solubility (mg/L)	Mass F. 1		
7.18	0.000		
11.50	0.100		
18.20	0.200		
36.30	0.300		
58.20	0.400		
399.0	0.500		
618.0	0.600		
5201.0	0.800		
212511.0	1.000		

Constants of Jouyban–Acree model

J_0	−938.257
J_1	−248.612
J_2	−608.990

Reference of data: [51].

Solute	Solvent 1	Solvent 2	T (°C)
Chlorpyrifos	Methanol	Water	35.0
Solubility (mg/L)	Mass F. 1		
9.07	0.000		
18.20	0.100		
32.10	0.200		
51.70	0.300		
90.40	0.400		
522.0	0.500		
865.0	0.600		
9302.0	0.800		
804651.0	1.000		

Constants of Jouyban–Acree model

J_0	−938.257
J_1	−248.612
J_2	−608.990

Reference of data: [51].

Solute	Solvent 1	Solvent 2	T (°C)
Clindamycin phosphate	Water	Ethanol	5
Solubility (mole F.)	Mole F. 1		
0.000006	0.000		
0.001370	0.150		
0.001700	0.309		
0.003800	0.450		
0.005810	0.615		
0.018480	0.752		
0.176080	1.000		

Constants of Jouyban–Acree model

J_0	286.837
J_1	−1107.518
J_2	1680.615

Reference of data: [52].

Solute	Solvent 1	Solvent 2	T (°C)
Clindamycin phosphate	Water	Ethanol	10
Solubility (mole F.)	Mole F. 1		
0.00009	0.000		
0.00148	0.150		
0.00257	0.309		
0.00454	0.450		
0.00678	0.615		
0.02142	0.752		
0.17986	1.000		

Constants of Jouyban–Acree model

J_0	286.837
J_1	−1107.518
J_2	1680.615

Reference of data: [52].

Solute	Solvent 1	Solvent 2	T (°C)
Clindamycin phosphate	Water	Ethanol	15
Solubility (mole F.)	Mole F. 1		
0.00009	0.000		
0.00171	0.150		
0.00361	0.309		
0.00519	0.450		
0.00753	0.615		
0.03433	0.752		
0.18589	1.000		

Constants of Jouyban–Acree model

J_0	286.837
J_1	−1107.518
J_2	1680.615

Reference of data: [52].

Solute	Solvent 1	Solvent 2	T (°C)
Clindamycin phosphate	Water	Ethanol	20
Solubility (mole F.)	Mole F. 1		
0.00011	0.000		
0.00284	0.150		
0.00374	0.309		
0.00608	0.450		
0.00940	0.615		
0.03693	0.752		
0.19957	1.000		

Constants of Jouyban–Acree model

J_0	286.837
J_1	−1107.518
J_2	1680.615

Reference of data: [52].

Solute	Solvent 1	Solvent 2	T (°C)
Clindamycin phosphate	Water	Ethanol	25
Solubility (mole F.)	Mole F. 1		
0.00014	0.000		
0.00322	0.150		
0.00410	0.309		
0.00876	0.450		
0.01360	0.615		
0.04320	0.752		
0.20292	1.000		

Constants of Jouyban–Acree model

J_0	286.837
J_1	−1107.518
J_2	1680.615

Reference of data: [52].

Solute	Solvent 1	Solvent 2	T (°C)
Clindamycin phosphate	Water	Ethanol	30
Solubility (mole F.)	Mole F. 1		
0.00024	0.000		
0.00418	0.150		
0.00572	0.309		
0.00917	0.450		
0.01512	0.615		
0.05089	0.752		
0.20942	1.000		

Constants of Jouyban–Acree model

J_0	286.837
J_1	−1107.518
J_2	1680.615

Reference of data: [52].

Solute	Solvent 1	Solvent 2	T (°C)
Clindamycin phosphate	Water	Ethanol	35
Solubility (mole F.)	Mole F. 1		
0.00031	0.000		
0.00457	0.150		
0.00616	0.309		
0.01299	0.450		
0.01949	0.615		
0.05127	0.752		
0.21665	1.000		

Constants of Jouyban–Acree model

J_0	286.837
J_1	−1107.518
J_2	1680.615

Reference of data: [52].

Solute	Solvent 1	Solvent 2	T (°C)
Clindamycin phosphate	Water	Ethanol	40
Solubility (mole F.)	Mole F. 1		
0.00034	0.000		
0.00500	0.150		
0.00737	0.309		
0.01407	0.450		
0.02086	0.615		
0.05343	0.752		
0.22286	1.000		

Constants of Jouyban–Acree model

J_0	286.837
J_1	−1107.518
J_2	1680.615

Reference of data: [52].

Solute	Solvent 1	Solvent 2	T (°C)
Clindamycin phosphate	Water	Ethanol	45
Solubility (mole F.)	Mole F. 1		
0.00042	0.000		
0.00544	0.150		
0.00798	0.309		
0.01728	0.450		
0.02574	0.615		
0.06087	0.752		
0.22787	1.000		

Constants of Jouyban–Acree model

J_0	286.837
J_1	−1107.518
J_2	1680.615

Reference of data: [52].

Solute	Solvent 1	Solvent 2	T (°C)
Clindamycin phosphate	Water	Ethanol	50
Solubility (mole F.)	Mole F. 1		
0.00045	0.000		
0.00555	0.150		
0.01103	0.309		
0.02452	0.450		
0.03561	0.615		
0.06374	0.752		
0.23177	1.000		

Constants of Jouyban–Acree model

J_0	286.837
J_1	−1107.518
J_2	1680.615

Reference of data: [52].

Solute	Solvent 1	Solvent 2	T (°C)
Clindamycin phosphate	Water	Ethanol	55
Solubility (mole F.)	Mole F. 1		
0.00047	0.000		
0.00673	0.150		
0.01473	0.309		
0.02558	0.450		
0.03764	0.615		
0.06681	0.752		
0.23495	1.000		

Constants of Jouyban–Acree model

J_0	286.837
J_1	−1107.518
J_2	1680.615

Reference of data: [52].

Solute	Solvent 1	Solvent 2	T (°C)
Clindamycin phosphate	Water	Ethanol	60
Solubility (mole F.)	Mole F. 1		
0.00056	0.000		
0.00762	0.150		
0.01612	0.309		
0.02723	0.450		
0.03982	0.615		
0.06905	0.752		
0.24071	1.000		

Constants of Jouyban–Acree model

J_0	286.837
J_1	−1107.518
J_2	1680.615

Reference of data: [52].

Solute	Solvent 1	Solvent 2	T (°C)
Clindamycin phosphate	Water	Ethanol	65
Solubility (mole F.)	Mole F. 1		
0.00074	0.000		
0.00872	0.150		
0.01974	0.309		
0.02897	0.450		
0.04291	0.615		
0.07650	0.752		
0.25187	1.000		

Constants of Jouyban–Acree model

J_0	286.837
J_1	−1107.518
J_2	1680.615

Reference of data: [52].

Solute	Solvent 1	Solvent 2	T (°C)
Clindamycin phosphate	Water	Ethanol	70
Solubility (mole F.)	Mole F. 1		
0.00112	0.000		
0.01140	0.150		
0.02179	0.309		
0.03451	0.450		
0.05063	0.615		
0.09147	0.752		
0.29239	1.000		

Constants of Jouyban–Acree model

J_0	286.837
J_1	−1107.518
J_2	1680.615

Reference of data: [52].

Solute	Solvent 1	Solvent 2	T (°C)
Clonazepam	Ethanol	Water	30
Solubility (mole F.)	Vol. F. 1		
0.00000089	0.000		
0.00000185	0.100		
0.00000431	0.200		
0.00001610	0.300		
0.00005150	0.400		
0.00014100	0.500		
0.00031500	0.600		
0.00058100	0.700		
0.00090500	0.800		
0.00105000	0.900		
0.00102000	1.000		

Constants of Jouyban–Acree model

J_0	807.830
J_1	669.687
J_2	−479.282

Reference of data: [50].

Solute	Solvent 1	Solvent 2	T (°C)
Clonazepam	Polyethylene glycol 200	Water	30
Solubility (mole F.)	Vol. F. 1		
0.00000084	0.000		
0.00000323	0.100		
0.00000820	0.200		
0.00002020	0.300		
0.00005010	0.400		
0.00013100	0.500		
0.00036900	0.600		
0.00103000	0.700		
0.00317000	0.800		
0.00870000	0.900		
0.01804000	1.000		

Constants of Jouyban–Acree model

J_0	21.219
J_1	−34.723
J_2	593.235

Reference of data: [50].

Solute	Solvent 1	Solvent 2	T (°C)
Clonazepam	Propylene glycol	Water	30
Solubility (mole F.)	Vol. F. 1		
0.00000088	0.000		
0.00000203	0.100		
0.00000348	0.200		
0.00000744	0.300		
0.00001950	0.400		
0.00005130	0.500		
0.00015000	0.600		
0.00022600	0.700		
0.00047300	0.800		
0.00079600	0.900		
0.00098100	1.000		

Constants of Jouyban–Acree model

J_0	287.624
J_1	486.255
J_2	0

Reference of data: [50].

Solute	Solvent 1	Solvent 2	T (°C)
Clonazepam	Ethanol	Water	25
Solubility (mole/L)	Vol. F. 1		
0.000095	0.00		
0.000128	0.10		
0.000249	0.20		
0.000628	0.30		
0.001897	0.40		
0.004524	0.50		
0.008358	0.60		
0.013360	0.70		
0.017880	0.80		
0.019567	0.90		
0.016192	1.00		

Constants of Jouyban–Acree model

J_0	644.074
J_1	781.166
J_2	−559.144

Reference of data: [53].

Solute	Solvent 1	Solvent 2	T (°C)
Clonazepam	N-Methylpyrrolidone	Water	25
Solubility (mole/L)	Vol. F. 1		
0.000095	0.00		
0.000348	0.10		
0.000745	0.20		
0.001732	0.30		
0.004480	0.40		
0.006770	0.50		
0.037871	0.60		
0.074333	0.70		
0.164334	0.80		
0.389755	0.90		
0.666715	1.00		

Constants of Jouyban–Acree model

J_0	185.185
J_1	0
J_2	0

Reference of data: [54].

Solute	Solvent 1	Solvent 2	T (°C)
Clonazepam	Propylene glycol	Water	25
Solubility (mole/L)	Vol. F. 1		
0.00010	0.00		
0.00011	0.10		
0.00018	0.20		
0.00028	0.30		
0.00058	0.40		
0.00098	0.50		
0.00274	0.60		
0.00469	0.70		
0.00890	0.80		
0.01173	0.90		
0.01854	1.00		

Constants of Jouyban–Acree model

J_0	−89.917
J_1	539.014
J_2	0

Reference of data: [55].

Solute	Solvent 1	Solvent 2	T (°C)
Dexamethasone sodium phosphate	Water	Ethanol	25.0
Solubility (mol F.)	Mass F. 1		
0.0252600	1.000		
0.0182700	0.926		
0.0139900	0.873		
0.0093250	0.736		
0.0060810	0.559		
0.0045380	0.479		
0.0034090	0.406		
0.0017820	0.241		
0.0006069	0.092		
0.0001751	0.000		

Constants of Jouyban–Acree model

J_0 429.097
J_1 −612.666
J_2 256.505

Reference of data: [56].

Solute	Solvent 1	Solvent 2	T (°C)
Dexamethasone sodium phosphate	Water	Methanol	25.0
Solubility (mol F.)	Mass F. 1		
0.025260	1.000		
0.018670	0.878		
0.012570	0.747		
0.007871	0.578		
0.005351	0.415		
0.003746	0.194		
0.003621	0.142		
0.003790	0.078		
0.004077	0.059		
0.004964	0.040		
0.006681	0.021		
0.009523	0.000		

Constants of Jouyban–Acree model

J_0 −412.603
J_1 647.360
J_2 −796.838

Reference of data: [56].

Solute	Solvent 1	Solvent 2	T (°C)
Diazepam	Ethanol	Water	30
Solubility (mole F.)	Vol. F. 1		
0.00000331	0.000		
0.00000785	0.100		
0.00002040	0.200		
0.00007380	0.300		
0.00028700	0.400		
0.00097600	0.500		
0.00219000	0.600		
0.00449000	0.700		
0.00679000	0.800		
0.00858000	0.900		
0.00754000	1.000		

Constants of Jouyban–Acree model

J_0	908.489
J_1	773.923
J_2	−453.683

Reference of data: [50].

Solute	Solvent 1	Solvent 2	T (°C)
Diazepam	Polyethylene glycol 200	Water	30
Solubility (mole F.)	Vol. F. 1		
0.00000350	0.000		
0.0000165	0.100		
0.0000443	0.200		
0.0001000	0.300		
0.0002330	0.400		
0.0005020	0.500		
0.0012300	0.600		
0.0033300	0.700		
0.0072900	0.800		
0.0168500	0.900		
0.0343800	1.000		

Constants of Jouyban–Acree model

J_0	209.221
J_1	−257.597
J_2	604.273

Reference of data: [50].

Solute	Solvent 1	Solvent 2	T (°C)
Diazepam	Propylene glycol	Water	30
Solubility (mole F.)	Vol. F. 1		
0.00000346	0.000		
0.00000700	0.100		
0.00001440	0.200		
0.00003220	0.300		
0.00007670	0.400		
0.00020000	0.500		
0.00049000	0.600		
0.00094800	0.700		
0.00191000	0.800		
0.00331000	0.900		
0.00515000	1.000		

Constants of Jouyban–Acree model

J_0	192.098
J_1	356.275
J_2	0

Reference of data: [50].

Solute	Solvent 1	Solvent 2	T (°C)
Diazepam	Ethanol	Water	25
Solubility (mole/L)	Vol. F. 1		
0.000152	0.00		
0.000390	0.10		
0.000737	0.20		
0.001989	0.30		
0.006657	0.40		
0.016729	0.50		
0.039249	0.60		
0.074858	0.70		
0.100714	0.80		
0.134731	0.90		
0.091580	1.00		

Constants of Jouyban–Acree model

J_0	763.592
J_1	731.827
J_2	0

Reference of data: [53].

Solute	Solvent 1	Solvent 2	T (°C)
Diazepam	N-Methylpyrrolidone	Water	25
Solubility (mole/L)	Vol. F. 1		
0.000155	0.00		
0.003660	0.10		
0.007372	0.20		
0.012592	0.30		
0.019376	0.40		
0.054339	0.50		
0.064200	0.60		
0.135020	0.70		
0.284932	0.80		
0.596340	0.90		
1.317100	1.00		

Constants of Jouyban–Acree model

J_0	494.280
J_1	−1264.279
J_2	1508.947

Reference of data: [54].

Solute	Solvent 1	Solvent 2	T (°C)
Diazepam	Propylene glycol	Water	25
Solubility (mole/L)	Vol. F. 1		
0.00015	0.00		
0.00044	0.10		
0.00052	0.20		
0.00094	0.30		
0.00171	0.40		
0.00285	0.50		
0.00552	0.60		
0.01168	0.70		
0.02164	0.80		
0.03494	0.90		
0.04282	1.00		

Constants of Jouyban–Acree model

J_0	161.839
J_1	0
J_2	0

Reference of data: [55].

Solute	Solvent 1	Solvent 2	T (°C)
Diclofenac (Na)	Ethanol	Water	25.0
Solubility (mole/L)	Vol. F. 1		
0.0674	0.000		
0.0735	0.100		
0.1073	0.200		
0.1761	0.300		
0.2980	0.400		
0.3995	0.500		
0.4766	0.600		
0.5300	0.670		
0.5495	0.700		
0.5917	0.770		
0.6303	0.800		
0.7267	0.830		
0.8517	0.850		
0.9827	0.870		
1.1851	0.900		
0.7907	0.930		
0.6439	0.950		
0.5544	0.970		
0.5202	1.000		

Constants of Jouyban–Acree model

J_0	478.937
J_1	151.290
J_2	0

Reference of data: [57].

Solute	Solvent 1	Solvent 2	T (°C)
Diclofenac (Na)	Methanol	Water	25.0
Solubility (mole/L)	Vol. F. 1		
0.0674	0.000		
0.0728	0.100		
0.0954	0.200		
0.0971	0.300		
0.1334	0.350		
0.1348	0.400		
0.1696	0.500		
0.2172	0.600		
0.2358	0.700		
0.5887	0.800		
1.5042	0.850		
1.4242	0.900		
1.3310	0.950		
1.0466	1.000		

Constants of Jouyban–Acree model

J_0	112.334
J_1	330.768
J_2	0

Reference of data: [57].

Solute	Solvent 1	Solvent 2	T (°C)
Diclofenac (Na)	Propanol (2)	Water	25.0
Solubility (mole/L)	Vol. F. 1		
0.0674	0.000		
0.1145	0.100		
0.1272	0.200		
0.2910	0.300		
0.3403	0.400		
0.4386	0.450		
0.8420	0.500		
0.4052	0.550		
0.3198	0.600		
0.3026	0.700		
0.1633	0.800		
0.1533	0.900		
0.0256	1.000		

Constants of Jouyban–Acree model

J_0	946.740
J_1	−195.318
J_2	0

Reference of data: [57].

Solute	Solvent 1	Solvent 2	T (°C)
Diglycine	Water	Dioxane	25.0
Solubility (g/100 g)	Mass F. 1		
22.75	1.000		
7.70	0.800		
1.54	0.600		
0.195	0.400		

Constants of Jouyban–Acree model

J_0	
J_1	
J_2	

Reference of data: [23].

Solute	Solvent 1	Solvent 2	T (°C)
Diglycine	Water	Ethanol	20.0
Solubility (mass F.)	Vol. F. 1		
18.45	1.000		
6.10	0.800		
0.94	0.500		

Constants of Jouyban–Acree model

J_0
J_1
J_2

Reference of data: [58].

Solute	Solvent 1	Solvent 2	T (°C)
Diglycine	Water	Ethanol	25.0
Solubility (mass F.)	Vol. F. 1		
20.5	1.000		
6.96	0.800		
1.02	0.500		

Constants of Jouyban–Acree model

J_0
J_1
J_2

Reference of data: [58].

Solute	Solvent 1	Solvent 2	T (°C)
Diglycine	Water	Ethanol	41.5
Solubility (mass F.)	Vol. F. 1		
26.4	1.000		
10.34	0.800		
1.62	0.500		

Constants of Jouyban–Acree model

J_0
J_1
J_2

Reference of data: [58].

Solute	Solvent 1	Solvent 2	T (°C)
Diglycine	Water	Ethanol	60.0
Solubility (mass F.)	Vol. F. 1		
34.36	1.000		
14.30	0.800		
2.63	0.500		

Constants of Jouyban–Acree model

J_0

J_1

J_2

Reference of data: [58].

Solute	Solvent 1	Solvent 2	T (°C)
Dihydroxyphenylalanine	Water	Ethanol	25.0
Solubility (g/100 g)	Mass F. 1		
0.38	1.000		
0.264	0.800		
0.189	0.600		
0.114	0.400		
0.039	0.200		

Constants of Jouyban–Acree model

J_0

J_1

J_2

Reference of data: [23].

Solute	Solvent 1	Solvent 2	T (°C)
Dodecyl p-aminobenzoate	Propylene glycol	Water	37.0
Solubility (mole F.)	Vol. F. 1		
0.000000000288	0.000		
0.000000003690	0.200		
0.000000044978	0.400		
0.000003388442	0.600		
0.000041114972	0.800		
0.001076465214	1.000		

Constants of Jouyban–Acree model

J_0	−230.979
J_1	0
J_2	0

Reference of data: [33].

Solute	Solvent 1	Solvent 2	T (°C)
Eflucimibe polymorph A	Ethanol	Heptane (n)	20
Solubility (mole F.)	Mass F. 1		
0.0005650	1.000		
0.0007940	0.960		
0.0009050	0.930		
0.0010100	0.880		
0.0012500	0.720		
0.0012500	0.670		
0.0011600	0.570		
0.0008400	0.460		
0.0004590	0.310		
0.0002660	0.220		
0.0001010	0.120		
0.0000094	0.000		

Constants of Jouyban–Acree model

J_0	1116.939
J_1	−377.971
J_2	674.422

Reference of data: [59].

Solute	Solvent 1	Solvent 2	T (°C)
Eflucimibe polymorph A	Ethanol	Heptane (n)	30
Solubility (mole F.)	Mass F. 1		
0.0010400	1.000		
0.0014100	0.960		
0.0015900	0.930		
0.0018000	0.880		
0.0020200	0.720		
0.0019900	0.670		
0.0017500	0.570		
0.0013400	0.460		
0.0007270	0.310		
0.0004000	0.220		
0.0001900	0.120		
0.0000272	0.000		

Constants of Jouyban–Acree model

J_0	1116.939
J_1	−377.971
J_2	674.422

Reference of data: [59].

Solute	Solvent 1	Solvent 2	T (°C)
Eflucimibe polymorph A	Ethanol	Heptane (n)	35
Solubility (mole F.)	Mass F. 1		
0.001470	1.000		
0.002020	0.960		
0.002370	0.930		
0.002680	0.880		
0.002940	0.720		
0.002810	0.670		
0.002440	0.570		
0.001910	0.460		
0.001070	0.310		
0.000625	0.220		
0.000260	0.120		
0.0000412	0.000		

Constants of Jouyban–Acree model

J_0	1116.939
J_1	−377.971
J_2	674.422

Reference of data: [59].

Solute	Solvent 1	Solvent 2	T (°C)
Eflucimibe polymorph A	Ethanol	Heptane (n)	40
Solubility (mole F.)	Mass F. 1		
0.0020700	1.000		
0.0027400	0.960		
0.0030500	0.930		
0.0036400	0.880		
0.0038000	0.720		
0.0036900	0.670		
0.0034300	0.570		
0.0027200	0.460		
0.0015000	0.310		
0.0010300	0.220		
0.0004730	0.120		
0.0000801	0.000		

Constants of Jouyban–Acree model

J_0	1116.939
J_1	−377.971
J_2	674.422

Reference of data: [59].

Solute	Solvent 1	Solvent 2	T (°C)
Eflucimibe polymorph A	Ethanol	Heptane (n)	50
Solubility (mole F.)	Mass F. 1		
0.004450	1.000		
0.005490	0.960		
0.006160	0.930		
0.007030	0.880		
0.007710	0.720		
0.007660	0.670		
0.006970	0.570		
0.005480	0.460		
0.003680	0.310		
0.002790	0.220		
0.000920	0.120		
0.000217	0.000		

Constants of Jouyban–Acree model

J_0	1116.939
J_1	−377.971
J_2	674.422

Reference of data: [59].

Solute	Solvent 1	Solvent 2	T (°C)
Eflucimibe polymorph A	Ethanol	Heptane (n)	54
Solubility (mole F.)	Mass F. 1		
0.005650	1.000		
0.006870	0.960		
0.007640	0.930		
0.008940	0.880		
0.010100	0.720		
0.010100	0.670		
0.009550	0.570		
0.008240	0.460		
0.005630	0.310		
0.004110	0.220		
0.001600	0.120		
0.000364	0.000		

Constants of Jouyban–Acree model

J_0	1116.939
J_1	−377.971
J_2	674.422

Reference of data: [59].

Solute	Solvent 1	Solvent 2	T (°C)
Eflucimibe polymorph B	Ethanol	Heptane (n)	20
Solubility (mole F.)	Mass F. 1		
0.00057700	1.000		
0.00082200	0.960		
0.00103000	0.880		
0.00135000	0.720		
0.00132000	0.570		
0.00053900	0.310		
0.00000953	0.000		

Constants of Jouyban–Acree model

J_0	1162.308
J_1	−426.772
J_2	784.757

Reference of data: [59].

Solute	Solvent 1	Solvent 2	T (°C)
Eflucimibe polymorph B	Ethanol	Heptane (n)	35
Solubility (mole F.)	Mass F. 1		
0.0016200	1.000		
0.0022800	0.960		
0.0031300	0.880		
0.0033000	0.720		
0.0025600	0.570		
0.0013400	0.310		
0.0000452	0.000		

Constants of Jouyban–Acree model

J_0	1162.308
J_1	−426.772
J_2	784.757

Reference of data: [59].

Solute	Solvent 1	Solvent 2	T (°C)
Eflucimibe polymorph B	Ethanol	Heptane (n)	40
Solubility (mole F.)	Mass F. 1		
0.00207	1.000		
0.00310	0.960		
0.00411	0.880		
0.00456	0.720		
0.00397	0.570		
0.00207	0.310		
0.00008	0.000		

Constants of Jouyban–Acree model

J_0	1162.308
J_1	−426.772
J_2	784.757

Reference of data: [59].

Solute	Solvent 1	Solvent 2	T (°C)
Eflucimibe polymorph B	Ethanol	Heptane (*n*)	50
Solubility (mole F.)	Mass F. 1		
0.004750	1.000		
0.006100	0.960		
0.008070	0.880		
0.008810	0.720		
0.007690	0.570		
0.004430	0.310		
0.000232	0.000		

Constants of Jouyban–Acree model

J_0	1162.308
J_1	−426.772
J_2	784.757

Reference of data: [59].

Solute	Solvent 1	Solvent 2	T (°C)
Eflucimibe polymorph B	Ethanol	Heptane (*n*)	54
Solubility (mole F.)	Mass F. 1		
0.006380	1.000		
0.007780	0.960		
0.010400	0.880		
0.011700	0.720		
0.010400	0.570		
0.006600	0.310		
0.000383	0.000		

Constants of Jouyban–Acree model

J_0	1162.308
J_1	−426.772
J_2	784.757

Reference of data: [59].

Solute	Solvent 1	Solvent 2	T (°C)
Erythromycin A dehydrate	Acetone	Water	20
Solubility (mole F.)	Mole F. 1		
0.00914000	1.000		
0.00600000	0.735		
0.00341000	0.554		
0.00181000	0.406		
0.00082400	0.317		
0.00036210	0.236		
0.00025010	0.173		
0.00019140	0.117		
0.00014340	0.071		
0.00005854	0.033		
0.00005400	0.000		

Constants of Jouyban–Acree model

J_0	1003.063
J_1	−371.496
J_2	0

Reference of data: [60].

Solute	Solvent 1	Solvent 2	T (°C)
Erythromycin A dehydrate	Acetone	Water	25
Solubility (mole F.)	Mole F. 1		
0.01310000	1.000		
0.00820000	0.735		
0.00471000	0.554		
0.00238900	0.406		
0.00117700	0.317		
0.00050210	0.236		
0.00033120	0.173		
0.00021660	0.117		
0.00014090	0.071		
0.00005098	0.033		
0.00004700	0.000		

Constants of Jouyban–Acree model

J_0	1003.063
J_1	−371.496
J_2	0

Reference of data: [60].

Solute	Solvent 1	Solvent 2	T (°C)
Erythromycin A dehydrate	Acetone	Water	30
Solubility (mole F.)	Mole F. 1		
0.02170000	1.000		
0.01264000	0.735		
0.00632000	0.554		
0.00349000	0.406		
0.00172100	0.317		
0.00073200	0.236		
0.00042980	0.173		
0.00023230	0.117		
0.00014070	0.071		
0.00004667	0.033		
0.00004200	0.000		

Constants of Jouyban–Acree model

J_0	1003.063
J_1	−371.496
J_2	0

Reference of data: [60].

Solute	Solvent 1	Solvent 2	T (°C)
Erythromycin A dehydrate	Acetone	Water	35
Solubility (mole F.)	Mole F. 1		
0.02930000	1.000		
0.01700000	0.735		
0.00828000	0.554		
0.00492000	0.406		
0.00243000	0.317		
0.00096120	0.236		
0.00053590	0.173		
0.00024830	0.117		
0.00014000	0.071		
0.00004186	0.033		
0.00004000	0.000		

Constants of Jouyban–Acree model

J_0	1003.063
J_1	−371.496
J_2	0

Reference of data: [60].

Solute	Solvent 1	Solvent 2	T (°C)
Erythromycin A dehydrate	Acetone	Water	40
Solubility (mole F.)	Mole F. 1		
0.04330000	1.000		
0.02464000	0.735		
0.01183000	0.554		
0.00638000	0.406		
0.00301000	0.317		
0.00119000	0.236		
0.00061820	0.173		
0.00027180	0.117		
0.00013880	0.071		
0.00003857	0.033		
0.00003800	0.000		

Constants of Jouyban–Acree model

J_0	1003.063
J_1	−371.496
J_2	0

Reference of data: [60].

Solute	Solvent 1	Solvent 2	T (°C)
Erythromycin A dehydrate	Acetone	Water	45
Solubility (mole F.)	Mole F. 1		
0.05420000	1.000		
0.03097000	0.735		
0.01489000	0.554		
0.00827000	0.406		
0.00389000	0.317		
0.00158000	0.236		
0.00073860	0.173		
0.00028990	0.117		
0.00013800	0.071		
0.00003623	0.033		
0.00003500	0.000		

Constants of Jouyban–Acree model

J_0	1003.063
J_1	−371.496
J_2	0

Reference of data: [60].

Solute	Solvent 1	Solvent 2	T (°C)
Erythromycin A dehydrate	Acetone	Water	50
Solubility (mole F.)	Mole F. 1		
0.08040000	1.000		
0.04536000	0.735		
0.02155000	0.554		
0.01069000	0.406		
0.00511000	0.317		
0.00201000	0.236		
0.00089530	0.173		
0.00032050	0.117		
0.00013760	0.071		
0.00003267	0.033		
0.00003400	0.000		

Constants of Jouyban–Acree model

J_0	1003.063
J_1	−371.496
J_2	0

Reference of data: [60].

Solute	Solvent 1	Solvent 2	T (°C)
Ethyl maltol	Ethanol	Water	20.0
Solubility (mole F.)	Mole F. 1		
0.0021	0.0000		
0.0115	0.1435		
0.0246	0.2811		
0.0462	0.4771		
0.0624	0.6100		
0.0762	0.7800		
0.0776	0.8800		
0.0715	0.9500		
0.0599	1.0000		

Constants of Jouyban–Acree model

J_0	953.176
J_1	−897.800
J_2	833.104

Reference of data: [61].

Solute	Solvent 1	Solvent 2	T (°C)
Ethyl maltol	Ethanol	Water	25.0
Solubility (mole F.)	Mole F. 1		
0.0023	0.0000		
0.0152	0.1435		
0.0373	0.2811		
0.0648	0.4771		
0.0799	0.6100		
0.0928	0.7800		
0.0930	0.8800		
0.0902	0.9500		
0.0798	1.0000		

Constants of Jouyban–Acree model

J_0	953.176
J_1	−897.800
J_2	833.104

Reference of data: [61].

Solute	Solvent 1	Solvent 2	T (°C)
Ethyl maltol	Ethanol	Water	30.0
Solubility (mole F.)	Mole F. 1		
0.0027	0.0000		
0.0216	0.1435		
0.0540	0.2811		
0.0847	0.4771		
0.1023	0.6100		
0.1142	0.7800		
0.1126	0.8800		
0.1097	0.9500		
0.0974	1.0000		

Constants of Jouyban–Acree model

J_0	953.176
J_1	−897.800
J_2	833.104

Reference of data: [61].

Solute	Solvent 1	Solvent 2	T (°C)
Ethyl maltol	Ethanol	Water	35.0
Solubility (mole F.)	Mole F. 1		
0.0033	0.0000		
0.0319	0.1435		
0.0804	0.2811		
0.1185	0.4771		
0.1341	0.6100		
0.1551	0.7800		
0.1620	0.8800		
0.1486	0.9500		
0.1329	1.0000		

Constants of Jouyban–Acree model

J_0	953.176
J_1	−897.800
J_2	833.104

Reference of data: [61].

Solute	Solvent 1	Solvent 2	T (°C)
Ethyl maltol	Ethanol	Water	40.0
Solubility (mole F.)	Mole F. 1		
0.0040	0.0000		
0.0466	0.1435		
0.1221	0.2811		
0.1672	0.4771		
0.1761	0.6100		
0.1987	0.7800		
0.1871	0.8800		
0.1811	0.9500		
0.1739	1.0000		

Constants of Jouyban–Acree model

J_0	953.176
J_1	−897.800
J_2	833.104

Reference of data: [61].

Solute	Solvent 1	Solvent 2	T (°C)
Ethyl maltol	Ethanol	Water	45.0
Solubility (mole F.)	Mole F. 1		
0.0051	0.0000		
0.0733	0.1435		
0.1878	0.2811		
0.2200	0.4771		
0.2271	0.6100		
0.2426	0.7800		
0.2445	0.8800		
0.2340	0.9500		
0.2272	1.0000		

Constants of Jouyban–Acree model

J_0	953.176
J_1	−897.800
J_2	833.104

Reference of data: [61].

Solute	Solvent 1	Solvent 2	T (°C)
Ethyl maltol	Ethanol	Water	50.0
Solubility (mole F.)	Mole F. 1		
0.0067	0.0000		
0.1182	0.1435		
0.2699	0.2811		
0.2856	0.4771		
0.2808	0.6100		
0.3024	0.7800		
0.3126	0.8800		
0.3020	0.9500		
0.2907	1.0000		

Constants of Jouyban–Acree model

J_0	953.176
J_1	−897.800
J_2	833.104

Reference of data: [61].

Solute	Solvent 1	Solvent 2	T (°C)
Ethyl maltol	Ethanol	Water	55.0
Solubility (mole F.)	Mole F. 1		
0.0089	0.0000		
0.1573	0.1435		
0.3636	0.2811		
0.3592	0.4771		
0.3577	0.6100		
0.3976	0.7800		
0.3968	0.8800		
0.3857	0.9500		
0.3669	1.0000		

Constants of Jouyban–Acree model

J_0	953.176
J_1	−897.800
J_2	833.104

Reference of data: [61].

Solute	Solvent 1	Solvent 2	T (°C)
Ethyl maltol	Ethanol	Water	60.0
Solubility (mole F.)	Mole F. 1		
0.0126	0.0000		
0.1976	0.1435		
0.4370	0.2811		
0.4495	0.4771		
0.4490	0.6100		
0.4799	0.7800		
0.4724	0.8800		
0.4649	0.9500		
0.4594	1.0000		

Constants of Jouyban–Acree model

J_0	953.176
J_1	−897.800
J_2	833.104

Reference of data: [61].

Solute	Solvent 1	Solvent 2	T (°C)
Ethyl p-aminobenzoate	Propylene glycol	Water	27.0
Solubility (mole F.)	Vol. F. 1		
0.0001073	0.000		
0.0001666	0.100		
0.0002639	0.200		
0.0004760	0.300		
0.0009780	0.400		
0.0023579	0.500		
0.0045620	0.600		
0.0095616	0.700		
0.0181334	0.800		
0.0320647	0.900		
0.0566989	1.000		

Constants of Jouyban–Acree model

J_0	−76.947
J_1	322.502
J_2	0

Reference of data: [32].

Solute	Solvent 1	Solvent 2	T (°C)
Ethyl p-aminobenzoate	Propylene glycol	Water	37.0
Solubility (mole F.)	Vol. F. 1		
0.00018450	0.000		
0.00046881	0.200		
0.00190546	0.400		
0.00460257	0.600		
0.01485936	0.800		
0.06902398	1.000		

Constants of Jouyban–Acree model

J_0	−150.490
J_1	0
J_2	0

Reference of data: [33].

Solute	Solvent 1	Solvent 2	T (°C)
Ethyl *p*-hydroxybenzoate	Propylene glycol	Water	27.0
Solubility (mole F.)	Vol. F. 1		
0.0001021	0.000		
0.0001463	0.100		
0.0002692	0.200		
0.0004760	0.300		
0.0008938	0.400		
0.0024788	0.500		
0.0060967	0.600		
0.0138427	0.700		
0.0254765	0.800		
0.0483156	0.900		
0.0780817	1.000		

Constants of Jouyban–Acree model

J_0	−77.744
J_1	488.107
J_2	0

Reference of data: [32].

Solute	Solvent 1	Solvent 2	T (°C)
Fluasterone	Ethanol	Water	Room temperature
Solubility (mole/L)	Vol. F. 1		
0.00000016	0.000		
0.00000037	0.063		
0.00000087	0.125		
0.00000207	0.188		
0.00000491	0.251		
0.00002770	0.376		
0.00016000	0.501		
0.00080000	0.627		
0.00354000	0.752		
0.01015000	0.877		

Constants of Jouyban–Acree model

J_0

J_1

J_2

Reference of data: [62].

Solute	Solvent 1	Solvent 2	T (°C)
Fluasterone	Methanol	Water	Room temperature
Solubility (mole/L)	Vol. F. 1		
0.00000016	0.000		
0.00000032	0.063		
0.00000066	0.127		
0.00000281	0.253		
0.00001190	0.379		
0.00003090	0.506		
0.00021000	0.632		
0.00104000	0.759		
0.00460000	0.886		

Constants of Jouyban–Acree model

J_0
J_1
J_2

Reference of data: [62].

Solute	Solvent 1	Solvent 2	T (°C)
Fluasterone	Propanol (1)	Water	Room temperature
Solubility (mole/L)	Vol. F. 1		
0.000000155	0.000		
0.000000164	0.003		
0.000000208	0.012		
0.000000324	0.031		
0.000000672	0.062		
0.000002910	0.124		
0.000057200	0.249		
0.000830000	0.373		
0.002660000	0.498		
0.006090000	0.622		
0.01083	0.746		

Constants of Jouyban–Acree model

J_0
J_1
J_2

Reference of data: [62].

Solute	Solvent 1	Solvent 2	T (°C)
Furosemide	Ethanol	Water	25
Solubility (mole F.)	Vol. F. 1		
0.00000231	0.000		
0.00000240	0.050		
0.00000293	0.100		
0.00000483	0.200		
0.00000805	0.250		
0.00001340	0.300		
0.00005222	0.400		
0.00014054	0.500		
0.00030659	0.600		
0.00076167	0.700		
0.00129402	0.800		
0.00200924	0.900		
0.00240549	1.000		

Constants of Jouyban–Acree model

J_0	297.529
J_1	908.075
J_2	−566.761

Reference of data: [63].

Solute	Solvent 1	Solvent 2	T (°C)
Furosemide	Propylene glycol	Water	25.0
Solubility (mole F.)	Vol. F. 1		
0.0000023	0.000		
0.0000026	0.050		
0.0000035	0.100		
0.0000052	0.200		
0.0000064	0.250		
0.0000083	0.300		
0.0000206	0.400		
0.0000502	0.500		
0.0001440	0.600		
0.0003360	0.700		
0.0010300	0.800		
0.0019600	0.900		
0.0037300	1.000		

Constants of Jouyban–Acree model

J_0	−323.918
J_1	548.855
J_2	233.181

Reference of data: [63].

Solute	Solvent 1	Solvent 2	T (°C)
Glucose (D)	Water	Ethanol	35.0
Solubility (mass F.)	Mass F. 1		
0.0042	1.000		
0.0219	0.880		
0.0736	0.741		
0.0885	0.695		
0.1488	0.586		
0.2238	0.466		
0.2855	0.357		
0.2885	0.346		
0.3770	0.263		
0.4323	0.199		
0.4790	0.156		
0.4976	0.130		
0.4920	0.113		
0.5202	0.099		
0.5514	0.067		
0.5620	0.066		
0.5818	0.000		
0.0042	1.000		

Constants of Jouyban–Acree model

J_0	884.250
J_1	−4460.462
J_2	0

Reference of data: [64].

Solute	Solvent 1	Solvent 2	T (°C)
Glutamine	Water	Dioxane	25.0
Solubility (g/100 g)	Mass F. 1		
4.150	1.000		
1.960	0.800		
0.671	0.600		
0.164	0.400		

Constants of Jouyban–Acree model

J_0
J_1
J_2

Reference of data: [23].

Solute	Solvent 1	Solvent 2	T (°C)
Glutamine	Water	Ethanol	25.0
Solubility (g/100 g)	Mass F. 1		
4.15	1.000		
1.87	0.800		
0.78	0.600		
0.26	0.400		

Constants of Jouyban–Acree model

J_0
J_1
J_2

Reference of data: [23].

Solute	Solvent 1	Solvent 2	T (°C)
Glycine	Water	Ethanol	25
Solubility (mole F.)	Vol. F. 1		
0.05669893	1.000		
0.02843882	0.800		
0.01227734	0.600		
0.00456197	0.400		
0.00105946	0.200		
0.00025354	0.100		
0.00002300	0.000		

Constants of Jouyban–Acree model

J_0 975.979
J_1 924.103
J_2 867.029

Reference of data: [12].

Solute	Solvent 1	Solvent 2	T (°C)
Glycine	Water	1-Propanol	25.0
Solubility (mole F.)	Mole F. 1		
0.05660	1.000		
0.04990	0.986		
0.03510	0.950		
0.02420	0.890		
0.01500	0.806		
0.00686	0.691		
0.00197	0.529		
0.00023	0.270		

Constants of Jouyban–Acree model

J_0
J_1
J_2

Reference of data: [14].

Solute	Solvent 1	Solvent 2	T (°C)
Glycine	Water	2-Propanol	25.0
Solubility (mole F.)	Mole F. 1		
0.056600	1.000		
0.048900	0.986		
0.031900	0.952		
0.017300	0.887		
0.009700	0.804		
0.003960	0.690		
0.000857	0.527		
0.000195	0.270		

Constants of Jouyban–Acree model

J_0
J_1
J_2

Reference of data: [14].

Solute	Solvent 1	Solvent 2	T (°C)
Glycine	Water	Dioxane	25.0
Solubility (g/100 g)	Mass F. 1		
25.160	1.000		
10.900	0.800		
6.350	0.700		
3.300	0.600		
0.640	0.400		
0.062	0.200		

Constants of Jouyban–Acree model

J_0
J_1
J_2

Reference of data: [23].

Solute	Solvent 1	Solvent 2	T (°C)
Glycine	Water	Ethanol	20.0
Solubility (g/100 g)	Vol. F. 1		
22.95	1.000		
9.29	0.800		
2.05	0.500		

Constants of Jouyban–Acree model

J_0

J_1

J_2

Reference of data: [58].

Solute	Solvent 1	Solvent 2	T (°C)
Glycine	Water	Ethanol	25.0
Solubility (g/100 g)	Vol. F. 1		
25.18	1.000		
10.78	0.800		
2.41	0.500		

Constants of Jouyban–Acree model

J_0

J_1

J_2

Reference of data: [58].

Solute	Solvent 1	Solvent 2	T (°C)
Glycine	Water	Ethanol	41.5
Solubility (g/100 g)	Vol. F. 1		
29.0	1.000		
15.50	0.800		
3.99	0.500		

Constants of Jouyban–Acree model

J_0

J_1

J_2

Reference of data: [58].

Solute	Solvent 1	Solvent 2	T (°C)
Glycine	Water	Ethanol	60.0
Solubility (g/100 g)	Vol. F. 1		
33.47	1.000		
19.66	0.800		
6.38	0.500		

Constants of Jouyban–Acree model

J_0

J_1

J_2

Reference of data: [58].

Solute	Solvent 1	Solvent 2	T (°C)
Glycine	Water	Ethanol	25.0
Solubility (g/100 g)	Mass F. 1		
25.160	1.000		
11.300	0.800		
6.830	0.700		
4.250	0.600		
2.570	0.500		
1.400	0.400		
0.243	0.200		
0.050	0.100		

Constants of Jouyban–Acree model

J_0

J_1

J_2

Reference of data: [23].

Solute	Solvent 1	Solvent 2	T (°C)
Glycine	Water	Ethanol	25.0
Solubility (g/L)	Vol. F. 1		
206.41	1.000		
162.51	0.900		
120.67	0.800		
76.58	0.700		
46.55	0.600		
26.62	0.500		
14.12	0.400		

Constants of Jouyban–Acree model

J_0

J_1

J_2

Reference of data: [65].

Solute	Solvent 1	Solvent 2	T (°C)
Glycine	Water	Methanol	15.0
Solubility (mole F.)	Mole F. 1		
3.0250	1.000		
2.8500	0.920		
2.6750	0.840		
1.1750	0.748		
1.0125	0.656		
0.8750	0.553		
0.1225	0.458		
0.8750	0.359		
0.0675	0.241		
0.0160	0.121		
0.0065	0.000		

Constants of Jouyban–Acree model

J_0	489.883
J_1	0
J_2	0

Reference of data: [15].

Solute	Solvent 1	Solvent 2	T (°C)
Glycine	Water	Methanol	20.0
Solubility (mole F.)	Mole F. 1		
3.180	1.000		
3.020	0.920		
2.860	0.840		
1.267	0.748		
1.115	0.656		
0.910	0.553		
0.140	0.458		
0.100	0.359		
0.080	0.241		
0.025	0.121		
0.009	0.000		

Constants of Jouyban–Acree model

J_0	489.883
J_1	0
J_2	0

Reference of data: [15].

Solute	Solvent 1	Solvent 2	T (°C)
Glycine	Water	Methanol	25.0
Solubility (mole F.)	Mole F. 1		
3.3250	1.000		
3.1750	0.920		
3.0250	0.840		
1.3625	0.748		
1.2250	0.656		
0.9500	0.553		
0.1575	0.458		
0.1150	0.359		
0.1050	0.241		
0.0400	0.121		
0.0115	0.000		

Constants of Jouyban–Acree model

J_0	489.883
J_1	0
J_2	0

Reference of data: [15].

Solute	Solvent 1	Solvent 2	T (°C)
Glycine	Water	Ethanol	25.0
Solubility (g/L)	Vol. F. 1		
206.41	1.000		
162.51	0.900		
120.67	0.800		
76.58	0.700		
46.55	0.600		
26.62	0.500		
14.12	0.400		

Constants of Jouyban–Acree model

J_0
J_1
J_2

Reference of data: [65].

Solute	Solvent 1	Solvent 2	T (°C)
Hexachlorobenzene	Ethanol	Water	23.0
Solubility (mole/L)	Vol. F. 1		
0.00000002	0.000		
0.00000009	0.200		
0.00000324	0.400		
0.00007079	0.600		
0.00045709	0.800		
0.00316228	1.000		

Constants of Jouyban–Acree model

J_0	432.657
J_1	866.238
J_2	−1561.730

Reference of data: [66].

Solute	Solvent 1	Solvent 2	T (°C)
Hexaglycine	Water	Ethanol	25.0
Solubility (g/L)	Vol. F. 1		
0.60	1.000		
0.42	0.900		
0.30	0.800		
0.22	0.700		
0.17	0.600		
0.13	0.500		
0.10	0.400		

Constants of Jouyban–Acree model

J_0

J_1

J_2

Reference of data: [65].

Solute	Solvent 1	Solvent 2	T (°C)
Hexyl p-aminobenzoate	Propylene glycol	Water	37.0
Solubility (mole F.)	Vol. F. 1		
0.00000193	0.000		
0.00000820	0.200		
0.00005834	0.400		
0.00076736	0.600		
0.00477529	0.800		
0.07030723	1.000		

Constants of Jouyban–Acree model

J_0	−375.784
J_1	0
J_2	0

Reference of data: [33].

Solute	Solvent 1	Solvent 2	T (°C)
Histidine	Water	Dioxane	25.0
Solubility (g/100 g)	Mass F. 1		
4.280	1.000		
2.470	0.800		
1.630	0.700		
1.000	0.600		
0.255	0.400		

Constants of Jouyban–Acree model

J_0
J_1
J_2

Reference of data: [23].

Solute	Solvent 1	Solvent 2	T (°C)
Histidine (L)	Water	Methanol	15.0
Solubility (mole F.)	Mole F. 1		
0.258000	1.000		
0.253000	0.920		
0.161000	0.840		
0.145000	0.748		
0.111000	0.656		
0.043700	0.553		
0.014500	0.458		
0.011300	0.359		
0.001150	0.241		
0.000824	0.121		
0.000483	0.000		

Constants of Jouyban–Acree model

J_0	438.789
J_1	700.286
J_2	−662.393

Reference of data: [15].

Solute	Solvent 1	Solvent 2	T (°C)
Histidine (L)	Water	Methanol	20.0
Solubility (mole F.)	Mole F. 1		
0.264000	1.000		
0.259000	0.920		
0.166000	0.840		
0.150000	0.748		
0.116000	0.656		
0.046800	0.553		

(continued)

Solute	Solvent 1	Solvent 2	T (°C)
0.016200	0.458		
0.013500	0.359		
0.001510	0.241		
0.001120	0.121		
0.000646	0.000		

Constants of Jouyban–Acree model

J_0	438.789
J_1	700.286
J_2	−662.393

Reference of data: [15].

Solute	Solvent 1	Solvent 2	T (°C)
Histidine (L)	Water	Methanol	25.0
Solubility (mole F.)	Mole F. 1		
0.2720	1.000		
0.2660	0.920		
0.1710	0.840		
0.1550	0.748		
0.1220	0.656		
0.0505	0.553		
0.0188	0.458		
0.0160	0.359		
0.0020	0.241		
0.0015	0.121		
0.0009	0.000		

Constants of Jouyban–Acree model

J_0	438.789
J_1	700.286
J_2	−662.393

Reference of data: [15].

Solute	Solvent 1	Solvent 2	T (°C)
Hydrocortisone	Propylene glycol	Water	25.0
Solubility (mole F.)	Vol. F. 1		
0.0000147	0.000		
0.0000410	0.200		
0.0001540	0.400		
0.0005200	0.600		
0.0016800	0.800		
0.0033800	1.000		

Constants of Jouyban–Acree model

J_0	129.808
J_1	277.835
J_2	0

Reference of data: [67].

Solute	Solvent 1	Solvent 2	T (°C)
Hydroxybenzoic acid (para)	Dioxane	Water	25.0
Solubility (mole F.)	Vol. F. 1		
0.0006	0.000		
0.0072	0.200		
0.0302	0.450		
0.0478	0.500		
0.0585	0.550		
0.0710	0.600		
0.0820	0.650		
0.0939	0.700		
0.1150	0.800		
0.1210	0.850		
0.1270	0.900		
0.1220	0.950		
0.0844	1.000		

Constants of Jouyban–Acree model

J_0	943.071
J_1	−80.781
J_2	576.336

Reference of data: [68].

Solute	Solvent 1	Solvent 2	T (°C)
Hydroxyphenyl glycine sulfate (DL-p)	Water	Acetone	30.0
Solubility (mass F.)	Mass F. 1		
0.0000218	0.000		
0.0000150	0.244		
0.0000322	0.392		
0.0000699	0.492		
0.0001293	0.563		
0.0002951	0.659		
0.0004964	0.721		
0.0006611	0.763		
0.0008303	0.801		
0.0010434	0.843		
0.0012366	0.890		
0.0013517	0.942		
0.0012842	1.000		

Constants of Jouyban–Acree model

J_0	−550.108
J_1	1245.713
J_2	−213.838

Reference of data: [69,70].

Solute	Solvent 1	Solvent 2	T (°C)
Hydroxyphenyl glycine sulfate (DL-p)	Water	Acetone	35.0
Solubility (mass F.)	Mass F. 1		
0.0000270	0.0000		
0.0000147	0.2437		
0.0000341	0.3919		
0.0000780	0.4916		
0.0001461	0.5631		
0.0003360	0.6591		
0.0005428	0.7205		
0.0007346	0.7632		
0.0009401	0.8011		
0.0011700	0.8430		
0.0014311	0.8896		
0.0016484	0.9415		
0.0017522	1.0000		

Constants of Jouyban–Acree model

J_0	−550.108
J_1	1245.713
J_2	−213.838

Reference of data: [69,70].

Solute	Solvent 1	Solvent 2	T (°C)
Hydroxyphenyl glycine sulfate (DL-p)	Water	Acetone	41.0
Solubility (mass F.)	Mass F. 1		
0.0000337	0.0000		
0.0000156	0.2437		
0.0000370	0.3919		
0.0000868	0.4916		
0.0001624	0.5631		
0.0003774	0.6591		
0.0006096	0.7205		
0.0008276	0.7632		
0.0010658	0.8011		
0.0013446	0.8430		
0.0016890	0.8896		
0.0020435	0.9415		
0.0023387	1.0000		

Constants of Jouyban–Acree model

J_0	−550.108
J_1	1245.713
J_2	−213.838

Reference of data: [69,70].

Solute	Solvent 1	Solvent 2	T (°C)
Hydroxyphenyl glycine sulfate (DL-*p*)	Water	Acetone	46.0
Solubility (mass F.)	Mass F. 1		
0.0000354	0.0000		
0.0000163	0.2437		
0.0000405	0.3919		
0.0000940	0.4916		
0.0001794	0.5631		
0.0004138	0.6591		
0.0006707	0.7205		
0.0009150	0.7632		
0.0011609	0.8011		
0.0014742	0.8430		
0.0018538	0.8896		
0.0022546	0.9415		
0.0026328	1.0000		

Constants of Jouyban–Acree model

J_0	−550.108
J_1	1245.713
J_2	−213.838

Reference of data: [69,70].

Solute	Solvent 1	Solvent 2	T (°C)
Hydroxyphenyl glycine sulfate (DL-*p*)	Water	Acetone	50.0
Solubility (mass F.)	Mass F. 1		
0.00004000	0.0000		
0.00001830	0.2437		
0.00004480	0.3919		
0.00010490	0.4916		
0.00019930	0.5631		
0.00045160	0.6591		
0.00073640	0.7205		
0.00098690	0.7632		
0.00125810	0.8011		
0.00158560	0.8430		
0.00200070	0.8896		
0.00245100	0.9415		
0.00288660	1.0000		

Constants of Jouyban–Acree model

J_0	−550.108
J_1	1245.713
J_2	−213.838

Reference of data: [69,70].

Solute	Solvent 1	Solvent 2	T (°C)
Hydroxyphenylglycine (D-p)	Water	Propanol (2)	20.1
Solubility (mole F.)	Vol. F. 1		
0.001960	1.000		
0.001631	0.968		
0.001343	0.930		
0.001090	0.886		
0.000862	0.834		
0.000667	0.770		
0.000506	0.690		
0.000340	0.589		
0.000194	0.455		

Constants of Jouyban–Acree model

J_0
J_1
J_2

Reference of data: [71].

Solute	Solvent 1	Solvent 2	T (°C)
Hydroxyphenylglycine (D-p)	Water	Propanol (2)	25.0
Solubility (mole F.)	Vol. F. 1		
0.002101	1.000		
0.001753	0.968		
0.001457	0.930		
0.001187	0.886		
0.000991	0.834		
0.000762	0.770		
0.000562	0.690		
0.000375	0.589		
0.000213	0.455		

Constants of Jouyban–Acree model

J_0
J_1
J_2

Reference of data: [71].

Solute	Solvent 1	Solvent 2	T (°C)
Hydroxyphenylglycine (D-p)	Water	Propanol (2)	30.0
Solubility (mole F.)	Vol. F. 1		
0.002237	1.000		
0.001899	0.968		
0.001652	0.930		
0.001339	0.886		
0.001125	0.834		
0.000858	0.770		
0.000640	0.690		
0.000416	0.589		
0.000234	0.455		

Constants of Jouyban–Acree model

J_0

J_1

J_2

Reference of data: [71].

Solute	Solvent 1	Solvent 2	T (°C)
Hydroxyphenylglycine (D-p)	Water	Propanol (2)	35.0
Solubility (mole F.)	Vol. F. 1		
0.002407	1.000		
0.002043	0.968		
0.001731	0.930		
0.001490	0.886		
0.001236	0.834		
0.000952	0.770		
0.000695	0.690		
0.000442	0.589		
0.000255	0.455		

Constants of Jouyban–Acree model

J_0

J_1

J_2

Reference of data: [71].

Solute	Solvent 1	Solvent 2	T (°C)
Hydroxyphenylglycine (D-p)	Water	Propanol (2)	39.9
Solubility (mole F.)	Vol. F. 1		
0.002548	1.000		
0.002216	0.968		
0.001929	0.930		
0.001634	0.886		
0.001361	0.834		
0.001047	0.770		
0.000774	0.690		
0.000488	0.589		
0.000277	0.455		

Constants of Jouyban–Acree model

J_0

J_1

J_2

Reference of data: [71].

Solute	Solvent 1	Solvent 2	T (°C)
Hydroxyphenylglycine (D-p)	Water	Propanol (2)	45.0
Solubility (mole F.)	Vol. F. 1		
0.002718	1.000		
0.002402	0.968		
0.002079	0.930		
0.001824	0.886		
0.001514	0.834		
0.001171	0.770		
0.000854	0.690		
0.000525	0.589		
0.000300	0.455		

Constants of Jouyban–Acree model

J_0

J_1

J_2

Reference of data: [71].

Solute	Solvent 1	Solvent 2	T (°C)
Hydroxyphenylglycine (D-p)	Water	Propanol (2)	50.0
Solubility (mole F.)	Vol. F. 1		
0.002982	1.000		
0.002592	0.968		
0.002246	0.930		
0.002004	0.886		
0.001645	0.834		
0.001259	0.770		
0.000940	0.690		
0.000570	0.589		
0.000323	0.455		

Constants of Jouyban–Acree model

J_0

J_1

J_2

Reference of data: [71].

Solute	Solvent 1	Solvent 2	T (°C)
Hydroxyphenylglycine (D-p)	Water	Propanol (2)	55.0
Solubility (mole F.)	Vol. F. 1		
0.003187	1.000		
0.002831	0.968		
0.002493	0.930		
0.002192	0.886		
0.001819	0.834		
0.001421	0.770		
0.001029	0.690		
0.000645	0.589		
0.000346	0.455		

Constants of Jouyban–Acree model

J_0
J_1
J_2

Reference of data: [71].

Solute	Solvent 1	Solvent 2	T (°C)
Hydroxyphenylglycine (D-p)	Water	Propanol (2)	60.0
Solubility (mole F.)	Vol. F. 1		
0.003439	1.000		
0.003076	0.968		
0.002724	0.930		
0.002417	0.886		
0.001928	0.834		
0.001541	0.770		
0.001152	0.690		
0.000699	0.589		
0.000371	0.455		

Constants of Jouyban–Acree model

J_0
J_1
J_2

Reference of data: [71].

Solute	Solvent 1	Solvent 2	T (°C)
Hydroxyphenylglycine (D-p)	Water	Propanol (2)	65.0
Solubility (mole F.)	Vol. F. 1		
0.003663	1.000		
0.003330	0.968		
0.002982	0.930		
0.002590	0.886		
0.002152	0.834		
0.001699	0.770		
0.001238	0.690		
0.000760	0.589		
0.000396	0.455		

Constants of Jouyban–Acree model

J_0
J_1
J_2

Reference of data: [71].

Solute	Solvent 1	Solvent 2	T (°C)
Hydroxyphenylglycine (D-p)	Water	Propanol (2)	69.9
Solubility (mole F.)	Vol. F. 1		
0.003910	1.000		
0.003601	0.968		
0.003259	0.930		
0.002786	0.886		
0.002345	0.834		
0.001863	0.770		
0.001325	0.690		
0.000812	0.589		
0.000421	0.455		

Constants of Jouyban–Acree model

J_0
J_1
J_2

Reference of data: [71].

Solute	Solvent 1	Solvent 2	T (°C)
Ibesartan (form B)	Water	Ethanol	10
Solubility (mole F.)	Vol. F. 1		
0.00001028	0.683		
0.00002002	0.581		
0.00002460	0.447		
0.00003845	0.264		

Constants of Jouyban–Acree model

J_0
J_1
J_2

Reference of data: [72].

Solute	Solvent 1	Solvent 2	T (°C)
Ibesartan (form B)	Water	Ethanol	15
Solubility (mole F.)	Vol. F. 1		
0.00001239	0.683		
0.00002295	0.581		
0.00002771	0.447		
0.00004771	0.264		

Constants of Jouyban–Acree model

J_0
J_1
J_2

Reference of data: [72].

Solute	Solvent 1	Solvent 2	T (°C)
Ibesartan (form B)	Water	Ethanol	20
Solubility (mole F.)	Vol. F. 1		
0.00000750	0.764		
0.00001645	0.683		
0.00003293	0.581		
0.00005743	0.447		
0.00006270	0.264		

Constants of Jouyban–Acree model

J_0
J_1
J_2

Reference of data: [72].

Solute	Solvent 1	Solvent 2	T (°C)
Ibesartan (form B)	Water	Ethanol	25
Solubility (mole F.)	Vol. F. 1		
0.00000999	0.764		
0.00002133	0.683		
0.00004255	0.581		
0.00005743	0.447		
0.00008555	0.264		

Constants of Jouyban–Acree model

J_0
J_1
J_2

Reference of data: [72].

Solute	Solvent 1	Solvent 2	T (°C)
Ibesartan (form B)	Water	Ethanol	30
Solubility (mole F.)	Vol. F. 1		
0.00001118	0.764		
0.00002999	0.683		
0.00005653	0.581		
0.00008345	0.447		
0.00011630	0.264		

Constants of Jouyban–Acree model

J_0
J_1
J_2

Reference of data: [72].

Solute	Solvent 1	Solvent 2	T (°C)
Ibesartan (form B)	Water	Ethanol	35
Solubility (mole F.)	Vol. F. 1		
0.00001672	0.764		
0.00004050	0.683		
0.00007354	0.581		
0.00011050	0.447		
0.00015590	0.264		

Constants of Jouyban–Acree model

J_0
J_1
J_2

Reference of data: [72].

Solute	Solvent 1	Solvent 2	T (°C)
Ibesartan (form B)	Water	Ethanol	40
Solubility (mole F.)	Vol. F. 1		
0.00002346	0.764		
0.00005740	0.683		
0.00009572	0.581		
0.00014520	0.447		
0.00020030	0.264		

Constants of Jouyban–Acree model

J_0
J_1
J_2

Reference of data: [72].

Solute	Solvent 1	Solvent 2	T (°C)
Ibesartan (form B)	Water	Ethanol	45
Solubility (mole F.)	Vol. F. 1		
0.00002958	0.764		
0.00007185	0.683		
0.00011970	0.581		
0.00019220	0.447		
0.00025990	0.264		

Constants of Jouyban–Acree model

J_0
J_1
J_2

Reference of data: [72].

Solute	Solvent 1	Solvent 2	T (°C)
Ibesartan (form B)	Water	Ethanol	50
Solubility (mole F.)	Vol. F. 1		
0.00003936	0.764		
0.00009804	0.683		
0.00015980	0.581		
0.00025590	0.447		
0.00033200	0.264		

Constants of Jouyban–Acree model

J_0
J_1
J_2

Reference of data: [72].

Solute	Solvent 1	Solvent 2	T (°C)
Ibesartan (form B)	Water	Ethanol	55
Solubility (mole F.)	Vol. F. 1		
0.00004581	0.764		
0.00012650	0.683		
0.00019770	0.581		
0.00033400	0.447		
0.00042010	0.264		

Constants of Jouyban–Acree model

J_0
J_1
J_2

Reference of data: [72].

Solute	Solvent 1	Solvent 2	T (°C)
Ibesartan (form B)	Water	Ethanol	60
Solubility (mole F.)	Vol. F. 1		
0.00005778	0.764		
0.00016180	0.683		
0.00025630	0.581		
0.00042500	0.447		
0.00053570	0.264		

Constants of Jouyban–Acree model

J_0

J_1

J_2

Reference of data: [72].

Solute	Solvent 1	Solvent 2	T (°C)
Ibuprofen	Ethanol	Propylene glycol	20
Solubility (mole F.)	Mass F. 1		
0.0747	0.000		
0.1170	0.200		
0.1600	0.400		
0.1840	0.600		
0.2020	0.800		
0.2050	1.000		

Constants of Jouyban–Acree model

J_0	94.559
J_1	0
J_2	0

Reference of data: [73].

Solute	Solvent 1	Solvent 2	T (°C)
Ibuprofen	Ethanol	Propylene glycol	25
Solubility (mole F.)	Mass F. 1		
0.0985	0.000		
0.1470	0.200		
0.1920	0.400		
0.2090	0.600		
0.2330	0.800		
0.2410	1.000		

Constants of Jouyban–Acree model

J_0	94.559
J_1	0
J_2	0

Reference of data: [73].

Solute	Solvent 1	Solvent 2	T (°C)
Ibuprofen	Ethanol	Propylene glycol	30
Solubility (mole F.)	Mass F. 1		
0.141	0.000		
0.183	0.200		
0.231	0.400		
0.250	0.600		
0.268	0.800		
0.284	1.000		

Constants of Jouyban–Acree model

J_0	94.559
J_1	0
J_2	0

Reference of data: [73].

Solute	Solvent 1	Solvent 2	T (°C)
Ibuprofen	Ethanol	Propylene glycol	35
Solubility (mole F.)	Mass F. 1		
0.188	0.000		
0.230	0.200		
0.276	0.400		
0.290	0.600		
0.316	0.800		
0.340	1.000		

Constants of Jouyban–Acree model

J_0	94.559
J_1	0
J_2	0

Reference of data: [73].

Solute	Solvent 1	Solvent 2	T (°C)
Ibuprofen	Ethanol	Propylene glycol	40
Solubility (mole F.)	Mass F. 1		
0.217	0.000		
0.260	0.200		
0.310	0.400		
0.337	0.600		
0.363	0.800		
0.367	1.000		

Constants of Jouyban–Acree model

J_0	94.559
J_1	0
J_2	0

Reference of data: [73].

Solute	Solvent 1	Solvent 2	T (°C)
Ibuprofen	Propylene glycol	Water	20
Solubility (mole F.)	Vol. F. 1		
0.00001199	0.000		
0.00001985	0.100		
0.00003070	0.200		
0.00006829	0.300		
0.00017910	0.400		
0.00056860	0.500		
0.00175100	0.600		
0.00567800	0.700		
0.01905000	0.800		
0.06966000	0.900		
0.20790000	1.000		

Constants of Jouyban–Acree model

J_0	−545.695
J_1	374.414
J_2	0

Reference of data: [74].

Solute	Solvent 1	Solvent 2	T (°C)
Ibuprofen	Propylene glycol	Water	20
Solubility (mole F.)	Mass F. 1		
0.0000112	0.000		
0.0000179	0.200		
0.0000878	0.400		
0.0004960	0.600		
0.0065100	0.800		
0.0747000	1.000		

Constants of Jouyban–Acree model

J_0	−851.074
J_1	0
J_2	0

Reference of data: [75].

Solute	Solvent 1	Solvent 2	T (°C)
Ibuprofen	Propylene glycol	Water	25
Solubility (mole F.)	Mass F. 1		
0.0000136	0.000		
0.0000218	0.200		
0.0001060	0.400		
0.0005690	0.600		
0.0077700	0.800		
0.0985000	1.000		

Constants of Jouyban–Acree model

J_0	−851.074
J_1	0
J_2	0

Reference of data: [75].

Solute	Solvent 1	Solvent 2	T (°C)
Ibuprofen	Propylene glycol	Water	30
Solubility (mole F.)	Mass F. 1		
0.0000155	0.000		
0.0000292	0.200		
0.0001250	0.400		
0.0007960	0.600		
0.0118000	0.800		
0.1410000	1.000		

Constants of Jouyban–Acree model

J_0	−851.074
J_1	0
J_2	0

Reference of data: [75].

Solute	Solvent 1	Solvent 2	T (°C)
Ibuprofen	Propylene glycol	Water	35
Solubility (mole F.)	Mass F. 1		
0.000019	0.000		
0.000036	0.200		
0.000171	0.400		
0.001080	0.600		
0.017800	0.800		
0.188000	1.000		

Constants of Jouyban–Acree model

J_0	−851.074
J_1	0
J_2	0

Reference of data: [75].

Solute	Solvent 1	Solvent 2	T (°C)
Ibuprofen	Propylene glycol	Water	40
Solubility (mole F.)	Mass F. 1		
0.0000224	0.000		
0.0000457	0.200		
0.0002250	0.400		
0.0013600	0.600		
0.0027300	0.800		
0.2170000	1.000		

Constants of Jouyban–Acree model

J_0	−851.074
J_1	0
J_2	0

Reference of data: [75].

Solute	Solvent 1	Solvent 2	T (°C)
Isoleucine (L)	Water	1-Propanol	25.0
Solubility (mole F.)	Mole F. 1		
0.00438	1.000		
0.00403	0.985		
0.00336	0.949		
0.00331	0.887		
0.00317	0.804		
0.00262	0.690		
0.00147	0.528		
0.000336	0.271		
0.000045	0.000		
0.004380	1.000		

Constants of Jouyban–Acree model

J_0	509.024
J_1	0
J_2	0

Reference of data: [14].

Solute	Solvent 1	Solvent 2	T (°C)
Isoleucine (L)	Water	2-Propanol	25.0
Solubility (mole F.)	Mole F. 1		
0.004380	1.000		
0.003970	0.986		
0.002990	0.950		
0.002280	0.885		
0.002050	0.802		
0.001210	0.689		
0.000808	0.527		
0.000297	0.270		
0.000019	0.000		

Constants of Jouyban–Acree model

J_0 520.170
J_1 −647.348
J_2 0

Reference of data: [14].

Solute	Solvent 1	Solvent 2	T (°C)
Isoproternol HCl	Water	Propylene glycol	22.0
Solubility (g/L)	Vol. F. 1		
389	1.000		
357.8	0.900		
335.1	0.800		
288.3	0.600		
243.3	0.400		
91.6	0.900		

Constants of Jouyban–Acree model

J_0 520.170
J_1 −647.348
J_2 0

Reference of data: [76].

Solute	Solvent 1	Solvent 2	T (°C)
Ketoprofen	Propylene glycol	Water	25.0
Solubility (g/L)	Vol. F. 1		
0.1073	0.000		
0.8930	0.100		
1.380	0.200		
4.530	0.300		
10.210	0.400		
15.340	0.500		
23.150	0.600		
48.710	0.700		
169.160	0.800		
247.910	0.900		
446.380	1.000		

Constants of Jouyban–Acree model

J_0 375.512
J_1 −493.781
J_2 947.356

Reference of data: [77].

Solute	Solvent 1	Solvent 2	T (°C)
Ketoprofen	Propylene glycol	Water	37.0
Solubility (g/L)	Vol. F. 1		
0.1308	0.000		
0.980	0.100		
1.480	0.200		
4.830	0.300		
11.030	0.400		
16.470	0.500		
28.170	0.600		
75.870	0.700		
215.980	0.800		
320.720	0.900		
515.070	1.000		

Constants of Jouyban–Acree model

J_0	375.512
J_1	−493.781
J_2	947.356

Reference of data: [77].

Solute	Solvent 1	Solvent 2	T (°C)
Lamivudine (polymorphs I and II)	Ethanol	Water	25.0
Solubility (g/L)	Vol. F. 1		
11.4	0.000		
28.5	0.050		
45.4	0.100		
60.3	0.120		
71.9	0.150		
85.8	0.180		
92.3	0.200		
171.8	0.400		
173	0.600		
108.30	0.800		
91.3	0.900		
84.9	1.000		

Constants of Jouyban–Acree model

J_0	897.913
J_1	−754.901
J_2	275.654

Reference of data: [78].

Solute	Solvent 1	Solvent 2	T (°C)
Lamotrigine	Ethanol	Water	25
Solubility (mole/L)	Vol. F. 1		
0.000729	0.00		
0.001869	0.10		
0.002311	0.20		
0.004388	0.30		
0.009685	0.40		
0.020323	0.50		
0.035616	0.60		
0.042416	0.70		
0.043858	0.80		
0.037616	0.90		
0.013975	1.00		

Constants of Jouyban–Acree model

J_0	955.731
J_1	773.516
J_2	0

Reference of data: [53].

Solute	Solvent 1	Solvent 2	T (°C)
Lamotrigine	N-Methylpyrrolidone	Water	25
Solubility (mole/L)	Vol. F. 1		
0.000729	0.00		
0.003205	0.10		
0.007324	0.20		
0.012431	0.30		
0.020131	0.40		
0.029460	0.50		
0.030614	0.60		
0.032826	0.70		
0.040228	0.80		
0.045758	0.90		
0.058526	1.00		

Constants of Jouyban–Acree model

J_0	742.968
J_1	−648.861
J_2	0

Reference of data: [54].

Solute	Solvent 1	Solvent 2	T (°C)
Lamotrigine	Propylene glycol	Water	25
Solubility (mole/L)	Vol. F. 1		
0.000730	0.00		
0.001250	0.10		
0.002010	0.20		
0.003110	0.30		
0.007320	0.40		
0.012670	0.50		
0.035330	0.60		
0.059940	0.70		
0.112190	0.80		
0.174380	0.90		
0.204190	1.00		

Constants of Jouyban–Acree model

J_0	112.240
J_1	472.388
J_2	0

Reference of data: [55].

Solute	Solvent 1	Solvent 2	T (°C)
Lorazepam	Ethanol	Water	30
Solubility (mole F.)	Vol. F. 1		
0.00000345	0.000		
0.00000737	0.100		
0.00001710	0.200		
0.00006630	0.300		
0.00033600	0.400		
0.00044900	0.500		
0.00095900	0.600		
0.00173000	0.700		
0.00253000	0.800		
0.00269000	0.900		
0.00196000	1.000		

Constants of Jouyban–Acree model

J_0	924.315
J_1	621.135
J_2	0

Reference of data: [50].

Solute	Solvent 1	Solvent 2	T (°C)
Lorazepam	Polyethylene glycol 200	Water	30
Solubility (mole F.)	Vol. F. 1		
0.00000348	0.000		
0.00001470	0.100		
0.00004630	0.200		
0.00011900	0.300		
0.00033400	0.400		
0.00086400	0.500		
0.00265000	0.600		
0.00825000	0.700		
0.02570000	0.800		
0.05770000	0.900		
0.09580000	1.000		

Constants of Jouyban–Acree model

J_0	335.570
J_1	0
J_2	0

Reference of data: [50].

Solute	Solvent 1	Solvent 2	T (°C)
Lorazepam	Propylene glycol	Water	30
Solubility (mole F.)	Vol. F. 1		
0.00000347	0.000		
0.00000684	0.100		
0.00001400	0.200		
0.00003720	0.300		
0.00010300	0.400		
0.00020500	0.500		
0.00039600	0.600		
0.00073500	0.700		
0.00133000	0.800		
0.00221000	0.900		
0.00479000	1.000		

Constants of Jouyban–Acree model

J_0	175.997
J_1	0
J_2	0

Reference of data: [50].

Solute	Solvent 1	Solvent 2	T (°C)
Losartan potassium	2-Propanol	Cyclohexane	20.0
Solubility (mole F.)	Mole F. 1		
0.00125000	0.583		
0.00023460	0.412		
0.00019650	0.359		
0.00009891	0.318		
0.00007683	0.286		
0.00006509	0.259		
0.00004691	0.219		
0.00003713	0.189		

Constants of Jouyban–Acree model

J_0
J_1
J_2

Reference of data: [79].

Solute	Solvent 1	Solvent 2	T (°C)
Losartan potassium	2-Propanol	Cyclohexane	30.0
Solubility (mole F.)	Mole F. 1		
0.00126800	0.583		
0.00031720	0.412		
0.00018800	0.359		
0.00011350	0.318		
0.00009254	0.286		
0.00007067	0.259		
0.00005393	0.219		
0.00005146	0.189		

Constants of Jouyban–Acree model

J_0
J_1
J_2

Reference of data: [79].

Solute	Solvent 1	Solvent 2	T (°C)
Losartan potassium	2-Propanol	Cyclohexane	40.0
Solubility (mole F.)	Mole F. 1		
0.00139600	0.583		
0.00035420	0.412		
0.00019500	0.359		
0.00013820	0.318		
0.00011310	0.286		
0.00008830	0.259		
0.00007350	0.219		
0.00006180	0.189		

Constants of Jouyban–Acree model

J_0

J_1

J_2

Reference of data: [79].

Solute	Solvent 1	Solvent 2	T (°C)
Losartan potassium	2-Propanol	Cyclohexane	50.0
Solubility (mole F.)	Mole F. 1		
0.0017220	0.583		
0.0004520	0.412		
0.0002240	0.359		
0.0001436	0.318		
0.0001178	0.286		
0.0000998	0.259		
0.0000899	0.219		
0.0000612	0.189		

Constants of Jouyban–Acree model

J_0

J_1

J_2

Reference of data: [79].

Solute	Solvent 1	Solvent 2	T (°C)
Losartan potassium	2-Propanol	Cyclohexane	60.0
Solubility (mole F.)	Mole F. 1		
0.00188600	0.583		
0.00050550	0.412		
0.00030190	0.359		
0.00021460	0.318		
0.00017140	0.286		
0.00013690	0.259		
0.00009980	0.219		
0.00007529	0.189		

Constants of Jouyban–Acree model

J_0

J_1

J_2

Reference of data: [79].

Solute	Solvent 1	Solvent 2	T (°C)
Losartan potassium	2-Propanol	Cyclohexane	70.0
Solubility (mole F.)	Mole F. 1		
0.00206100	0.583		
0.00053130	0.412		
0.00038920	0.359		
0.00031240	0.318		
0.00022560	0.286		
0.00017890	0.259		
0.00011980	0.219		
0.00008842	0.189		

Constants of Jouyban–Acree model

J_0

J_1

J_2

Reference of data: [79].

Solute	Solvent 1	Solvent 2	T (°C)
Losartan potassium	Water	2-Propanol	20.0
Solubility (mole F.)	Mole F. 1		
0.01485	0.094		
0.02144	0.122		
0.02881	0.150		
0.03278	0.176		
0.04034	0.201		
0.04425	0.225		
0.04842	0.248		
0.05415	0.271		

Constants of Jouyban–Acree model

J_0

J_1

J_2

Reference of data: [79].

Solute	Solvent 1	Solvent 2	T (°C)
Losartan potassium	Water	2-Propanol	30.0
Solubility (mole F.)	Mole F. 1		
0.01832	0.094		
0.02462	0.122		
0.02985	0.150		
0.03378	0.176		
0.04351	0.201		
0.05085	0.225		
0.05273	0.248		
0.06021	0.271		

Constants of Jouyban–Acree model

J_0

J_1

J_2

Reference of data: [79].

Solute	Solvent 1	Solvent 2	T (°C)
Losartan potassium	Water	2-Propanol	40.0
Solubility (mole F.)	Mole F. 1		
0.02014	0.094		
0.02627	0.122		
0.03418	0.150		
0.03791	0.176		
0.04574	0.201		
0.05149	0.225		
0.05604	0.248		
0.06325	0.271		

Constants of Jouyban–Acree model

J_0

J_1

J_2

Reference of data: [79].

Solute	Solvent 1	Solvent 2	T (°C)
Losartan potassium	Water	2-Propanol	50.0
Solubility (mole F.)	Mole F. 1		
0.02345	0.094		
0.03037	0.122		
0.03885	0.150		
0.04394	0.176		
0.04920	0.201		
0.05474	0.225		
0.05671	0.248		
0.06478	0.271		

Constants of Jouyban–Acree model

J_0

J_1

J_2

Reference of data: [79].

Solute	Solvent 1	Solvent 2	T (°C)
Losartan potassium	Water	2-Propanol	60.0
Solubility (mole F.)	Mole F. 1		
0.03149	0.094		
0.03832	0.122		
0.04629	0.150		
0.04928	0.176		
0.06023	0.201		
0.06280	0.225		
0.06920	0.248		
0.07590	0.271		

Constants of Jouyban–Acree model

J_0
J_1
J_2

Reference of data: [79].

Solute	Solvent 1	Solvent 2	T (°C)
Losartan potassium	Water	2-Propanol	70.0
Solubility (mole F.)	Mole F. 1		
0.03580	0.094		
0.03941	0.122		
0.04770	0.150		
0.05591	0.176		
0.06243	0.201		
0.06887	0.225		
0.07422	0.248		
0.08014	0.271		

Constants of Jouyban–Acree model

J_0
J_1
J_2

Reference of data: [79].

Solute	Solvent 1	Solvent 2	T (°C)
Lovastatin	Acetone	Water	5.0
Solubility (mole F.)	Mole F. 1		
0.0000468	0.200		
0.0003150	0.300		
0.0011500	0.400		
0.0020000	0.500		
0.0041600	0.600		
0.0055400	0.700		
0.0058200	0.800		
0.0060100	0.900		
0.0063750	1.000		

Constants of Jouyban–Acree model

J_0

J_1

J_2

Reference of data: [80].

Solute	Solvent 1	Solvent 2	T (°C)
Lovastatin	Acetone	Water	10.0
Solubility (mole F.)	Mole F. 1		
0.0000656	0.200		
0.0004100	0.300		
0.0014600	0.400		
0.0025000	0.500		
0.0050700	0.600		
0.0066900	0.700		
0.0070700	0.800		
0.0073000	0.900		
0.0074160	1.000		

Constants of Jouyban–Acree model

J_0

J_1

J_2

Reference of data: [80].

Solute	Solvent 1	Solvent 2	T (°C)
Lovastatin	Acetone	Water	15.0
Solubility (mole F.)	Mole F. 1		
0.0000905	0.200		
0.0005260	0.300		
0.0018300	0.400		
0.0030600	0.500		
0.0061600	0.600		
0.0080200	0.700		
0.0082300	0.800		
0.0084200	0.900		
0.0086920	1.000		

Constants of Jouyban–Acree model

J_0

J_1

J_2

Reference of data: [80].

Solute	Solvent 1	Solvent 2	T (°C)
Lovastatin	Acetone	Water	20.0
Solubility (mole F.)	Mole F. 1		
0.0001235	0.200		
0.0006960	0.300		
0.0022900	0.400		
0.0039100	0.500		
0.0074400	0.600		
0.0096600	0.700		
0.0098400	0.800		
0.0100300	0.900		
0.0320600	1.000		

Constants of Jouyban–Acree model

J_0

J_1

J_2

Reference of data: [80].

Solute	Solvent 1	Solvent 2	T (°C)
Lovastatin	Acetone	Water	25.0
Solubility (mole F.)	Mole F. 1		
0.0001666	0.200		
0.0009120	0.300		
0.0028900	0.400		
0.0048800	0.500		
0.0088300	0.600		
0.0114000	0.700		
0.0117600	0.800		
0.0119700	0.900		
0.0104300	1.000		

Constants of Jouyban–Acree model

J_0

J_1

J_2

Reference of data: [80].

Solute	Solvent 1	Solvent 2	T (°C)
Lovastatin	Acetone	Water	30.0
Solubility (mole F.)	Mole F. 1		
0.0002243	0.200		
0.0011600	0.300		
0.0036000	0.400		
0.0061800	0.500		
0.0105200	0.600		
0.0136100	0.700		
0.0152400	0.800		
0.0154700	0.900		
0.0123300	1.000		

Constants of Jouyban–Acree model

J_0
J_1
J_2

Reference of data: [80].

Solute	Solvent 1	Solvent 2	T (°C)
Lovastatin	Acetone	Water	35.0
Solubility (mole F.)	Mole F. 1		
0.0003021	0.200		
0.0015000	0.300		
0.0045900	0.400		
0.0076200	0.500		
0.0127800	0.600		
0.0162500	0.700		
0.0175300	0.800		
0.0178700	0.900		
0.0157300	1.000		

Constants of Jouyban–Acree model

J_0
J_1
J_2

Reference of data: [80].

Solute	Solvent 1	Solvent 2	T (°C)
Lovastatin	Acetone	Water	40.0
Solubility (mole F.)	Mole F. 1		
0.0004046	0.200		
0.0020200	0.300		
0.0057800	0.400		
0.0094700	0.500		
0.0155700	0.600		
0.0195200	0.700		
0.0213800	0.800		
0.0220400	0.900		
0.0183200	1.000		

Constants of Jouyban–Acree model

J_0
J_1
J_2

Reference of data: [80].

Solute	Solvent 1	Solvent 2	T (°C)
Lovastatin	Acetone	Water	45.0
Solubility (mole F.)	Mole F. 1		
0.0005433	0.200		
0.0025600	0.300		
0.0072300	0.400		
0.0116600	0.500		
0.0190900	0.600		
0.0232300	0.700		
0.0249000	0.800		
0.0263600	0.900		
0.0225900	1.000		

Constants of Jouyban–Acree model

J_0
J_1
J_2

Reference of data: [80].

Solute	Solvent 1	Solvent 2	T (°C)
Lovastatin	Acetone	Water	50.0
Solubility (mole F.)	Mole F. 1		
0.0007401	0.200		
0.0033500	0.300		
0.0089900	0.400		
0.0146300	0.500		
0.0235500	0.600		
0.0279500	0.700		
0.0301000	0.800		
0.0318800	0.900		
0.0271600	1.000		

Constants of Jouyban–Acree model

J_0
J_1
J_2

Reference of data: [80].

Solute	Solvent 1	Solvent 2	T (°C)
Meloxicam	Ethanol	Glycerol	25
Solubility (g/L)	Vol. F. 1		
0.138	0.000		
0.298	0.100		
0.392	0.200		
0.403	0.400		
0.484	0.600		
0.407	0.800		
0.372	0.900		
0.354	1.000		

Constants of Jouyban–Acree model

J_0	358.542
J_1	−345.732
J_2	347.931

Reference of data: [49].

Solute	Solvent 1	Solvent 2	T (°C)
Meloxicam	Ethanol	Water	25
Solubility (g/L)	Vol. F. 1		
0.012	0.000		
0.042	0.100		
0.051	0.200		
0.061	0.400		
0.134	0.600		
0.255	0.800		
0.298	0.900		
0.354	1.000		

Constants of Jouyban–Acree model

J_0	307.299
J_1	0
J_2	0

Reference of data: [49].

Solute	Solvent 1	Solvent 2	T (°C)
Meloxicam	Polyethylene glycol 400	Ethanol	25
Solubility (g/L)	Vol. F. 1		
0.354	0.000		
0.783	0.200		
1.610	0.400		
2.730	0.600		
3.842	0.800		
4.023	0.900		
3.763	1.000		

Constants of Jouyban–Acree model

J_0	325.343
J_1	113.568
J_2	0

Reference of data: [49].

Solute	Solvent 1	Solvent 2	T (°C)
Metharbital	Ethanol	Water	25
Solubility (g/L)	Mass F. 1		
41.90	1.000		
43.70	0.975		
46.10	0.950		
47.90	0.925		
50.00	0.900		
50.70	0.875		
51.20	0.850		
50.90	0.825		
50.30	0.800		
49.30	0.775		
48.00	0.750		
46.00	0.725		
44.70	0.700		
43.40	0.675		
39.00	0.650		
36.60	0.625		
36.20	0.600		
33.40	0.575		
28.90	0.550		
26.20	0.525		
23.50	0.500		
21.20	0.475		
18.70	0.450		
16.20	0.425		
14.00	0.400		
12.10	0.375		
10.20	0.350		
8.60	0.325		
7.40	0.300		
6.30	0.275		
5.30	0.250		
4.50	0.225		
4.00	0.200		
3.50	0.175		
3.20	0.150		
2.90	0.125		
2.70	0.100		
2.50	0.075		
2.30	0.050		
2.20	0.025		
2.00	0.000		

Constants of Jouyban–Acree model

J_0	482.872
J_1	454.204
J_2	−307.083

Reference of data: [20].

Solute	Solvent 1	Solvent 2	T (°C)
Methyl p-aminobenzoate	Propylene glycol	Water	27.0
Solubility (mole F.)	Vol. F. 1		
0.00016010	0.000		
0.00023170	0.100		
0.00034220	0.200		
0.00055310	0.300		
0.00109170	0.400		
0.00213350	0.500		
0.00373500	0.600		
0.00687410	0.700		
0.01191450	0.800		
0.01984110	0.900		
0.03206470	1.000		

Constants of Jouyban–Acree model

J_0	−70.459
J_1	253.632
J_2	0

Reference of data: [32].

Solute	Solvent 1	Solvent 2	T (°C)
Methyl p-hydroxybenzoate	Propylene glycol	Water	27.0
Solubility (mole F.)	Vol. F. 1		
0.00030350	0.000		
0.00043940	0.100		
0.00066880	0.200		
0.00119450	0.300		
0.00231120	0.400		
0.00535350	0.500		
0.01240070	0.600		
0.02351770	0.700		
0.04242570	0.800		
0.06720550	0.900		
0.09442020	1.000		

Constants of Jouyban–Acree model

J_0	−1.178
J_1	500.058
J_2	0

Reference of data: [32].

Solute	Solvent 1	Solvent 2	T (°C)
Metronidazole	Dioxane	Water	35
Solubility (mole F.)	Vol. F. 1		
0.01894	1.000		
0.02701	0.900		
0.02112	0.800		
0.01894	0.700		
0.01843	0.600		
0.01745	0.500		
0.01562	0.400		
0.01394	0.300		
0.01394	0.200		
0.01305	0.100		
0.01000	0.000		

Constants of Jouyban–Acree model

J_0	88.323
J_1	69.779
J_2	383.294

Reference of data: [81].

Solute	Solvent 1	Solvent 2	T (°C)
Nalidixic acid	Dioxane	Water	10
Solubility (mole F.)	Vol. F. 1		
0.00000180	0.000		
0.00000650	0.100		
0.00001050	0.200		
0.00002530	0.300		
0.00005600	0.400		
0.00012740	0.500		
0.00027250	0.600		
0.00065550	0.700		
0.00096510	0.800		
0.00106910	0.850		
0.00104850	0.900		
0.00069740	1.000		

Constants of Jouyban–Acree model

J_0	723.869
J_1	5951.546
J_2	639.195

Reference of data: [1].

Solute	Solvent 1	Solvent 2	T (°C)
Nalidixic acid	Dioxane	Water	20
Solubility (mole F.)	Vol. F. 1		
0.00000210	0.000		
0.00000740	0.100		
0.00001230	0.200		
0.00003290	0.300		
0.00007170	0.400		
0.00016290	0.500		
0.00037510	0.600		
0.00088440	0.700		
0.00129630	0.800		
0.00142510	0.850		
0.00153220	0.900		
0.00099320	1.000		

Constants of Jouyban–Acree model

J_0	723.869
J_1	5951.546
J_2	639.195

Reference of data: [1].

Solute	Solvent 1	Solvent 2	T (°C)
Nalidixic acid	Dioxane	Water	25
Solubility (mole F.)	Vol. F. 1		
0.00000240	0.000		
0.00000860	0.100		
0.00001560	0.200		
0.00003710	0.300		
0.00008570	0.400		
0.00020890	0.500		
0.00044430	0.600		
0.00108820	0.700		
0.00147720	0.800		
0.00177650	0.850		
0.00189390	0.900		
0.00116200	1.000		

Constants of Jouyban–Acree model

J_0	723.869
J_1	5951.546
J_2	639.195

Reference of data: [1].

Solute	Solvent 1	Solvent 2	T (°C)
Nalidixic acid	Dioxane	Water	30
Solubility (mole F.)	Vol. F. 1		
0.00000290	0.000		
0.00000950	0.100		
0.00001850	0.200		
0.00004920	0.300		
0.00011460	0.400		
0.00025680	0.500		
0.00058390	0.600		
0.00122260	0.700		
0.00184600	0.800		
0.00227300	0.850		
0.00234780	0.900		
0.00134000	1.000		

Constants of Jouyban–Acree model

J_0	723.869
J_1	5951.546
J_2	639.195

Reference of data: [1].

Solute	Solvent 1	Solvent 2	T (°C)
Nalidixic acid	Dioxane	Water	35
Solubility (mole F.)	Vol. F. 1		
0.00000360	0.000		
0.00001110	0.100		
0.00002540	0.200		
0.00006400	0.300		
0.00013760	0.400		
0.00033840	0.500		
0.00075200	0.600		
0.00142860	0.700		
0.00240320	0.800		
0.00277020	0.850		
0.00275560	0.900		
0.00150650	1.000		

Constants of Jouyban–Acree model

J_0	723.869
J_1	5951.546
J_2	639.195

Reference of data: [1].

Solute	Solvent 1	Solvent 2	T (°C)
Nalidixic acid	Dioxane	Water	40
Solubility (mole F.)	Vol. F. 1		
0.00000500	0.000		
0.00001340	0.100		
0.00002940	0.200		
0.00008680	0.300		
0.00019470	0.400		
0.00047950	0.500		
0.00095620	0.600		
0.00180240	0.700		
0.00279170	0.800		
0.00349560	0.850		
0.00319990	0.900		
0.00166450	1.000		

Constants of Jouyban–Acree model

J_0	723.869
J_1	5951.546
J_2	639.195

Reference of data: [1].

Solute	Solvent 1	Solvent 2	T (°C)
Naproxen	Ethanol	Propylene glycol	20
Solubility (mole F.)	Mass F. 1		
0.00614	0.000		
0.00848	0.200		
0.01040	0.400		
0.01170	0.600		
0.01250	0.800		
0.01240	1.000		

Constants of Jouyban–Acree model

J_0	137.563
J_1	−26.667
J_2	0

Reference of data: [73].

Solute	Solvent 1	Solvent 2	T (°C)
Naproxen	Ethanol	Propylene glycol	25
Solubility (mole F.)	Mass F. 1		
0.00726	0.000		
0.00986	0.200		
0.01260	0.400		
0.01410	0.600		
0.01490	0.800		
0.01490	1.000		

Constants of Jouyban–Acree model

J_0	137.563
J_1	−26.667
J_2	0

Reference of data: [73].

Solute	Solvent 1	Solvent 2	T (°C)
Naproxen	Ethanol	Propylene glycol	30
Solubility (mole F.)	Mass F. 1		
0.00861	0.000		
0.01250	0.200		
0.01590	0.400		
0.01790	0.600		
0.01990	0.800		
0.01980	1.000		

Constants of Jouyban–Acree model

J_0	137.563
J_1	−26.667
J_2	0

Reference of data: [73].

Solute	Solvent 1	Solvent 2	T (°C)
Naproxen	Ethanol	Propylene glycol	35
Solubility (mole F.)	Mass F. 1		
0.01060	0.000		
0.01490	0.200		
0.01870	0.400		
0.02070	0.600		
0.02210	0.800		
0.02320	1.000		

Constants of Jouyban–Acree model

J_0	137.563
J_1	−26.667
J_2	0

Reference of data: [73].

Solute	Solvent 1	Solvent 2	T (°C)
Naproxen	Ethanol	Propylene glycol	40
Solubility (mole F.)	Mass F. 1		
0.01250	0.000		
0.01790	0.200		
0.02310	0.400		
0.02710	0.600		
0.02930	0.800		
0.02840	1.000		

Constants of Jouyban–Acree model

J_0	137.563
J_1	−26.667
J_2	0

Reference of data: [73].

Solute	Solvent 1	Solvent 2	T (°C)
Naproxen	Ethanol	Water	20.0
Solubility (mole F.)	Mass F. 1		
0.00000436	0.000		
0.00000523	0.100		
0.00001300	0.200		
0.00005638	0.300		
0.00025090	0.400		
0.00073650	0.500		
0.00175800	0.600		
0.00343300	0.700		
0.00574700	0.800		
0.00895000	0.900		
0.01242000	1.000		

Constants of Jouyban–Acree model

J_0	683.895
J_1	620.328
J_2	−1026.657

Reference of data: [82].

Solute	Solvent 1	Solvent 2	T (°C)
Naproxen	Ethanol	Water	25.0
Solubility (mole F.)	Mass F. 1		
0.00000513	0.000		
0.00000701	0.100		
0.00001824	0.200		
0.00008842	0.300		
0.00032800	0.400		
0.00095100	0.500		

(continued)

Solute	Solvent 1	Solvent 2	T (°C)
0.00216000	0.600		
0.00422200	0.700		
0.00730000	0.800		
0.01100000	0.900		
0.01494000	1.000		

Constants of Jouyban–Acree model

J_0	683.895
J_1	620.328
J_2	−1026.657

Reference of data: [82].

Solute	Solvent 1	Solvent 2	T (°C)
Naproxen	Ethanol	Water	30.0
Solubility (mole F.)	Mass F. 1		
0.00000591	0.000		
0.00000924	0.100		
0.00002440	0.200		
0.00012350	0.300		
0.00044100	0.400		
0.00121300	0.500		
0.00283200	0.600		
0.00527000	0.700		
0.00905000	0.800		
0.01357000	0.900		
0.01976000	1.000		

Constants of Jouyban–Acree model

J_0	683.895
J_1	620.328
J_2	−1026.657

Reference of data: [82].

Solute	Solvent 1	Solvent 2	T (°C)
Naproxen	Ethanol	Water	35.0
Solubility (mole F.)	Mass F. 1		
0.00000660	0.000		
0.00001240	0.100		
0.00003400	0.200		
0.00015780	0.300		
0.00058500	0.400		
0.00153200	0.500		
0.00348400	0.600		
0.00652600	0.700		
0.01138000	0.800		
0.01700000	0.900		
0.02323000	1.000		

Constants of Jouyban–Acree model

J_0	683.895
J_1	620.328
J_2	−1026.657

Reference of data: [82].

Solute	Solvent 1	Solvent 2	T (°C)
Naproxen	Ethanol	Water	40.0
Solubility (mole F.)	Mass F. 1		
0.00000768	0.000		
0.00001443	0.100		
0.00004340	0.200		
0.00021000	0.300		
0.00066400	0.400		
0.00186100	0.500		
0.00451000	0.600		
0.00774000	0.700		
0.01330000	0.800		
0.01985000	0.900		
0.02836000	1.000		

Constants of Jouyban–Acree model

J_0	683.895
J_1	620.328
J_2	−1026.657

Reference of data: [82].

Solute	Solvent 1	Solvent 2	T (°C)
Naproxen	Propylene glycol	Water	20
Solubility (mole F.)	Mass F. 1		
0.00000431	0.000		
0.00000900	0.200		
0.00003270	0.400		
0.00016200	0.600		
0.00114000	0.800		
0.00614000	1.000		

Constants of Jouyban–Acree model

J_0	−365.112
J_1	249.368
J_2	0

Reference of data: [75].

Solute	Solvent 1	Solvent 2	T (°C)
Naproxen	Propylene glycol	Water	25
Solubility (mole F.)	Mass F. 1		
0.00000510	0.000		
0.00001160	0.200		
0.00004240	0.400		
0.00022600	0.600		
0.00131000	0.800		
0.00726000	1.000		

Constants of Jouyban–Acree model

J_0	−365.112
J_1	249.368
J_2	0

Reference of data: [75].

Solute	Solvent 1	Solvent 2	T (°C)
Naproxen	Propylene glycol	Water	30
Solubility (mole F.)	Mass F. 1		
0.00000591	0.000		
0.00001380	0.200		
0.00004870	0.400		
0.00027100	0.600		
0.00159000	0.800		
0.00861000	1.000		

Constants of Jouyban–Acree model

J_0	−365.112
J_1	249.368
J_2	0

Reference of data: [75].

Solute	Solvent 1	Solvent 2	T (°C)
Naproxen	Propylene glycol	Water	35
Solubility (mole F.)	Mass F. 1		
0.00000660	0.000		
0.00001610	0.200		
0.00006600	0.400		
0.00032900	0.600		
0.00187000	0.800		
0.01060000	1.000		

Constants of Jouyban–Acree model

J_0	−365.112
J_1	249.368
J_2	0

Reference of data: [75].

Solute	Solvent 1	Solvent 2	T (°C)
Naproxen	Propylene glycol	Water	40
Solubility (mole F.)	Mass F. 1		
0.00000768	0.000		
0.00001950	0.200		
0.00007520	0.400		
0.00037500	0.600		
0.00215000	0.800		
0.01250000	1.000		

Constants of Jouyban–Acree model

J_0	−365.112
J_1	249.368
J_2	0

Reference of data: [75].

Solute	Solvent 1	Solvent 2	T (°C)
Nifedipine	Polyethylene glycol 400	Dimethyl isosorbide	37
Solubility (mg/g)	Mole F. 1		
18.50	0.000		
18.70	0.050		
19.10	0.130		
19.80	0.300		
21.50	0.570		
22.70	0.800		
23.80	1.000		

Constants of Jouyban–Acree model

J_0	
J_1	
J_2	

Reference of data: [83].

Solute	Solvent 1	Solvent 2	T (°C)
Nifedipine	Propylene glycol	Dimethyl isosorbide	37
Solubility (mg/g)	Mole F. 1		
18.50	0.000		
20.50	0.200		
22.80	0.430		
25.60	0.700		
26.00	0.870		
27.30	0.950		
28.10	1.000		

Constants of Jouyban–Acree model

J_0
J_1
J_2

Reference of data: [83].

Solute	Solvent 1	Solvent 2	T (°C)
Niflumic acid	Ethanol	Water	10
Solubility (mole F.)	Vol. F. 1		
0.00000492	0.000		
0.00000750	0.100		
0.00001090	0.200		
0.00001960	0.300		
0.00003260	0.400		
0.00006440	0.500		
0.00043500	0.700		
0.00360000	0.900		
0.01030000	1.000		

Constants of Jouyban–Acree model

J_0	−438.769
J_1	265.979
J_2	179.837

Reference of data: [36].

Solute	Solvent 1	Solvent 2	T (°C)
Niflumic acid	Ethanol	Water	15
Solubility (mole F.)	Vol. F. 1		
0.00000505	0.000		
0.00000783	0.100		
0.00001170	0.200		
0.00002130	0.300		
0.00003870	0.400		
0.00008500	0.500		
0.00059600	0.700		
0.00455000	0.900		
0.01190000	1.000		

Constants of Jouyban–Acree model

J_0	−438.769
J_1	265.979
J_2	179.837

Reference of data: [36].

Solute	Solvent 1	Solvent 2	T (°C)
Niflumic acid	Ethanol	Water	20
Solubility (mole F.)	Vol. F. 1		
0.00000525	0.000		
0.00000827	0.100		
0.00001250	0.200		
0.00002350	0.300		
0.00004510	0.400		
0.00011400	0.500		
0.00074000	0.700		
0.00552000	0.900		
0.01350000	1.000		

Constants of Jouyban–Acree model

J_0	−438.769
J_1	265.979
J_2	179.837

Reference of data: [36].

Solute	Solvent 1	Solvent 2	T (°C)
Niflumic acid	Ethanol	Water	25
Solubility (mole F.)	Vol. F. 1		
0.00000545	0.000		
0.00000875	0.100		
0.00001340	0.200		
0.00002610	0.300		
0.00005410	0.400		
0.00014600	0.500		
0.00097100	0.700		
0.00684000	0.900		
0.01630000	1.000		

Constants of Jouyban–Acree model

J_0	−438.769
J_1	265.979
J_2	179.837

Reference of data: [36].

Solute	Solvent 1	Solvent 2	T (°C)
Niflumic acid	Ethanol	Water	30
Solubility (mole F.)	Vol. F. 1		
0.00000563	0.000		
0.00000915	0.100		
0.00001460	0.200		
0.00002880	0.300		
0.00006710	0.400		
0.00018200	0.500		
0.00119000	0.700		
0.00900000	0.900		
0.01850000	1.000		

Constants of Jouyban–Acree model

J_0	−438.769
J_1	265.979
J_2	179.837

Reference of data: [36].

Solute	Solvent 1	Solvent 2	T (°C)
Niflumic acid	Ethanol	Water	35
Solubility (mole F.)	Vol. F. 1		
0.00000587	0.000		
0.00000974	0.100		
0.00001570	0.200		
0.00003200	0.300		
0.00008250	0.400		
0.00023100	0.500		
0.00164000	0.700		
0.01050000	0.900		
0.02250000	1.000		

Constants of Jouyban–Acree model

J_0	−438.769
J_1	265.979
J_2	179.837

Reference of data: [36].

Solute	Solvent 1	Solvent 2	T (°C)
Niflumic acid	Ethyl acetate	Ethanol	10
Solubility (mole F.)	Vol. F. 1		
0.01030	0.000		
0.01490	0.100		
0.02370	0.300		
0.03700	0.500		
0.04290	0.600		
0.04720	0.700		
0.05070	0.800		
0.03440	0.900		
0.02220	1.000		

Constants of Jouyban–Acree model

J_0	446.973
J_1	198.059
J_2	215.889

Reference of data: [36].

Solute	Solvent 1	Solvent 2	T (°C)
Niflumic acid	Ethyl acetate	Ethanol	15
Solubility (mole F.)	Vol. F. 1		
0.01190	0.000		
0.01790	0.100		
0.02730	0.300		
0.04000	0.500		
0.04600	0.600		
0.05000	0.700		
0.05130	0.800		
0.03550	0.900		
0.02310	1.000		

Constants of Jouyban–Acree model

J_0	446.973
J_1	198.059
J_2	215.889

Reference of data: [36].

Solute	Solvent 1	Solvent 2	T (°C)
Niflumic acid	Ethyl acetate	Ethanol	20
Solubility (mole F.)	Vol. F. 1		
0.01350	0.000		
0.02060	0.100		
0.03120	0.300		
0.04280	0.500		
0.05040	0.600		
0.05270	0.700		
0.05190	0.800		
0.03650	0.900		
0.02440	1.000		

Constants of Jouyban–Acree model

J_0	446.973
J_1	198.059
J_2	215.889

Reference of data: [36].

Solute	Solvent 1	Solvent 2	T (°C)
Niflumic acid	Ethyl acetate	Ethanol	25
Solubility (mole F.)	Vol. F. 1		
0.01630	0.000		
0.02400	0.100		
0.03520	0.300		
0.04820	0.500		
0.05420	0.600		
0.05500	0.700		
0.05240	0.800		
0.03740	0.900		
0.02540	1.000		

Constants of Jouyban–Acree model

J_0	446.973
J_1	198.059
J_2	215.889

Reference of data: [36].

Solute	Solvent 1	Solvent 2	T (°C)
Niflumic acid	Ethyl acetate	Ethanol	30
Solubility (mole F.)	Vol. F. 1		
0.01850	0.000		
0.02660	0.100		
0.04000	0.300		
0.05100	0.500		
0.05700	0.600		
0.05800	0.700		
0.05280	0.800		
0.03860	0.900		
0.02640	1.000		

Constants of Jouyban–Acree model

J_0	446.973
J_1	198.059
J_2	215.889

Reference of data: [36].

Solute	Solvent 1	Solvent 2	T (°C)
Niflumic acid	Ethyl acetate	Ethanol	35
Solubility (mole F.)	Vol. F. 1		
0.02250	0.000		
0.03100	0.100		
0.04560	0.300		
0.05580	0.500		
0.06070	0.600		
0.06080	0.700		
0.05330	0.800		
0.03970	0.900		
0.02730	1.000		

Constants of Jouyban–Acree model

J_0	446.973
J_1	198.059
J_2	215.889

Reference of data: [36].

Solute	Solvent 1	Solvent 2	T (°C)
Nimesulide	Ethanol	Glycerol	25
Solubility (g/L)	Mole F. 1		
0.2180	0.000		
0.4160	0.100		
0.6910	0.200		
1.6930	0.400		
2.7490	0.600		
3.4200	0.800		
4.0400	0.900		
3.3200	1.000		

Constants of Jouyban–Acree model

J_0	504.751
J_1	0
J_2	0

Reference of data: [49].

Solute	Solvent 1	Solvent 2	T (°C)
Nimesulide	Ethanol	Water	25
Solubility (g/L)	Vol. F. 1		
0.0140	0.000		
0.0620	0.100		
0.1010	0.200		
0.1250	0.400		
0.6420	0.600		
2.6280	0.800		
3.5600	0.900		
3.3200	1.000		

Constants of Jouyban–Acree model

J_0	390.050
J_1	0
J_2	0

Reference of data: [49].

Solute	Solvent 1	Solvent 2	T (°C)
Nimesulide	Polyethylene glycol 400	Ethanol	25
Solubility (g/L)	Vol. F. 1		
3.320	0.000		
9.900	0.200		
24.640	0.400		
36.00	0.600		
56.5120	0.800		
65.6000	0.900		
63.1200	1.000		

Constants of Jouyban–Acree model

J_0	395.790
J_1	0
J_2	0

Reference of data: [49].

Solute	Solvent 1	Solvent 2	T (°C)
Nitroaniline (*ortho*)	Dimethyl formamide	Water	30.0
Solubility (mole F.)	Mole F. 1		
0.00013	0.000		
0.00032	0.007		
0.00045	0.012		
0.00074	0.024		
0.00091	0.030		
0.00129	0.040		
0.00168	0.053		
0.00299	0.082		
0.00463	0.104		
0.00722	0.118		
0.01467	0.136		
0.03576	0.189		
0.08826	0.259		
0.14538	0.315		
0.19906	0.402		
0.30533	0.495		
0.34698	0.560		
0.40278	0.677		
0.47406	0.816		
0.54325	1.000		

Constants of Jouyban–Acree model

J_0	1804.916
J_1	−1793.276
J_2	1728.360

Reference of data: [16].

Solute	Solvent 1	Solvent 2	T (°C)
Nitroaniline (*para*)	Dimethyl formamide	Water	30.0
Solubility (mole F.)	Mole F. 1		
0.000110	0.000		
0.000460	0.007		
0.000720	0.012		
0.001180	0.024		
0.003360	0.040		
0.004580	0.065		
0.006250	0.082		
0.007750	0.104		
0.008990	0.118		
0.011200	0.135		
0.015230	0.189		
0.030120	0.244		
0.059380	0.315		
0.122130	0.402		
0.160840	0.472		
0.236970	0.560		
0.314810	0.677		
0.372080	0.816		
0.412140	1.000		

Constants of Jouyban–Acree model

J_0	1129.764
J_1	−304.106
J_2	723.548

Reference of data: [16].

Solute	Solvent 1	Solvent 2	T (°C)
Octyl *p*-aminobenzoate	Propylene glycol	Water	37.0
Solubility (mole F.)	Vol. F. 1		
0.00000007	0.000		
0.00000068	0.200		
0.00000569	0.400		
0.00010304	0.600		
0.00109396	0.800		
0.02051162	1.000		

Constants of Jouyban–Acree model

J_0	−268.764
J_1	0
J_2	0

Reference of data: [33].

Solute	Solvent 1	Solvent 2	T (°C)
Oxolinic acid	Ethanol	Water	20
Solubility (mole F.)	Vol. F. 1		
0.00000087	0.000		
0.00000264	0.100		
0.00000478	0.200		
0.00000661	0.300		
0.00000884	0.400		
0.00001070	0.500		
0.00001412	0.600		
0.00001713	0.700		
0.00001738	0.800		
0.00001407	0.900		
0.00000683	1.000		

Constants of Jouyban–Acree model

J_0	743.954
J_1	169.755
J_2	695.164

Reference of data: [84].

Solute	Solvent 1	Solvent 2	T (°C)
Oxolinic acid	Ethanol	Water	25
Solubility (mole F.)	Vol. F. 1		
0.00000108	0.000		
0.00000292	0.100		
0.00000540	0.200		
0.00000741	0.300		
0.00000997	0.400		
0.00001266	0.500		
0.00001634	0.600		
0.00002017	0.700		
0.00002099	0.800		
0.00001688	0.900		
0.00000824	1.000		

Constants of Jouyban–Acree model

J_0	743.954
J_1	169.755
J_2	695.164

Reference of data: [84].

Solute	Solvent 1	Solvent 2	T (°C)
Oxolinic acid	Ethanol	Water	30
Solubility (mole F.)	Vol. F. 1		
0.00000134	0.000		
0.00000331	0.100		
0.00000603	0.200		
0.00000854	0.300		
0.00001164	0.400		
0.00001541	0.500		
0.00001971	0.600		
0.00002325	0.700		
0.00002584	0.800		
0.00002111	0.900		
0.00001056	1.000		

Constants of Jouyban–Acree model

J_0	743.954
J_1	169.755
J_2	695.164

Reference of data: [84].

Solute	Solvent 1	Solvent 2	T (°C)
Oxolinic acid	Ethanol	Water	35
Solubility (mole F.)	Vol. F. 1		
0.00000163	0.000		
0.00000369	0.100		
0.00000693	0.200		
0.00000964	0.300		
0.00001348	0.400		
0.00001830	0.500		
0.00002390	0.600		
0.00002952	0.700		
0.00003284	0.800		
0.00002654	0.900		
0.00001291	1.000		

Constants of Jouyban–Acree model

J_0	743.954
J_1	169.755
J_2	695.164

Reference of data: [84].

Solute	Solvent 1	Solvent 2	T (°C)
Oxolinic acid	Ethanol	Water	40
Solubility (mole F.)	Vol. F. 1		
0.00000202	0.000		
0.00000404	0.100		
0.00000762	0.200		
0.00001109	0.300		
0.00001587	0.400		
0.00002213	0.500		
0.00002934	0.600		
0.00003549	0.700		
0.00004094	0.800		
0.00003212	0.900		
0.00001630	1.000		

Constants of Jouyban–Acree model

J_0	743.954
J_1	169.755
J_2	695.164

Reference of data: [84].

Solute	Solvent 1	Solvent 2	T (°C)
Oxolinic acid	Ethyl acetate	Ethanol	20
Solubility (mole F.)	Vol. F. 1		
0.00000683	0.000		
0.00000987	0.100		
0.00001345	0.200		
0.00001821	0.300		
0.00002969	0.500		
0.00003292	0.600		
0.00003475	0.700		
0.00002998	0.900		
0.00002010	1.000		

Constants of Jouyban–Acree model

J_0	452.982
J_1	135.792
J_2	92.308

Reference of data: [84].

Solute	Solvent 1	Solvent 2	T (°C)
Oxolinic acid	Ethyl acetate	Ethanol	25
Solubility (mole F.)	Vol. F. 1		
0.00000824	0.000		
0.00001178	0.100		
0.00001653	0.200		
0.00002256	0.300		
0.00003489	0.500		
0.00003812	0.600		
0.00004159	0.700		
0.00003428	0.900		
0.00002450	1.000		

Constants of Jouyban–Acree model

J_0	452.982
J_1	135.792
J_2	92.308

Reference of data: [84].

Solute	Solvent 1	Solvent 2	T (°C)
Oxolinic acid	Ethyl acetate	Ethanol	30
Solubility (mole F.)	Vol. F. 1		
0.00001056	0.000		
0.00001496	0.100		
0.00002075	0.200		
0.00002785	0.300		
0.00004118	0.500		
0.00004604	0.600		
0.00004912	0.700		
0.00004077	0.900		
0.00002980	1.000		

Constants of Jouyban–Acree model

J_0	452.982
J_1	135.792
J_2	92.308

Reference of data: [84].

Solute	Solvent 1	Solvent 2	T (°C)
Oxolinic acid	Ethyl acetate	Ethanol	35
Solubility (mole F.)	Vol. F. 1		
0.00001291	0.000		
0.00001912	0.100		
0.00002654	0.200		
0.00003406	0.300		
0.00005124	0.500		
0.00005395	0.600		
0.00005889	0.700		
0.00004908	0.900		
0.00003546	1.000		

Constants of Jouyban–Acree model

J_0	452.982
J_1	135.792
J_2	92.308

Reference of data: [84].

Solute	Solvent 1	Solvent 2	T (°C)
Oxolinic acid	Ethyl acetate	Ethanol	40
Solubility (mole F.)	Vol. F. 1		
0.00001630	0.000		
0.00002390	0.100		
0.00003290	0.200		
0.00004280	0.300		
0.00006339	0.500		
0.00006447	0.600		
0.00006931	0.700		
0.00005839	0.900		
0.00004229	1.000		

Constants of Jouyban–Acree model

J_0	452.982
J_1	135.792
J_2	92.308

Reference of data: [84].

Solute	Solvent 1	Solvent 2	T (°C)
Propyl p-aminobenzoate	Propylene glycol	Water	27.0
Solubility (mole F.)	Vol. F. 1		
0.00004770	0.000		
0.00007490	0.100		
0.00012970	0.200		
0.00025100	0.300		
0.00061740	0.400		
0.00154920	0.500		
0.00318280	0.600		
0.00848040	0.700		
0.01984110	0.800		
0.03955750	0.900		
0.08374320	1.000		

Constants of Jouyban–Acree model

J_0	−1813225
J_1	343.897
J_2	0

Reference of data: [32].

Solute	Solvent 1	Solvent 2	T (°C)
Propyl p-hydroxybenzoate	Propylene glycol	Water	27.0
Solubility (mole F.)	Vol. F. 1		
0.00004360	0.000		
0.00006770	0.100		
0.00010950	0.200		
0.00020350	0.300		
0.00051570	0.400		
0.00161250	0.500		
0.00392650	0.600		
0.01046210	0.700		
0.02447750	0.800		
0.05669890	0.900		
0.08629360	1.000		

Constants of Jouyban–Acree model

J_0	−143.496
J_1	630.835
J_2	0

Reference of data: [32].

Solute	Solvent 1	Solvent 2	T (°C)
Paroxetine HCl hemihydrate (A)	Acetone	Water	20
Solubility (mole F.)	Mole F. 1		
0.00003410	1.000		
0.00001827	0.688		
0.00001527	0.495		
0.00001441	0.364		
0.00001374	0.269		
0.00001244	0.197		
0.00001092	0.141		
0.00000977	0.095		
0.00000827	0.053		
0.00000704	0.027		
0.00000598	0.000		

Constants of Jouyban–Acree model

J_0	42.583
J_1	−233.516
J_2	110.519

Reference of data: [85].

Solute	Solvent 1	Solvent 2	T (°C)
Paroxetine HCl hemihydrate (A)	Acetone	Water	25
Solubility (mole F.)	Mole F. 1		
0.00004030	1.000		
0.00002137	0.688		
0.00001676	0.495		
0.00001602	0.364		
0.00001517	0.269		
0.00001429	0.197		
0.00001283	0.141		
0.00001149	0.095		
0.00000910	0.053		
0.00000804	0.027		
0.00000698	0.000		

Constants of Jouyban–Acree model

J_0	42.583
J_1	−233.516
J_2	110.519

Reference of data: [85].

Solute	Solvent 1	Solvent 2	T (°C)
Paroxetine HCl hemihydrate (A)	Acetone	Water	30
Solubility (mole F.)	Mole F. 1		
0.00004586	1.000		
0.00002689	0.688		
0.00002200	0.495		
0.00002063	0.364		
0.00001895	0.269		
0.00001786	0.197		
0.00001585	0.141		
0.00001453	0.095		
0.00001247	0.053		
0.00001053	0.027		
0.00000898	0.000		

Constants of Jouyban–Acree model

J_0	42.583
J_1	−233.516
J_2	110.519

Reference of data: [85].

Solute	Solvent 1	Solvent 2	T (°C)
Paroxetine HCl hemihydrate (A)	Acetone	Water	35
Solubility (mole F.)	Mole F. 1		
0.00005022	1.000		
0.00003198	0.688		
0.00002811	0.495		
0.00002575	0.364		
0.00002336	0.269		
0.00002076	0.197		
0.00001833	0.141		
0.00001710	0.095		
0.00001605	0.053		
0.00001452	0.027		
0.00001336	0.000		

Constants of Jouyban–Acree model

J_0	42.583
J_1	−233.516
J_2	110.519

Reference of data: [85].

Solute	Solvent 1	Solvent 2	T (°C)
Paroxetine HCl hemihydrate (A)	Acetone	Water	40
Solubility (mole F.)	Mole F. 1		
0.00005648	1.000		
0.00003590	0.688		
0.00003274	0.495		
0.00002968	0.364		
0.00002714	0.269		
0.00002451	0.197		
0.00002200	0.141		
0.00001997	0.095		
0.00001809	0.053		
0.00001702	0.027		
0.00001560	0.000		

Constants of Jouyban–Acree model

J_0	42.583
J_1	−233.516
J_2	110.519

Reference of data: [85].

Solute	Solvent 1	Solvent 2	T (°C)
Paroxetine HCl hemihydrate (A)	Acetone	Water	45
Solubility (mole F.)	Mole F. 1		
0.00006124	1.000		
0.00004012	0.688		
0.00003745	0.495		
0.00003360	0.364		
0.00003105	0.269		
0.00002821	0.197		
0.00002443	0.141		
0.00002264	0.095		
0.00002106	0.053		
0.00001962	0.027		
0.00001856	0.000		

Constants of Jouyban–Acree model

J_0	42.583
J_1	−233.516
J_2	110.519

Reference of data: [85].

Solute	Solvent 1	Solvent 2	T (°C)
Paroxetine HCl hemihydrate (A)	Acetone	Water	50
Solubility (mole F.)	Mole F. 1		
0.00006568	1.000		
0.00004553	0.688		
0.00004357	0.495		
0.00003764	0.364		
0.00003490	0.269		
0.00003114	0.197		
0.00002760	0.141		
0.00002574	0.095		
0.00002380	0.053		
0.00002237	0.027		
0.00002102	0.000		

Constants of Jouyban–Acree model

J_0	42.583
J_1	−233.516
J_2	110.519

Reference of data: [85].

Solute	Solvent 1	Solvent 2	T (°C)
Pentachlorobenzene	Ethanol	Water	23.0
Solubility (mole/ L)	Vol. F. 1		
0.00000076	0.000		
0.00000214	0.200		
0.00014791	0.400		
0.00239883	0.600		
0.01412538	0.800		
0.08317638	1.000		

Constants of Jouyban–Acree model

J_0	557.595
J_1	1106.861
J_2	−2371.516

Reference of data: [66].

Solute	Solvent 1	Solvent 2	T (°C)
Pentaglycine	Water	Ethanol	25.0
Solubility (g/L)	Vol. F. 1		
1.50	1.000		
1.07	0.900		
0.73	0.800		
0.50	0.700		
0.32	0.600		
0.21	0.500		
0.12	0.400		

Constants of Jouyban–Acree model

J_0

J_1

J_2

Reference of data: [65].

Solute	Solvent 1	Solvent 2	T (°C)
Pentobarbital	Ethanol	Water	25
Solubility (g/L)	Mass F. 1		
250.40	1.000		
250.40	0.975		
246.20	0.950		
243.30	0.925		
236.10	0.900		
226.40	0.875		
218.60	0.850		
208.80	0.825		
193.10	0.800		
180.40	0.775		
169.30	0.750		
157.10	0.725		
137.10	0.700		
123.40	0.675		
110.60	0.650		
98.30	0.625		
85.00	0.600		
72.00	0.575		
62.50	0.550		
51.00	0.525		
40.00	0.500		
34.50	0.475		
31.30	0.450		
24.30	0.425		
20.60	0.400		
15.50	0.375		
13.20	0.350		
10.30	0.300		
7.80	0.275		
5.50	0.250		
4.00	0.225		
3.10	0.200		
2.70	0.175		
2.10	0.150		
1.70	0.125		
1.40	0.100		
0.90	0.075		
0.60	0.050		
0.50	0.025		
0.50	0.000		

Constants of Jouyban–Acree model

J_0	711.985
J_1	227.209
J_2	−135.466

Reference of data: [20].

Solute	Solvent 1	Solvent 2	T (°C)
Phenacetin	Dioxane	Water	20
Solubility (mole F.)	Vol. F. 1		
0.000067	0.000		
0.000215	0.100		
0.000317	0.150		
0.000511	0.200		
0.000816	0.250		
0.001238	0.300		
0.002605	0.400		
0.006410	0.500		
0.014101	0.600		
0.024935	0.700		
0.037407	0.800		
0.039785	0.900		
0.014118	1.000		

Constants of Jouyban–Acree model

J_0	1007.296
J_1	680.457
J_2	693.953

Reference of data: [86].

Solute	Solvent 1	Solvent 2	T (°C)
Phenacetin	Dioxane	Water	25
Solubility (mole F.)	Vol. F. 1		
0.000082	0.000		
0.000248	0.100		
0.000404	0.150		
0.000648	0.200		
0.001016	0.250		
0.001549	0.300		
0.003510	0.400		
0.008353	0.500		
0.017785	0.600		
0.029056	0.700		
0.043801	0.800		
0.047080	0.900		
0.017158	1.000		

Constants of Jouyban–Acree model

J_0	1007.296
J_1	680.457
J_2	693.953

Reference of data: [86].

Solute	Solvent 1	Solvent 2	T (°C)
Phenacetin	Dioxane	Water	30
Solubility (mole F.)	Vol. F. 1		
0.000103	0.000		
0.000293	0.100		
0.000520	0.150		
0.000811	0.200		
0.001310	0.250		
0.001820	0.300		
0.004470	0.400		
0.010154	0.500		
0.022050	0.600		
0.036604	0.700		
0.053575	0.800		
0.054916	0.900		
0.021419	1.000		

Constants of Jouyban–Acree model

J_0	1007.296
J_1	680.457
J_2	693.953

Reference of data: [86].

Solute	Solvent 1	Solvent 2	T (°C)
Phenacetin	Dioxane	Water	35
Solubility (mole F.)	Vol. F. 1		
0.000118	0.000		
0.000402	0.100		
0.000622	0.150		
0.000980	0.200		
0.001661	0.250		
0.002614	0.300		
0.005490	0.400		
0.012659	0.500		
0.026184	0.600		
0.042360	0.700		
0.062000	0.800		
0.065257	0.900		
0.026415	1.000		

Constants of Jouyban–Acree model

J_0	1007.296
J_1	680.457
J_2	693.953

Reference of data: [86].

Solute	Solvent 1	Solvent 2	T (°C)
Phenacetin	Dioxane	Water	40
Solubility (mole F.)	Vol. F. 1		
0.000145	0.000		
0.000461	0.100		
0.000737	0.150		
0.001275	0.200		
0.002040	0.250		
0.003187	0.300		
0.007022	0.400		
0.016022	0.500		
0.030878	0.600		
0.051965	0.700		
0.071001	0.800		
0.076358	0.900		
0.030675	1.000		

Constants of Jouyban–Acree model

J_0	1007.296
J_1	680.457
J_2	693.953

Reference of data: [86].

Solute	Solvent 1	Solvent 2	T (°C)
Phenacetin	Ethanol	Water	25
Solubility (mole F.)	Vol. F. 1		
0.00010	0.000		
0.00020	0.100		
0.00050	0.200		
0.00070	0.300		
0.00160	0.400		
0.00580	0.500		
0.00820	0.600		
0.01260	0.700		
0.01500	0.800		
0.01720	0.900		
0.01450	1.000		

Constants of Jouyban–Acree model

J_0	637.700
J_1	490.862
J_2	0

Reference of data: [11].

Solute	Solvent 1	Solvent 2	T (°C)
Phenacetin	Ethyl acetate	Ethanol	25
Solubility (mole F.)	Vol. F. 1		
0.01450	0.000		
0.02050	0.100		
0.02330	0.200		
0.02690	0.300		
0.03220	0.400		
0.03030	0.500		
0.02680	0.600		
0.02350	0.700		
0.01980	0.800		
0.01750	0.900		
0.01540	1.000		

Constants of Jouyban–Acree model

J_0	341.278
J_1	−154.550
J_2	0

Reference of data: [11].

Solute	Solvent 1	Solvent 2	T (°C)
Phenobarbital	Ethanol	Water	25
Solubility (g/L)	Mass F. 1		
118.40	1.000		
122.60	0.975		
127.80	0.950		
131.10	0.925		
132.30	0.900		
130.60	0.875		
126.40	0.850		
120.60	0.825		
112.30	0.800		
104.00	0.775		
97.70	0.750		
90.20	0.725		
82.40	0.700		
78.20	0.675		

(continued)

(continued)

Solute	Solvent 1	Solvent 2	T (°C)
70.30	0.650		
62.10	0.625		
52.00	0.600		
46.60	0.575		
41.10	0.550		
36.30	0.525		
30.60	0.500		
25.40	0.475		
21.20	0.450		
18.00	0.425		
15.00	0.400		
11.50	0.375		
10.00	0.350		
7.90	0.325		
6.20	0.300		
5.10	0.275		
4.50	0.250		
4.00	0.225		
3.00	0.200		
2.70	0.175		
2.50	0.150		
2.30	0.125		
1.90	0.100		
1.70	0.075		
1.50	0.050		
1.30	0.025		
1.20	0.000		

Constants of Jouyban–Acree model

J_0	462.444
J_1	537.461
J_2	−176.940

Reference of data: [20].

Solute	Solvent 1	Solvent 2	T (°C)
Phenylalanine (L)	Water	Methanol	15.0
Solubility (mole/L)	Mass F. 1		
0.17400	1.000		
0.16500	0.920		
0.07940	0.840		
0.06870	0.748		
0.02950	0.656		
0.02600	0.553		
0.02180	0.458		
0.00730	0.359		
0.00608	0.241		
0.00440	0.121		
0.00114	0.000		

Constants of Jouyban–Acree model

J_0	103.264
J_1	−247.914
J_2	518.213

Reference of data: [15].

Solute	Solvent 1	Solvent 2	T (°C)
Phenylalanine (L)	Water	Methanol	20.0
Solubility (mole/L)	Mass F. 1		
0.17700	1.000		
0.16700	0.920		
0.07960	0.840		
0.07030	0.748		
0.03080	0.656		
0.02750	0.553		
0.02340	0.458		
0.00831	0.359		
0.00692	0.241		
0.00513	0.121		
0.00141	0.000		

Constants of Jouyban–Acree model

J_0	103.264
J_1	−247.914
J_2	518.213

Reference of data: [15].

Solute	Solvent 1	Solvent 2	T (°C)
Phenylalanine (L)	Water	Methanol	25.0
Solubility (mole/L)	Mass F. 1		
0.18000	1.000		
0.16900	0.920		
0.08000	0.840		
0.07200	0.748		
0.03200	0.656		
0.02900	0.553		
0.02600	0.458		
0.00920	0.359		
0.00820	0.241		
0.00620	0.121		
0.00180	0.000		

Constants of Jouyban–Acree model

J_0	103.264
J_1	−247.914
J_2	518.213

Reference of data: [15].

Solute	Solvent 1	Solvent 2	T (°C)
Phenobarbital	Propylene glycol	Water	25
Solubility (mole/L)	Vol. F. 1		
0.005330	0.00		
0.009120	0.10		
0.011270	0.20		
0.018950	0.30		
0.032910	0.40		
0.056800	0.50		
0.089030	0.60		
0.166420	0.70		
0.251610	0.80		
0.511250	0.90		
0.642260	1.00		

Constants of Jouyban–Acree model

J_0	−24.980
J_1	161.670
J_2	0

Reference of data: [55].

Solute	Solvent 1	Solvent 2	T (°C)
Phenyl salicylate	Ethanol	Water	25
Solubility (mass F.)	Mass F. 1		
0.000150	0.000		
0.000200	0.200		
0.002200	0.400		
0.007600	0.500		
0.021000	0.600		
0.044000	0.700		
0.077000	0.800		
0.140000	0.900		
0.177000	0.923		
0.350000	1.000		

Constants of Jouyban–Acree model

J_0	22.351
J_1	898.348
J_2	−1421.393

Reference of data: [10].

Solute	Solvent 1	Solvent 2	T (°C)
Phenylalanine	Water	Dioxane	25
Solubility (g/100 g)	Mass F. 1		
2.790	1.000		
2.470	0.800		
2.240	0.700		
1.900	0.600		
1.090	0.400		
0.265	0.200		
0.050	0.100		

Constants of Jouyban–Acree model

J_0
J_1
J_2

Reference of data: [23].

Solute	Solvent 1	Solvent 2	T (°C)
Phenylalanine	Water	Ethanol	25
Solubility (g/100 g)	Mass F. 1		
2.790	1.000		
1.860	0.800		
1.560	0.700		
1.480	0.600		
1.380	0.500		
1.230	0.400		
0.602	0.200		
0.245	0.100		

Constants of Jouyban–Acree model

J_0
J_1
J_2

Reference of data: [23].

Solute	Solvent 1	Solvent 2	T (°C)
Phenylalanine (L)	Water	1-Propanol	25
Solubility (mole F.)	Mole F. 1		
0.00290	1.000		
0.00280	0.968		
0.00300	0.931		
0.00340	0.886		
0.00369	0.834		
0.00369	0.770		
0.00337	0.691		
0.00262	0.589		
0.00157	0.455		
0.000557	0.271		
0.000066	0.000		

Constants of Jouyban–Acree model

J_0	770.362
J_1	0
J_2	0

Reference of data: [14].

Solute	Solvent 1	Solvent 2	T (°C)
Phenylalanine (L)	Water	2-Propanol	25
Solubility (mole F.)	Mole F. 1		
0.002900	1.000		
0.002540	0.968		
0.002160	0.930		
0.002210	0.886		
0.002370	0.834		
0.002220	0.770		
0.001840	0.690		
0.001330	0.589		
0.000699	0.455		
0.000218	0.271		
0.000022	0.000		

Constants of Jouyban–Acree model

J_0	616.966
J_1	0
J_2	0

Reference of data: [14].

Solute	Solvent 1	Solvent 2	T (°C)
Phenytoin	1,3-Butanediol	Water	25
Solubility (mg/L)	Vol. F. 1		
20.3	0.000		
45.0	0.100		
80.0	0.200		
173.0	0.300		
395.0	0.400		
870.0	0.500		
1780.0	0.600		
3200.0	0.700		
5560.0	0.800		
8520.0	0.900		
11500.0	1.000		

Constants of Jouyban–Acree model

J_0	287.782
J_1	266.678
J_2	0

Reference of data: [87].

Solute	Solvent 1	Solvent 2	T (°C)
Phenytoin	Ethanol	Water	25
Solubility (mg/L)	Vol. F. 1		
20.3	0.000		
38.5	0.100		
83.3	0.200		
251.0	0.300		
879.0	0.400		
2430.0	0.500		
5170.0	0.600		
8780.0	0.700		
12500.0	0.800		
15500.0	0.900		
14800.0	1.000		

Constants of Jouyban–Acree model

J_0	741.655
J_1	687.822
J_2	−560.595

Reference of data: [87].

Solute	Solvent 1	Solvent 2	T (°C)
Phenytoin	Glycerin	Water	25
Solubility (mg/L)	Vol. F. 1		
20.3	0.000		
24.7	0.100		
33.9	0.200		
47.2	0.300		
70.2	0.400		
97.5	0.500		
146.0	0.600		
228.0	0.700		
382.0	0.800		
613.0	0.900		
1010.0	1.000		

Constants of Jouyban–Acree model

J_0	−193.723
J_1	33.535
J_2	0

Reference of data: [87].

Solute	Solvent 1	Solvent 2	T (°C)
Phenytoin	Methanol	Water	25
Solubility (mg/L)	Vol. F. 1		
20.3	0.000		
36.5	0.100		
69.6	0.200		
155.0	0.300		
364.0	0.400		
928.0	0.500		
1960.0	0.600		
4670.0	0.700		
8670.0	0.800		
14700.0	0.900		
23800.0	1.000		

Constants of Jouyban–Acree model

J_0	114.881
J_1	388.429
J_2	0

Reference of data: [87].

Solute	Solvent 1	Solvent 2	T (°C)
Phenytoin	Polyethylene glycol 200	Water	25
Solubility (mg/L)	Vol. F. 1		
20.3	0.000		
51.9	0.100		
123.0	0.200		
294.0	0.300		
735.0	0.400		
1920.0	0.500		
5060.0	0.600		
13200.0	0.700		
29300.0	0.800		
54100.0	0.900		
73900.0	1.000		

Constants of Jouyban–Acree model

J_0	233.390
J_1	378.994
J_2	342.695

Reference of data: [87].

Solute	Solvent 1	Solvent 2	T (°C)
Phenytoin	Polyethylene glycol 400	Water	25
Solubility (mg/L)	Vol. F. 1		
20.3	0.000		
67.7	0.100		
144.0	0.200		
358.0	0.300		
941.0	0.400		
2640.0	0.500		
7790.0	0.600		
20300.0	0.700		
45900.0	0.800		
70900.0	0.900		
71800.0	1.000		

Constants of Jouyban–Acree model

J_0	402.072
J_1	546.091
J_2	639.967

Reference of data: [87].

Solute	Solvent 1	Solvent 2	T (°C)
Phenytoin	Propylene glycol	Water	25
Solubility (mg/L)	Vol. F. 1		
20.3	0.000		
36.4	0.100		
54.4	0.200		
121.0	0.300		
243.0	0.400		
625.0	0.500		
1480.0	0.600		
2930.0	0.700		
5440.0	0.800		
10300.0	0.900		
17600.0	1.000		

Constants of Jouyban–Acree model

J_0	0
J_1	0
J_2	0

Reference of data: [87].

Solute	Solvent 1	Solvent 2	T (°C)
Phenytoin	Water	Sorbitol	25
Solubility (mg/L)	Vol. F. 1		
20.3	0.000		
18.4	0.100		
18.4	0.200		
19.5	0.300		
17.5	0.400		
17.0	0.500		
17.0	0.600		
17.5	0.700		
18.0	0.800		
15.4	0.900		
17.0	1.000		

Constants of Jouyban–Acree model

J_0	−32.816
J_1	0
J_2	0

Reference of data: [87].

Solute	Solvent 1	Solvent 2	T (°C)
Phthalic acid	Water	Acetic acid	25.2
Solubility (g/100 g)	Mass F. 1		
1.990	0.200		
2.260	0.150		
2.630	0.100		
2.850	0.050		
2.960	0.000		

Constants of Jouyban–Acree model

J_0

J_1

J_2

Reference of data: [25].

Solute	Solvent 1	Solvent 2	T (°C)
Phthalic acid	Water	Acetic acid	35.3
Solubility (g/100 g)	Mass F. 1		
3.740	0.200		
3.980	0.150		
4.230	0.100		
4.360	0.050		
4.220	0.000		

Constants of Jouyban–Acree model

J_0

J_1

J_2

Reference of data: [25].

Solute	Solvent 1	Solvent 2	T (°C)
Phthalic acid	Water	Acetic acid	45.2
Solubility (g/100 g)	Mass F. 1		
6.490	0.200		
6.610	0.150		
6.940	0.100		
6.820	0.050		
6.290	0.000		

Constants of Jouyban–Acree model

J_0

J_1

J_2

Reference of data: [25].

Solute	Solvent 1	Solvent 2	T (°C)
Phthalic acid	Water	Acetic acid	55.1
Solubility (g/100 g)	Mass F. 1		
10.60	0.200		
10.70	0.150		
10.59	0.100		
9.57	0.050		
8.78	0.000		

Constants of Jouyban–Acree model

J_0

J_1

J_2

Reference of data: [25].

Solute	Solvent 1	Solvent 2	T (°C)
Phthalic acid	Water	Acetic acid	64.9
Solubility (g/100 g)	Mass F. 1		
16.390	0.200		
16.060	0.150		
15.130	0.100		
13.620	0.050		
12.040	0.000		

Constants of Jouyban–Acree model

J_0
J_1
J_2

Reference of data: [25].

Solute	Solvent 1	Solvent 2	T (°C)
Phthalic acid	Water	Acetic acid	75.2
Solubility (g/100 g)	Mass F. 1		
23.73	0.200		
23.14	0.150		
22.32	0.100		
19.72	0.050		
16.33	0.000		

Constants of Jouyban–Acree model

J_0
J_1
J_2

Reference of data: [25].

Solute	Solvent 1	Solvent 2	T (°C)
Phthalic acid	Water	Acetic acid	85.5
Solubility (g/100 g)	Mass F. 1		
35.04	0.200		
32.15	0.150		
30.03	0.100		
26.24	0.050		
21.53	0.000		

Constants of Jouyban–Acree model

J_0
J_1
J_2

Reference of data: [25].

Solute	Solvent 1	Solvent 2	T (°C)
Phthalic acid	Water	Acetic acid	94.8
Solubility (g/100 g)	Mass F. 1		
44.78	0.200		
41.58	0.150		
37.87	0.100		
33.46	0.050		
27.52	0.000		

Constants of Jouyban–Acree model

J_0
J_1
J_2

Reference of data: [25].

Solute	Solvent 1	Solvent 2	T (°C)
Physostigmine	2-Propanol	Isopropyl myristate	25
Solubility (mole F.)	Vol. F. 1		
0.0090	0.000		
0.0280	0.100		
0.0360	0.300		
0.0450	0.500		
0.0460	0.700		
0.0470	0.900		
0.0510	1.000		

Constants of Jouyban–Acree model

J_0	426.536
J_1	−530.599
J_2	0

Reference of data: [88].

Solute	Solvent 1	Solvent 2	T (°C)
Potassium clavulanate	Water	2-Propanol	0
Solubility (mole F.)	Mole F. 1		
0.00004940	0.000		
0.00010340	0.050		
0.00023140	0.090		
0.00051600	0.150		
0.00131480	0.200		

Constants of Jouyban–Acree model

J_0
J_1
J_2

Reference of data: [89].

Solute	Solvent 1	Solvent 2	T (°C)
Potassium clavulanate	Water	2-Propanol	4
Solubility (mole F.)	Mole F. 1		
0.00005450	0.000		
0.00011050	0.050		
0.00023710	0.090		
0.00057780	0.150		
0.00143270	0.200		

Constants of Jouyban–Acree model

J_0
J_1
J_2

Reference of data: [89].

Solute	Solvent 1	Solvent 2	T (°C)
Potassium clavulanate	Water	2-Propanol	8
Solubility (mole F.)	Mole F. 1		
0.00006050	0.000		
0.00012220	0.050		
0.00026260	0.090		
0.00065350	0.150		
0.00160720	0.200		

Constants of Jouyban–Acree model

J_0
J_1
J_2

Reference of data: [89].

Solute	Solvent 1	Solvent 2	T (°C)
Potassium clavulanate	Water	2-Propanol	12
Solubility (mole F.)	Mole F. 1		
0.00006810	0.000		
0.00013910	0.050		
0.00028440	0.090		
0.00073070	0.150		
0.00179230	0.200		

Constants of Jouyban–Acree model

J_0
J_1
J_2

Reference of data: [89].

Solute	Solvent 1	Solvent 2	T (°C)
Potassium clavulanate	Water	2-Propanol	16
Solubility (mole F.)	Mole F. 1		
0.00007470	0.000		
0.00015250	0.050		
0.00031650	0.090		
0.00078690	0.150		
0.00205380	0.200		

Constants of Jouyban–Acree model

J_0
J_1
J_2

Reference of data: [89].

Solute	Solvent 1	Solvent 2	T (°C)
Potassium clavulanate	Water	2-Propanol	20
Solubility (mole F.)	Mole F. 1		
0.00008130	0.000		
0.00017420	0.050		
0.00035460	0.090		
0.00092020	0.150		
0.00227880	0.200		

Constants of Jouyban–Acree model

J_0
J_1
J_2

Reference of data: [89].

Solute	Solvent 1	Solvent 2	T (°C)
Potassium clavulanate	Water	2-Propanol	24
Solubility (mole F.)	Mole F. 1		
0.00009170	0.000		
0.00019820	0.050		
0.00041280	0.090		
0.00100850	0.150		
0.00256310	0.200		

Constants of Jouyban–Acree model

J_0
J_1
J_2

Reference of data: [89].

Solute	Solvent 1	Solvent 2	T (°C)
Potassium clavulanate	Water	2-Propanol	28
Solubility (mole F.)	Mole F. 1		
0.00010310	0.000		
0.00022240	0.050		
0.00049630	0.090		
0.00115520	0.150		
0.00285170	0.200		

Constants of Jouyban–Acree model

J_0

J_1

J_2

Reference of data: [89].

Solute	Solvent 1	Solvent 2	T (°C)
Potassium clavulanate	Water	2-Propanol	32
Solubility (mole F.)	Mole F. 1		
0.00012310	0.000		
0.00024900	0.050		
0.00059410	0.090		
0.00133690	0.150		
0.00295730	0.200		

Constants of Jouyban–Acree model

J_0

J_1

J_2

Reference of data: [89].

Solute	Solvent 1	Solvent 2	T (°C)
Resveratrol (*trans*)	Ethanol	Water	0.1
Solubility (mole F.)	Mole F. 1		
0.000220	0.200		
0.001150	0.300		
0.002360	0.400		
0.003720	0.500		
0.005720	0.600		
0.007540	0.700		
0.008690	0.800		
0.009400	0.900		
0.010060	1.000		

Constants of Jouyban–Acree model

J_0

J_1

J_2

Reference of data: [90].

Solute	Solvent 1	Solvent 2	T (°C)
Resveratrol (*trans*)	Ethanol	Water	10.1
Solubility (mole F.)	Mole F. 1		
0.000038	0.100		
0.000380	0.200		
0.001660	0.300		
0.003360	0.400		
0.005050	0.500		
0.007210	0.600		
0.009440	0.700		
0.010930	0.800		
0.011920	0.900		
0.012740	1.000		

Constants of Jouyban–Acree model

J_0

J_1

J_2

Reference of data: [90].

Solute	Solvent 1	Solvent 2	T (°C)
Resveratrol (*trans*)	Ethanol	Water	20.1
Solubility (mole F.)	Mole F. 1		
0.0000029	0.000		
0.0000708	0.100		
0.0006800	0.200		
0.0024400	0.300		
0.0046200	0.400		
0.0071300	0.500		
0.0096300	0.600		
0.0120100	0.700		
0.0135200	0.800		
0.0146500	0.900		
0.0156200	1.000		

Handbook of Solubility Data for Pharmaceuticals

Constants of Jouyban–Acree model

J_0

J_1

J_2

Reference of data: [90].

Solute	Solvent 1	Solvent 2	T (°C)
Resveratrol (*trans*)	Ethanol	Water	30.1
Solubility (mole F.)	Mole F. 1		
0.00000457	0.000		
0.00010460	0.100		
0.00098000	0.200		
0.00292000	0.300		
0.00522000	0.400		
0.00804000	0.500		
0.01059000	0.600		
0.01354000	0.700		
0.01541000	0.800		
0.01700000	0.900		
0.01823000	1.000		

Constants of Jouyban–Acree model

J_0

J_1

J_2

Reference of data: [90].

Solute	Solvent 1	Solvent 2	T (°C)
Resveratrol (*trans*)	Ethanol	Water	40.1
Solubility (mole F.)	Mole F. 1		
0.00000646	0.000		
0.00015130	0.100		
0.00143000	0.200		
0.00327000	0.300		
0.00607000	0.400		
0.00922000	0.500		
0.01222000	0.600		
0.01533000	0.700		
0.01797000	0.800		
0.01961000	0.900		
0.02081000	1.000		

Constants of Jouyban–Acree model

J_0

J_1

J_2

Reference of data: [90].

Solute	Solvent 1	Solvent 2	T (°C)
Resveratrol (*trans*)	Ethanol	Water	50.1
Solubility (mole F.)	Mole F. 1		
0.0000093	0.000		
0.0002212	0.100		
0.0019000	0.200		
0.0038400	0.300		
0.0070600	0.400		
0.0104100	0.500		
0.0137400	0.600		
0.0175300	0.700		
0.0203100	0.800		
0.0222000	0.900		
0.0236300	1.000		

Constants of Jouyban–Acree model

J_0
J_1
J_2

Reference of data: [90].

Solute	Solvent 1	Solvent 2	T (°C)
Rapamycin	Butyrolactone (γ)	Water	25.0
Solubility (g/L)	Vol. F. 1		
0.0026	0.000		
0.0630	0.200		
0.5780	0.300		
3.7540	0.400		

Constants of Jouyban–Acree model

J_0
J_1
J_2

Reference of data: [91].

Solute	Solvent 1	Solvent 2	T (°C)
Rapamycin	Dimethyl isosorbide	Water	25.0
Solubility (g/L)	Vol. F. 1		
0.0026	0.000		
0.0160	0.200		
0.0480	0.300		
0.2970	0.400		

Constants of Jouyban–Acree model

J_0

J_1

J_2

Reference of data: [91].

Solute	Solvent 1	Solvent 2	T (°C)
Rapamycin	Ethanol	Water	25.0
Solubility (g/L)	Vol. F. 1		
0.0026	0.000		
0.0090	0.200		

Constants of Jouyban–Acree model

J_0

J_1

J_2

Reference of data: [91].

Solute	Solvent 1	Solvent 2	T (°C)
Rapamycin	Ethoxy diglycol (Transcutol)	Water	25.0
Solubility (g/L)	Vol. F. 1		
0.0026	0.000		
0.0150	0.200		
0.0270	0.300		
0.1580	0.400		

Constants of Jouyban–Acree model

J_0

J_1

J_2

Reference of data: [91].

Solute	Solvent 1	Solvent 2	T (°C)
Rapamycin	Glycerol formal	Water	25.0
Solubility (g/L)	Vol. F. 1		
0.0026	0.000		
0.0140	0.200		
0.0320	0.300		
0.1760	0.400		

Constants of Jouyban–Acree model

J_0

J_1

J_2

Reference of data: [91].

Solute	Solvent 1	Solvent 2	T (°C)
Rapamycin	N-Methyl-2-pyrrolidinone	Water	25.0
Solubility (g/L)	Vol. F. 1		
0.0026	0.000		
0.0250	0.200		
0.0470	0.300		
0.2610	0.400		

Constants of Jouyban–Acree model

J_0

J_1

J_2

Reference of data: [91].

Solute	Solvent 1	Solvent 2	T (°C)
Rapamycin	Triethylene glycol dimethyl ether (Triglyme)	Water	25.0
Solubility (g/L)	Vol. F. 1		
0.0026	0.000		
0.0110	0.200		
0.0170	0.300		
0.0800	0.400		

Constants of Jouyban–Acree model

J_0

J_1

J_2

Reference of data: [91].

Solute	Solvent 1	Solvent 2	T (°C)
Rifapentine	Acetic acid	n-Octanol	5.0
Solubility (mole F.)	Mole F. 1		
0.0523	1.000		
0.0467	0.900		
0.0414	0.799		
0.0353	0.700		
0.0276	0.600		
0.0203	0.499		
0.0140	0.401		

Constants of Jouyban–Acree model

J_0

J_1

J_2

Reference of data: [92].

Solute	Solvent 1	Solvent 2	T (°C)
Rifapentine	Acetic acid	*n*-Octanol	10.0
Solubility (mole F.)	Mole F. 1		
0.0558	1.000		
0.0494	0.900		
0.0427	0.799		
0.0360	0.700		
0.0281	0.600		
0.0210	0.499		
0.0145	0.401		

Constants of Jouyban–Acree model

J_0

J_1

J_2

Reference of data: [92].

Solute	Solvent 1	Solvent 2	T (°C)
Rifapentine	Acetic acid	*n*-Octanol	15.0
Solubility (mole F.)	Mole F. 1		
0.0595	1.000		
0.0524	0.900		
0.0440	0.799		
0.0367	0.700		
0.0287	0.600		
0.0217	0.499		
0.0151	0.401		

Constants of Jouyban–Acree model

J_0

J_1

J_2

Reference of data: [92].

Solute	Solvent 1	Solvent 2	T (°C)
Rifapentine	Acetic acid	n-Octanol	20.0
Solubility (mole F.)	Mole F. 1		
0.0632	1.000		
0.0556	0.900		
0.0455	0.799		
0.0374	0.700		
0.0293	0.600		
0.0224	0.499		
0.0157	0.401		

Constants of Jouyban–Acree model

J_0

J_1

J_2

Reference of data: [92].

Solute	Solvent 1	Solvent 2	T (°C)
Rifapentine	Acetic acid	n-Octanol	24.0
Solubility (mole F.)	Mole F. 1		
0.0670	1.000		
0.0589	0.900		
0.0471	0.799		
0.0381	0.700		
0.0300	0.600		
0.0231	0.499		
0.0164	0.401		

Constants of Jouyban–Acree model

J_0

J_1

J_2

Reference of data: [92].

Solute	Solvent 1	Solvent 2	T (°C)
Rifapentine	Acetic acid	n-Octanol	30.0
Solubility (mole F.)	Mole F. 1		
0.0707	1.000		
0.0623	0.900		
0.0487	0.799		
0.0389	0.700		
0.0307	0.600		
0.0236	0.499		
0.0171	0.401		

Constants of Jouyban–Acree model

J_0

J_1

J_2

Reference of data: [92].

Solute	Solvent 1	Solvent 2	T (°C)
Rifapentine	Acetic acid	n-Octanol	35.0
Solubility (mole F.)	Mole F. 1		
0.0755	1.000		
0.0656	0.900		
0.0503	0.799		
0.0396	0.700		
0.0315	0.600		
0.0244	0.499		
0.0178	0.401		

Constants of Jouyban–Acree model

J_0

J_1

J_2

Reference of data: [92].

Solute	Solvent 1	Solvent 2	T (°C)
Rifapentine	Acetic acid	n-Octanol	40.0
Solubility (mole F.)	Mole F. 1		
0.0805	1.000		
0.0691	0.900		
0.0520	0.799		
0.0404	0.700		
0.0323	0.600		
0.0252	0.499		
0.0187	0.401		

Constants of Jouyban–Acree model

J_0

J_1

J_2

Reference of data: [92].

Solute	Solvent 1	Solvent 2	T (°C)
Rifapentine	Acetic acid	n-Octanol	45.0
Solubility (mole F.)	Mole F. 1		
0.0858	1.000		
0.0727	0.900		
0.0538	0.799		
0.0412	0.700		
0.0331	0.600		
0.0261	0.499		
0.0195	0.401		

Constants of Jouyban–Acree model

J_0

J_1

J_2

Reference of data: [92].

Solute	Solvent 1	Solvent 2	T (°C)
Rifapentine	Acetic acid	n-Octanol	50.0
Solubility (mole F.)	Mole F. 1		
0.0913	1.000		
0.0762	0.900		
0.0557	0.799		
0.0421	0.700		
0.0340	0.600		
0.0270	0.499		
0.0204	0.401		

Constants of Jouyban–Acree model

J_0

J_1

J_2

Reference of data: [92].

Solute	Solvent 1	Solvent 2	T (°C)
Rofecoxib	Ethanol	Glycerol	25
Solubility (g/L)	Vol. F. 1		
0.108	0.000		
0.124	0.100		
0.203	0.200		
0.544	0.400		
0.685	0.600		
0.759	0.800		
0.739	0.900		
0.683	1.000		

Constants of Jouyban–Acree model

J_0	456.262
J_1	139.576
J_2	−454.658

Reference of data: [49].

Solute	Solvent 1	Solvent 2	T (°C)
Rofecoxib	Ethanol	Water	25
Solubility (g/L)	Vol. F. 1		
0.009	0.000		
0.016	0.100		
0.032	0.200		
0.168	0.400		
0.548	0.600		
1.018	0.800		
1.041	0.900		
0.683	1.000		

Constants of Jouyban–Acree model

J_0	713.440
J_1	578.394
J_2	0

Reference of data: [49].

Solute	Solvent 1	Solvent 2	T (°C)
Rofecoxib	Polyethylene glycol 400	Ethanol	25
Solubility (g/L)	Vol. F. 1		
0.683	0.000		
1.005	0.200		
2.228	0.400		
3.964	0.600		
8.143	0.800		
8.973	0.900		
11.234	1.000		

Constants of Jouyban–Acree model

J_0	0
J_1	0
J_2	0

Reference of data: [49].

Solute	Solvent 1	Solvent 2	T (°C)
Rofecoxib	Ethanol	Water	25
Solubility (mg/L)	Vol. F. 1		
8.190	0.000		
25.260	0.200		
228.7	0.400		
633.5	0.600		
879.2	0.800		
390.5	1.000		

Constants of Jouyban–Acree model

J_0	1130.129
J_1	611.409
J_2	−584.590

Reference of data: [93].

Solute	Solvent 1	Solvent 2	T (°C)
Rofecoxib	Ethanol	Water	30
Solubility (mg/L)	Vol. F. 1		
9.360	0.000		
42.440	0.200		
306.6	0.400		
865.3	0.600		
1187	0.800		
506	1.000		

Constants of Jouyban–Acree model

J_0	1130.129
J_1	611.409
J_2	−584.590

Reference of data: [93].

Solute	Solvent 1	Solvent 2	T (°C)
Rofecoxib	Ethanol	Water	35
Solubility (mg/L)	Vol. F. 1		
11.180	0.000		
55.610	0.200		
418.7	0.400		
1183	0.600		
1381	0.800		
614.4	1.000		

Constants of Jouyban–Acree model

J_0	1130.129
J_1	611.409
J_2	−584.590

Reference of data: [93].

Solute	Solvent 1	Solvent 2	T (°C)
Rofecoxib	Methanol	Water	25
Solubility (mg/L)	Vol. F. 1		
8.190	0.000		
24.3	0.200		
132.3	0.400		
538.3	0.600		
1220	0.800		
1400	1.000		

Constants of Jouyban–Acree model

J_0	536.269
J_1	476.318
J_2	−385.797

Reference of data: [93].

Solute	Solvent 1	Solvent 2	T (°C)
Rofecoxib	Methanol	Water	30
Solubility (mg/L)	Vol. F. 1		
9.360	0.000		
29.650	0.200		
156.2	0.400		
730.3	0.600		
1509	0.800		
1961	1.000		

Constants of Jouyban–Acree model

J_0	536.269
J_1	476.318
J_2	−385.797

Reference of data: [93].

Solute	Solvent 1	Solvent 2	T (°C)
Rofecoxib	Methanol	Water	35
Solubility (mg/L)	Vol. F. 1		
11.180	0.000		
44.880	0.200		
245.2	0.400		
973.5	0.600		
1977	0.800		
2751	1.000		

Constants of Jouyban–Acree model

J_0	536.269
J_1	476.318
J_2	−385.797

Reference of data: [93].

Solute	Solvent 1	Solvent 2	T (°C)
Rofecoxib (S)	Ethanol	Water	25.0
Solubility (mg/L)	Mass F. 1		
8.2	0.000		
23.4	0.100		
33.3	0.200		
82.1	0.300		
228.7	0.400		
395.5	1.000		

Constants of Jouyban–Acree model

J_0	1554.698
J_1	3189.580
J_2	3349.072

Reference of data: [94].

Solute	Solvent 1	Solvent 2	T (°C)
Rofecoxib (S)	Ethanol	Water	30.0
Solubility (mg/L)	Mass F. 1		
9.4	0.000		
35.3	0.100		
54	0.200		
121.4	0.300		
306.6	0.400		
510.4	1.000		

Constants of Jouyban–Acree model

J_0	1554.698
J_1	3189.580
J_2	3349.072

Reference of data: [94].

Solute	Solvent 1	Solvent 2	T (°C)
Rofecoxib (S)	Ethanol	Water	35.0
Solubility (mg/L)	Mass F. 1		
11.2	0.000		
44.2	0.100		
82	0.200		
197.3	0.300		
418.7	0.400		
624.3	1.000		

Constants of Jouyban–Acree model

J_0	1554.698
J_1	3189.580
J_2	3349.072

Reference of data: [94].

Solute	Solvent 1	Solvent 2	T (°C)
Rofecoxib (S)	Glycerol	Water	25.0
Solubility (mg/L)	Mass F. 1		
8.2	0.000		
12.1	0.100		
13.8	0.200		
15.8	0.300		
18.4	0.400		
70.6	1.000		

Constants of Jouyban–Acree model

J_0	−104.819
J_1	−250.020
J_2	0

Reference of data: [94].

Solute	Solvent 1	Solvent 2	T (°C)
Rofecoxib (S)	Glycerol	Water	30.0
Solubility (mg/L)	Mass F. 1		
9.4	0.000		
13.3	0.100		
15.2	0.200		
18.1	0.300		
22.1	0.400		
95.3	1.000		

Constants of Jouyban–Acree model

J_0	−104.819
J_1	−250.020
J_2	0

Reference of data: [94].

Solute	Solvent 1	Solvent 2	T (°C)
Rofecoxib (S)	Glycerol	Water	35.0
Solubility (mg/L)	Mass F. 1		
11.2	0.000		
15	0.100		
17.4	0.200		
21.4	0.300		
25.5	0.400		
118.1	1.000		

Constants of Jouyban–Acree model

J_0	−104.819
J_1	−250.020
J_2	0

Reference of data: [94].

Solute	Solvent 1	Solvent 2	T (°C)
Rofecoxib (S)	Mannitol	Water	25.0
Solubility (mg/L)	Mass F. 1		
8.20	0.000		
13.70	0.010		
14.00	0.020		
14.20	0.050		
14.30	0.100		

Constants of Jouyban–Acree model

J_0

J_1

J_2

Reference of data: [95].

Solute	Solvent 1	Solvent 2	T (°C)
Rofecoxib (S)	Mannitol	Water	30.0
Solubility (mg/L)	Mass F. 1		
9.50	0.000		
14.30	0.010		
14.40	0.020		
14.50	0.050		
14.70	0.100		

Constants of Jouyban–Acree model

J_0

J_1

J_2

Reference of data: [95].

Solute	Solvent 1	Solvent 2	T (°C)
Rofecoxib (S)	Mannitol	Water	35.0
Solubility (mg/L)	Mass F. 1		
11.30	0.000		
14.50	0.010		
14.60	0.020		
14.70	0.050		
14.80	0.100		

Constants of Jouyban–Acree model

J_0

J_1

J_2

Reference of data: [95].

Solute	Solvent 1	Solvent 2	T (°C)
Rofecoxib (S)	Poly(vinylpyrrolidone) K30	Water	25.0
Solubility (mg/L)	Mass F. 1		
8.20	0.000		
18.10	0.010		
23.80	0.020		
60.10	0.050		
90.30	0.100		

Constants of Jouyban–Acree model

J_0

J_1

J_2

Reference of data: [95].

Solute	Solvent 1	Solvent 2	T (°C)
Rofecoxib (S)	Poly(vinylpyrrolidone) K30	Water	30.0
Solubility (mg/L)	Mass F. 1		
9.50	0.000		
19.60	0.010		
25.30	0.020		
70.30	0.050		
108.00	0.100		

Constants of Jouyban–Acree model

J_0

J_1

J_2

Reference of data: [95].

Solute	Solvent 1	Solvent 2	T (°C)
Rofecoxib (S)	Poly(vinylpyrrolidone) K30	Water	35.0
Solubility (mg/L)	Mass F. 1		
11.30	0.000		
20.70	0.010		
28.50	0.020		
84.30	0.050		
126.20	0.100		

Constants of Jouyban–Acree model

J_0

J_1

J_2

Reference of data: [95].

Solute	Solvent 1	Solvent 2	T (°C)
Rofecoxib (S)	Polyethylene glycol 4000	Water	25.0
Solubility (mg/L)	Mass F. 1		
8.20	0.000		
17.10	0.010		
20.20	0.020		
24.60	0.050		
46.90	0.100		

Constants of Jouyban–Acree model

J_0

J_1

J_2

Reference of data: [95].

Solute	Solvent 1	Solvent 2	T (°C)
Rofecoxib (S)	Polyethylene glycol 4000	Water	30.0
Solubility (mg/L)	Mass F. 1		
9.50	0.000		
18.20	0.010		
21.60	0.020		
26.80	0.050		
57.50	0.100		

Constants of Jouyban–Acree model

J_0

J_1

J_2

Reference of data: [95].

Solute	Solvent 1	Solvent 2	T (°C)
Rofecoxib (S)	Polyethylene glycol 4000	Water	35.0
Solubility (mg/L)	Mass F. 1		
11.30	0.000		
19.10	0.010		
23.20	0.020		
29.60	0.050		
67.20	0.100		

Constants of Jouyban–Acree model

J_0

J_1

J_2

Reference of data: [95].

Solute	Solvent 1	Solvent 2	T (°C)
Rofecoxib (S)	Polyethylene glycol 6000	Water	25.0
Solubility (mg/L)	Mass F. 1		
8.20	0.000		
14.50	0.010		
15.30	0.020		
18.10	0.050		
20.00	0.100		

Constants of Jouyban–Acree model

J_0
J_1
J_2

Reference of data: [95].

Solute	Solvent 1	Solvent 2	T (°C)
Rofecoxib (S)	Polyethylene glycol 6000	Water	30.0
Solubility (mg/L)	Mass F. 1		
9.50	0.000		
15.50	0.010		
16.40	0.020		
19.20	0.050		
21.30	0.100		

Constants of Jouyban–Acree model

J_0
J_1
J_2

Reference of data: [95].

Solute	Solvent 1	Solvent 2	T (°C)
Rofecoxib (S)	Polyethylene glycol 6000	Water	35.0
Solubility (mg/L)	Mass F. 1		
11.30	0.000		
17.50	0.010		
18.90	0.020		
22.30	0.050		
25.30	0.100		

Constants of Jouyban–Acree model

J_0
J_1
J_2

Reference of data: [95].

Solute	Solvent 1	Solvent 2	T (°C)
Rofecoxib (S)	Propylene glycol	Water	25.0
Solubility (mole F.)	Mass F. 1		
8.20	0.000		
20.30	0.100		
30.80	0.200		
48.60	0.300		
70.10	0.400		
170.80	1.000		

Constants of Jouyban–Acree model

J_0	659.086
J_1	912.253
J_2	1493.225

Reference of data: [94].

Solute	Solvent 1	Solvent 2	T (°C)
Rofecoxib (S)	Propylene glycol	Water	30.0
Solubility (mole F.)	Mass F. 1		
9.40	0.000		
25.10	0.100		
36.00	0.200		
60.00	0.300		
85.20	0.400		
190.60	1.000		

Constants of Jouyban–Acree model

J_0	659.086
J_1	912.253
J_2	1493.225

Reference of data: [94].

Solute	Solvent 1	Solvent 2	T (°C)
Rofecoxib (S)	Propylene glycol	Water	35.0
Solubility (mole F.)	Mass F. 1		
11.20	0.000		
28.80	0.100		
42.10	0.200		
66.30	0.300		
97.50	0.400		
226.50	1.000		

Constants of Jouyban–Acree model

J_0	659.086
J_1	912.253
J_2	1493.225

Reference of data: [94].

Solute	Solvent 1	Solvent 2	T (°C)
Rofecoxib (S)	Urea	Water	25.0
Solubility (mg/L)	Mass F. 1		
8.20	0.000		
19.20	0.010		
38.70	0.020		
61.50	0.050		
95.60	0.100		

Constants of Jouyban–Acree model

J_0
J_1
J_2

Reference of data: [95].

Solute	Solvent 1	Solvent 2	T (°C)
Rofecoxib (S)	Urea	Water	30.0
Solubility (mg/L)	Mass F. 1		
9.50	0.000		
23.40	0.010		
45.60	0.020		
80.60	0.050		
120.20	0.100		

Constants of Jouyban–Acree model

J_0
J_1
J_2

Reference of data: [95].

Solute	Solvent 1	Solvent 2	T (°C)
Rofecoxib (S)	Urea	Water	35.0
Solubility (mg/L)	Mass F. 1		
11.30	0.000		
28.60	0.010		
58.60	0.020		
115.70	0.050		
155.80	0.100		

Constants of Jouyban–Acree model

J_0
J_1
J_2

Reference of data: [95].

Solute	Solvent 1	Solvent 2	T (°C)
Salicylic acid	1,4-Dioxane	Water	25.0
Solubility (mole F.)	Mole F. 1		
0.0003	0.000		
0.0008	0.024		
0.0018	0.051		
0.0066	0.083		
0.0221	0.125		
0.0436	0.173		
0.0784	0.239		
0.1212	0.329		
0.1616	0.457		
0.2173	0.657		
0.2610	1.000		

Constants of Jouyban–Acree model

J_0	1489.552
J_1	−1496.403
J_2	2186.382

Reference of data: [96].

Solute	Solvent 1	Solvent 2	T (°C)
Salicylic acid	Dioxane	Water	10
Solubility (mole/L)	Vol. F. 1		
0.0080	0.000		
0.0210	0.100		
0.0490	0.200		
0.1110	0.300		
0.5700	0.400		
0.8410	0.500		
1.3810	0.600		
2.1110	0.700		
2.4110	0.800		
2.9060	0.900		
2.5960	1.000		

Constants of Jouyban–Acree model

J_0	946.596
J_1	49.281
J_2	−181.601

Reference of data: [24].

Solute	Solvent 1	Solvent 2	T (°C)
Salicylic acid	Dioxane	Water	20
Solubility (mole/L)	Vol. F. 1		
0.0110	0.000		
0.0290	0.100		
0.0710	0.200		
0.2580	0.300		
0.7060	0.400		
0.9760	0.500		
1.5170	0.600		
2.4110	0.700		
2.8080	0.800		
3.7790	0.900		
3.0650	1.000		

Constants of Jouyban–Acree model

J_0	946.596
J_1	49.281
J_2	−181.601

Reference of data: [24].

Solute	Solvent 1	Solvent 2	T (°C)
Salicylic acid	Dioxane	Water	25
Solubility (mole/L)	Vol. F. 1		
0.0120	0.000		
0.0390	0.100		
0.0870	0.200		
0.3730	0.300		
2.8960	0.700		
3.1680	0.800		
4.5820	0.900		
3.6800	1.000		

Constants of Jouyban–Acree model

J_0	946.596
J_1	49.281
J_2	−181.601

Reference of data: [24].

Solute	Solvent 1	Solvent 2	T (°C)
Salicylic acid	Dioxane	Water	30
Solubility (mole/L)	Vol. F. 1		
0.0150	0.000		
0.0410	0.100		
0.1070	0.200		
0.5590	0.300		
3.1270	0.700		
3.6050	0.800		
5.3320	0.900		
3.9740	1.000		

Constants of Jouyban–Acree model

J_0	946.596
J_1	49.281
J_2	−181.601

Reference of data: [24].

Solute	Solvent 1	Solvent 2	T (°C)
Salicylic acid	Dioxane	Water	35
Solubility (mole/L)	Vol. F. 1		
0.0170	0.000		
0.0470	0.100		
0.1270	0.200		
0.8320	0.300		
3.4120	0.700		
4.1090	0.800		
6.1830	0.900		
4.3400	1.000		

Constants of Jouyban–Acree model

J_0	946.596
J_1	49.281
J_2	−181.601

Reference of data: [24].

Solute	Solvent 1	Solvent 2	T (°C)
Salicylic acid	Dioxane	Water	40
Solubility (mole/L)	Vol. F. 1		
0.0200	0.000		
0.0580	0.100		
0.1580	0.200		
1.1660	0.300		
3.8370	0.700		
4.6890	0.800		
7.3740	0.900		
4.6590	1.000		

Constants of Jouyban–Acree model

J_0	946.596
J_1	49.281
J_2	−181.601

Reference of data: [24].

Solute	Solvent 1	Solvent 2	T (°C)
Salicylic acid	Ethanol	Ethyl acetate	25.0
Solubility (mole F.)	Mole F. 1		
0.13830	0.000		
0.16530	0.161		
0.17670	0.299		
0.18100	0.420		
0.18080	0.526		
0.18390	0.628		
0.17380	0.713		
0.17540	0.795		
0.16160	0.870		
0.14460	0.939		
0.14500	1.000		

Constants of Jouyban–Acree model

J_0	135.504
J_1	0
J_2	0

Reference of data: [96].

Solute	Solvent 1	Solvent 2	T (°C)
Salicylic acid	Ethanol	Ethyl acetate	25
Solubility (mole F.)	Vol. F. 1		
0.13970	1.000		
0.14580	0.900		
0.15790	0.800		
0.16530	0.700		
0.17490	0.600		
0.18300	0.500		
0.18930	0.400		
0.18540	0.300		
0.16580	0.200		
0.13090	0.100		
0.11360	0.000		

Constants of Jouyban–Acree model

J_0	196.085
J_1	−117.499
J_2	0

Reference of data: [11].

Solute	Solvent 1	Solvent 2	T (°C)
Salicylic acid	Ethanol	Water	25.0
Solubility (mole F.)	Mole F. 1		
0.00030	0.000		
0.00040	0.033		
0.00070	0.073		
0.00200	0.118		
0.006300	0.171		
0.016100	0.235		
0.026000	0.273		
0.029400	0.290		
0.034200	0.318		
0.044900	0.367		
0.054100	0.416		
0.066600	0.477		
0.078100	0.549		
0.098100	0.631		
0.117800	0.7407		
0.129500	0.8457		
0.145000	1.000		

Constants of Jouyban–Acree model

J_0	1294.915
J_1	−759.489
J_2	0

Reference of data: [96].

Solute	Solvent 1	Solvent 2	T (°C)
Salicylic acid	Ethanol	Water	25
Solubility (mole F.)	Vol. F. 1		
0.00020	0.000		
0.00050	0.100		
0.00070	0.200		
0.00190	0.300		
0.00750	0.400		
0.01610	0.500		
0.03680	0.600		
0.06790	0.700		
0.08520	0.800		
0.12780	0.900		
0.13970	1.000		

Constants of Jouyban–Acree model

J_0	522.590
J_1	543.601
J_2	0

Reference of data: [11].

Solute	Solvent 1	Solvent 2	T (°C)
Salicylic acid	Ethanol	Water	25
Solubility (g/L)	Vol. F. 1		
1.890	0.000		
2.390	0.100		
4.470	0.200		
11.170	0.300		
29.000	0.400		
66.480	0.500		
132.820	0.600		
191.280	0.700		
254.350	0.800		
305.320	0.900		
291.300	1.000		

Constants of Jouyban–Acree model

J_0	535.267
J_1	683.480
J_2	−611.474

Reference of data: [97].

Solute	Solvent 1	Solvent 2	T (°C)
Salicylic acid	Methanol	Water	25.0
Solubility (mole F.)	Mole F. 1		
0.00030	0.000		
0.00060	0.090		
0.00210	0.194		
0.01130	0.360		
0.02520	0.458		
0.06500	0.692		
0.12230	1.000		

Constants of Jouyban–Acree model

J_0	884.935
J_1	−52.128
J_2	−870.693

Reference of data: [96].

Solute	Solvent 1	Solvent 2	T (°C)
Salicylic acid	Polyethylene glycol 300	Water	25.0
Solubility (mole F.)	Mole F. 1		
0.00030	0.000		
0.00070	0.015		
0.00150	0.025		
0.00410	0.039		
0.01490	0.057		
0.03500	0.083		
0.07290	0.123		
0.17120	0.194		
0.26700	0.351		
0.49310	1.000		

Constants of Jouyban–Acree model

J_0	3262.590
J_1	6984.018
J_2	13529.026

Reference of data: [96].

Solute	Solvent 1	Solvent 2	T (°C)
Salicylic acid	Propylene glycol	Water	25
Solubility (g/L)	Vol. F. 1		
1.890	0.000		
2.400	0.100		
3.240	0.200		
5.240	0.300		
9.290	0.400		
20.360	0.500		
40.390	0.600		
75.520	0.700		
126.550	0.800		
187.630	0.900		
248.630	1.000		

Constants of Jouyban–Acree model

J_0	−51.052
J_1	510.802
J_2	0

Reference of data: [97].

Solute	Solvent 1	Solvent 2	T (°C)
Salmeterol xinafoate	Dioxane	Water	19
Solubility (mole F.)	Vol. F. 1		
0.00000279	0.000		
0.00000964	0.100		
0.00003643	0.200		
0.00013369	0.300		
0.00051056	0.400		
0.00185476	0.500		
0.00621991	0.600		
0.02004050	0.700		
0.03111703	0.800		
0.03404745	0.900		
0.02192780	0.950		
0.00429630	1.000		

Constants of Jouyban–Acree model

J_0	1402.663
J_1	1693.171
J_2	1456.366

Reference of data: [98].

Solute	Solvent 1	Solvent 2	T (°C)
Serine (L)	Water	Methanol	10
Solubility (g/L)	Vol. F. 1		
227.150	1.000		
189.785	0.900		
115.840	0.800		
64.150	0.700		
42.050	0.600		
18.035	0.400		
8.510	0.200		
6.500	0.000		

Constants of Jouyban–Acree model

J_0	−135.052
J_1	491.099
J_2	0

Reference of data: [99].

Solute	Solvent 1	Solvent 2	T (°C)
Serine (L)	Water	Methanol	30
Solubility (g/L)	Vol. F. 1		
394.040	1.000		
334.680	0.900		
243.280	0.800		
168.870	0.700		
90.035	0.600		
26.205	0.400		
6.955	0.200		
7.030	0.000		

Constants of Jouyban–Acree model

J_0	−135.052
J_1	491.099
J_2	0

Reference of data: [99].

Solute	Solvent 1	Solvent 2	T (°C)
Serine (L)	Water	Methanol	15
Solubility (mol/L)	Mole F. 1		
0.453	1.000		
0.427	0.920		
0.243	0.840		
0.205	0.748		
0.1245	0.656		
0.0913	0.553		
0.0525	0.458		
0.0186	0.359		
0.0073	0.241		
0.0026	0.121		
0.0008	0.000		

Constants of Jouyban–Acree model

J_0	516.939
J_1	0
J_2	0

Reference of data: [15].

Solute	Solvent 1	Solvent 2	T (°C)
Serine (L)	Water	Methanol	20
Solubility (mol/L)	Mole F. 1		
0.461	1.000		
0.432	0.920		
0.248	0.840		
0.210	0.748		
0.131	0.656		
0.0975	0.553		

(continued)

Solute	Solvent 1	Solvent 2	T (°C)
0.0575	0.458		
0.0209	0.359		
0.00851	0.241		
0.00309	0.121		
0.00105	0.000		

Constants of Jouyban–Acree model

J_0	516.939
J_1	0
J_2	0

Reference of data: [15].

Solute	Solvent 1	Solvent 2	T (°C)
Serine (L)	Water	Methanol	25
Solubility (mol/L)	Mole F. 1		
0.47	1.000		
0.438	0.920		
0.253	0.840		
0.217	0.748		
0.136	0.656		
0.103	0.553		
0.063	0.458		
0.024	0.359		
0.01	0.241		
0.00375	0.121		
0.00135	0.000		

Constants of Jouyban–Acree model

J_0	516.939
J_1	0
J_2	0

Reference of data: [15].

Solute	Solvent 1	Solvent 2	T (°C)
Silybin	Polyethylene glycol 6000	Water	20
Solubility (mass F.)	Mass F. 1		
0.00004240	0.000		
0.00006260	0.001		
0.00007250	0.002		
0.00008660	0.005		
0.00010100	0.008		
0.00011200	0.010		
0.00012300	0.012		
0.00014200	0.015		
0.00017400	0.020		

Constants of Jouyban–Acree model

J_0

J_1

J_2

Reference of data: [100].

Solute	Solvent 1	Solvent 2	T (°C)
Silybin	Polyethylene glycol 6000	Water	25
Solubility (mass F.)	Mass F. 1		
0.0000540	0.000		
0.0000780	0.001		
0.0000975	0.002		
0.0001110	0.005		
0.0001270	0.008		
0.0001500	0.010		
0.0001530	0.012		
0.0001700	0.015		
0.0002130	0.020		

Constants of Jouyban–Acree model

J_0

J_1

J_2

Reference of data: [100].

Solute	Solvent 1	Solvent 2	T (°C)
Silybin	Polyethylene glycol 6000	Water	30
Solubility (mass F.)	Mass F. 1		
0.0000691	0.000		
0.0000916	0.001		
0.0001190	0.002		
0.0001400	0.005		
0.0001610	0.008		
0.0001960	0.010		
0.0001950	0.012		
0.0002100	0.015		
0.0002550	0.020		

Constants of Jouyban–Acree model

J_0

J_1

J_2

Reference of data: [100].

Solute	Solvent 1	Solvent 2	T (°C)
Silybin	Polyethylene glycol 6000	Water	35
Solubility (mass F.)	Mass F. 1		
0.0000833	0.000		
0.0001160	0.001		
0.0001500	0.002		
0.0001850	0.005		
0.0002130	0.008		
0.0002360	0.010		
0.0002360	0.012		
0.0002510	0.015		
0.0003020	0.020		

Constants of Jouyban–Acree model

J_0
J_1
J_2

Reference of data: [100].

Solute	Solvent 1	Solvent 2	T (°C)
Silybin	Polyethylene glycol 6000	Water	40
Solubility (mass F.)	Mass F. 1		
0.0000997	0.000		
0.0001520	0.001		
0.0001820	0.002		
0.0002290	0.005		
0.0002670	0.008		
0.0002790	0.010		
0.0002980	0.012		
0.0003120	0.015		
0.0003530	0.020		

Constants of Jouyban–Acree model

J_0
J_1
J_2

Reference of data: [100].

Solute	Solvent 1	Solvent 2	T (°C)
Stavudine (form 1)	Methanol	1,2-Dimethylmethoxyethane	24.8
Solubility (g/100 g)	Mass F. 1		
2.260	0.000		
4.450	0.092		
8.060	0.314		
8.390	0.348		
9.770	0.619		
9.120	0.868		
9.250	0.904		
9.080	0.925		
8.720	1.000		

Constants of Jouyban–Acree model

J_0	380.412
J_1	−305.984
J_2	229.392

Reference of data: [101].

Solute	Solvent 1	Solvent 2	T (°C)
Stavudine (form 1)	Methanol	2-Propanol	24.8
Solubility (g/100 g)	Mass F. 1		
1.440	0.000		
1.560	0.047		
1.710	0.116		
1.970	0.191		
2.330	0.279		
2.770	0.351		
3.440	0.481		
4.280	0.576		
6.770	0.842		
7.270	0.883		
7.760	0.915		
8.110	0.954		
8.720	1.000		

Constants of Jouyban–Acree model

J_0	11.339
J_1	46.368
J_2	0

Reference of data: [101].

Solute	Solvent 1	Solvent 2	T (°C)
Sulfamethizole	Dioxane	Water	25.0
Solubility (mole F.)	Vol. F. 1		
0.00003546	0.000		
0.00014637	0.100		
0.00023462	0.200		
0.00049242	0.300		
0.00095949	0.400		
0.00212180	0.500		
0.00410677	0.600		
0.00587480	0.700		
0.00717210	0.750		
0.00687120	0.770		
0.00818520	0.800		
0.00769610	0.850		
0.00773090	0.870		
0.00706420	0.900		
0.00622600	0.920		
0.00495460	0.940		
0.00389690	0.960		
0.00225500	0.980		
0.00096418	1.000		

Constants of Jouyban–Acree model

J_0	1130.091
J_1	1008.982
J_2	1845.299

Reference of data: [102].

Solute	Solvent 1	Solvent 2	T (°C)
Sulfanilamide	Dioxane	Water	25.0
Solubility (mole F.)	Vol. F. 1		
0.00064461	0.000		
0.00183270	0.100		
0.00579950	0.200		
0.01360000	0.300		
0.03110000	0.400		
0.06050000	0.500		
0.09180000	0.600		
0.12710000	0.700		
0.15350000	0.750		
0.18770000	0.800		
0.19470000	0.850		
0.18130000	0.900		
0.17470000	0.940		
0.15470000	0.960		
0.13000000	0.980		
0.08030000	1.000		

Constants of Jouyban–Acree model

J_0	1029.151
J_1	412.354
J_2	641.871

Reference of data: [103].

Solute	Solvent 1	Solvent 2	T (°C)
Sulfapyridine	Dioxane	Water	25.0
Solubility (mole F.)	Vol. F. 1		
0.00001768	0.000		
0.00005207	0.200		
0.00016017	0.400		
0.00027515	0.500		
0.00045279	0.600		
0.00060976	0.650		
0.00074579	0.700		
0.00082768	0.750		
0.00092980	0.800		
0.00103350	0.850		
0.00107430	0.870		
0.00099777	0.900		
0.00096181	0.920		
0.00086300	0.940		
0.00072208	0.960		
0.00057877	0.980		
0.00033935	1.000		

Constants of Jouyban–Acree model

J_0	592.700
J_1	770.122
J_2	1065.399

Reference of data: [104].

Solute	Solvent 1	Solvent 2	T (°C)
Sulfisomidine	Dioxane	Water	25.0
Solubility (mole F.)	Vol. F. 1		
0.0000895	0.000		
0.0002740	0.100		
0.0007130	0.200		
0.0015750	0.300		
0.0023490	0.350		
0.0030310	0.400		
0.0037890	0.450		
0.0059830	0.550		
0.0074760	0.600		
0.0083900	0.650		
0.0102600	0.700		
0.0112100	0.750		

(continued)

Solute	Solvent 1	Solvent 2	T (°C)
0.0117800	0.770		
0.0118900	0.800		
0.0116000	0.850		
0.0101200	0.900		
0.0095610	0.920		
0.0079310	0.940		
0.0064310	0.960		
0.0047130	0.980		
0.0025080	1.000		

Constants of Jouyban–Acree model

J_0	1160.072
J_1	531.431
J_2	1150.568

Reference of data: [105].

Solute	Solvent 1	Solvent 2	T (°C)
Sulfadimidine	Dioxane	Water	25.0
Solubility (mole F.)	Vol. F. 1		
0.00000303	0.000		
0.00000663	0.100		
0.00002435	0.200		
0.00007570	0.300		
0.00018845	0.400		
0.00043172	0.500		
0.00096605	0.600		
0.00141971	0.650		
0.00175631	0.700		
0.00241824	0.750		
0.00286682	0.800		
0.00347616	0.850		
0.00358261	0.900		
0.00358509	0.920		
0.00353102	0.940		
0.00309314	0.960		
0.00274600	0.980		
0.00206300	0.990		
0.00149486	1.000		

Constants of Jouyban–Acree model

J_0	923.600
J_1	796.920
J_2	749.482

Reference of data: [106].

Solute	Solvent 1	Solvent 2	T (°C)
Sulfadiazine	Dimethyl formamide	Water	20.0
Solubility (mole F.)	Mass F. 1		
0.00000327	0.000		
0.00000354	0.005		
0.00000378	0.010		
0.00000437	0.020		
0.00000500	0.030		
0.00000644	0.050		
0.00001330	0.100		
0.00003350	0.200		
0.00007910	0.300		
0.00040800	0.500		
0.00294000	0.700		
0.00775000	0.780		
0.02320000	0.890		
0.06020000	1.000		

Constants of Jouyban–Acree model

J_0	116.596
J_1	−195.519
J_2	0

Reference of data: [107].

Solute	Solvent 1	Solvent 2	T (°C)
Sulfadiazine	Dimethyl formamide	Water	30.0
Solubility (mole F.)	Mass F. 1		
0.00000547	0.000		
0.00000592	0.005		
0.00000639	0.010		
0.00000721	0.020		
0.00000818	0.030		
0.00001060	0.050		
0.00001970	0.100		
0.00004870	0.200		
0.00011400	0.300		
0.00058400	0.500		
0.00373000	0.700		
0.00884000	0.780		
0.02600000	0.890		
0.06290000	1.000		

Constants of Jouyban–Acree model

J_0	116.596
J_1	−195.519
J_2	0

Reference of data: [107].

Solute	Solvent 1	Solvent 2	T (°C)
Sulfadiazine	Dimethyl formamide	Water	40.0
Solubility (mole F.)	Mass F. 1		
0.00000929	0.000		
0.00000997	0.005		
0.00001070	0.010		
0.00001210	0.020		
0.00001380	0.030		
0.00001740	0.050		
0.00003190	0.100		
0.00008210	0.200		
0.00017500	0.300		
0.00088100	0.500		
0.00550000	0.700		
0.01140000	0.780		
0.02900000	0.890		
0.06570000	1.000		

Constants of Jouyban–Acree model

J_0	116.596
J_1	−195.519
J_2	0

Reference of data: [107].

Solute	Solvent 1	Solvent 2	T (°C)
Sulfadiazine	Dioxane	Water	25.0
Solubility (mole F.)	Vol. F. 1		
0.00000440	0.000		
0.00001446	0.100		
0.00006076	0.200		
0.00013687	0.300		
0.00020994	0.400		
0.00055246	0.500		
0.00084140	0.600		
0.00126445	0.700		
0.00161622	0.750		
0.00176035	0.800		
0.00180634	0.850		
0.00176767	0.870		
0.00162405	0.900		
0.00136144	0.940		
0.00108543	0.960		
0.00077268	0.980		
0.00049671	1.000		

Constants of Jouyban–Acree model

J_0	1177.367
J_1	490.025
J_2	1089.688

Reference of data: [106].

Solute	Solvent 1	Solvent 2	T (°C)
Sulfamethazine	Ethanol	Water	25.0
Solubility (mole F.)	Vol. F. 1		
0.00000299	0.000		
0.00018971	0.300		
0.00063620	0.500		
0.00100779	0.600		
0.00124328	0.700		
0.00137404	0.750		
0.00148848	0.800		
0.00144449	0.850		
0.00126840	0.900		
0.00137404	0.950		
0.00074659	1.000		

Constants of Jouyban–Acree model

J_0	1334.613
J_1	−239.454
J_2	763.784

Reference of data: [108].

Solute	Solvent 1	Solvent 2	T (°C)
Sulfamethoxazol	Dioxane	Water	25.0
Solubility (mole F.)	Vol. F. 1		
0.00002317	0.000		
0.00010060	0.100		
0.00044167	0.200		
0.00175146	0.300		
0.00303529	0.400		
0.00801309	0.500		
0.01658441	0.600		
0.02800271	0.700		
0.04177342	0.750		
0.05297854	0.800		
0.06129146	0.850		
0.06084151	0.900		
0.05561603	0.940		
0.05207149	0.960		
0.03000544	1.000		

Constants of Jouyban–Acree model

J_0	1140.273
J_1	303.584
J_2	869.058

Reference of data: [106].

Solute	Solvent 1	Solvent 2	T (°C)
Sulfamethoxypyridazine	Dioxane	Water	25.0
Solubility (mole F.)	Vol. F. 1		
0.00003726	0.000		
0.00016581	0.100		
0.00147096	0.300		
0.00380365	0.400		
0.01132922	0.500		
0.01570363	0.550		
0.01854385	0.570		
0.02547417	0.600		
0.03855671	0.700		
0.05388898	0.750		
0.06105201	0.800		
0.07087613	0.850		
0.07003256	0.860		
0.06423918	0.900		
0.05495409	0.940		
0.05065240	0.960		
0.04189865	0.980		
0.02372466	1.000		

Constants of Jouyban–Acree model

J_0	1213.119
J_1	640.581
J_2	895.956

Reference of data: [106].

Solute	Solvent 1	Solvent 2	T (°C)
Sulfamethoxypridazine	Ethanol	Water	25.0
Solubility (mole F.)	Vol. F. 1		
0.00003723	1.000		
0.00028555	0.700		
0.00110049	0.500		
0.00164535	0.400		
0.00209228	0.350		
0.00246787	0.300		
0.00265131	0.250		
0.00271327	0.200		
0.00259726	0.150		
0.00213306	0.100		
0.00211014	0.070		
0.00191278	0.050		
0.00144680	0.000		

Constants of Jouyban–Acree model

J_0	787.712
J_1	505.115
J_2	0

Reference of data: [106].

Solute	Solvent 1	Solvent 2	T (°C)
Sulfamethoxypridazine	Ethyl acetate	Ethanol	25.0
Solubility (mole F.)	Vol. F. 1		
0.00144680	0.000		
0.00167072	0.050		
0.00197024	0.100		
0.00275924	0.200		
0.00412867	0.350		
0.00446494	0.400		
0.00483294	0.500		
0.00508429	0.600		
0.00534283	0.700		
0.00474198	0.800		
0.00423234	0.830		
0.00357496	0.870		
0.00326989	0.900		
0.00205825	1.000		

Constants of Jouyban–Acree model

J_0	562.669
J_1	197.111
J_2	0

Reference of data: [106].

Solute	Solvent 1	Solvent 2	T (°C)
Sulfamethoxypridazine	Ethyl acetate	n-Hexane	25.0
Solubility (mole F.)	Vol. F. 1		
0.00205825	1.000		
0.00007363	0.500		
0.00000773	0.300		
0.00000038	0.000		

Constants of Jouyban–Acree model

J_0	408.069
J_1	0
J_2	0

Reference of data: [106].

Solute	Solvent 1	Solvent 2	T (°C)
Sulfanilamide	Ethanol	Water	25.0
Solubility (mole F.)	Vol. F. 1		
0.00064259	0.000		
0.00115923	0.100		
0.00200924	0.200		
0.00337959	0.300		
0.00609675	0.400		
0.00775048	0.500		
0.00995182	0.600		
0.01088902	0.650		
0.01265124	0.700		
0.01270195	0.800		
0.01083471	0.900		
0.00755912	1.000		

Constants of Jouyban–Acree model

J_0	668.002
J_1	214.891
J_2	0

Reference of data: [108].

Solute	Solvent 1	Solvent 2	T (°C)
Terfenadine	Ethanol	Water	25.0
Solubility (mole/kg)	Mass F. 1		
0.071	1.000		
0.049	0.855		
0.035	0.736		
0.0221	0.637		
0.0135	0.553		

Constants of Jouyban–Acree model

J_0
J_1
J_2

Reference of data: [109].

Solute	Solvent 1	Solvent 2	T (°C)
Tetraglycine	Water	Ethanol	25.0
Solubility (g/L)	Vol. F. 1		
5.250	1.000		
3.180	0.900		
1.740	0.800		
0.980	0.700		
0.730	0.600		
0.460	0.500		
0.360	0.400		

Constants of Jouyban–Acree model

J_0
J_1
J_2

Reference of data: [65].

Solute	Solvent 1	Solvent 2	T (°C)
Theobromine	Dioxane	Water	25.0
Solubility (mole F.)	Vol. F. 1		
0.000033	0.000		
0.000076	0.100		
0.000164	0.200		
0.000280	0.300		
0.000425	0.400		
0.000560	0.500		
0.000674	0.600		
0.000753	0.700		
0.000747	0.800		
0.000594	0.900		
0.000444	1.000		

Constants of Jouyban–Acree model

J_0	802.001
J_1	−35.151
J_2	90.174

Reference of data: [110].

Solute	Solvent 1	Solvent 2	T (°C)
Theophylline	Dioxane	Water	25.0
Solubility (mole F.)	Vol. F. 1		
0.00074140	0.000		
0.00106680	0.050		
0.00155830	0.100		
0.00210460	0.150		
0.00278310	0.200		
0.00351580	0.250		
0.00468180	0.300		
0.00567830	0.350		
0.00668560	0.400		
0.00832950	0.450		
0.01144490	0.550		
0.01254110	0.600		
0.01314360	0.620		

(continued)

Solute	Solvent 1	Solvent 2	T (°C)
0.01438030	0.660		
0.01429260	0.700		
0.01427110	0.750		
0.01425920	0.770		
0.01387360	0.800		
0.01317700	0.850		
0.01170740	0.900		
0.00259590	1.000		

Constants of Jouyban–Acree model

J_0	987.475
J_1	591.156
J_2	708.897

Reference of data: [111].

Solute	Solvent 1	Solvent 2	T (°C)
Theophylline	Ethylene glycol	Water	25.0
Solubility (mole/L)	Mole F. 1		
0.0340	0.000		
0.0350	0.008		
0.0366	0.017		
0.0356	0.026		
0.0360	0.035		
0.0367	0.044		
0.0393	0.054		
0.0381	0.075		
0.04010	0.097		
0.0434	0.122		
0.0447	0.148		
0.0476	0.177		
0.0540	0.244		
0.0669	0.327		
0.0658	0.430		
0.0632	0.564		
0.0547	1.000		

Constants of Jouyban–Acree model

J_0	224.099
J_1	−262.707
J_2	−366.879

Reference of data: [112].

Solute	Solvent 1	Solvent 2	T (°C)
Theophylline	Water	Acetonitrile	25.0
Solubility (mole/L)	Mole F. 1		
0.0340	1.000		
0.0386	0.990		
0.0469	0.977		
0.0556	0.967		
0.0678	0.950		
0.0923	0.927		
0.1040	0.922		
0.1130	0.911		
0.1375	0.888		
0.1494	0.868		
0.1908	0.772		
0.1716	0.638		
0.1291	0.494		
0.0675	0.295		
0.0326	0.134		
0.0195	0.068		
0.0081	0.000		

Constants of Jouyban–Acree model

J_0	1038.482
J_1	354.774
J_2	1075.038

Reference of data: [112].

Solute	Solvent 1	Solvent 2	T (°C)
Theophylline	Water	Methanol	25.0
Solubility (mole/L)	Mole F. 1		
0.0340	1.000		
0.0370	0.971		
0.0394	0.942		
0.0431	0.910		
0.0595	0.806		
0.0763	0.727		
0.0899	0.641		
0.0992	0.541		
0.0958	0.488		
0.0794	0.372		
0.0598	0.239		
0.0396	0.081		
0.0332	0.000		

Constants of Jouyban–Acree model

J_0	546.249
J_1	89.476
J_2	−319.516

Reference of data: [112].

Solute	Solvent 1	Solvent 2	T (°C)
Theophylline anhydrate	Propylene glycol	Water	30.0
Solubility (g/L)	Vol. F. 1		
8.557	0.000		
11.17	0.200		
16.22	0.400		
22.19	0.600		
26.15	0.700		
22.89	0.800		
18.95	0.900		
12.60	1.000		

Constants of Jouyban–Acree model

J_0	348.633
J_1	363.099
J_2	0

Reference of data: [113].

Solute	Solvent 1	Solvent 2	T (°C)
Theophylline hydrate	Propylene glycol	Water	30.0
Solubility (g/L)	Vol. F. 1		
8.705	0.000		
11.42	0.200		
16.19	0.400		
21.79	0.600		
25.7	0.700		
23.82	0.800		
19.19	0.900		
12.92	1.000		

Constants of Jouyban–Acree model

J_0	336.493
J_1	357.120
J_2	0

Reference of data: [113].

Solute	Solvent 1	Solvent 2	T (°C)
Thiamylal	Ethanol	Water	25.0
Solubility (g/L)	Mass F. 1		
160.8	1.000		
149.9	0.975		
135.4	0.950		
124.6	0.925		
112.7	0.900		
102.0	0.875		
93.2	0.850		

(continued)

(continued)

Solute	Solvent 1	Solvent 2	T (°C)
82.3	0.825		
71.8	0.800		
61.9	0.775		
54.9	0.750		
43.3	0.725		
37.7	0.700		
32.3	0.675		
28.7	0.650		
23.1	0.625		
18.1	0.600		
15.4	0.575		
13.0	0.550		
10.3	0.525		
8.2	0.500		
6.5	0.475		
4.7	0.450		
3.4	0.425		
2.5	0.400		
1.96	0.375		
1.41	0.350		
0.73	0.300		
0.51	0.275		
0.35	0.250		
0.23	0.225		
0.19	0.200		
0.15	0.175		
0.12	0.150		
0.10	0.125		
0.09	0.100		
0.07	0.075		
0.06	0.050		
0.06	0.025		
0.05	0.000		

Constants of Jouyban–Acree model

J_0	514.044
J_1	655.945
J_2	−740.301

Reference of data: [20].

Solute	Solvent 1	Solvent 2	T (°C)
Thiopental	Ethanol	Water	25.0
Solubility (g/L)	Mass F. 1		
56.3	1.000		
62.3	0.975		

(continued)			
Solute	**Solvent 1**	**Solvent 2**	**T (°C)**
74.2	0.950		
97.1	0.925		
94.9	0.900		
86.6	0.875		
79.9	0.850		
71.6	0.825		
63.7	0.800		
55.4	0.775		
50.4	0.750		
41.1	0.725		
36.3	0.700		
31.0	0.675		
38.0	0.650		
23.5	0.625		
18.8	0.600		
16.3	0.575		
14.0	0.550		
11.2	0.525		
9.1	0.500		
7.3	0.475		
5.4	0.450		
4.5	0.425		
3.2	0.400		
2.4	0.375		
2.0	0.350		
0.9	0.300		
0.7	0.275		
0.5	0.250		
0.3	0.225		
0.28	0.200		
0.23	0.175		
0.19	0.150		
0.17	0.125		
0.15	0.100		
0.12	0.075		
0.11	0.050		
0.09	0.025		
0.08	0.000		

Constants of Jouyban–Acree model

J_0	713.894
J_1	948.072
J_2	−222.494

Reference of data: [20].

Solute	Solvent 1	Solvent 2	T (°C)
trans-Stilbene	2,2,4-Trimethyl pentane	1-Butanol	25.0
Solubility (mole F.)	Mole F. 1		
0.00803	1.000		
0.00852	0.819		
0.00857	0.672		
0.00815	0.459		
0.00771	0.356		
0.00739	0.270		
0.00645	0.132		
0.00595	0.061		
0.00533	0.000		

Constants of Jouyban–Acree model

J_0	119.421
J_1	−25.668
J_2	39.947

Reference of data: [114].

Solute	Solvent 1	Solvent 2	T (°C)
trans-Stilbene	2,2,4-Trimethyl pentane	1-Propanol	25.0
Solubility (mole F.)	Mole F. 1		
0.00803	1.000		
0.00845	0.868		
0.00799	0.641		
0.00691	0.403		
0.00624	0.306		
0.00576	0.232		
0.00484	0.105		
0.00442	0.049		
0.00403	0.000		

Constants of Jouyban–Acree model

J_0	141.446
J_1	0
J_2	0

Reference of data: [115].

Solute	Solvent 1	Solvent 2	T (°C)
trans-Stilbene	2,2,4-Trimethyl pentane	2-Butanol	25.0
Solubility (mole F.)	Mole F. 1		
0.00803	1.000		
0.00857	0.816		
0.00820	0.684		
0.00714	0.459		
0.00658	0.355		
0.00600	0.269		
0.00493	0.123		
0.00442	0.064		
0.00382	0.000		

Constants of Jouyban–Acree model

J_0	147.503
J_1	−8.374
J_2	75.320

Reference of data: [116].

Solute	Solvent 1	Solvent 2	T (°C)
trans-Stilbene	2,2,4-Trimethyl pentane	2-Propanol	25.0
Solubility (mole F.)	Mole F. 1		
0.00803	1.000		
0.00800	0.786		
0.00747	0.636		
0.00613	0.414		
0.00539	0.314		
0.00470	0.238		
0.00366	0.115		
0.00318	0.053		
0.00279	0.000		

Constants of Jouyban–Acree model

J_0	182.834
J_1	−21.667
J_2	0

Reference of data: [117].

Solute	Solvent 1	Solvent 2	T (°C)
trans-Stilbene	Cyclohexane	1-Butanol	25.0
Solubility (mole F.)	Mole F. 1		
0.01374	1.000		
0.01414	0.866		
0.01341	0.767		
0.01161	0.565		
0.01029	0.456		
0.00934	0.356		
0.00740	0.178		
0.00640	0.092		
0.00533	0.000		

Constants of Jouyban–Acree model

J_0	122.322
J_1	13.385
J_2	63.845

Reference of data: [114].

Solute	Solvent 1	Solvent 2	T (°C)
trans-Stilbene	Cyclohexane	1-Propanol	25.0
Solubility (mole F.)	Mole F. 1		
0.01374	1.000		
0.01386	0.840		
0.01318	0.732		
0.01102	0.506		
0.00959	0.405		
0.00817	0.312		
0.00571	0.140		
0.00470	0.062		
0.00403	0.000		

Constants of Jouyban–Acree model

J_0	196.111
J_1	0
J_2	0

Reference of data: [115].

Solute	Solvent 1	Solvent 2	T (°C)
trans-Stilbene	Cyclohexane	2-Butanol	25.0
Solubility (mole F.)	Mole F. 1		
0.01374	1.000		
0.01385	0.884		
0.01337	0.779		
0.01097	0.567		
0.00951	0.461		
0.00804	0.363		
0.00567	0.180		
0.00468	0.092		
0.00382	0.000		

Constants of Jouyban–Acree model

J_0	168.840
J_1	40334
J_2	0

Reference of data: [116].

Solute	Solvent 1	Solvent 2	T (°C)
trans-Stilbene	Cyclohexane	2-Propanol	25.0
Solubility (mole F.)	Mole F. 1		
0.01374	1.000		
0.01351	0.843		
0.01267	0.737		
0.00979	0.525		
0.00829	0.413		
0.00676	0.316		
0.00459	0.165		
0.00363	0.083		
0.00279	0.000		

Constants of Jouyban–Acree model

J_0	224.737
J_1	0
J_2	0

Reference of data: [117].

Solute	Solvent 1	Solvent 2	T (°C)
trans-Stilbene	Methyl cyclohexane	1-Butanol	25.0
Solubility (mole F.)	Mole F. 1		
0.01413	1.000		
0.01405	0.855		
0.01327	0.724		
0.01135	0.514		
0.01058	0.406		
0.00952	0.317		
0.00756	0.166		
0.00651	0.082		
0.00533	0.000		

Constants of Jouyban–Acree model

J_0	140.115
J_1	−35.641
J_2	46.342

Reference of data: [114].

Solute	Solvent 1	Solvent 2	T (°C)
trans-Stilbene	Methyl cyclohexane	1-Propanol	25.0
Solubility (mole F.)	Mole F. 1		
0.01413	1.000		
0.01383	0.834		
0.01259	0.695		
0.00972	0.460		
0.00860	0.371		
0.00744	0.280		
0.00561	0.132		
0.00495	0.072		
0.00403	0.000		

Constants of Jouyban–Acree model

J_0	156.239
J_1	−7.069
J_2	56.755

Reference of data: [115].

Solute	Solvent 1	Solvent 2	T (°C)
trans-Stilbene	Methyl cyclohexane	2-Butanol	25.0
Solubility (mole F.)	Mole F. 1		
0.01413	1.000		
0.01382	0.857		
0.01296	0.729		
0.01094	0.522		
0.00949	0.417		
0.00821	0.318		
0.00619	0.158		
0.00505	0.083		
0.00382	0.000		

Constants of Jouyban–Acree model

J_0	186.521
J_1	−57.895
J_2	70.669

Reference of data: [116].

Solute	Solvent 1	Solvent 2	T (°C)
trans-Stilbene	Methyl cyclohexane	2-Propanol	25.0
Solubility (mole F.)	Mole F. 1		
0.01413	1.000		
0.01368	0.838		
0.01238	0.702		
0.00973	0.485		
0.00839	0.394		
0.00658	0.281		
0.00448	0.138		
0.00355	0.067		
0.00279	0.000		

Constants of Jouyban–Acree model

J_0	239.409
J_1	−42.567
J_2	0

Reference of data: [117].

Solute	Solvent 1	Solvent 2	T (°C)
trans-Stilbene	*n*-Heptane	1-Butanol	25.0
Solubility (mole F.)	Mole F. 1		
0.01085	1.000		
0.01149	0.840		
0.01083	0.705		
0.00961	0.483		
0.00874	0.383		
0.00833	0.297		
0.00674	0.133		
0.00613	0.073		
0.00533	0.000		

Constants of Jouyban–Acree model

J_0	135.063
J_1	0
J_2	0

Reference of data: [114].

Solute	Solvent 1	Solvent 2	T (°C)
trans-Stilbene	*n*-Heptane	1-Propanol	25.0
Solubility (mole F.)	Mole F. 1		
0.01085	1.000		
0.01091	0.789		
0.01035	0.672		
0.00846	0.434		
0.00761	0.336		
0.00693	0.254		
0.00529	0.113		
0.00467	0.056		
0.00403	0.000		

Constants of Jouyban–Acree model

J_0	163.123
J_1	−25.442
J_2	57.561

Reference of data: [115].

Solute	Solvent 1	Solvent 2	T (°C)
trans-Stilbene	*n*-Heptane	2-Butanol	25.0
Solubility (mole F.)	Mole F. 1		
0.01085	1.000		
0.01119	0.839		
0.01059	0.704		
0.00901	0.488		
0.00820	0.394		
0.00739	0.299		
0.00546	0.140		
0.00467	0.074		
0.00382	0.000		

Constants of Jouyban–Acree model

J_0	183.624
J_1	−32.140
J_2	50.863

Reference of data: [116].

Solute	Solvent 1	Solvent 2	*T* (°C)
trans-Stilbene	*n*-Heptane	2-Propanol	25.0
Solubility (mole F.)	Mole F. 1		
0.01085	1.000		
0.01094	0.814		
0.01021	0.674		
0.00789	0.438		
0.00685	0.345		
0.00595	0.262		
0.00410	0.118		
0.00340	0.059		
0.00279	0.000		

Constants of Jouyban–Acree model

J_0	232.572
J_1	−38.814
J_2	38.076

Reference of data: [117].

Solute	Solvent 1	Solvent 2	*T* (°C)
trans-Stilbene	*n*-Hexane	1-Butanol	25.0
Solubility (mole F.)	Mole F. 1		
0.00960	1.000		
0.01035	0.842		
0.01021	0.730		
0.00906	0.512		
0.00831	0.409		
0.00758	0.303		
0.00647	0.151		
0.00594	0.083		
0.00533	0.000		

Constants of Jouyban–Acree model

J_0	117.207
J_1	43.686
J_2	35.738

Reference of data: [114].

Solute	Solvent 1	Solvent 2	T (°C)
trans-Stilbene	*n*-Hexane	1-Propanol	25.0
Solubility (mole F.)	Mole F. 1		
0.00960	1.000		
0.00955	0.825		
0.00909	0.685		
0.00791	0.464		
0.00731	0.366		
0.00654	0.279		
0.00518	0.127		
0.00458	0.058		
0.00403	0.000		

Constants of Jouyban–Acree model

J_0	143.722
J_1	−28.822
J_2	0

Reference of data: [115].

Solute	Solvent 1	Solvent 2	T (°C)
trans-Stilbene	*n*-Hexane	2-Butanol	25.0
Solubility (mole F.)	Mole F. 1		
0.00960	1.000		
0.01002	0.855		
0.00976	0.728		
0.00849	0.512		
0.00771	0.417		
0.00678	0.319		
0.00513	0.155		
0.00444	0.079		
0.00382	0.000		

Constants of Jouyban–Acree model

J_0	169.741
J_1	18.626
J_2	0

Reference of data: [116].

Solute	Solvent 1	Solvent 2	T (°C)
trans-Stilbene	*n*-Hexane	2-Propanol	25.0
Solubility (mole F.)	Mole F. 1		
0.00960	1.000		
0.00960	0.829		
0.00920	0.696		
0.00746	0.462		
0.00667	0.392		
0.00560	0.285		
0.00398	0.131		
0.00333	0.062		
0.00279	0.000		

Constants of Jouyban–Acree model

J_0	209.807
J_1	−18.585
J_2	0

Reference of data: [117].

Solute	Solvent 1	Solvent 2	T (°C)
trans-Stilbene	*n*-Octane	1-Butanol	25.0
Solubility (mole F.)	Mole F. 1		
0.01241	1.000		
0.01233	0.814		
0.01156	0.683		
0.01008	0.466		
0.00910	0.356		
0.00835	0.270		
0.00707	0.126		
0.00619	0.057		
0.00533	0.000		

Constants of Jouyban–Acree model

J_0	117.486
J_1	−35.277
J_2	94.570

Reference of data: [114].

Solute	Solvent 1	Solvent 2	T (°C)
trans-Stilbene	*n*-Octane	1-Propanol	25.0
Solubility (mole F.)	Mole F. 1		
0.01241	1.000		
0.01218	0.782		
0.01121	0.640		
0.00899	0.409		
0.00785	0.309		
0.00693	0.231		
0.00532	0.103		
0.00464	0.049		
0.00403	0.000		

Constants of Jouyban–Acree model

J_0	176.276
J_1	−31.062
J_2	41.246

Reference of data: [115].

Solute	Solvent 1	Solvent 2	T (°C)
trans-Stilbene	*n*-Octane	2-Butanol	25.0
Solubility (mole F.)	Mole F. 1		
0.01241	1.000		
0.01235	0.821		
0.01151	0.686		
0.00973	0.456		
0.00875	0.359		
0.00776	0.267		
0.00572	0.127		
0.00478	0.064		
0.00382	0.000		

Constants of Jouyban–Acree model

J_0	199.982
J_1	−82.636
J_2	75.309

Reference of data: [116].

Solute	Solvent 1	Solvent 2	T (°C)
trans-Stilbene	*n*-Octane	2-Propanol	25.0
Solubility (mole F.)	Mole F. 1		
0.01241	1.000		
0.01179	0.795		
0.01061	0.638		
0.00826	0.416		
0.00709	0.324		
0.00590	0.234		
0.00040	0.104		
0.00344	0.054		
0.00279	0.000		

Constants of Jouyban–Acree model

J_0	0
J_1	0
J_2	0

Reference of data: [117].

Solute	Solvent 1	Solvent 2	T (°C)
Triglycine	Water	Dioxane	25.0
Solubility (g/100 g)	Mass F. 1		
6.500	1.000		
2.000	0.800		
0.420	0.600		
0.045	0.400		

Constants of Jouyban–Acree model

J_0
J_1
J_2

Reference of data: [23].

Solute	Solvent 1	Solvent 2	T (°C)
Triglycine	Water	Ethanol	20.0
Solubility (g/100 g)	Vol. F. 1		
6.87	1.000		
1.89	0.800		
0.25	0.500		

Constants of Jouyban–Acree model

J_0
J_1
J_2

Reference of data: [58].

Solute	Solvent 1	Solvent 2	T (°C)
Triglycine	Water	Ethanol	25.1
Solubility (g/100 g)	Mass F. 1		
6.45	1.000		
2.14	0.800		
0.08	0.600		
0.105	0.400		

Constants of Jouyban–Acree model

J_0

J_1

J_2

Reference of data: [23].

Solute	Solvent 1	Solvent 2	T (°C)
Triglycine	Water	Ethanol	25.0
Solubility (g/100 g)	Vol. F. 1		
7.11	1.000		
2.00	0.800		
0.25	0.500		

Constants of Jouyban–Acree model

J_0

J_1

J_2

Reference of data: [58].

Solute	Solvent 1	Solvent 2	T (°C)
Triglycine	Water	Ethanol	41.5
Solubility (g/100 g)	Vol. F. 1		
11.08	1.000		
3.51	0.800		
0.47	0.500		

Constants of Jouyban–Acree model

J_0

J_1

J_2

Reference of data: [58].

Solute	Solvent 1	Solvent 2	T (°C)
Triglycine	Water	Ethanol	60.0
Solubility (mole F.)	Vol. F. 1		
16.72	1.000		
5.52	0.800		
0.84	0.500		

Constants of Jouyban–Acree model

J_0

J_1

J_2

Reference of data: [58].

Solute	Solvent 1	Solvent 2	T (°C)
Triglycine	Water	Ethanol	25.0
Solubility (g/L)	Vol. F. 1		
72.35	1.000		
41.70	0.900		
22.31	0.800		
12.30	0.700		
6.02	0.600		
3.12	0.500		
1.59	0.400		

Constants of Jouyban–Acree model

J_0

J_1

J_2

Reference of data: [65].

Solute	Solvent 1	Solvent 2	T (°C)
Trimethoprim	Dioxane	Water	25.0
Solubility (mole F.)	Vol. F. 1		
0.000021	0.000		
0.000117	0.100		
0.000364	0.200		
0.000530	0.300		
0.000790	0.350		
0.000880	0.400		
0.001325	0.450		
0.001510	0.500		
0.001720	0.550		

(*continued*)

(continued)

Solute	Solvent 1	Solvent 2	T (°C)
0.002300	0.600		
0.003082	0.650		
0.003320	0.700		
0.003751	0.750		
0.004232	0.800		
0.004816	0.830		
0.004499	0.870		
0.004515	0.900		
0.003738	0.940		
0.003234	0.970		
0.003029	1.000		

Constants of Jouyban–Acree model

J_0	922.267
J_1	−246.512
J_2	888.774

Reference of data: [118].

Solute	Solvent 1	Solvent 2	T (°C)
Tryptophan	Water	Dioxane	25.0
Solubility (g/100 g)	Mass F. 1		
1.28	1.000		
2.13	0.800		
2.59	0.700		
2.97	0.600		
2.80	0.400		
0.873	0.200		
0.136	0.100		

Constants of Jouyban–Acree model

J_0

J_1

J_2

Reference of data: [23].

Solute	Solvent 1	Solvent 2	T (°C)
Tryptophan	Water	Ethanol	25.0
Solubility (g/100 g)	Mass F. 1		
1.38	1.000		
1.13	0.800		
1.11	0.700		
1.25	0.600		
1.42	0.500		
1.40	0.400		
0.78	0.200		
0.33	0.100		

Constants of Jouyban–Acree model

J_0
J_1
J_2

Reference of data: [23].

Solute	Solvent 1	Solvent 2	T (°C)
Tyrosine	Water	Dioxane	25.0
Solubility (g/100 g)	Mass F. 1		
0.0452	1.000		
0.0457	0.800		
0.0391	0.600		
0.0209	0.400		
0.0408	0.200		
0.00065	0.100		

Constants of Jouyban–Acree model

J_0
J_1
J_2

Reference of data: [23].

Solute	Solvent 1	Solvent 2	T (°C)
Tyrosine	Water	Ethanol	25.0
Solubility (g/100 g)	Mass F. 1		
0.0452	1.000		
0.0318	0.800		
0.0262	0.600		
0.0192	0.400		
0.0078	0.200		
0.0029	0.100		
0.00052	0.000		

Constants of Jouyban–Acree model

J_0	805.301
J_1	−858.034
J_2	452.159

Reference of data: [23].

Solute	Solvent 1	Solvent 2	T (°C)
Valdecoxib	Ethanol	Water	25.0
Solubility (mg/L)	Vol. F. 1		
10.25	0.000		
125	0.200		
3092	0.400		
6639	0.600		
14289	0.800		
9645	1.000		

Constants of Jouyban–Acree model

J_0	1172.185
J_1	554.640
J_2	0

Reference of data: [119].

Solute	Solvent 1	Solvent 2	T (°C)
Valdecoxib	Ethanol	Water	30.0
Solubility (mg/L)	Vol. F. 1		
10.4	0.000		
223	0.200		
4180	0.400		
9791	0.600		
16133	0.800		
13557	1.000		

Constants of Jouyban–Acree model

J_0	1172.185
J_1	554.640
J_2	0

Reference of data: [119].

Solute	Solvent 1	Solvent 2	T (°C)
Valdecoxib	Ethanol	Water	35.0
Solubility (mg/L)	Vol. F. 1		
11.1	0.000		
248	0.200		
5698	0.400		
14334	0.600		
20973	0.800		
15491	1.000		

Constants of Jouyban–Acree model

J_0	1172.185
J_1	554.640
J_2	0

Reference of data: [119].

Solute	Solvent 1	Solvent 2	T (°C)
Valdecoxib	Ethanol	Water	37.0
Solubility (mg/L)	Mass F. 1		
14264.82	1.000		
4392.62	0.500		
2870.76	0.400		
1251.61	0.300		
177.69	0.200		
34.96	0.100		
11.19	0.000		

Constants of Jouyban–Acree model

J_0	1275.691
J_1	−2242.948
J_2	−3896.886

Reference of data: [120].

Solute	Solvent 1	Solvent 2	T (°C)
Valdecoxib	Glycerol	Water	37.0
Solubility (mg/L)	Mass F. 1		
60.75	1.000		
43.08	0.500		
39.82	0.400		
37.83	0.300		
24.23	0.200		
12.98	0.100		
11.19	0.000		

Constants of Jouyban–Acree model

J_0	246.450
J_1	−1055.904
J_2	−1571.960

Reference of data: [120].

Solute	Solvent 1	Solvent 2	T (°C)
Valdecoxib	Methanol	Water	37.0
Solubility (mg/L)	Mass F. 1		
66546.85	1.000		
3846.39	0.500		
1123.81	0.400		
372.35	0.300		
67.95	0.200		
14.74	0.100		
11.19	0.000		

Constants of Jouyban–Acree model

J_0	899.895
J_1	1402.328
J_2	0

Reference of data: [120].

Solute	Solvent 1	Solvent 2	T (°C)
Valdecoxib (S)	Glycerol	Water	25.0
Solubility (mg/L)	Mass F. 1		
10.25	0.000		
12.2	0.100		
16.2	0.200		
21.8	0.300		
50.4	1.000		

Constants of Jouyban–Acree model

J_0	532.806
J_1	539.693
J_2	0

Reference of data: [121].

Solute	Solvent 1	Solvent 2	T (°C)
Valdecoxib (S)	Glycerol	Water	30.0
Solubility (mg/L)	Mass F. 1		
10.4	0.000		
13.3	0.100		
19.4	0.200		
29.7	0.300		
56.7	1.000		

Constants of Jouyban–Acree model

J_0	532.806
J_1	539.693
J_2	0

Reference of data: [121].

Solute	Solvent 1	Solvent 2	T (°C)
Valdecoxib (S)	Glycerol	Water	35.0
Solubility (mg/L)	Mass F. 1		
11.1	0.000		
13.8	0.100		
23.7	0.200		
36.3	0.300		
60.3	1.000		

Constants of Jouyban–Acree model

J_0	532.806
J_1	539.693
J_2	0

Reference of data: [121].

Solute	Solvent 1	Solvent 2	T (°C)
Valdecoxib (S)	Polyethylene glycol 10000	Water	25.0
Solubility (mg/L)	Mass F. 1		
10.25	0.000		
15.4	0.010		
19.1	0.020		
22.4	0.050		
27.3	0.100		

Constants of Jouyban–Acree model

J_0

J_1

J_2

Reference of data: [122].

Solute	Solvent 1	Solvent 2	T (°C)
Valdecoxib (S)	Polyethylene glycol 10000	Water	30.0
Solubility (mg/L)	Mass F. 1		
10.43	0.000		
20.7	0.010		
27.8	0.020		
32.3	0.050		
37.9	0.100		

Constants of Jouyban–Acree model

J_0

J_1

J_2

Reference of data: [122].

Solute	Solvent 1	Solvent 2	T (°C)
Valdecoxib (S)	Polyethylene glycol 10000	Water	35.0
Solubility (mg/L)	Mass F. 1		
11.0	0.000		
22.4	0.010		
33.1	0.020		
40.3	0.050		
50.8	0.100		

Constants of Jouyban–Acree model

J_0
J_1
J_2

Reference of data: [122].

Solute	Solvent 1	Solvent 2	T (°C)
Valdecoxib (S)	Polyethylene glycol 400	Water	25.0
Solubility (mg/L)	Mass F. 1		
10.25	0.000		
35.2	0.150		
107.7	0.300		
276.9	0.450		
734.5	0.600		
2135.4	1.000		

Constants of Jouyban–Acree model

J_0 795.104
J_1 0
J_2 0

Reference of data: [121].

Solute	Solvent 1	Solvent 2	T (°C)
Valdecoxib (S)	Polyethylene glycol 400	Water	30.0
Solubility (mg/L)	Mass F. 1		
10.4	0.000		
66.2	0.150		
220.1	0.300		
551.3	0.450		
1516.2	0.600		
2987.3	1.000		

Constants of Jouyban–Acree model

J_0 795.104
J_1 0
J_2 0

Reference of data: [121].

Solute	Solvent 1	Solvent 2	T (°C)
Valdecoxib (S)	Polyethylene glycol 400	Water	35.0
Solubility (mg/L)	Mass F. 1		
11.1	0.000		
101.4	0.150		
298.8	0.300		
1142.3	0.450		
2805.5	0.600		
4456.6	1.000		

Constants of Jouyban–Acree model

J_0	795.104
J_1	0
J_2	0

Reference of data: [121].

Solute	Solvent 1	Solvent 2	T (°C)
Valdecoxib (S)	Polyethylene glycol 4000	Water	25.0
Solubility (mg/L)	Mass F. 1		
10.25	0.000		
23.0	0.010		
27.6	0.020		
31.0	0.050		
38.4	0.100		

Constants of Jouyban–Acree model

J_0	
J_1	
J_2	

Reference of data: [122].

Solute	Solvent 1	Solvent 2	T (°C)
Valdecoxib (S)	Polyethylene glycol 4000	Water	30.0
Solubility (mg/L)	Mass F. 1		
10.43	0.000		
27.7	0.010		
36.5	0.020		
43.1	0.050		
51.4	0.100		

Constants of Jouyban–Acree model

J_0	
J_1	
J_2	

Reference of data: [122].

Solute	Solvent 1	Solvent 2	T (°C)
Valdecoxib (S)	Polyethylene glycol 4000	Water	35.0
Solubility (mg/L)	Mass F. 1		
11.0	0.000		
30.1	0.010		
41.2	0.020		
55.2	0.050		
68.6	0.100		

Constants of Jouyban–Acree model

J_0
J_1
J_2

Reference of data: [122].

Solute	Solvent 1	Solvent 2	T (°C)
Valdecoxib (S)	Polyethylene glycol 6000	Water	25.0
Solubility (mg/L)	Mass F. 1		
10.25	0.000		
20.1	0.010		
25.2	0.020		
28.5	0.050		
35.8	0.100		

Constants of Jouyban–Acree model

J_0
J_1
J_2

Reference of data: [122].

Solute	Solvent 1	Solvent 2	T (°C)
Valdecoxib (S)	Polyethylene glycol 6000	Water	30.0
Solubility (mg/L)	Mass F. 1		
10.43	0.000		
24.1	0.010		
33.1	0.020		
37.6	0.050		
48.3	0.100		

Constants of Jouyban–Acree model

J_0
J_1
J_2

Reference of data: [122].

Solute	Solvent 1	Solvent 2	T (°C)
Valdecoxib (S)	Polyethylene glycol 6000	Water	35.0
Solubility (mg/L)	Mass F. 1		
11.0	0.000		
27.6	0.010		
38.2	0.020		
50.2	0.050		
60.9	0.100		

Constants of Jouyban–Acree model

J_0
J_1
J_2

Reference of data: [122].

Solute	Solvent 1	Solvent 2	T (°C)
Valdecoxib (S)	Polyethylene glycol 8000	Water	25.0
Solubility (mg/L)	Mass F. 1		
10.25	0.000		
17.3	0.010		
21.4	0.020		
25.6	0.050		
31.1	0.100		

Constants of Jouyban–Acree model

J_0
J_1
J_2

Reference of data: [122].

Solute	Solvent 1	Solvent 2	T (°C)
Valdecoxib (S)	Polyethylene glycol 8000	Water	30.0
Solubility (mg/L)	Mass F. 1		
10.43	0.000		
22.1	0.010		
30.2	0.020		
35.2	0.050		
42.8	0.100		

Constants of Jouyban–Acree model

J_0
J_1
J_2

Reference of data: [122].

Solute	Solvent 1	Solvent 2	T (°C)
Valdecoxib (S)	Polyethylene glycol 8000	Water	35.0
Solubility (mg/L)	Mass F. 1		
11.0	0.000		
25.1	0.010		
35.3	0.020		
45.6	0.050		
56.1	0.100		

Constants of Jouyban–Acree model

J_0
J_1
J_2

Reference of data: [122].

Solute	Solvent 1	Solvent 2	T (°C)
Valdecoxib (S)	Propylene glycol	Water	25.0
Solubility (mg/L)	Mass F. 1		
10.25	0.000		
41.3	0.200		
187.6	0.400		
650.2	0.600		
1038.6	0.800		
1434.4	1.000		

Constants of Jouyban–Acree model

J_0 729.470
J_1 0
J_2 0

Reference of data: [121].

Solute	Solvent 1	Solvent 2	T (°C)
Valdecoxib (S)	Propylene glycol	Water	30.0
Solubility (mg/L)	Mass F. 1		
10.40	0.000		
76.2	0.200		
391.2	0.400		
1027.0	0.600		
1765.6	0.800		
1925.7	1.000		

Constants of Jouyban–Acree model

J_0 729.470
J_1 0
J_2 0

Reference of data: [121].

Solute	Solvent 1	Solvent 2	T (°C)
Valdecoxib (S)	Propylene glycol	Water	35.0
Solubility (mg/L)	Mass F. 1		
11.1	0.000		
118.3	0.200		
506.7	0.400		
1267.8	0.600		
2524.5	0.800		
2936.8	1.000		

Constants of Jouyban–Acree model

J_0	729.470
J_1	0
J_2	0

Reference of data: [121].

Solute	Solvent 1	Solvent 2	T (°C)
Valine (DL)	Water	Ethanol	25.0
Solubility (mole F.)	Vol. F. 1		
0.01078068	1.000		
0.00667090	0.800		
0.00404611	0.600		
0.00252883	0.400		
0.00107010	0.200		
0.00042221	0.100		
0.00007485	0.000		

Constants of Jouyban–Acree model

J_0	654.249
J_1	779.621
J_2	682.002

Reference of data: [12].

Solute	Solvent 1	Solvent 2	T (°C)
Valine (DL)	Water	Ethanol	0.0
Solubility (mole F.)	Vol. F. 1		
0.00366	0.751		
0.00159	0.499		
0.000695	0.258		
0.0000974	0.049		
0.0000534	0.000		

Constants of Jouyban–Acree model

J_0	
J_1	
J_2	

Reference of data: [13].

Solute	Solvent 1	Solvent 2	T (°C)
Valine (DL)	Water	Ethanol	25.0
Solubility (mole F.)	Vol. F. 1		
0.00575	0.747		
0.00317	0.490		
0.00147	0.257		
0.00020	0.049		

Constants of Jouyban–Acree model

J_0

J_1

J_2

Reference of data: [13].

Solute	Solvent 1	Solvent 2	T (°C)
Valine (DL)	Water	Ethanol	45.0
Solubility (mole F.)	Vol. F. 1		
0.00884	0.755		
0.00564	0.498		
0.00259	0.257		
0.000345	0.049		

Constants of Jouyban–Acree model

J_0

J_1

J_2

Reference of data: [13].

Solute	Solvent 1	Solvent 2	T (°C)
Valine (DL)	Water	Ethanol	65.0
Solubility (mole F.)	Vol. F. 1		
0.0128	0.755		
0.00922	0.499		
0.00417	0.258		
0.000585	0.049		

Constants of Jouyban–Acree model

J_0

J_1

J_2

Reference of data: [13].

Solute	Solvent 1	Solvent 2	T (°C)
Valine (L)	Water	Methanol	15.0
Solubility (mole/L)	Mass F. 1		
0.718	1.000		
0.676	0.920		
0.632	0.840		
0.493	0.748		
0.213	0.656		
0.196	0.553		
0.162	0.458		
0.0269	0.359		
0.0182	0.241		
0.000398	0.121		
0.000229	0.000		

Constants of Jouyban–Acree model

J_0	1094.049
J_1	0
J_2	0

Reference of data: [15].

Solute	Solvent 1	Solvent 2	T (°C)
Valine (L)	Water	Methanol	20.0
Solubility (mole/L)	Mass F. 1		
0.736	1.000		
0.689	0.920		
0.641	0.840		
0.499	0.748		
0.216	0.656		
0.203	0.553		
0.182	0.458		
0.0302	0.359		
0.0209	0.241		
0.000501	0.121		
0.000309	0.000		

Constants of Jouyban–Acree model

J_0	1094.049
J_1	0
J_2	0

Reference of data: [15].

Solute	Solvent 1	Solvent 2	T (°C)
Valine (L)	Water	Methanol	25.0
Solubility (mole/L)	Mass F. 1		
0.755	1.000		
0.70	0.920		
0.65	0.840		
0.505	0.748		
0.22	0.656		
0.21	0.553		
0.195	0.458		
0.035	0.359		
0.0245	0.241		
0.000625	0.121		
0.000425	0.000		

Constants of Jouyban–Acree model

J_0	1094.049
J_1	0
J_2	0

Reference of data: [15].

Solute	Solvent 1	Solvent 2	T (°C)
Valsartan	Ethanol	Water	20.0
Solubility (g/L)	Vol. F. 1		
556.0	0.000		
727.0	0.150		
1029.0	0.200		
1540.0	0.250		
2368.0	0.300		

Constants of Jouyban–Acree model

J_0
J_1
J_2

Reference of data: [123].

Solute	Solvent 1	Solvent 2	T (°C)
Valsartan	Polyethylene glycol 400	Water	25.0
Solubility (g/L)	Vol. F. 1		
613.0	0.000		
701.0	0.100		
1491.0	0.200		
2442.0	0.300		
4699.0	0.400		
7500.0	0.500		

Constants of Jouyban–Acree model

J_0

J_1

J_2

Reference of data: [124].

Solute	Solvent 1	Solvent 2	T (°C)
Valsartan	Propylene glycol	Water	20.0
Solubility (g/L)	Vol. F. 1		
556.0	0.000		
682.0	0.150		
923.0	0.200		
1200.0	0.250		
1599.0	0.300		

Constants of Jouyban–Acree model

J_0

J_1

J_2

Reference of data: [123].

Solute	Solvent 1	Solvent 2	T (°C)
Vanillin	Ethanol	Water	4.4
Solubility (g/L)	Mass F. 1		
3.00	0.000		
4.00	0.050		
4.50	0.100		
5.00	0.150		
6.00	0.200		
9.00	0.250		
14.00	0.300		

Constants of Jouyban–Acree model

J_0

J_1

J_2

Reference of data: [10].

Solute	Solvent 1	Solvent 2	T (°C)
Vanillin	Ethanol	Water	15.6
Solubility (g/L)	Mass F. 1		
5.2	0.000		
6.0	0.050		
7.5	0.100		
10.0	0.150		
13.5	0.200		
20.0	0.250		
33.0	0.300		

Constants of Jouyban–Acree model

J_0

J_1

J_2

Reference of data: [10].

Solute	Solvent 1	Solvent 2	T (°C)
Vanillin	Ethanol	Water	23.9
Solubility (g/L)	Mass F. 1		
9.0	0.000		
11.0	0.050		
13.5	0.100		
19.0	0.150		
28.0	0.200		
47.0	0.250		

Constants of Jouyban–Acree model

J_0

J_1

J_2

Reference of data: [10].

Solute	Solvent 1	Solvent 2	T (°C)
Vinbarbital	Ethanol	Water	25.0
Solubility (g/L)	Mass F. 1		
62.3	1.000		
62.7	0.975		
63.1	0.950		
63.3	0.925		
63.0	0.900		
62.2	0.875		
61.0	0.850		
59.2	0.825		
56.9	0.800		
54.2	0.775		

(continued)

Solute	Solvent 1	Solvent 2	T (°C)
51.3	0.750		
48.1	0.725		
44.8	0.700		
41.3	0.675		
37.9	0.650		
34.4	0.625		
31.1	0.600		
27.8	0.575		
24.7	0.550		
21.8	0.525		
19.0	0.500		
16.4	0.475		
14.0	0.450		
11.8	0.425		
9.8	0.400		
8.1	0.375		
6.6	0.350		
5.3	0.325		
4.2	0.300		
3.3	0.275		
2.6	0.250		
2.1	0.225		
1.7	0.200		
1.4	0.175		
1.2	0.150		
1.1	0.125		
1.0	0.100		
0.9	0.075		
0.8	0.050		
0.7	0.025		
0.7	0.000		

Constants of Jouyban–Acree model

J_0	541.462
J_1	491.893
J_2	−538.947

Reference of data: [20].

Solute	Solvent 1	Solvent 2	T (°C)
Vitamin E	Ethanol	Water	33
Solubility (mole F.)	Mass F. 1		
0.00000087	0.000		
0.00000137	0.100		
0.00000490	0.200		
0.00001391	0.300		
0.00002121	0.400		
0.00003222	0.480		

(*continued*)

(continued)

Solute	Solvent 1	Solvent 2	T (°C)
0.00004512	0.520		
0.00008255	0.570		
0.00015272	0.620		
0.00032919	0.660		
0.00069802	0.710		

Constants of Jouyban–Acree model

J_0
J_1
J_2

Reference of data: [125].

Solute	Solvent 1	Solvent 2	T (°C)
Vitamin K3	Ethanol	Water	33
Solubility (mole F.)	Mass F. 1		
0.000016	0.000		
0.000029	0.090		
0.000065	0.180		
0.000156	0.270		
0.000395	0.360		
0.000954	0.450		
0.001651	0.540		
0.002613	0.630		
0.004848	0.720		
0.005840	0.810		
0.006757	0.900		

Constants of Jouyban–Acree model

J_0
J_1
J_2

Reference of data: [125].

REFERENCES

1. Bustamante, P., Romero, S., Pena, A., Escalera, B., and Reillo, A., Enthalpy–entropy compensations for the solubility of drugs in solvent mixtures; paracetamol, acetanilide and nalidixic acid in dioxane–water. *Journal of Pharmaceutical Sciences*, 1998. 87: 1590–1596.
2. Jimenez, J.A. and Martinez, F., Temperature dependence of the solubility of acetaminophen in propylene glycol + ethanol mixtures. *Journal of Solution Chemistry*, 2005. 35: 335–352.
3. Martinez, F., Aplicacion del metodo extendido de Hildebrand al estudio de la solubilidad del acetaminofen en mezclas etanol–propilenoglicol. *Acta Farmaceutica Bonaerense*, 2005. 24: 215–224.
4. Prakongpan, S. and Nagai, T., Solubility of acetaminophen in cosolvents. *Chemical and Pharmaceutical Bulletin*, 1984. 32: 340–343.
5. Bustamante, P., Romero, S., and Reillo, A., Thermodynamics of paracetamol in amphiprotic and amphiprotic–aprotic solvent mixtures. *Pharmaceutical Sciences*, 1995. 1: 505–507.

6. Hojjati, H. and Rohani, S., Measurement and prediction of solubility of paracetamol in water–isopropanol solution. Part 1. Measurement and data analysis. *Organic Process Research and Development*, 2006. 10: 1101–1109.

7. Jimenez, J.A. and Martinez, F., Thermodynamic study of the solubility of acetaminophen in propylene glycol + water cosolvent mixtures. *Journal of Brazilian Chemical Society*, 2006. 17: 125–134.

8. Jouyban, A., Chan, H.K., Chew, N.Y.K., Khoubnasabjafari, M., and Acree, W.E. Jr., Solubility prediction of paracetamol in binary and ternary solvent mixtures using Jouyban–Acree model. *Chemical and Pharmaceutical Bulletin*, 2006. 54: 428–431.

9. Etman, M.A. and Naggar, V.F., Thermodynamics of paracetamol solubility in sugar–water cosolvent systems. *International Journal of Pharmaceutics*, 1990. 58: 177–184.

10. Stephen, H. and Stephen, T., *Solubilities of Inorganic and Organic Compounds*. 1964: Pergamon Press, Oxford.

11. Pena, M.A., Reillo, A., Escalera, B., and Bustamante, P., Solubility parameter of drugs for predicting the solubility profile type within a wide polarity range in solvent mixtures. *International Journal of Pharmaceutics*, 2006. 321: 155–161.

12. Greenstein, J.P. and Winitz, M., *Chemistry of Amino Acids*. 1961: John Wiley and Sons, New York.

13. Dunn, M.S. and Ross, F.J., Quantitative investigations of amino acids and peptides. IV. The solubilities of the amino acids in water–ethyl alcohol mixtures. *Journal of Biological Chemistry*, 1938. 125: 309–332.

14. Orella, C.J. and Kirwan, D.J., Correlation of amino acid solubilities in aqueous aliphatic alcohol solutions. *Industrial Engineering and Chemical Research*, 1991. 30: 1040–1045.

15. Dey, B.P. and Lahiri, S.C., Solubilities of amino acids in methanol + water mixtures at different temperatures. *Indian Journal of Chemistry A*, 1987. 27: 297–302.

16. Phatak, P.V. and Gaikar, V.G., Solubilities of *o*-chlorobenzoic acid and *o*- and *p*-nitroaniline in *n,n*-dimethylformamide + water. *Journal of Chemical Engineering Data*, 1996. 41: 1052–1054.

17. Leiterman, R.V., Mulski, M.J., and Connors, K.A., Solvent effect on chemical processes. 10. Solubility of alfa-cyclodextrin in binary aqueous-organic solvents: Relationship to solid phase composition. *Journal of Pharmacy and Pharmacology*, 1995. 84: 1272–1275.

18. Yalkowsky, S.H. and Flynn, G.L., Correlation and prediction of mass transport across membranes II: Influence of vehicle polarity on flux from solutions and suspensions. *Journal of Pharmaceutical Sciences*, 1975. 63: 1276–1280.

19. Paruta, A.N., Solubility profiles for antipyrine and aminopyrine in hydroalcoholic solutions. *Journal of Pharmaceutical Sciences*, 1967. 56: 1565–1569.

20. Breon, T.L. and Paruta, A.N., Solubility profiles for several barbiturates in hydroalcoholic mixtures. *Journal of Pharmaceutical Sciences*, 1970. 59: 1306–1313.

21. Karanth, H. and Josyula, V.R., Studies on solubility parameter of amoxycillin trihydrate: Influence on in vitro release and antibacterial activity. *Indian Journal of Pharmaceutical Sciences*, 2005. 67: 342–345.

22. Wang, L.H., Song, Y.T., Chen, Y., and Cheng, Y.Y., Solubility of artemisinin in ethanol + water from (278.15 to 343.15) K. *Journal of Chemical and Engineering Data*, 2007. 53: 757–758.

23. Nozaki, Y. and Tanford, C., The solubility of amino acids and two glycine peptides in aqueous ethanol and dioxane solutions. *Journal of Biological Chemistry*, 1971. 246: 2211–2217.

24. Pena, M.A., Bustamante, P., Escalera, B., and Reillo, A., Solubility and phase separation of benzocaine and salicylic acid in 1,4-dioxane–water mixtures at several temperatures. *Journal of Pharmaceutical and Biomedical Analysis*, 2004. 36: 571–578.

25. Wang, Q., Hou, L., Cheng, Y., and Li, X., Solubility of benzoic acid and phthalic acid in acetic acid + water solvent mixtures. *Journal of Chemical and Engineering Data*, 2007. 52: 936–940.

26. Pal, A. and Lahiri, S.C., Solubility and the thermodynamics of transfer of benzoic acid in mixed solvents. *Indian Journal of Chemistry A*, 1989. 28: 276–279.

27. Yurquina, A., Manzur, M.E., Brito, P., Manzo, R., and Molina, M.A.A., Solubility and dielectric properties of benzoic acid in a binary solvent: Water–ethylene glycol. *Journal of Molecular Liquids*, 2003. 108: 119–133.

28. Chellquist, M. and Gorman, W.G., Benzoyl peroxide solubility and stability in hydric solvents. *Pharmaceutical Research*, 1992. 9: 1341–1346.

29. Lu, Y.C., Lin, Q., Luo, G.Sh., and Dai, Y.Y., Solubility of berberine chloride in various solvents. *Journal of Chemical and Engineering Data*, 2006. 51: 642–644.

30. Acree, W.E. Jr., Solubility of biphenyl in binary solvent mixtures. *International Journal of Pharmaceutics*, 1984. 18: 47–52.

31. Khossravi, D. and Connors, K.A., Solvent effects on chemical processes. V: Hydrophobic and solvation effects on the solubilities of substituted biphenyls in methanol/water mixtures. *Journal of Pharmaceutical Sciences*, 1993. 82: 817–820.

32. Rubino, J.T. and Obeng, E.K., Influence of solute structure on deviation from log-linear solubility equation in propylene glycol: Water mixtures. *Journal of Pharmaceutical Sciences*, 1991. 80: 479–483.

33. Yalkowsky, S.H., Amidon, G.L., Zografi, G., and Flynn, G.L., Solubility of nonelectrolytes in polar solvents III: Alkyl *p*-aminobenzoates in polar and mixed solvents. *Journal of Pharmaceutical Sciences*, 1975. 64: 48–52.

34. Herrador, M.A. and Gonzalez, A.G., Solubility prediction of caffeine in aqueous *N,N*-dimethylformamide mixtures using the extended Hildebrand solubility approach. *International Journal of Pharmaceutics*, 1997. 156: 239–244.

35. Adjei, A., Newburger, J., and Martin, A., Extended Hildebrand approach. Solubility of caffeine in dioxane–water mixtures. *Journal of Pharmaceutical Sciences*, 1980. 69: 659–661.

36. Bustamante, P., Navarro, J., Romero, S., and Escalera, B., Thermodynamic origin of the solubility profile of drugs showing one or two maxima against the polarity of aqueous and nonaqueous mixtures: Niflumic acid and caffeine. *Journal of Pharmaceutical Sciences*, 2002. 91: 874–883.

37. Molzon, J.A., Lausier, J.M., and Paruta, A.N., Solubility of calcium oxalate in 1-alkanols and ethanol–water mixtures. *Journal of Pharmaceutical Sciences*, 1978. 67: 733–735.

38. Subrahmanyam, C.V.S. and Sarasija, S., Solubility behaviour of carbamazepine in binary solvents: Extended Hildebrand solubility approach to obtain solubility and other parameters. *Die Pharmazie*, 1997. 52: 939–942.

39. Treszczanowicz, T., Treszczanowicz, A.J., Kasparzycka-Guttmans, T., and Kulesza, A., Solubility of beta-carotene in binary solvents formed by some hydrocarbons with ketones. *Journal of Chemical Engineering Data*, 2001. 46: 792–794.

40. Treszczanowicz, T., Kasparzycka-Guttmans, T., and Treszczanowicz, A.J., Solubility of beta-carotene in binary solvents formed by some hydrocarbons with dibutyl ether and 1,2-dimethoxyethane. *Journal of Chemical Engineering Data*, 2003. 48: 1517–1520.

41. Treszczanowicz, T., Kasparzycka-Guttmans, T., and Treszczanowicz, A.J., Solubility of beta-carotene in binary solvents formed by some hydrocarbons with cyclohexane and 1-octanol. *Journal of Chemical Engineering Data*, 2001. 46: 1494–1496.

42. Treszczanowicz, T., Kasparzycka-Guttmans, T., and Treszczanowicz, A.J., Solubility of beta-carotene in binary solvents formed by some hydrocarbons with *tert*-butyl methyl ether and with *tert*-amyl methyl ether. *Journal of Chemical Engineering Data*, 2005. 50: 973–976.

43. Dong, H., Fu, R., Yan, W., and Zhu, M., Solubility and density of the disodium salt hemiheptahydrate of ceftriaxone in (acetone + water) at *T* = (298.15, 308.15, and 318.15) K. *Journal of Chemical Thermodynamics*, 2004. 36: 155–159.

44. Wu, J., Wang, J., Zhang, M., and Qian, Y., Solubility of cefazolin sodium pentahydrate in binary system of ethanol + water mixtures. *Journal of Chemical and Engineering Data*, 2006. 51: 1404–1405.

45. Wu, J., Wang, J., Zhang, M., and Qian, Y., Erratum: Solubility of cefazolin sodium pentahydrate in binary system of ethanol + water mixtures. *Journal of Chemical and Engineering Data*, 2006. 51: 2275.

46. Wu, J. and Wang, J., Solubility of cefazolin sodium pentahydrate in aqueous 2-propanol mixtures. *Journal of Chemical Engineering Data*, 2005. 50: 980–982.

47. Zhu, M., Solubility and density of the disodium salt hemiheptahydrate of ceftriaxone in water + ethanol mixtures. *Journal of Chemical and Engineering Data*, 2001. 46: 175–176.

48. Fu, R., Yan, W., and Zhu, M., Solubility and density of the disodium salt hemiheptahydrate of ceftriaxone in water + methanol + mixtures. *Journal of Chemical and Engineering Data*, 2004. 49: 262–263.

49. Seedher, N. and Bhatia, S., Solubility enhancement of cox-2 inhibitors using various solvent systems. *AAPS Pharmaceutical Sciences*, 2003. 4: Article 33.

50. Shokri, J., Solubilization by cosolvency and skin penetration of benzodiazepines, PhD dissertation, 2002, Tabriz University (Medical Sciences), Tabriz, Iran.

51. Kulkarni, A.R., Soppimath, K.S., Dave, A.M., Mehta, M.H., and Aminabhavi, T.M., Solubility study of hazardous pesticide (chlorpyrifos) by gas chromatography. *Journal of Hazardous Materials A*, 2000. 80: 9–13.

52. Chen, Y. and Wang, J.K., Solubility of clindamycin phosphate in binary water–ethanol solvent. *Journal of Chemical and Engineering Data*, 2007. 52: 1908–1910.

53. Shayanfar, A., Fakhree, M.A.A., Acree, W.E. Jr., and Jouyban, A., Solubility of lamotrigine, diazepam and clonazepam in ethanol + water mixtures at 298.15 K, *Journal of Chemical Engineering Data*, 2009. 54: 1107–1109.

54. Shayanfar, A., Acree, W.E. Jr., and Jouyban, A., Solubility of lamotrigine, diazepam and clonazepam in *N*-methylpyrrolidone + water mixtures at 298.15 K. *Journal of Chemical and Engineering Data*, 2009: Accepted for publication.

55. Shayanfar, A., Acree, W.E. Jr., and Jouyban, A., Solubility of lamotrigine, diazepam, clonazepam and phenobarbital in propylene glycol + water mixtures at 298.15 K. *Journal of Chemical and Engineering Data*, 2009. 54: 1153–1157

56. Hao, H.-X., Wang, J.-K., and Wang, Y.-L., Solubility of dexamethasone sodium phosphate in different solvents. *Journal of Chemical and Engineering Data*, 2004. 49: 1697–1698.

57. Saei, A.A., Jabbaribar, F., Fakhree, M.A.A., Acree, W.E. Jr., and Jouyban, A., Solubility of sodium diclofenac in binary water + alcohol solvent mixtures at 25°C. *Journal of Drug Delivery Science and Technology*, 2008. 18: 149–151.

58. Conio, G., Curletto, L., and Patrone, E., On the temperature coefficient of the solubility of some glycyl peptides in water–ethanol mixtures. *Journal of Biological Chemistry*, 1973. 248: 5448–5450.

59. Teychene, S., Autret, J.M., and Biscans, B., Determination of solubility profiles of eflucimibe polymorphs: Experimental and modeling. *Journal of Pharmaceutical Sciences*, 2005. 95: 871–882.

60. Wang, Z., Wang, J., Zhang, M., and Dang, L., Solubility of Erythromycin A dihydrate in different pure solvents and acetone + water binary mixtures between 293 K and 323 K. *Journal of Chemical Engineering Data*, 2006. 51: 1062–1065.

61. Liu, B.S., Liu, R.J., Hu, Y.Q., and Hu, Q.F., Solubility of ethyl maltol in aqueous ethanol mixtures. *Journal of Chemical and Engineering Data*, 2009. 53: 2712–2714, 54: 1171.

62. He, Y., Li, P., and Yalkowsky, S.H., Solubilization of fluasterone in cosolvent/cyclodextrin combinations. *International Journal of Pharmaceutics*, 2003. 264: 25–34.

63. Jouyban-Gharamaleki, A., Dastmalchi, S., Chan, H.K., Hanaee, J., Javanmard, A., and Barzegar-Jalali, M., Solubility prediction for furosemide in water–cosolvent mixtures using the minimum number of experiments. *Drug Development and Industrial Pharmacy*, 2001. 27: 577–583.

64. Bockstanz, G.L., Buffa, M., and Lira, C.T., Solubilities of alfa-anhydrous glucose in ethanol/water mixtures. *Journal of Chemical and Engineering Data*, 1989. 34: 426–429.

65. Lu, J., Wang, X.-J., Yang, X. and Ching, C.-B., Solubilities of glycine and its oligopeptides in aqueous solutions. *Journal of Chemical and Engineering Data*, 2006. 51: 1593–1596.

66. Li, A. and Yalkowsky, S.H., Solubility of organic solutes in ethanol–water mixtures. *Journal of Pharmaceutical Sciences*, 1994. 83: 1735–1740.

67. Hagen, T.A. and Flynn, G.L., Solubility of hydrocortisone in organic and aqueous media: Evidence for regular solution behaviour in apolar solvents. *Journal of Pharmaceutical Sciences*, 1983. 72: 409–415.

68. Wu, P.L. and Martin, A., Extended Hildebrand solubility approach: *p*-Hydroxybenzoic acid in mixtures of dioxane and water. *Journal of Pharmaceutical Sciences*, 1983. 72: 587–595.

69. Ren, G.B., Wang, J.K., Yin, Q.X., and Zhang, M.J., Solubility of DL-*p*-hydroxyphenylglycine sulfate in binary acetone + water solvent mixtures. *Journal of Chemical Engineering Data*, 2004. 49: 1376–1378.

70. Ren, G.B., Wang, J.K., Yin, Q.X., Zhang, and M.J., Erratum: Solubility of DL-*p*-hydroxyphenylglycine sulfate in binary acetone + water solvent mixtures. *Journal of Chemical Engineering Data*, 2005. 50: 298.

71. Zhou, Z., Qu, Y., Wei, H., and Chen, L., Solubility of D-(*p*-hydroxy)phenylglycine in water + 2-propanol from (293 to 343) K. *Journal of Chemical and Engineering Data*, 2008. 53: 2900–2901.

72. Gao, Y. and Tian, J., Solubility of irbesartan form B in an aqueous ethanol mixture. *Journal of Chemical and Engineering Data*, 2008. 53: 535–537.

73. Pacheco, D.P., Manrique, Y.J., and Martinez, F., Thermodynamic study of the solubility of ibuprofen and naproxen in some ethanol + propylene glycol mixtures. *Fluid Phase Equilibria*, 2007. 262: 23–31.

74. Herkenne, C., Naik, A., Kalia, Y.N., Hadgraft, J., and Guy, R.H., Effect of propylene glycol on ibuprofen absorption into human skin in vivo. *Journal of Pharmaceutical Sciences*, 2008. 97: 185–197.

75. Manrique, Y.J., Pacheco, D.P., and Martinez, F., Thermodynamics of mixing and solvation of ibuprofen and naproxen in propylene glycol + water cosolvent mixtures. *Journal of Solution Chemistry*, 2008. 37, 165–181.

76. Patel, R.A. and Vasavada, R.C., Transdermal delivery of isoproternol HCl: An investigation of stability, solubility, partition coefficient, and vehicle effects. *Pharmaceutical Research*, 1988. 5: 116–119.

77. Singhai, A.K., Jain, S., and Jain, N.K., Cosolvent solubilization and formulation of an aqueous injection of ketoprofen. *Die Pharmazie*, 1996. 51: 737–740.

78. Jozwiakowski, M.J., Nguyen, N.-A.T., Sisco, J.M., and Spancake, C.W., Solubility behaviour of lamivudine crystal forms in recrystallization solvents. *Journal of Pharmaceutical Sciences*, 1996. 85: 193–199.

79. Guo, K., Yin, Q., Zhang, M., and Wang, J., Solubility of losartan potassium in different binary systems from (293.15 to 343.15) K. *Journal of Chemical and Engineering Data*, 2008. 53: 1138–1140.
80. Sun, H. and Wang, J., Solubility of lovastatin in acetone + water solvent mixtures. *Journal of Chemical and Engineering Data*, 2008. 53: 1335–1337.
81. Mourya, V.K., Yadav, S.K., and Saini, T.R., Solubility studies of metronidazole in binary solvent blends. *Indian Journal of Pharmaceutical Sciences*, 1997. 59: 200–202.
82. Pacheco, D.P. and Martínez, F., Thermodynamic analysis of the solubility of naproxen in ethanol + water cosolvent mixtures. *Physics and Chemistry of Liquids*, 2007. 45: 581–595.
83. Squillante, E., Needham, T., and Zia, H., Solubility and in vitro transdermal permeation of nifedipine. *International Journal of Pharmaceutics*, 1997. 159: 171–180.
84. Jouyban-Gharamaleki, A., Romero, S., Bustamante, P., and Clark, B.J., Multiple solubility maxima of oxolinic acid in mixed solvents and a new extension of Hildebrand solubility approach. *Chemical and Pharmaceutical Bulletin*, 2000. 48: 175–178.
85. Ren, G., Wang, J., and Li, G., Solubility of proxetine hydrochloride hemi-hydrate in (water + acetone). *Journal of Chemical Thermodynamics*, 2005. 37: 860–865.
86. Bustamante, C. and Bustamante, P., Nonlinear enthalpy–entropy compensation for the solubility of phenacetin in dioxane–water solvent mixtures. *Journal of Pharmaceutical Sciences*, 1996. 85: 1109–1111.
87. Rubino, J.T., Blanchard, J., and Yalkowsky, S.H., Solubilization by cosolvents II: Phenytoin in binary and ternary solvents. *Journal of Parenteral Science and Technology*, 1984. 38: 215–221.
88. Pardo, A., Shiri, Y., and Cohen, S., Partial molal volumes and solubilities of physostigmine in isopropanol: Isopropyl myristate solvents in relation to skin penetrability. *Journal of Pharmaceutical Sciences*, 1991. 80: 567–572.
89. Liu, B.-S., Wang, J.-K., and Sun, H., Solubility of potassium clavulanate in aqueous 2-propanol mixtures. *Journal of Chemical and Engineering Data*, 2006. 51: 291–293.
90. Sun, X., Shao, Y., and Yan, W., Measurement and correlation of solubilities of *trans*-resveratrol in ethanol + water and acetone + water mixed solvents at different temperatures. *Journal of Chemical and Engineering Data*, 2008. 53: 2562–2566.
91. Simamora, P., Alvarez, J.M., and Yalkowsky, S.H., Solubilization of rapamycin. *International Journal of Pharmaceutics*, 2001. 213: 25–29.
92. Zhou, K., Li, J., and Zheng, D.Sh., Solubility of Rifapentine in the binary system of acetic acid and *n*-octanol solvent mixtures. *Journal of Chemical and Engineering Data*, 2008. 53: 1978–1979.
93. Desai, K.G.H., Kulkarni, A.R., and Aminabhavi, T.M., Solubility of refecoxib in the presence of methanol, ethanol, and sodium lauryl sulfate at (298.15, 303.15, and 308.15) K. *Journal of Chemical Engineering Data*, 2003. 48: 942–945.
94. Liu, C., Desai, K.G.H., Tang, X., and Chen, X., Solubility of rofecoxib in the presence of aqueous solutions of glycerol, propylene glycol, ethanol, Span 20, Tween 80, and sodium lauryl sulfate at (298.15, 303.15, and 308.15) K. *Journal of Chemical and Engineering Data*, 2005. 50: 2061–2064.
95. Liu, C., Desai, K.G.H., and Liu, C., Solubility of rofecoxib in the presence of mannitol, poly(vinylpyrrolidone) K30, urea, polyethylene glycol 4000, and polyethylene glycol 6000 at (298.15, 303.15, and 308.15) K. *Journal of Chemical and Engineering Data*, 2005. 50: 661–665.
96. Matsuda, H., Kaburagi, K., Matsumoto, Sh., Kurihara, K., Tochigi, K., and Tomono, K., Solubilities of salicylic acid in pure solvents and binary mixtures containing cosolvent. *Journal of Chemical and Engineering Data*, 2008. 54: 480–484.
97. Jouyban, A., Chew, N.Y.K., Chan, H.K., Khoubnasabjafari, M., and Acree, W.E. Jr., Solubility prediction of salicylic acid in water–ethanol–propylene glycol mixtures using the Jouyban–Acree model. *Die Pharmazie*, 2006. 61: 417–419.
98. Jouyban-Gharamaleki, A., York, P., Hanna, M., and Clark, B.J., Solubility prediction of salmeterol xinafoate in water–dioxane mixtures. *International Journal of Pharmaceutics*, 2001. 216: 33–41.
99. Charmolue, H. and Rousseau, R.W., L-Serine obtained by methanol addition in batch crystallization. *AIChE Journal*, 1991. 37: 1121–1128.
100. Bai, T.C., Yan, G.B. Hu, J., Zhang, H.L., and Huang, C.G., Solubility of silybin in aqueous poly (ethylene glycol) solution. *International Journal of Pharmaceutics*, 2006. 308: 100–106.
101. Mirmehrabi, M., Rohani, S., and Perry, L., Thermodynamic modeling of activity coefficient and prediction of solubility: Part 2. Semipredictive or semiempirical models. *Journal of Pharmaceutical Sciences*, 2006. 95: 798–809.
102. Reillo, A., Cordoba, M., Escalera, B., Selles, E., and Cordoba, M. Jr., Prediction of sulfamethiazole solubility in dioxane–water mixtures. *Die Pharmazie*, 1995. 50: 472–475.

103. Reillo, A., Escalera, B., and Selles, E., Prediction of sulfanilamide solubility in dioxane–water mixtures. *Die Pharmazie*, 1993. 48: 904–907.

104. Reillo, A., Bustamante, P., Escalera, B., Jimenez, M.M., and Selles, E., Solubility parameter-based methods for predicting the solubility of sulfapyridine in solvent mixtures. *Drug Development and Industrial Pharmacy*, 1995. 21: 2073–2084.

105. Martin, A., Wu, P.L., and Velasquez, T., Extended Hildebrand solubility approach. Sulfonamides in binary and ternary solvents. *Journal of Pharmaceutical Sciences*, 1985. 74: 277–282.

106. Bustamante, P., Escalera, B., Martin, A., and Selles, E., A modification of the extended Hildebrand approach to predict the solubility of structurally related drugs in solvent mixtures. *Journal of Pharmacy and Pharmacology*, 1993. 45: 253–257.

107. Elworthy, P.H. and Worthington, E.C., The solubility of sulphadiazine in water–dimethylformamide mixtures. *Journal of Pharmacy and Pharmacology*, 1968. 20: 830–835.

108. Bustamante, P., Ochoa, R., Reillo, A., and Escalera, J.B., Chameleonic effect of sulfanilamide and sulfamethazine in solvent mixtures. Solubility curves with two maxima. *Chemical and Pharmaceutical Bulletin*, 1994. 42: 1129–1133.

109. Canotiho, J., Costa, F.S., Sousa, A.T., Redinho, J.S., and Leitao, M.L.P., Enthalpy of solution of terfenadine in ethanol/water mixtures. *Thermochimica Acta*, 2000. 344: 9–13.

110. Martin, A., Paruta, A.N., and Adjei, A., Extended-Hildebrand solubility approach: Methylxanthines in mixed solvents. *Journal of Pharmaceutical Sciences*, 1981. 70: 1115–1120.

111. Martin, A., Newburger, J., and Adjei, A., Extended Hildebrand solubility approach: Solubility of theophylline in polar binary solvents. *Journal of Pharmaceutical Sciences*, 1980. 69: 487–491.

112. Khossravi, D. and Connors, K.A., Solvent effects on chemical processes. I: Solubility of aromatic and heterocyclic compounds in binary aqueous-organic solvents. *Journal of Pharmaceutical Sciences*, 1992. 81: 371–379.

113. Gould, P.L., Howard, J.R., and Oldershaw, G.A., The effect of hydrate formation on the solubility of theophylline in binary aqueous cosolvent system. *International Journal of Pharmaceutics*, 1989. 51: 195–202.

114. De Fina, K.M., Hernandez, C.E., and Acree, W.E. Jr., Solubility of *trans*-stilbene in binary alkane + 1-butanol solvent mixtures at 298.2 K. *Physics and Chemistry of Liquids*, 2000. 38: 211–216.

115. Hernandez, C.E., Roy, L.E., Sharp, T.L., Childress, S.D., De Fina, K.M., Deng, T.H., and Acree, W.E. Jr., Solubility of *trans*-stilbene in binary alkane + 1-propanol solvent mixtures at 298.2 K. *Physics and Chemistry of Liquids*, 1999. 37: 757–763.

116. Deng, T.H., Childress, S.D., De Fina, K.M., Hernandez, C.E., Roy, L.E., Sharp, T.L., McKethan, B., Rubio, A., Sanchez, M., Wright, D., and Acree, W.E. Jr., Solubility of *trans*-stilbene in binary alkane + 2-butanol solvent mixtures at 298.15 K. *Physics and Chemistry of Liquids*, 1999. 37: 735–740.

117. De Fina, K.M., Hernandez, C.E., Sharp, T.L., and Acree, W.E. Jr., Solubility of *trans*-stilbene in binary alkane + 2-propanol solvent mixtures at 298.2 K. *Physics and Chemistry of Liquids*, 2000. 38: 89–94.

118. Subrahmanyam, C.V.S., Ravi Prakash, K., and Gundu Rao, P., Estimation of the solubility parameter of trimethoprim by current method. *Pharmaceutica Acta Helvetia*, 1996. 71: 175–183.

119. Liu, C., Desai, K.G.H., and Liu, C., Solubility of valdecoxib in the presence of ethanol and sodium lauryl sulfate at (298.15, 303.15, and 308.15) K. *Journal of Chemical Engineering Data*, 2004. 49: 1847–1850.

120. Desai, H.K.G. and Park, H.J., Solubility studies on valdecoxib in the presence of carriers, cosolvents, and surfactants. *Drug Development Research*, 2004. 62: 41–48.

121. Liu, C., Desai, K.G.H., Chen, X., and Tang, X., Solubility of valdecoxib in the presence of glycerol, propylene glycol, and polyethylene glycol 400 at (298.15, 303.15, and 308.15) K. *Journal of Chemical and Engineering Data*, 2005. 50: 1736–1739.

122. Liu, C., Desai, K.G.H., and Liu, C., Solubility of valdecoxib in the presence of polyethylene glycol 4000, polyethylene glycol 6000, polyethylene glycol 8000, and polyethylene glycol 10000 at (298.15, 303.15, and 308.15) K. *Journal of Chemical and Engineering Data*, 2005. 50: 278–282.

123. Mbah, C.J., Physicochemical properties of valsartan and the effect of ethyl alcohol, propylene glycol and pH on its solubility. *Die Pharmazie*, 2005. 60: 849–850.

124. Mbah, C.J., Solubilization of valsartan by aqueous glycerol, polyethylene glycol and micellar solutions. *Die Pharmazie*, 2006. 61: 322–324.

125. Dubbs, M. and Gupta, R.B., Solubility of vitamin E (alfa-tocopherol) and vitamin K_3 (menadione) in ethanol–water mixture. *Journal of Chemical and Engineering Data*, 1998. 43: 590–591.

4 Solubility Data in Ternary Solvent Mixtures

This chapter reports the solubility data of pharmaceutically interested solutes in aqueous and non-aqueous ternary mixtures collected from the literature. The solvent compositions and the solubilities in mixed solvents were reported using different units which included in the tables concerning the originally reported data. The solubility data in monosolvents are also listed in the tables collected form the original reference and from other references. It should be noted that when the solubility data in monosolvents taken from a different reference, some differences in solubilities are expected.

The model constants of the Jouyban–Acree model for predicting the solubility at all solvent compositions are reported when the solubility data of the drug in monosolvents were available and the minimum number of data points ($N = 10$) in mixed solvents was available to calculate the statistically significant model constants. The model for ternary solvent mixtures at various temperatures could be written as the following equation.

$$\log X_{m,T} = f_1 \log X_{1,T} + f_2 \log X_{2,T} + f_3 \log X_{3,T}$$

$$+ \frac{J_{0\text{-}12} f_1 f_2}{T} + \frac{J_{1\text{-}12} f_1 f_2 (f_1 - f_2)}{T} + \frac{J_{2\text{-}12} f_1 f_2 (f_1 - f_2)^2}{T}$$

$$+ \frac{J_{0\text{-}13} f_1 f_3}{T} + \frac{J_{1\text{-}13} f_1 f_3 (f_1 - f_3)}{T} + \frac{J_{2\text{-}13} f_1 f_3 (f_1 - f_3)^2}{T}$$

$$+ \frac{J_{0\text{-}23} f_2 f_3}{T} + \frac{J_{1\text{-}23} f_2 f_3 (f_2 - f_3)}{T} + \frac{J_{2\text{-}23} f_2 f_3 (f_2 - f_3)^2}{T}$$

where
    X is the solute solubility
    f denotes the fraction of the solvent in the mixtures in the absence of the solute
    T is the absolute temperature
    J terms are the model constants

Subscripts m, 1, 2, and 3 are the solvent mixture and neat solvents 1–3.

Solute	Solvent 1	Solvent 2	Solvent 3	T (°C)
Acetaminophen	Acetone	Water	Toluene	0
Solubility (g/kg)	% w/w 1	% w/w 2	% w/w 3	
98.000	3.00	67.90	29.10	
118.100	3.00	82.45	14.55	
128.000	3.00	92.15	4.85	
178.000	7.00	65.10	27.90	
199.000	7.00	79.05	13.95	
209.500	7.00	88.35	4.65	
281.400	15.00	72.25	12.75	
299.700	15.00	80.75	4.25	
300.600	30.00	66.50	3.50	

X_1	X_2	X_3
55.600	7.210	0.220

Constants of Jouyban–Acree model

J_{0-12}	−9.560
J_{1-12}	0.147
J_{2-12}	0
J_{0-13}	13.790
J_{1-13}	−1.442
J_{2-13}	0
J_{0-23}	−59.120
J_{1-23}	2.148
J_{2-23}	−0.022

Reference of data: [1].

Solute	Solvent 1	Solvent 2	Solvent 3	T (°C)
Acetaminophen	Acetone	Water	Toluene	5
Solubility (g/kg)	% w/w 1	% w/w 2	% w/w 3	
103.800	3.00	67.90	29.10	
125.400	3.00	82.45	14.55	
137.100	3.00	92.15	4.85	
182.300	7.00	65.10	27.90	
210.400	7.00	79.05	13.95	
222.800	7.00	88.35	4.65	
298.500	15.00	72.25	12.75	
317.500	15.00	80.75	4.25	
324.540	30.00	66.50	3.50	

X_1	X_2	X_3
62.300	8.210	0.270

Constants of Jouyban–Acree model

$J_{0\text{-}12}$	−9.560
$J_{1\text{-}12}$	0.147
$J_{2\text{-}12}$	0
$J_{0\text{-}13}$	13.790
$J_{1\text{-}13}$	−1.442
$J_{2\text{-}13}$	0
$J_{0\text{-}23}$	−59.120
$J_{1\text{-}23}$	2.148
$J_{2\text{-}23}$	−0.022

Reference of data: [1].

Solute	Solvent 1	Solvent 2	Solvent 3	T (°C)
Acetaminophen	Acetone	Water	Toluene	10
Solubility (g/kg)	% w/w 1	% w/w 2	% w/w 3	
110.800	3.00	67.90	29.10	
134.700	3.00	82.45	14.55	
147.700	3.00	92.15	4.85	
191.200	7.00	65.10	27.90	
223.500	7.00	79.05	13.95	
237.500	7.00	88.35	4.65	
317.700	15.00	72.25	12.75	
337.600	15.00	80.75	4.25	
350.900	30.00	66.50	3.50	

X_1	X_2	X_3
69.600	9.440	0.320

Constants of Jouyban–Acree model

$J_{0\text{-}12}$	−9.560
$J_{1\text{-}12}$	0.147
$J_{2\text{-}12}$	0
$J_{0\text{-}13}$	13.790
$J_{1\text{-}13}$	−1.442
$J_{2\text{-}13}$	0
$J_{0\text{-}23}$	−59.120
$J_{1\text{-}23}$	2.148
$J_{2\text{-}23}$	−0.022

Reference of data: [1].

Solute	Solvent 1	Solvent 2	Solvent 3	T (°C)
Acetaminophen	Acetone	Water	Toluene	15
Solubility (g/kg)	% w/w 1	% w/w 2	% w/w 3	
118.900	3.00	67.90	29.10	
145.400	3.00	82.45	14.55	
159.600	3.00	92.15	4.85	
204.400	7.00	65.10	27.90	
238.000	7.00	79.05	13.95	
253.900	7.00	88.35	4.65	
338.900	15.00	72.25	12.75	
359.600	15.00	80.75	4.25	
380.400	30.00	66.50	3.50	

X_1	X_2	X_3
78.500	10.970	0.360

Constants of Jouyban–Acree model

J_{0-12}	−9.560
J_{1-12}	0.147
J_{2-12}	0
J_{0-13}	13.790
J_{1-13}	−1.442
J_{2-13}	0
J_{0-23}	−59.120
J_{1-23}	2.148
J_{2-23}	−0.022

Reference of data: [1].

Solute	Solvent 1	Solvent 2	Solvent 3	T (°C)
Acetaminophen	Acetone	Water	Toluene	20
Solubility (g/kg)	% w/w 1	% w/w 2	% w/w 3	
128.300	3.00	67.90	29.10	
157.300	3.00	82.45	14.55	
173.400	3.00	92.15	4.85	
217.300	7.00	65.10	27.90	
254.100	7.00	79.05	13.95	
272.600	7.00	88.35	4.65	
362.700	15.00	72.25	12.75	
384.300	15.00	80.75	4.25	
412.100	30.00	66.50	3.50	

X_1	X_2	X_3
88.100	12.780	0.370

Constants of Jouyban–Acree model

J_{0-12}	−9.560
J_{1-12}	0.147
J_{2-12}	0
J_{0-13}	13.790
J_{1-13}	−1.442
J_{2-13}	0
J_{0-23}	−59.120
J_{1-23}	2.148
J_{2-23}	−0.022

Reference of data: [1].

Solute	Solvent 1	Solvent 2	Solvent 3	T (°C)
Acetaminophen	Acetone	Water	Toluene	25
Solubility (g/kg)	% w/w 1	% w/w 2	% w/w 3	
138.600	3.00	67.90	29.10	
171.600	3.00	82.45	14.55	
189.100	3.00	92.15	4.85	
244.200	7.00	65.10	27.90	
272.500	7.00	79.05	13.95	
292.600	7.00	88.35	4.65	
387.800	15.00	72.25	12.75	
410.100	15.00	80.75	4.25	
448.000	30.00	66.50	3.50	

X_1	X_2	X_3
99.800	14.900	0.370

Constants of Jouyban–Acree model

J_{0-12}	−9.560
J_{1-12}	0.147
J_{2-12}	0
J_{0-13}	13.790
J_{1-13}	−1.442
J_{2-13}	0
J_{0-23}	−59.120
J_{1-23}	2.148
J_{2-23}	−0.022

Reference of data: [1].

Solute	Solvent 1	Solvent 2	Solvent 3	T (°C)
Acetaminophen	Acetone	Water	Toluene	30
Solubility (g/kg)	% w/w 1	% w/w 2	% w/w 3	
150.100	3.00	67.90	29.10	
185.800	3.00	82.45	14.55	
205.200	3.00	92.15	4.85	
264.900	7.00	65.10	27.90	
290.300	7.00	79.05	13.95	
314.500	7.00	88.35	4.65	
415.000	15.00	72.25	12.75	
437.800	15.00	80.75	4.25	
487.600	30.00	66.50	3.50	

X_1	X_2	X_3
111.700	17.390	0.340

Constants of Jouyban–Acree model

$J_{0\text{-}12}$	−9.560
$J_{1\text{-}12}$	0.147
$J_{2\text{-}12}$	0
$J_{0\text{-}13}$	13.790
$J_{1\text{-}13}$	−1.442
$J_{2\text{-}13}$	0
$J_{0\text{-}23}$	−59.120
$J_{1\text{-}23}$	2.148
$J_{2\text{-}23}$	−0.022

Reference of data: [1].

Solute	Solvent 1	Solvent 2	Solvent 3	T (°C)
Acetaminophen	Ethanol	Propylene glycol	Water	25
Solubility (mol/L)	Vol. F. 1	Vol. F. 2	Vol. F. 3	
0.6791	0.1	0.5	0.5	
1.2576	0.5	0.1	0.5	
1.2024	0.5	0.5	0.1	

X_1	X_2	X_3
0.937	0.660	0.099

Constants of Jouyban–Acree model[a]

$J_{0\text{-}12}$	581.214
$J_{1\text{-}12}$	−298.354
$J_{2\text{-}12}$	−225.247
$J_{0\text{-}13}$	269.570
$J_{1\text{-}13}$	−260.891
$J_{2\text{-}13}$	−80.228
$J_{0\text{-}23}$	44.227
$J_{1\text{-}23}$	−69.793
$J_{2\text{-}23}$	−78.739

Reference of Data: [2].

[a] The constants were computed employing subbinary data ($N = 33$).

Solute	Solvent 1	Solvent 2	Solvent 3	T (°C)
Acetaminophen	Ethanol	Propylene glycol	Water	25
Solubility (mol/L)	Vol. F. 1	Vol. F. 2	Vol. F. 3	
0.174	0.1	0.1	0.8	
0.350	0.2	0.1	0.7	
0.543	0.3	0.1	0.6	
0.741	0.4	0.1	0.5	
1.073	0.5	0.1	0.4	
1.246	0.6	0.1	0.3	
1.522	0.7	0.1	0.2	
1.568	0.8	0.1	0.1	
0.242	0.1	0.2	0.7	
0.432	0.2	0.2	0.6	
0.631	0.3	0.2	0.5	
0.990	0.4	0.2	0.4	
1.116	0.5	0.2	0.3	
1.462	0.6	0.2	0.2	
1.537	0.7	0.2	0.1	
0.375	0.1	0.3	0.6	
0.583	0.2	0.3	0.5	
0.712	0.3	0.3	0.4	
1.083	0.4	0.3	0.3	
1.323	0.5	0.3	0.2	
1.448	0.6	0.3	0.1	
0.517	0.1	0.4	0.5	
0.664	0.2	0.4	0.4	
0.942	0.3	0.4	0.3	
1.209	0.4	0.4	0.2	
1.284	0.5	0.4	0.1	
0.607	0.1	0.5	0.4	
0.878	0.2	0.5	0.3	
1.062	0.3	0.5	0.2	
1.477	0.4	0.5	0.1	
0.819	0.1	0.6	0.3	
1.051	0.2	0.6	0.2	
1.116	0.3	0.6	0.1	
1.010	0.1	0.7	0.2	
1.051	0.2	0.7	0.1	
1.020	0.1	0.8	0.1	

X_1	X_2	X_3
0.85	0.72	0.09

Constants of Jouyban–Acree model[a]

$J_{0\text{-}12}$	577.279
$J_{1\text{-}12}$	−237.822
$J_{2\text{-}12}$	−46.033
$J_{0\text{-}13}$	291.176
$J_{1\text{-}13}$	−123.601
$J_{2\text{-}13}$	11.385
$J_{0\text{-}23}$	227.375
$J_{1\text{-}23}$	116.270
$J_{2\text{-}23}$	430.752

Reference of data: [3].

[a] The model constants were computed using subbinary and ternary solvent data at 25°C and 30°C.

Solute	Solvent 1	Solvent 2	Solvent 3	T (°C)
Acetaminophen	Ethanol	Propylene glycol	Water	30
Solubility (mol/L)	Vol. F. 1	Vol. F. 2	Vol. F. 3	
0.237	0.1	0.1	0.8	
0.399	0.2	0.1	0.7	
0.613	0.3	0.1	0.6	
0.970	0.4	0.1	0.5	
1.162	0.5	0.1	0.4	
1.284	0.6	0.1	0.3	
1.600	0.7	0.1	0.2	
1.751	0.8	0.1	0.1	
0.364	0.1	0.2	0.7	
0.502	0.2	0.2	0.6	
0.795	0.3	0.2	0.5	
1.020	0.4	0.2	0.4	
1.259	0.5	0.2	0.3	
1.537	0.6	0.2	0.2	
1.682	0.7	0.2	0.1	
0.436	0.1	0.3	0.6	
0.631	0.2	0.3	0.5	
0.990	0.3	0.3	0.4	
1.162	0.4	0.3	0.3	
1.522	0.5	0.3	0.2	
1.492	0.6	0.3	0.1	
0.613	0.1	0.4	0.5	
0.887	0.2	0.4	0.4	
1.073	0.3	0.4	0.3	
1.616	0.4	0.4	0.2	
1.477	0.5	0.4	0.1	
0.763	0.1	0.5	0.4	
1.020	0.2	0.5	0.3	
1.105	0.3	0.5	0.2	
1.323	0.4	0.5	0.1	
0.835	0.1	0.6	0.3	
1.062	0.2	0.6	0.2	
1.073	0.3	0.6	0.1	
0.961	0.1	0.7	0.2	
1.062	0.2	0.7	0.1	
1.051	0.1	0.8	0.1	

X_1	X_2	X_3
0.85	0.72	0.09

Constants of Jouyban–Acree model[a]

J_{0-12}	577.279
J_{1-12}	−237.822
J_{2-12}	−46.033
J_{0-13}	291.176
J_{1-13}	−123.601
J_{2-13}	11.385
J_{0-23}	227.375
J_{1-23}	116.270
J_{2-23}	430.752

Reference of data: [3].

[a] The model constants were computed using subbinary and ternary solvent data at 25°C and 30°C.

Solute	Solvent 1	Solvent 2	Solvent 3	T (°C)
Biphenyl dimethyl dicarboxylate	Dimethylacetamide	Ethanol	Water	25
Solubility (g/kg)	Vol. ratio 1	Vol. ratio 2	Vol. ratio 3	
0.930	5.0	1.0	4.0	
4.420	5.0	3.0	2.0	
4.430	5.0	3.0	2.0	
9.020	5.0	4.0	1.0	
11.780	5.0	4.5	0.5	
7.580	5.5	3.0	1.5	
9.060	5.5	3.5	1.0	
3.810	6.0	1.0	3.0	
4.200	6.0	1.5	2.5	
10.540	6.0	2.0	2.0	
11.910	6.0	2.5	1.5	
12.850	6.0	3.0	1.0	
7.400	6.0	4.0	1.0	
5.360	6.0	4.0	2.0	
11.080	6.5	1.5	2.0	
12.910	6.5	2.0	1.5	
17.050	6.5	2.5	1.0	
17.120	6.5	3.0	0.5	
11.760	7.0	1.0	2.0	
14.640	7.0	1.5	1.5	
19.270	7.0	2.0	1.0	
6.210	7.0	2.0	2.0	
20.660	7.0	2.5	0.5	
7.830	7.0	3.0	2.0	
10.830	7.0	3.0	1.0	
22.510	7.5	1.5	1.0	
23.750	7.5	2.0	0.5	
30.210	8.0	1.0	1.0	
36.240	8.0	1.5	0.5	
37.280	8.5	1.0	0.5	

X_1	X_2	X_3
90.060	0.810	0.002

Constants of Jouyban–Acree model

J_{0-12}	−183.917
J_{1-12}	36.457
J_{2-12}	−8.375
J_{0-13}	−543.029
J_{1-13}	105.840
J_{2-13}	75.082
J_{0-23}	902.107
J_{1-23}	−304.251
J_{2-23}	−7.847

Reference of data: [4].

Solute	Solvent 1	Solvent 2	Solvent 3	T (°C)
Diazepam	Ethanol	Propylene glycol	Water	20
Solubility (g/L)	Vol. F. 1	Vol. F. 2	Vol. F. 3	
0.408	0.2	0.2	0.7	
13.000	0.2	0.7	0.2	
9.520	0.3	0.3	0.3	
6.020	0.5	0.0	0.5	
28.000	0.7	0.2	0.2	

X_1	X_2	X_3
27.800	7.420	0.048

Constants of Jouyban–Acree model

J_{0-12}
J_{1-12}
J_{2-12}
J_{0-13}
J_{1-13}
J_{2-13}
J_{0-23}
J_{1-23}
J_{2-23}

Reference of data: [5].

Solute	Solvent 1	Solvent 2	Solvent 3	T (°C)
Phenobarbital	Glycerol	Propylene glycol	Water	32
Solubility (g/kg)	Vol. F. 1	Vol. F. 2	Vol. F. 3	
2.640	0.1	0.1	0.8	
3.670	0.1	0.2	0.7	
5.910	0.1	0.3	0.6	
18.250	0.1	0.5	0.4	
4.090	0.1	0.2	0.7	
3.110	0.2	0.1	0.7	
4.420	0.2	0.2	0.6	
7.280	0.2	0.3	0.5	
23.220	0.2	0.5	0.3	
4.680	0.2	0.2	0.6	
3.470	0.3	0.1	0.6	
5.540	0.3	0.2	0.5	
10.030	0.3	0.3	0.4	
31.500	0.3	0.5	0.2	
6.140	0.3	0.2	0.5	
7.090	0.3	0.2	0.5	
42.240	0.3	0.5	0.2	
4.570	0.4	0.1	0.5	
14.100	0.4	0.3	0.3	
45.540	0.4	0.5	0.1	
4.810	0.4	0.1	0.5	
9.760	0.4	0.2	0.4	
6.870	0.5	0.1	0.4	
11.320	0.5	0.2	0.3	
21.160	0.5	0.3	0.2	
24.360	0.6	0.3	0.1	
13.390	0.6	0.2	0.2	
17.860	0.6	0.2	0.2	
8.900	0.7	0.1	0.2	
22.760	0.7	0.2	0.1	
10.960	0.7	0.1	0.2	

X_1	X_2	X_3
NA	NA	1.290

Constants of Jouyban–Acree model

J_{0-12}
J_{1-12}
J_{2-12}
J_{0-13}
J_{1-13}
J_{2-13}
J_{0-23}
J_{1-23}
J_{2-23}

Reference of data: [6].

Solute	Solvent 1	Solvent 2	Solvent 3	T (°C)
Phenytion	Propylene glycol	Ethanol	Water	25
Solubility (mg/L)	Vol. F. 1	Vol. F. 2	Vol. F. 3	
601.00	0.2	0.2	0.6	
4770.00	0.2	0.4	0.4	
12300.00	0.2	0.6	0.2	
1520.00	0.4	0.2	0.4	
12100.00	0.4	0.4	0.2	
9700.00	0.6	0.2	0.2	

X_1	X_2	X_3
17600.00	14800.00	20.30

Constants of Jouyban–Acree model

$J_{0\text{-}12}$
$J_{1\text{-}12}$
$J_{2\text{-}12}$
$J_{0\text{-}13}$
$J_{1\text{-}13}$
$J_{2\text{-}13}$
$J_{0\text{-}23}$
$J_{1\text{-}23}$
$J_{2\text{-}23}$

Reference of data: [5].

Solute	Solvent 1	Solvent 2	Solvent 3	T (°C)
Phenytion	Propylene glycol	Polyethylene glycol 400	Water	25
Solubility (mg/L)	Vol. F. 1	Vol. F. 2	Vol. F. 3	
520.00	0.2	0.2	0.6	
4950.00	0.2	0.4	0.4	
28100.00	0.2	0.6	0.2	
2730.00	0.4	0.2	0.4	
17200.00	0.4	0.4	0.2	
11100.00	0.6	0.2	0.2	

X_1	X_2	X_3
17600.00	71800.00	20.30

Constants of Jouyban–Acree model

$J_{0\text{-}12}$
$J_{1\text{-}12}$
$J_{2\text{-}12}$
$J_{0\text{-}13}$
$J_{1\text{-}13}$
$J_{2\text{-}13}$
$J_{0\text{-}23}$
$J_{1\text{-}23}$
$J_{2\text{-}23}$

Reference of data: [5].

REFERENCES

1. Granberg, R.A. and Rasmuson, A.C., Solubility of paracetamol in binary and ternary mixtures of water + acetone + toluene. *Journal of Chemical Engineering Data*, 2000. 45: 478–483.
2. Jouyban, A., Chan, H.K., Chew, N.Y.K., Koubnasabjafari, M., and Acree, W.E. Jr., Solubility prediction of paracetamol in binary and ternary solvent mixtures using Jouyban–Acree model. *Chemical and Pharmaceutical Bulletin*, 2006. 54: 127–133.
3. Jouyban, A., Azarmir, O., Mirzaei, S., Hassanzadeh, D., Ghafourian, T., Acree, W.E. Jr., and Nokhodchi, A., Solubility prediction of paracetamol in water–ethanol–propylene glycol mixtures at 25 and 30°C using practical approaches. *Chemical and Pharmaceutical Bulletin*, 2008. 56: 602–606.
4. Han, S.K., Kim, G.Y., and Park, Y.H., Solubilization of biphenyl dimethyl dicarboxylate by cosolvency. *Drug Development and Industrial Pharmacy*, 1999. 25: 1193–1197.
5. Rubino, J.T., Blanchard, J., and Yalkowsky, S.H., Solubilization by cosolvents II: Phenytion in binary and ternary solvents. *Journal of Parenteral Science and Technology*, 1984. 38: 215–224.
6. Belloto, R.J. Jr., Dean, A.M., Moustafa, M.A., Molonika, A.M., Gouda, M.W., and Sokoloski, T.D., Statistical techniques applied to solubility predictions and pharmaceutical formulations: An approach to problem solving using mixture response surface methodology. *International Journal of Pharmaceutics*, 1985. 23: 195–207.

Subject Index

A

Acree, W.E. Jr., 30–31, 37, 40–45, 49–50, 55, 203, 507
Amidon, G.L., 37, 50

C

Connors, K.A., 38
Cosolvency, 26, 30–31, 50, 53–54
Cosolvent, 1, 28, 30–39, 41–45, 49–50, 53, 55–57, 59

H

Hildebrand equation, 1, 36
Hildebrand solubility parameter, 1, 16, 31–32

J

Jouyban, A., 30–31, 37, 40–45, 49–50, 55, 203, 507
Jouyban–Acree model, 30–31, 37, 40–45, 49–50, 55, 203, 507

K

Khossravi, D., 38
Khossravi–Connors model, 38
Kinetic solubility, 2–4

L

Log-linear, 23, 32–36, 38, 44, 49, 55–57, 59

M

Margules equations, 49
Mixture response surface, 38

P

Partial solubility parameter, 21, 32
Partition coefficient, 4, 7–8, 14, 23, 33, 35
Pharmaceutical formulation, 2, 49
Pharmaceutical industry, 2, 30, 59
Polymorph, 3, 28, 32, 37
Prediction, 1–2, 4–7, 14–15, 23, 25–26, 28–31, 33–37, 41–42, 45, 49, 55–56, 59

Q

Quantitative structure–property relationship (QSPR), 13, 42–43, 54–56

S

Solubility parameter, 1, 15–19, 21–22, 31–32, 36–37, 54
Solubilization, 1, 30, 33, 35, 49, 59

T

Thermodynamic solubility, 3

W

Williams–Amidon model, 37, 50
Williams, N.A., 37, 50

Y

Yalkowsky, 1, 4–5, 15, 25–26, 32, 55, 57, 59

Drug Index

Solvent Index

CAS Number Index

Chemical Formula Index